Worlds within worlds
AN INTRODUCTION TO BIOLOGY

Worlds within worlds

AN INTRODUCTION TO BIOLOGY

Thomas C. Emmel
University of Florida

HARCOURT BRACE JOVANOVICH, INC.

New York   Chicago   San Francisco   Atlanta

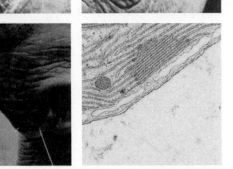

This book is dedicated to my father,
Edward F. Emmel, who has spent a lifetime pointing
out worlds within worlds to me.

*"This little group of islands [the Galápagos]
is like a world within a world."*

CHARLES DARWIN

ISBN: 0-15-597672-9

Library of Congress Catalog Card Number: 76-43226

DESIGN AND ART DIRECTION BY BETTY BINNS GRAPHICS

PRINTED IN THE UNITED STATES OF AMERICA

Preface

This innovative textbook is intended for the undergraduate student who wants a *comprehensive* introduction to the nature and functioning of living systems within the biosphere on earth. Rather than moving from the chemical/molecular level up through the cell and individual organism to ecology, this book *begins* with an analysis of the entire biosphere as an integral unit and then successively investigates in finer and finer detail how the parts of this biosphere function, arriving finally at the submolecular level. This approach has proved to be the most stimulating, relevant, and readily accepted method of presenting biology in actual teaching with thousands of University of Florida students.

There are two important features of this organization. First, it allows students to explore initially the subjects in biology that are intrinsically the most absorbing and exciting, especially in light of current world problems. From a broadly based ecoevolutionary approach entailing a panoramic view of biological interactions in the biosphere, students are urged by their own aroused curiosity to pursue the details of how these phenomena function down through the "lower" levels of organization to the interacting parts of the living individual organism. Hence, in effect, students are encouraged to probe for answers to certain questions in much the same way that scientists actually pursue their own research.

Our study centers first on the *biosphere*, with the large-scale cyclic mechanisms of life, energy, and materials explained in detail; yet at every point examples emphasize the significance of these cycles to human existence and the recent new role of human beings in affecting the balance of these ecological relationships in various ecosystems. We give full consideration to marine, fresh-water, terrestrial, and atmospheric communities and their components. Pollution and other problems in applied ecology are emphasized appropriately and used to illustrate contemporary processes in the biosphere. The ecoevolutionary interactions instrumental in shaping the structure of a biotic community — including such topics as predation, symbiosis, and competition — are treated in detail.

Moving from the biome-community level, subsequent major units of the book treat the *population* level of organization, including growth and regulation of population size and density, biogeographic and ecogeographic factors affecting distribution, and genetics and evolution at the population level. The necessary genetic discussion at this point anticipates, but does not depend on, more extensive genetic coverage later. The mechanisms of evolutionary change that may be observed in modern populations of organisms are treated in substantial detail, continuing and expanding on the thread of evolutionary theory that was initially introduced in the discussion of ecological relationships within the biosphere level. The historical evidence of evolution in the fossil record is reviewed and given proper emphasis in the story of the development of life on earth. At this point, we develop the concepts of species and race with the introduction of taxonomic principles and the classification of plant and animal divisions. Behavior at the population level — the interactions of organisms and their reactions to environmental stimuli — is treated in several detailed chapters in the final portion of the population unit, serving as a transition into our consideration of the individual organism.

As we progress from the population level of organization within the biosphere to the *individual* level, we at last begin to study in necessary detail the complex physiological and chemical mechanisms that are involved in the smooth functioning of the biosphere and its component units. But at no point do we lose sight of the integrated whole of life; for instance, there are no depressingly sterile sections on topics such as "orbital shells" and "chemical bonds" out of context from their biological manifestations and significance. Detailed consideration is given to cellular structure and functioning, and in turn to how these cellular interactions are melded into the tissue and organ levels of physiological homeostasis. Genetic mechanisms and genetic control of cellular synthetic processes are taken up in detail, with due attention being paid to Mendelian genetics, gene interactions, the chemical and physical nature of the gene, and the action of the gene in controlling protein synthesis. As throughout the book, examples relating to human beings are frequently used and emphasized. Photosynthesis and cellular respiration are treated as integral parts of the overall problems of nutrition and metabolism. In addition to processes involving both synthesis and catabolysis, ultrastructure and other appropriate concepts of modern molecular biology are introduced. Hormonal and nervous integration, of key significance in the organism's behavioral adaptation and adjustment to environmental changes, are emphasized in separate chapters. Growth and developmental processes in both plants and animals are treated, including an extensive section on human development. Reproductive physiology ends this sequence of chapters, emphasizing the continuity of life and the essential similarity and unity at this level of all organisms in the earth's biosphere.

While we feel this overall approach—going from the grand scale of biosphere biome to the community, population, individual, and, ultimately, the physicochemical nature of the biological organization of that individual—will be most effective with students, the text does offer the instructor considerable flexibility. Each major chapter grouping of the book is self-contained in explanatory definitions and can be used in the instructor's preferred order.

The instructor and student alike will find a number of useful accessory aids in using this book. A complete glossary reviews definitions of all important terms introduced in the text. Color inserts provide pictorial essays on especially interesting subjects, and special inserts in more detail. These inserts do not depend on the text and are optional for inclusion in reading assignments. A Study Guide, which includes extensive review material and study questions for each of the 34 chapters in this text, is available for students. An Instructor's Manual contains useful suggestions for lecture organization, for teaching aids such as 16mm films and their sources, and for testing approaches. A Laboratory Manual written especially for students using this text is also available.

Every effort has been made by the author and publisher to present in this text and its associated material as clear and yet detailed an introduction to the exciting worlds of biology as is possible to achieve.

We would like to acknowledge with great gratitude the aid of the following reviewers of principal portions of this text: Guy Bush, University of Texas at Austin; Karl Grossenbacher, Mills College; David Kieffer, Montgomery College; Georgia Lesh-Laurie, Case Western Reserve University; William Luke, Louisiana State University; Diane Wagner-Merner, University of South Florida; Gerald Posner, The City College (CUNY); Douglas Pratt, University of Minnesota. In preparing this book, I have had the great fortune to work with the best possible editor, Marilyn Davis of Harcourt Brace Jovanovich. From the inception of this textbook, she has provided constant guidance and encouragement which have resulted in a more comprehensive, harmoniously unified work. Betty Binns and Martin Lubin have done an outstanding job on the illustrations and total design of this book. Sylvia Newman exercised her exceptionally fine editorial pencil to improve the basic text and the flow of concepts through the book. Gail Lemkowitz and the editorial staff at Harcourt Brace Jovanovich provided unusually careful and expert attention to every detail involved in bringing this project to completion. Hal M. Ingman, Jr. offered considerable, indeed invaluable, assistance in the final days of completion of this project. Steven P. Brisko handled the overall compilation of the index with dedicated effort. Donald E. Goodman and Jill Jordan wrote the excellent laboratory manual which accompanies this text, and Marjorie Goldstein and Madeline Liff Simon prepared an outstanding study guide which will undoubtedly engender profound thanks from a great many college students. To all who helped, I am most grateful.

THOMAS C. EMMEL

Contents

PART ONE

The earth as a biosphere

Beginning with a look at the planet Mars and the place of our own planet in the universe, in Chapter 1 we explore our living world from the perspective of evolutionary change and ecological relationships and introduce the concept of the biosphere. In Chapter 2 we delve into the general organization of life in the biosphere and the major environments of the world.

In Chapter 3 we examine the sources and flow of energy in the organic world. The flow of materials in the great biogeochemical cycles on earth occupies our attention in Chapter 4. Finally, in Chapter 5, we investigate the alteration of the natural landscape and the destruction of our own human ecological relationships through environmental degradation and pollution.

1   Our living world

Standing on the reddish powdery soil by the side of the spaceship Beagle II, the first human visitor to Mars gazed in awe across the mighty chasm toward the 15-mile-high volcano Nix Olympica. Jumbled uplifts and dusty desert basins were being whipped by fierce winds under the late afternoon sun. High above the frost-rimmed craters a wisp of whitish cloud fled across the pink sky. And on the rock in front of the scientist's eager eyes lay the sole surviving form of higher life on the dying planet, a melancholy cockroach on a bit of lichen.

The July 20, 1976 landing of the unmanned Viking 1 probe on the surface of the red planet resulted in worldwide headlines twelve days later that life may have been discovered for the first time elsewhere in the solar system. The scientists' initial elation turned to caution as the telemetered data began to indicate the possibility of signals resulting from a chemical reaction within the soil sample retrieved by the landing craft, rather than the action of Martian microorganisms. Yet no one could say for certain exactly what had really occurred in the depths of the Viking's reaction chamber, and appetites were keenly whetted for a future manned voyage to our sister planet which could provide definitive answers on the existence of life there.

The actual scenario of the discovery of advanced life on the planet Mars has yet to be played out, but when the first biologists physically visit the fourth planet out from the sun, what can they expect to find? Undoubtedly not a cockroach, though it has been one of the most successful organisms on the earth in both an ecological and an evolutionary sense. Cockroaches of one kind or another are found from equatorial jungles to quonset huts in the Arctic, and according to the fossil record they have been around with almost the same body plan for over 250 million years. On Mars, however, they would have great biological difficulty in even surviving at high noon on the equator, where the average surface temperature ranges from near 32°F in the early afternoon to some 135° below zero just before sunrise (Figure 1–1). The fierce Martian dust storms known to obscure the planet's surface for months on end would likely blast even a lichen off an exposed volcanic substrate. The point is that while we will not know for sure about the character of life on Mars until we land there, from our experience on earth in extreme environments we can make reasonable predictions about what Martian organisms are *not* likely to be.

We also know today that our planet Earth is only an infinitesimal dot in the vastness of the universe, and we might reasonably predict that life could develop along other lines elsewhere. In fact, our home galaxy is far out from the center of the universe, and our solar system itself is located far out on the periphery of our own Milky Way galaxy. The scale of distances involved is almost incomprehensible. The disk alone of our home galaxy is some 80,000 light *years* across and 10,000 light years thick (light travels at 186,000 miles per second,

FIGURE 1–1

The planet Mars.

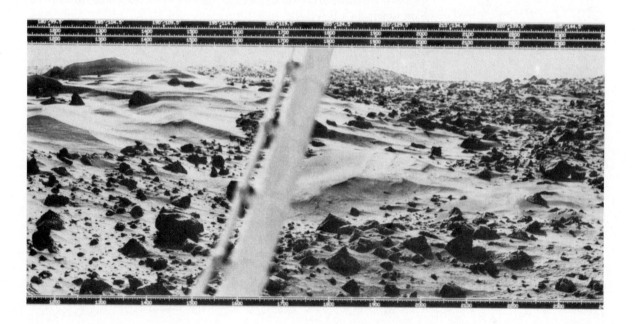

FIGURE 1–2

The planet Earth from space (Apollo 17 flight). The Mediterranean Sea and the Arabian Peninsula are in view at the top, most of Africa is covered by clouds, and Antarctica with the South Pole ice cap is visible at the bottom.

and a light year represents the distance light can travel over a year's time). Even the closest star system to earth in our galaxy, Alpha Centauri, is a distance of 4.3 light years. And to travel to the Andromeda nebula, the *next* neighboring galaxy of the 17 in our little cluster, would require crossing a distance of about 680,000 light years. About a *billion* galaxies lie within the range of our most powerful optical telescopes, and the total extent of the universe is unknown and perhaps beyond human comprehension. We have no idea where in the universe other experiments with the creation of living forms from nonliving matter may be going on, though astronomers have recently detected the presence of planetary bodies around one or more of the nearest stars; and it may be presumed that millions of planets probably exist throughout even the known panorama of the universe. While our nature as intelligent and reasoning beings leads us to raise questions about the existence of life elsewhere in the universe, inevitably our main focus of interest falls where we can personally observe events—on the globe that we call home, the planet Earth (Figure 1–2).

The biosphere

With over a million years of human proprietorship on this unique planet, we still know surprisingly little about it. We know that we presently share its surface with about 2 million other species of organisms and that in former times — since the creation of the earth, in fact — about 400 species of plants and animals have lived and died for every one that exists today. We know that these organisms are marvelously adapted to live in arctic winters, hot and arid desert valleys, humid jungles, shallow alpine lakes, and the deepest ocean troughs. Yet the total relief of the earth's surface measures less than 13 miles from the greatest ocean depths to the highest mountain, and it is this narrow "skin" of an 8,000-mile-diameter planet that contains the trillions of terrene organisms. Biologists have called this hollow sphere of life the **biosphere** — the subject that will occupy us in this book as well as affect each of us in multitudinous ways for the rest of our lives.

The levels of life that we can explore on earth, then, range from the gargantuan concept of the biosphere down to the submicroscopic level where the business of cellular maintenance takes place. Biologists, like everyone else, tend to categorize things and usually feel most comfortable starting with a small unit (such as chemical structure) and working upward to more and more complex situations in the organic world. Often this approach merely enables one to see the *trees* as collections of various chemical processes, but somehow obscures the *forest*, or "big picture," by the time one arrives at that stage. We are embarking in this book on a new kind of approach, where we shall start with the "big picture" and gradually focus with finer and finer detail on the processes and events at work in the natural world. As an initial step, let us look at the great unifying principles of life on earth that have been contributed by biologists.

Themes in the organic world

Since the time of Aristotle (384–322 B.C.), naturalists and biologists have attempted to synthesize general principles from data collected via direct observation and experimentation. From their innovative efforts and discoveries have emerged ordering concepts for the biological world that are analogous in influence on contempory scientific endeavor to those natural laws discovered in physics and astronomy by such workers as Sir Isaac Newton, Copernicus, and Albert Einstein. A consideration of some of these biological principles will best be taken up in detail in later chapters. The *theory of innate behavior*, for instance, developed largely by Konrad Lorenz and

Niko Tinbergen, states that many behavioral acts are genetically programed into the individual from birth rather than requiring learning through life's experiences. We can understand this concept better after a broad introduction to behavior itself (Chapters 18 and 19) and the evolutionary and ecological factors involved. The *theory of the organizer and induction in embryonic development*, first proposed by Hans Spemann in 1921, states that certain embryonic tissues have the capacity to induce or channel the development of surrounding tissues into particular pathways, integrating the development and maturation of all required tissue types into a properly organized embryonic body. We can understand the significance and universality of this concept better with a background in genetics (Chapters 22 and 23) and development (Chapters 31 and 32).

The other major biological concepts, which fall under two of the three great unifying themes of life on earth: evolutionary change and ecological relationships, shall be our point of departure. The third great theme—the cellular and chemical similarity of life—will occupy us in the last 14 chapters. From the perspective these themes provide, we gain an appeciation and understanding of how our living world functions and the biological significance of our place in it.

The theme of evolutionary change

While a number of eighteenth- and early nineteenth-century biologists had proposed that the living world does not remain static but that new forms arise over time, it remained for Charles Darwin (1809–1882) to propose a plausible mechanism that made possible the scientific acceptance of such a theory of evolution. In 1859 his book *On the Origin of Species by Means of Natural Selection* started a tidal wave of scientific and popular reaction that today influences much of the way we view our world and human knowledge. The concept of evolutionary change is now familiarly embraced on a chemical and physical level by astronomers studying the origin of new suns and solar systems and even the beginnings of the universe. Cell biologists use it to describe the origins of the minute structures they study under the powerful magnifications of the electron microscope. Darwin's original concept of **natural selection** as the mechanism of **evolution** has been supplemented by twentieth-century work in biology, but its basic premise still underlies what is believed to account for long-term organic change: Organisms that are best suited by their heredity for survival under particular environmental conditions will tend to leave more offspring than less favored organisms in the same population. Thus the hereditary make-up of a population changes from generation to generation, and over long periods of time new species will develop.

The basic theories of genetic inheritance and the nature of the gene, developed by biologists since Darwin's time, describe the hereditary units with which evolutionary mechanisms work, and the theory of mutation describes the ultimate source of variability for the genes on which natural selection operates. The theme of evolutionary change and the genetic theories involved in it begin to explain the sweeping panorama of life, from its earliest beginnings in the primordial seas of the unfinished planet to the incredible diversity of 2 million species of terrene life today.

The theme of ecological relationships

Just as all organisms share a common theme of evolutionary change through inheritance and natural selection operating on organisms with differential survival ability, we know today that all organisms share interrelationships with other organisms and their physical environment. The science of **ecology** is the study of these interrelationships. The complex web of ecological interrelationships between animals, plants, and microorganisms and their physical and chemical surroundings literally determines the perpetuation of the biotic (living) world as we know it. The pathways of energy flow and the cycling of materials between organisms and their environment have only been uncovered in detail in the past 40 years. Yet their extreme importance is shown almost daily as environmental pollution, burgeoning populations, and habitat destruction reap inevitable results in the terms of disastrous effects on human beings. The tremendous urgency of the problems our world faces in these areas has catapulted ecologists into the forefront of biological research and into the political arena as well.

Perhaps the greatest benefit that an exposure to biology provides is the concept of an ecological consciousness — an understanding of the interrelationships and interdependence of all forms of life on our earth — and an appreciation, both esthetic and intellectual, for the diversity of organic assemblages around the world, from the equatorial tropics of Africa, South America, and Asia, with their teeming rainforests and coral reefs, to the cooler temperate forests, grasslands, and mountains, and to the barren, almost lifeless, arctic polar regions of North America and Europe. We shall explore contemporary life on all levels, from giant marine lizards that feed on seaweed to tiny spiders and other organisms that spend most of their lives miles above the earth in the atmosphere; but we shall focus primarily on the human species as a biological organism affected by its surroundings.

We shall penetrate with an ecologist's insight the astonishing annual cycles of nature and see how organisms make provision for the

A   B

FIGURE 1–3

(A) The California desert is dry and shows little signs of life during much of the year. (B) In the spring, following sufficient rainfall, perennial plants, such as cactus, and millions of annual wildflowers spring forth to bloom; countless insects and other animal species that depend on these plants rush through a few hectic weeks of activity.

changing seasons. Thus, for instance, we shall see how a desert in southern California may lie scorched and almost devoid of green plants for 11 months of the year and then spring into spectacular verdant splendor with hundreds of annual wildflower species sprouting after the late winter rains (Figure 1–3). The orderliness of even the daily cycle of life, a response to ecological and evolutionary factors that have programed the organism's schedule of activities, will represent a fascinating and relevant discovery as we see how "biological clocks" operate in organisms ranging from roadside weeds to human beings.

As we penetrate deeper into the "forest" of our biological world, individual "trees" will become clearer, and we will be able to interpret the marvelous functioning of the individual organism and its constituent cells in the light of our ecological and evolutionary perspectives. Each species' remarkable adaptations to use the particular resources it requires can raise a feeling of awe in even the most blasé persons when they understand fully what is involved. Ultimately, we shall delve into the single cell and its chemical processes and rearrangements of matter and energy. On the molecular level, we shall see how the energy of the sun (ultimate source of all biologically useful energy) is captured by plants and transformed into a useful state that can be passed along to animals. We shall study how animals receive this energy and utilize it for growth processes and ultimately for reproduction of their own kind. Here we shall see the essential unity of life, extending from a sharing of common organic molecular forms to a universal "genetic code" and similar metabolic machinery. It is on this level that the tremendous complexity and interrelationships within the biosphere become universally understandable.

As we begin this adventure through the biosphere and on down through the levels and concepts of biological science, let us keep in mind the two great unifying themes of life on earth: evolutionary change and ecological relationships. A biological consciousness through seeing the unity and spendor of life in these themes can become an enriching and indeed vital part of the human experience and, with the interpretive bases these concepts provide, make the natural world an exceptionally exciting place to explore.

Summary

The earth is a tiny bit of matter in the vastness of the universe, yet to our present very limited knowledge it is only on this planet that life has developed. About 2 million species of organisms exist over the nearly 13-mile relief of the earth's surface, from the deepest ocean canyons to the highest mountains and the adjacent atmospheric area. This hollow sphere of life over the earth's surface is called the biosphere. We shall start at the level of the biosphere in this book and then focus in closer and closer detail on the processes and events at work in the world of life. The two great themes of evolutionary change and ecological relationships encompass many of the ordering concepts developed by biologists over the centuries. The theme of evolutionary change states that organisms best suited by their heredity for survival under particular ecological conditions will tend to be more successful at leaving offspring than less favored organisms in the same population (natural selection), and hence the hereditary make-up of the population will gradually change (evolution). The theme of ecological relationships states that all organisms are involved in interrelationships between other organisms and their physical environment. Ecology is the scientific study of these interrelationships, which ultimately involve pathways of energy flow and the cycling of materials such as chemical elements. The perspective provided by these **ecoevolutionary** themes will increase our appreciation and comprehension of the functioning of our living world and the biological significance of our place in it.

2   The organization of the biosphere

Ecology is the study of the relationships that organisms have with one another and with their environment. The word itself is derived from the Greek *oikos*, which means "house," that is, the environment around us. In recent years, we have come to associate ecology with human environmental problems, particularly those of pollution and conservation. Yet the science actually encompasses the entire functioning structure of the biosphere, and we must grasp the ways in which the physical environment and life in general are interrelated before we can really understand the problems and possible solutions involved in human environmental alteration.

The organization of life on earth

Ecologists view life in the biosphere at a series of levels of ecological organization. The *biosphere*, as we have seen earlier, encompasses the thin surface layer of the earth and its associated lower atmosphere, in which all living things exist. The biosphere may be broken up into about a dozen major **biomes,** broad groupings of plants and animals that spread across the earth at characteristic latitudes and altitudes. Our coniferous forests in northern North America form one

FIGURE 2-1

The hierarchy of organization in the natural world. From the whole living world (biosphere) to the smallest subatomic particles making up a single atom, each level may be explained in terms of the structure and functioning of the parts that compose it.

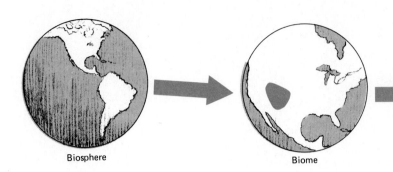

Biosphere   Biome

such biome, the taiga, which is essentially duplicated across northern Europe, Asia, and into Siberia south of the Arctic Circle. The great deserts of the world form another biome. We shall look at all major biome groupings of organisms later in this chapter.

We can break up a biome into a series of constituent ecosystems and communities. A **community** is an assemblage of plant and animal species living in a particular area; an **ecosystem** is composed of this biotic community and all the physical and environmental factors affecting it. Thus within the pine-spruce forests of the taiga biome we may find pond communities, willow bog communities, open meadow communities, and deep forest communities. Each community will have a different set of environmental factors impinging on it. The temperature range, amount of daily sunlight, soil moisture, quantity of humus and leaf litter in the soil, and other climatic and physical parameters will change in each subregion and hence each community will be a part of a different ecosystem.

A number of populations of plants and animals representing different species compose a biological community. A **population** consists of an aggregation of individuals of the same species and exhibits properties that individuals alone cannot—especially growth in numbers, a particular distributional pattern, density (degree of crowding), and a certain distribution of newborn through mature animals in the population.

Ultimately, the *individual* organism forms a basis for all our studies in ecology. Its reactions to environmental changes determine population fluctuations, community balance, and the flow of energy and materials through the ecosystem. The individual as a representative of a particular species helps to establish the unique biological character of a community and, on a broader scale, of a biome and the biosphere (Figure 2–1).

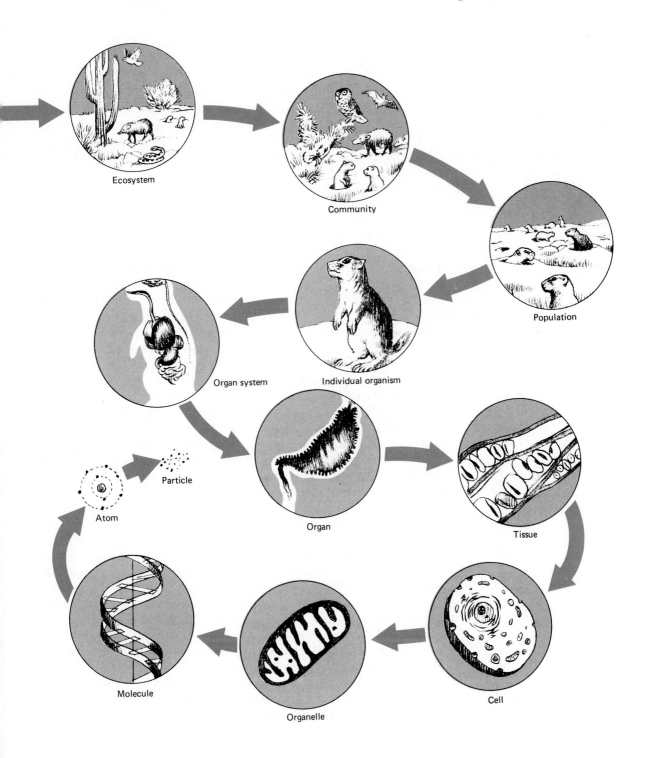

Environments on earth

Living organisms are known to occur in marine, terrestrial, fresh-water, and atmospheric environments, but are most successful in the first two areas. They do not occur (as far as is known to date) in deeper geological formations than surface soils and caves. Let us look briefly at the general characteristics of each of these four environments and perhaps determine why marine and terrestrial organisms have diversified to the greatest extent.

While there is some biochemical support for the fresh-water origin of living cells, the marine environment was the cradle of diversification for life several billion years ago, according to the geological fossil record. Today, it remains the greatest reservoir of both living organisms and the vital elements needed by life in all other environments. Thus the sea must be considered a major force in shaping the ecology of our world and the basic conditions for life on land as well as in fresh water. The sea covers 70.8 percent of the earth's surface and as much as 86 percent of this coverage is over a mile deep. Some of the ocean depths plunge to more than 4 1/2 miles (over 7,000 meters) from the surface. We know that all this water is inhabited, if only sparsely in the more extreme depths; thus there are no abiotic (nonliving) zones in the ocean. Yet on land (less than one-third of the earth's surface) quite an appreciable portion is either frozen polar ice, desert, or nearly barren high mountain; the whole Antarctic continent, for example, is a barren frozen waste. Even in the inhabited land, the zone of life extends from treetop height (100 meters, or 330 feet, maximum) to only a few feet below the surface of the soil. When we think again of the immense area and thickness of the oceans, we begin to grasp that the total mass of living things in the ocean is far greater than the combined mass of life from land and fresh water.

The seven seas—the North and South Atlantic, the North and South Pacific, the Indian, the Southern (or Antarctic), and the Arctic oceans—are continuously connected over all the world, unlike land and fresh-water habitats (Figure 2–2). However, troughs of great depth within the deep ocean basins are separated as effectively as the tops of different mountains on the land. Thus differences in depth, associated water pressures, water temperature, and salinity are the chief barriers to free movements of marine organisms. The sea is in continuous circulation because of strong winds and the earth's rotation; hence nutrients and other materials (even pesticides) are carried around the globe in water currents. The daily rhythm of tides produced by the pull of the moon and sun influences physical conditions in shore zones especially.

In land environments, available water becomes a major limiting factor for organisms. Terrestrial plants and animals must have mechanisms to avoid dehydration and to withstand rapid air-temperature changes. Likewise, air does not offer the physical support as a medium that water does. Strong skeletal support must be evolved to adapt to the terrestrial environment. Light in the sea is restricted to the sunlit surface layers (at 25 meters, it is only about 1 percent of the intensity of light just below the surface), whereas on land it penetrates to the deepest layers of the forest and into the top of the soil. Only deeper, soil-burrowing animals and cave-dwelling species live in a world that is entirely without light, comparable to the deep sea animals.

Land animals and plants do have the advantage of a relatively constant atmospheric content of oxygen and carbon dioxide, essential gases that may vary considerably in concentration in marine environments. Soil becomes the primary source of nutrients such as nitrates and phosphates, whereas the water medium carries the bulk of the free nutrient load in the oceans. Geographic barriers break up the continuity of land areas into continents and subcontinental areas. In general, then, the nature of terrestrial communities and ecosystems is determined by climate (moisture, temperature, light levels, etc.) and physical substrate (physiography of the terrain, soil character, etc.). The physiological demands of occupying the land environment have led to the development of the most complex and specialized groups of both the plant and animal kingdoms, especially the seed plants, insects, and warm-blooded vertebrates (birds and mammals).

Fresh-water environments are much more limited in geographic extent than terrestrial or marine regions. Standing-water, or lentic (*lenis*, calm), habitats include lakes, ponds, swamps, or bogs, whereas running-water, or lotic (*lotus*, washed), habitats encompass rivers, streams, brooks, and springs. Variation in temperature in fresh water is less extreme than in the air for terrestrial communities; however, the thermal variation in northern temperate lentic and lotic habitats may be far greater (because of winter cold and summer heat) than for similar environments in the tropics or the seas. These temperature changes also produce important patterns of circulation and stratification that greatly influence aquatic life. The lack of salts in the water creates a major problem for fresh-water organisms. Water is constantly entering fresh-water animals and must be excreted or pumped out again. Many marine groups that live where the salt concentrations are about the same in the outside environment as

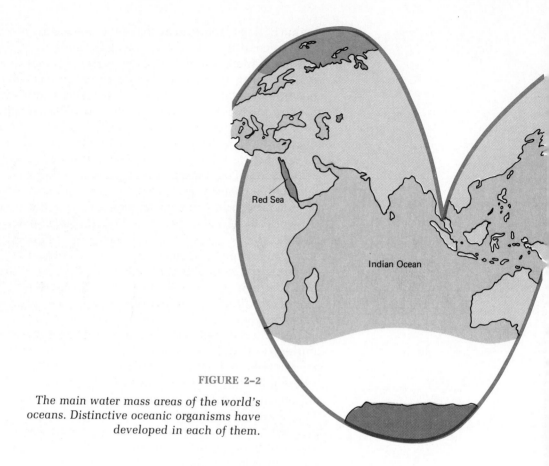

FIGURE 2–2

The main water mass areas of the world's oceans. Distinctive oceanic organisms have developed in each of them.

inside their cells have been unable to solve this difficulty and invade the fresh-water environment.

Atmospheric organisms were unknown until recently, when it was discovered that tiny algae and bacterial cells inhabit water droplets that are carried by wind currents for months at a time, high in the atmosphere. During such periods, these organisms survive and reproduce. Environmental barriers to extensive colonization of the atmosphere by organisms include very low temperatures; wind evaporation of the fluid medium surrounding the cells; lack of nutrients except by chance encounters of the water droplets with dust, pollutants, and other solid atmospheric particles; the high irradiation by ultraviolet and other wavelengths; and the pull of gravity on the water droplets and atmospheric changes (cloud formation) that eventually may result in the precipitation of the carrying medium onto the earth's surface.

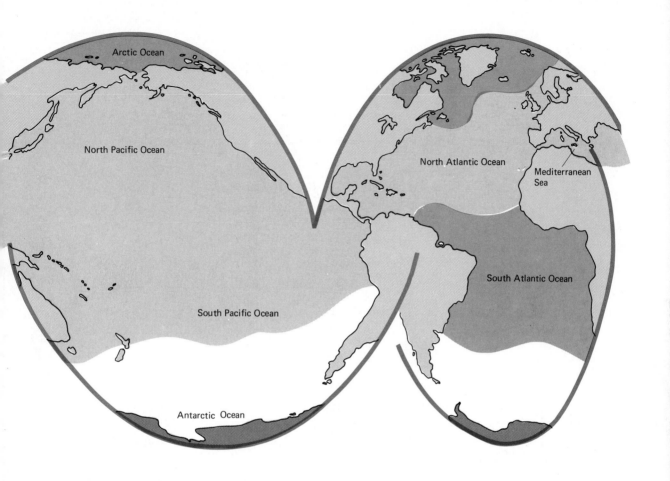

Looking across the spectrum of the earth's available environments, life has been remarkably successful in colonizing physically diverse areas. Let us now look at the major biomes across the earth and their organic diversity.

The biomes of the earth

Biomes are general groupings of similar plant and animal communities. These groupings are based on the distinctive life forms of the dominant species in each community, and in terrestrial regions dominant vegetational types are usually used to characterize biomes. Because climatic zones of particular temperature and rainfall patterns change in a relatively uniform pattern toward either pole from

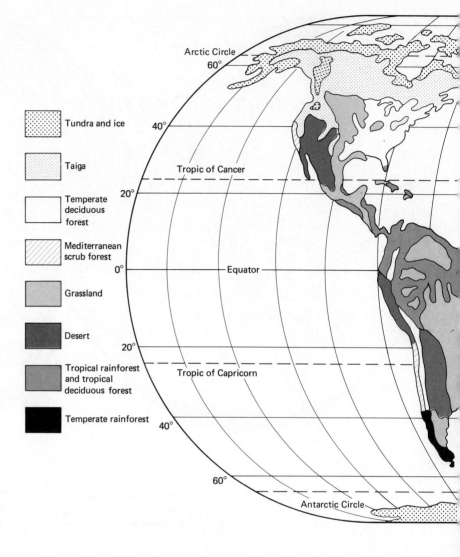

Tundra and ice

Taiga

Temperate deciduous forest

Mediterranean scrub forest

Grassland

Desert

Tropical rainforest and tropical deciduous forest

Temperate rainforest

Arctic Circle
60°
40°
Tropic of Cancer
20°
0°
Equator
20°
Tropic of Capricorn
40°
60°
Antarctic Circle

FIGURE 2–3

The major biomes of the world, as depicted in the distribution of original native vegetation.

the equator, the earth's biomes form more or less continuous latitudinal bands around the globe (Figure 2–3). We shall start at the equator in our brief survey of biomes.

Tropical rainforest

In a number of species, the tropical rainforests of the Old and New Worlds represent the richest biome on earth. There are more different kinds of organisms in this region than in any other biome;

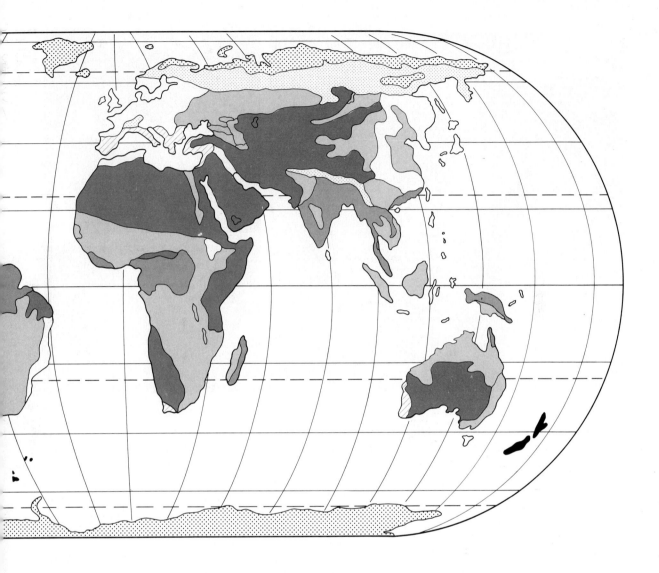

hence interactions between species such as predation and competition are very important limiting factors on population growth. There tend to be fewer individuals of any one species per unit area than in other biomes. Neither temperature nor water is a limiting factor on plant or animal growth, and there is less chance of extinction of a tropical population due to climatic vagaries. Rainfall, which may range from 100 inches to 200 inches or even 400 inches a year in some rainforest areas, tends to be more or less evenly distributed throughout the year. The mean annual temperature lies between

A   C

FIGURE 2–4

Creatures of the tropical rainforest. (A) Swarms of army ants, such as this Eciton hamatum *species in Costa Rica, overpower other insects and even vertebrates as they raid the forest in search of booty. (B) A delicate ithomiine butterfly,* Scada batesi, *feeds on a tiny composite flower near the forest floor in Ecuador. (C) A tapir, often hunted for food by forest Indians as well as jaguars, rests in the Colombian rainforest.*

about 68°F and 82°F, with very little seasonal fluctuation. In fact, the change from daytime to nighttime temperatures in a rainforest is greater than the annual mean temperature fluctuation.

Typical mammals of the Central and South American rainforests include monkeys, jaguars, ocelots, sloths, opossums, tapirs, coatis, and deer of several species. Parrots, macaws, tanagers, manakins, and other exotic birds fly about in the forest canopy and the several layers of understory trees. Impressive swarms of raiding army ants, hundreds of species of colorful butterflies, giant damselflies, and other insects fill the forest openings from dawn to dusk when even more insect species with nocturnal habits begin their activity. Arboreal orchids and bromeliads are particularly prominent features of the canopy flora (Figure 2–4; plate 2).

Tropical deciduous forest

To the north and south of the equatorial tropical rainforest biome but still within the Tropics of Cancer (lat. 23°27′ N) and Capricorn (lat. 23°27′ S) lies the tropical deciduous forest biome. With uniformly warm temperatures all year, the growth of the trees and herbaceous plants of this tropical forest type is affected only by seasonal distribution of the rainfall. A dry season of variable length, depending on location, alternates with a wet season. During the annual dry season such forests lose most leaves and then leaf out again at the onset of the rainy season. Both plants and animals exhibit special adaptations for avoiding dryness and water stress, often passing the dry season in a state of physiological inactivity in much the same manner as temperate-zone species pass the cold winter. The reproductive periods of most insects and vertebrates are keyed to the sudden initiation of the rainy season each year (Figure 2–5).

Tropical savanna

Tropical grasslands with scattered trees constitute the savanna biome. Rainfall is insufficient to support the thick growth of forests, and the poor character of the soils may also prevent the establishment of many trees. In many areas human inhabitants maintain savanna formations by setting fire to the dry grasses annually. The usual trees in tropical savannas are acacias in Africa and Central America and palms in South America. In Africa large herds of grazing hoofed mammals such as antelope and gazelles characterize the savanna fauna. In tropical Australian savannas these are replaced by marsupials such as kangaroo and wallaby species (Figure 2–6; plate 3).

FIGURE 2–5

The howler monkeys are widely distributed from Mexico to South America and are found in both tropical deciduous forest and evergreen rainforest.

FIGURE 2–6

A herd of wildebeest alertly watches for hunting lions on the Amboseli savanna below Mt. Kilimanjaro in East Africa.

Desert

Just to the north of the Tropic of Cancer and to the south of the Tropic of Capricorn is the worldwide desert biome. It is characterized by very low rainfall (generally less than 10 inches a year), a high evaporation rate, high diurnal temperatures, and relatively low nocturnal temperatures. Desert shrubs and other plants are generally low-growing, highly seasonal in leaf production and flowering, and have reduced, waxy, drought-resistant leaves. Cacti have virtually lost their leaves and instead use chlorophyll in the surface layers of their expanded stems for photosynthesis. Relatively few species of animals are found in deserts; most of the residents are nocturnal and have other behavioral and physiological adaptations to prevent water loss through evaporation or excretion. Because of the cool nights, dew is a significant source of moisture for desert organisms, which lack the availability of surface water or frequent rain showers. Many desert plants can absorb dew droplets and translocate the moisture from their leaves to their stems and roots (Figure 2–7; plate 6).

A

B   C

FIGURE 2–7

(A) The organ pipe cactus, Tachycerus marginatus, has photosynthetic pigments distributed throughout its expanded stems and lacks regular leaves, an adaptation to life in a land of constant drying winds. (B) An evening primrose, genus Oenothera, lifts its bouquet of bright flowers in the brief desert spring to attract a passing insect pollinator; many species bloom only in the early morning or evening hours when their nectar will not be evaporated as rapidly as at midday. (C) Among the inhabitants of the Arizona desert is the gila monster, Heloderma suspectum, one of two poisonous lizards in the world.

A

B

FIGURE 2–8

(A) Yucca plants and (B) the black-tailed prairie dog are characteristic inhabitants of the sweeping grasslands of the Great Plains of North America.

Grassland

Across the extensive central plains areas of temperate North America, Eurasia, and Australia, there is sufficient precipitation to support a thick growth of annual and perennial (the same plant continues to grow year after year) herbaceous plants, especially bunch grasses and spreading grasses with underground stems. Trees are absent except along the infrequent rivers and seasonally wet watercourses. Burrowing rodents are common. On the North American Great Plains, these include several species of prairie dogs, ground squirrels, and many mice. Large herds of grazing mammals such as buffalo and pronghorn were typically found in grasslands until recently. Rainfall patterns led to various ecological subdivisions of the temperate grasslands, such as the short-grass plains, long-grass plains, and prairie areas (Figure 2–8; plate 7).

Chaparral

The chaparral biome is relatively small in extent, but it occurs repeatedly in temperate parts of the world having a so-called Mediterranean climate, principally southern Europe, southern Australia, and southern California and adjacent areas of Arizona and Baja California. This climatic type is characteristically dry for most of the year, with low total precipitation in the form of winter rains and occasional summer thundershowers. The low, shrubby vegetation in the chaparral has thick, evergreen leaves whose toughness, abundance of resins, and waxy coatings make them resistant to prolonged periods of drought and moisture stress (Figure 2–9; plate 1).

Fires periodically sweep through chaparral, clearing out dead material and old growth and allowing reproduction of the various fire-adapted species of shrubs at the expense of trees. After a fire passes, the first rains of the next growing season cause the chaparral species to sprout from their fire-blackened stumps or from seeds lying dormant in the soil. However, if fires do not race through the chaparral every 10 to 20 years, the community becomes overly mature and growth of new seedlings is prevented by the densely growing older plants. When a fire finally occurs from lightning or events of human origin, it will be extremely destructive because of the amounts of accumulated fuel (dead branches, leaves, etc.) under the mature plants. Even the stumps and roots of many chaparral shrubs will be killed by the intense heat. When shrubs have been destroyed by such a fire, tree seeds floating in from adjacent forest areas can germinate and

FIGURE 2–9

Chaparral animal life: (A) California ground squirrel; (B) Anná's hummingbird; (C) mule deer; (D) a manzanita shrub.

become established in the area before the chaparral can recover. Thus chaparral is an example of a fire-maintained climax community (Chapter 6).

Temperate deciduous forest

The temperate deciduous forest biome is characterized by the broad-leaved trees and shrubs found in the great hardwood forests of eastern North America, with counterparts in eastern Asia and Europe. It is an ancient biome, believed to have been nearly continuous across the continents during Cretaceous times (136 to 165 million years ago). Since this forest type occurs in the temperate zone, the species are subject to considerable temperature fluctuations and variation in photosynthetic production, which is high in the summer months and low in the winter months. Thus the deciduous forest includes broad-leaved trees such as oaks, beeches, hickories, ashes, basswoods, maples, and other species that lose their leaves during the late fall and remain leafless during the winter. In North America, characteristic animals include Virginia white-tailed deer, black bears, raccoons, opossums, red foxes, squirrels, many mice and other rodents, and insectivores such as shrews (Figure 2–10; plate 4).

A   B

FIGURE 2-10

(A) Temperate deciduous forest: a maple woods in the northeast. (B) Gray squirrel (Sciurus carolinensis leucotis), a common denizen of eastern woodlands.

In California, the seasonally dry summer and fall months cause a peculiar variation of the temperate deciduous forest to develop, with scattered oak trees spread across rolling grassland. This plant formation is commonly called woodland. It receives considerable rainfall in winter, stimulating a rich growth of annual and perennial grasses. Formerly, it was an important grazing area for the Tule elk, antelope, and deer; today it is largely ranchland or is put to agricultural use for wheat or barley.

Taiga

The Russian word *taiga* has been adopted by ecologists to describe the wide coniferous forest belt south of the Arctic Circle around the world; it is continuous except for the interruptions of the Bering Sea and the North Atlantic. This evergreen needle-leaved forest is also present below the arctic-alpine zones of the highest mountains. Although these mountainous islands of taiga to the south of the continuous Siberian and Canadian taiga may vary somewhat in species diversity, the temperate evergreen forests are usually grouped with the taiga biome. In North America, the north-south orientation of the major mountain ranges allows great southward peninsular exten-

FIGURE 2–11

*(A) Clarke's nutcracker
(Nucifraga columbiana), a
common bird in the
coniferous forests of the
western United States. (B)
Coniferous forest and
meadows in Colorado. (C)
Bristlecone pine (Pinus
aristata), a conifer found in
the mountains of both
Colorado and eastern
California; some living
individuals of this species
have reached ages of over
4,000 years.*

sions of the taiga, particularly through the Pacific Northwest to Baja California and the Rocky Mountain states to Mexico (Figure 2–11; plate 7).

The northern taiga is dominated by spruces, firs, short-needled pines, and larches, whereas to the south long-needled pines, beeches, and aspens may occur. Most conspicuous among the larger animals of this biome are the moose, whose range includes Europe, Asia, and North America, deer, elk, weasels, ermines, and other fur-bearing mammals of the family Mustelidae, and a host of rodents and rabbits. The taiga, like the tundra to be examined next, is highly seasonal in animal activity. Its insects, spiders, and many vertebrate species become dormant during the winter. The spring influx and the autumn departure of migratory birds affect community structure and metabolism in a major way.

Tundra

Tundra is the Russian word for the zone north of timberline at the Arctic Circle; in other words, the treeless Arctic plains. The term is also applied to alpine grasslands and meadows on high mountains in the temperate zones and to the high treeless grasslands and paramos of the tropics. Tundra vegetation is characteristically low-growing, cold-adapted, and almost always perennial. Representative

FIGURE 2–12

(A) The arctic-alpine zone of tundra vegetation at an elevation of over 12,000 feet in Colorado, high above timberline. (B) Columbines nestled among the rocks in the Colorado tundra. (C) Caribou feeding on the Arctic tundra in Alaska.

plants are grasses, mosses, lichens, sedges, and shrubs, the latter growing in tight mats in sheltered places on the drier hillsides or along streams. The northern circumpolar tundra is the most continuous and in some respects the most sharply defined of all the biomes. It extends across Eurasia from Kamchatka to Lapland, and from Labrador across North America to Alaska. In this polar tundra, as well as in alpine communities on the higher mountains, the deeper layers of the soil may be permanently frozen (the *permafrost*), and only the surface soil thaws sufficiently to permit biological activity during the summer. The permafrost layer is impervious to water and hence in late spring the depressions of the rolling tundra plains are occupied by lakes, ponds, and bogs. This biome is also characterized by a short growing season (June–July), extremely low temperatures during the winter and nightly freezing temperatures even during the brief summer, and few animal species (Figure 2–12; plate 5).

Most of the mammals are burrowing species, such as lemmings and mice. Some of the larger herbivorous (plant-eating) mammals, such as musk oxen, reindeer, or caribou, must migrate annually between summer and winter ranges. In the Arctic oceans the floating polar ice forms an appendage to the adjacent land areas, and semi-aquatic mammals such as seals and walrus use the ice floes as resting platforms.

Orderliness and evolution in the biosphere

When we think about these different biomes and the complexities that exist in their organization and daily functioning, it becomes evident that their component species must be well adapted to each other and to the factors making up the physical environment or the whole system would soon collapse. Our own experience in nature tells us that natural communities and ecosystems manage to persist very well unless upset by some unusual outside factor such as a pesticide, a hurricane, or a clear-cutting forestry operation. Thus each of the ecosystems in the biosphere seems to be organized into a system analogous in many ways to the organization of an individual organism. The interrelation among organs in the body, or components of an ecosystem, is not haphazard. It has a particular spatial orientation, an involvement of time in operation of the system, an involvement of specific energetic sequences (food is consumed, processed, and the nutrient molecules reutilized), and a definite history of development. Orderliness in an ecosystem, then, can be characterized as spatial, temperal, metabolic, and evolutionary.

Spatial orderliness

Organisms of various species are adapted to different kinds of environments. Each plant or animal lives in a particular place — its physical **habitat** — and carries on a set of activities unique to its species that defines the way it survives and makes its living, an activity syndrome that can be called the organism's **adaptive strategy.** Each species, then, has its own ecological **niche,** composed of a place to live (the habitat) and a particular means (the adaptive strategy) of acquiring energy and maintaining itself. The niche is more than a physical space; it is a way of life and is believed to be unique for each species on earth.

Since each species of plant or animal has its own niche, the placement of individuals in the ecosystem is specific for different species. Some species populations are evenly spaced; others are clumped or are nongregarious in seemingly random patterns. Spatial orderliness is also often shown as **stratification** in an ecosystem (Figure 2–13). The living organisms in the ocean, a lake, or a terrestrial community seem to be arranged in vertical layers. Thus in a rainforest, one passes through wholly different sets of species as one goes from the soil up through the ground-level and lower-story vegetation to the uppermost forest canopy.

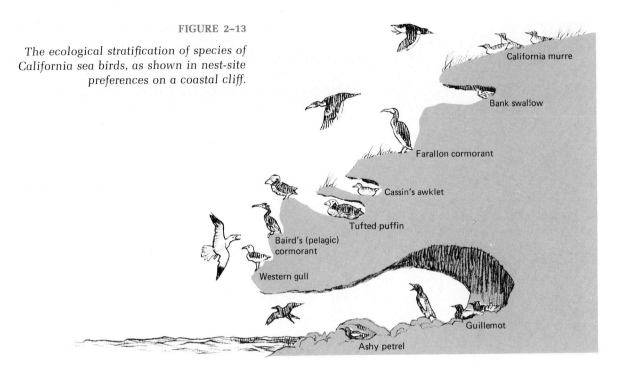

FIGURE 2–13

The ecological stratification of species of California sea birds, as shown in nest-site preferences on a coastal cliff.

Temporal orderliness

Time is reflected in the orderliness of ecosystems in many ways. Each species has a particular time of maximum activity during the 24-hour day, referred to as **periodicity** (Figure 2–14). Periodicity may be taken in a general sense, as nocturnal versus diurnal activity, or in a specific sense, as, for example, the number of minutes a particular desert flower is open and receptive to pollinators before the sun rises each morning. Many species can share the same area in a way that would not be possible if all species carried on their activities at one particular time or if their activity periods all coincided. On a longer time scale, virtually all organisms in biotic communities utilize particular times of the year to reproduce, and these breeding seasons represent adaptations to maximizing reproductive effort at the most advantageous times.

In animals, females will normally bear young at the time of the year when the maximum amount of food is available for their rearing. In plants, the dispersal of fruits and seeds or the seasonal appearance of climatic conditions promoting growth may be foremost in importance.

Midday

FIGURE 2–14

Periodicity in a coral reef ecosystem, as demonstrated in the activity and locations of fish species at midday and midnight.

Metabolic orderliness

Within the ecosystem, as within the body, transformations of energy and materials follow definite, orderly paths. Chemical elements cycle through the atmosphere, organisms, soil, and other parts of the physical environment. Because the elements tend to be used over and over again, and are not permanently lost from the community, transformations of matter are said to be cyclical. On the other hand, since some energy is lost at every step in the series of orderly transfers within ecosystems, the transfer of energy is not cyclical. Nonetheless, the intricate adaptations shown by plants and animals in capturing and utilizing energy reflect tremendous development of ordered interrelationships in the biosphere. The biological transfer of energy will be considered further on Chapter 3.

Evolutionary orderliness

Each community and ecosystem has a particular biotic composition, in terms of both number of individuals and number of species. The species of plants and animals living in that region are the most

Midnight (no moon)

recent products of evolutionary change. Each is adapted or adjusted to the particular environment it inhabits through **differential reproduction,** which results in what we call evolution, that is, change in the characteristics of a species or even the development of a new species. Darwin's chief contribution to biology was to show how this might occur through selection or removal of maladapted types. Differential reproduction simply means that the best-adapted individuals of a species tend to leave more offspring on the average than do less well-adapted genetic variants. We call this result *natural selection* or *survival of the fittest individuals.*

When we consider the whole biotic community in an ecosystem or even a biome, we see that the plant and animal species have not evolved independently of each other. Instead of a miscellaneous collection of adaptively unrelated organisms, the species composition of an ecosystem is *ordered.* There has been a coevolution of the plant and animal components of an ecosystem; mutual adjustments of species have taken place through natural selection. Thus the biotic parts of the ecosystem form a cohesive whole because they have shared a common recent evolutionary history, adjusting to each other and to the special set of physical environmental factors prevailing in that terrestrial or aquatic area. The evolutionary concept will provide one of the unifying themes in our examination of the world of life.

Summary

Life in the biosphere is organized into a series of levels of ecological organization: biomes, ecosystems, communities, populations, and individual organisms. It is found in terrestrial, marine, fresh-water, and atmospheric environments, each set of physical conditions requiring special physiological adaptations on the part of plants and animals. Major biomes in the terrestrial environment are tropical rainforest, tropical deciduous forest, tropical savanna, desert, grassland, chaparral, temperate deciduous forest, taiga, and tundra. Ecosystems, composed of biotic communities and their physical environment, exhibit spatial, temporal, metabolic, and evolutionary orderliness. Evolutionary change and adaptation play an important role in establishing the structure and functioning of species in a biotic community and provide a unifying theme for the study of biology in general.

3   The flow of energy

All organisms are dependent on energy in some form for their existence. Energy moves through plant and animal populations in an ecosystem in an orderly but unidirectional sequence. Energy flow is not cyclical; only materials cycle. The ultimate source of this unidirectional flow of energy for the earth's biosphere and its component ecosystems is light from the sun. Organisms use this light energy in a series of energy transfers initiated by the primary producers, the plants, which utilize photosynthesis to capture light energy and store it in the new form of chemical bonds. The light energy not captured by green plants is reflected by the earth's surface as invisible heat radiation. The small but vitally important fraction of the light energy converted into chemical bond energy is passed along to higher levels of consumer organisms in ecological systems. Some useful energy is lost as heat at every step in such a change of events. The behavior of energy during these transformations is described by two physical laws: the first and second laws of thermodynamics.

The laws of thermodynamics

The *first law of thermodynamics*, which involves the conservation of matter and energy, states that energy cannot be created or destroyed, only changed from one form to another. Thus, for instance, visible light energy may be changed by the process of photosynthesis in green plants into chemical energy in the form of chemical bonds present in the glucose sugar molecule. Later energy transformations in the plant may involve the breaking up of this glucose molecule for food in the process of cellular respiration, and at this point the chemical energy present in the molecular bonds is released as heat. The energy itself is never destroyed, only changed form.

The *second law of thermodynamics*, which deals with the tendency of energy to become dissipated in the universe, states that some useful energy — that is, energy available for doing work — is converted into heat energy at every transformation step. Thus energy transfers ultimately tend to disperse energy in the form of heat energy throughout the universe. When all molecular motion finally ceases, a uniform temperature of absolute zero will result. Any work, then, involves the escape of heat energy into the surrounding environment. When an animal ingests chemical potential energy in the form of plant food and converts it in a series of energy transfers in its cells, much of the useful energy is lost as body heat in order to enable a small part of the energy to be converted into the potential energy of chemical bonds in new protoplasm. In other words, energy transformations in the biological as well as the physical world are considerably less than 100 percent efficient. Since heat energy cannot be used easily to do work in organisms, more energy must be constantly supplied to a biological system from outside in order to counterbalance the inevitable loss of energy as heat. Ultimately, of course, this energy is derived from the processes of fusion of hydrogen atoms into helium in the depths of the sun.

Because energy transformations are one-way pathways in contrast to the cyclical behavior of materials, the behavior of energy in ecosystems is commonly termed **energy flow.** When plants are eaten by **herbivores** (plant-eating animals), the chemical bonds are broken and reformed into new molecular bonds within the animal's cell. Energy is lost at this stage. When **carnivores** (animals that eat plant-eating animals) eat herbivores, still another sequence of breaking and reforming of chemical bonds occurs and further heat energy escapes unused by the carnivore. The simplest description of this type of energy transfer is the food chain concept.

Food chains

A **food chain** simply involves the transfer of food energy from a given source through a series of species, each of which eats the next lower organism in the chain (Figure 3–1). This repeated series of nutritional relationships is always initiated with green plants that receive their energy from the sun. A very simple food chain is represented by the following:

Grass——→Deer——→Human Being

At each transfer of energy, a large proportion of the potential energy present in the chemical bonds of the food species is lost as heat, commonly 80 to 90 percent; therefore, the number of links in a food chain is usually limited to four or five. The shorter the food chain or the nearer the organism is to the beginning of the chain, the greater the available potential energy that can be converted into **biomass** (weight of living material) and utilized in cellular processes. Thus grazing herbivores that eat green plants tend to have a greater base of energy available to them than carnivores, which must feed on the energy successfully transformed from the feeding of herbivores into herbivore flesh. Thus the working of the two laws of thermodynamics is intimately involved in food chains. Energy is converted from one form into another, with loss of heat energy at each step.

Food chains may be of several types. In grazing food chains, the green plant base is eaten by grazing herbivores, which in turn may be eaten by carnivores. From the viewpoint of the end organism, this food chain could be equally well called a predatory food chain. In some food chains larger organisms are eaten by smaller species, as is the case frequently in parasitic food chains. For instance, a deer may be infested by intestinal tapeworms or a human being may have microscopic malarial parasites in his blood. In detritus food chains, sometimes also called saprophytic food chains, fungi, such as mushrooms, and microorganisms, including bacteria and yeasts, feed on dead organic matter.

If food chains were isolated sequences in nature, the transfer of energy in an ecosystem would be very simple to visualize and handle. However, normally an animal may eat several different foods, and in fact as a general rule in any biotic community food chains are interlocked in a complex pattern of feeding relations called a food web.

Hawk

Mockingbird

Caterpillar

Plant

FIGURE 3–1

A four-step food chain: a producer (plant) is eaten by a herbivore (caterpillar); the herbivore in turn is eaten by a primary carnivore (mockingbird), which may be eaten by a top carnivore (hawk).

Food webs

A **food web** is simply the total collection of feeding relationships in a biotic community (Figure 3–2). The many interlocking food chains tend to promote stability for organisms at the higher levels, providing them with alternative food sources should one or more of the prey species become less abundant. The more components involved, then, the more stable the system will probably be. The arctic foxes may feed on mice or lemmings throughout the year. During the summer there are a number of additional migratory bird species in the area, both young and adults, that can serve as prey. When the birds migrate south in the fall, foxes are not forced to die or move out but can feed on other species such as the mice that are in the same nutritional relationship to them.

FIGURE 3–2

A moderately simple food web composed of a number of alternate food chains. The top carnivores (owl, hawk, coyote) have a variety of prey species available as food; if one decreases in abundance, the predators can switch to other food chains. At each lower trophic level in this community, these options also exist.

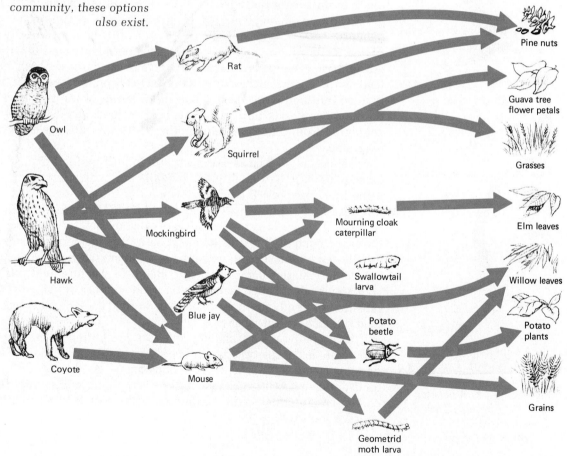

feeder") base on which the heterotrophic ("other-feeder") component of the community rests.

The **herbivores** represent the second trophic level. They are capable of converting energy stored in plant tissue into animal tissue. Because they are the first organismic level to feed on other organisms, their trophic level is sometimes called the primary consumer trophic level. Herbivores have a central role in the community—namely, that of providing food for the higher trophic levels, for only the herbivores such as cows and deer are adapted to live on a diet high in plant cellulose. Herbivores have specialized teeth with broad grinding surfaces, complicated stomach organs to process refractory plant cell walls, long intestines for absorption, and a well-developed symbiotic flora and fauna in their intestines that enable them to break up and utilize plant tissue.

The third trophic level is composed of the **carnivores.** They belong to the secondary consumer trophic level in cases where the carnivore feeds directly on the herbivores. The **top carnivores** in a food web are animals that feed on other carnivores. They represent the fourth, sometimes even the fifth, trophic level and may also be called tertiary or quaternary consumers. Generally the top carnivores are larger and more fierce in behavior than the first level of carnivores, which are second-level consumers. Notable adaptations among the carnivores include the familiar canine teeth for biting and piercing, the sheering teeth for ripping flesh, and talons and claws for holding prey as well as tearing flesh.

The **decomposers** consume plants and animals from all trophic levels as they feed on dead organic matter. They perform the instrumental role of breaking down dead plant and animal bodies, reducing them finally to inorganic nutrients that are able to disperse into the soil. Decomposers are represented by insects, millipedes, and other arthropods in the early stages of most decomposition scenarios, but in the later stages fungi and microorganisms such as bacteria and yeast are chiefly involved.

A species can occupy more than one trophic level. We do it daily as we eat meats and grains simultaneously. An East African baboon may normally graze on plants and yet occasionally take a small impala antelope for meat, thus being in the second and third trophic levels simultaneously (while eating producers and herbivores). This simultaneous existence in more than one trophic level, or feeding group, is characteristic of the general class of consumers called **omnivores.** Most organisms, however, can be classified by ecologists into one of the previously delineated trophic levels on the basis of their primary source of energy, which is generally done for simplicity of analysis of a community.

In natural communities all organisms that share the ⟋ source of nutrition are said to be at the same **trophic lev** 3). Each step in a food chain may be counted as being a c⟍ ber of levels away from plants, which are the ultimate sou⟍ energy. This trophic-level concept implies that these organ⟍ tain food through the same number of steps from plants ii⟍ chain.

The first trophic level in ecosystems, represented by green ⟋ and photosynthetic bacteria, consists of **producers.** These organ⟍ photosynthesize and are able to convert light energy from the su⟍ chemical bond energy through manufacturing food from sim⟍ inorganic substances. Producers represent the autotrophic ("se⟍

FIGURE 3–3

The trophic levels in the natural world: (A) producers: a giant tree and tiny mosses on its trunk, in the rainforest of eastern Ecuador; (B) herbivores: an elephant (Loxodonta africana) *feeding on a tree in Kenya; (C) carnivores: an East African lioness with a freshly killed Thomson gazelle; (D) decomposers: fungi on a rotting tree stump in the rainforest of Costa Rica.*

A

B

C

D

Ecological pyramids

Since this regular sequence of trophic levels occurs in the typical community, it is useful to express trophic structure graphically in the form of an *ecological pyramid* (Figure 3–4). The first trophic level, or producer level, forms the base, and successive levels form the narrowing tiers that lead to the apex. There are three general types of ecological pyramids.

In the *Eltonian pyramid*, the number of individual organisms for each trophic level is depicted. Successive levels of carnivores decrease very rapidly until there are very few carnivores at the top. In the *pyramid of biomass*, the total dry weight or other assessment of the total amount of living material present at any one time in any particular level is shown. Because loss of energy as well as materials occurs between each level, the total mass supported at each successive level is generally smaller than it was at the preceding level. In the *pyramid of energy*, the rate of energy flow and productivity (rate of organic production) of successive trophic levels is depicted. Obviously, less energy is transferred from each level than was put into

FIGURE 3–4

A food pyramid of five trophic levels in the sea.

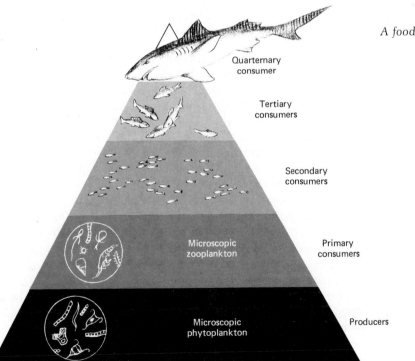

Quarternary consumer

Tertiary consumers

Secondary consumers

Microscopic zooplankton

Primary consumers

Microscopic phytoplankton

Producers

it, in accord with the second law of thermodynamics, and hence we have a sharply sloping pyramid toward the apex.

Given this pattern of trophic levels, what is the pathway of energy flow in an ecosystem? In a general sense, only a tiny fraction of the tremendous quantities of solar energy falling on an area each year is actually utilized by the plants and animals living there. On the average, in energy relationships in ecosystems around the world, more than 98 percent of the available light energy from the sun is lost and less than 2 percent is trapped by plants and stored as chemical bond energy. At each successive transfer between trophic levels, about 90 percent of the available potential energy is lost via respiration and heat. This happens in aquatic ecosystems as well as terrestrial ecosystems, in corn fields as well as forests.

The ecological efficiency of each trophic level in utilizing the amount of energy available to it is of considerable ecological interest to us. Understanding how this is controlled by the organisms involved can do much to help us plan the sensible management and intelligent use of existing major communities. Let us now look at this problem of the quantification of energy flow in more detail.

The quantification of energy flow

In an energy-flow analysis on the ecosystem level, one attempts to determine how much of the potential food supply each trophic level is using. In other words, how efficient is a particular level in exploiting the total amount of food available to it? All organisms in the ecosystem are assigned to one or more trophic levels and energy movement is estimated by establishing the energy values of populations operated at the same trophic level. In spite of the complexity of natural communities, such a task can be accomplished if conditions are quite stable in the ecological system. Fresh-water springs offer such conditions in areas with mild climates.

In a pioneering study of Silver Springs, a popular Florida tourist site, Howard T. Odum measured the number of units of energy captured by each trophic level over a year's time in this aquatic ecosystem. The continual imput of fresh water from an underground source included a nearly constant water temperature, even over a two-year period. Likewise, compared with most terrestrial systems the aquatic spring was relatively isolated from outside influences. The amount of energy entering the system could be accurately estimated, as well as the amount of energy leaving as heat energy or stored energy. The overall results of his study are depicted in Figure 3–5.

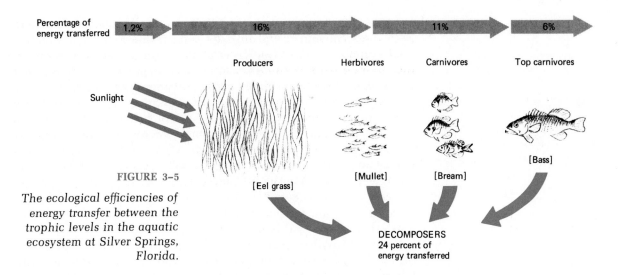

Percentage of energy transferred   1.2%   16%   11%   6%

Producers   Herbivores   Carnivores   Top carnivores

Sunlight

[Eel grass]   [Mullet]   [Bream]   [Bass]

DECOMPOSERS
24 percent of
energy transferred

FIGURE 3–5

The ecological efficiencies of energy transfer between the trophic levels in the aquatic ecosystem at Silver Springs, Florida.

Odum determined the ecological efficiencies of energy transfer by calculating the ratios between energy flow at different points along the food chain of trophic levels. At this particular latitude in Florida, Odum found that 1,700,000 kilocalories[1] were falling on each square meter of surface at Silver Springs over 365 days a year. Of this potentially available quantity, only 28,810 Kcal of energy per year were captured by photosynthesis in the green plant producers present within each square meter. By our definition of the ecological efficiency of a trophic level, only 1.2 percent of the solar energy present (20,810/1,700,000 Kcal) was actually trapped by chlorophyll molecules and converted into organic compounds by the producers. This is partly due to inefficiency in the plants themselves, including the complex processes of converting light to chemical energy, and partly to the fact that only about one-half of the sun's energy is represented by wavelengths in the useful visible spectrum, 4,000 to 7,000 angstroms.[2] The leaves of aquatic plants and the algal plant bodies floating in the water did not cover all of the available surface area of the bottom of the spring; hence some light reaching the spring never impinged on a plant. Light was also lost between chlorophyll molecules in the individual alga cell or leaf or was absorbed in the cell wall material and other nonphotosynthetic areas.

[1] The small calorie (cal) is used when small quantities of energy are involved and represents the amount of heat needed to raise one gram (or one milliliter) of water one degree centigrade at a pressure of one atmosphere (sea level), One kilocalorie (Cal or Kcal) represents the quantity of heat needed at a pressure of one atmosphere to raise the temperature of one thousand grams of water one degree centigrade.

[2] An angstrom is a minute unit of length equal to one ten-thousandth of a micron or one ten-thousand-millionth of a meter.

At the first consumer level in the trophic diagram (Figure 3–3), herbivores consumed and utilized some 3,368 Kcal of the 20,810 Kcal available in the plant material across an average square meter over the course of a year. This represents a 16 percent efficiency; that is, 84 percent of the available potential energy was lost to the environment. Part of this improvement over producer efficiency (1.2 percent) occurred because of the locomotory behavior of herbivores while finding their food. In the two higher consumer levels of Silver Springs — namely, predatory fish such as gar and bass — efficiency of energy utilization remains at a relatively low level. This is true despite the fact that the searching behavior of predators greatly increases their chances of finding suitable energy in the form of prey, in contrast to the restricted ability of sessile (nonmotile) plants to move their leaves and branches to trap sunlight energy.

Compared with higher levels, the decomposers that feed on all trophic levels are quite efficient. Odum did not separately quantify the amount of energy entering the decomposer populations from each trophic level but gave a total figure for the decomposers. These organisms utilize about 24 percent of the potential energy originally available in the chemical bonds of the producers. As Figure 3–3 shows, very little energy flow reaches the highest trophic levels, which are usually made up of fairly large organisms (fish and turtles in this spring) living at a low density per square meter. By the fourth trophic level in Silver Springs, potential energy has been reduced to such an extent that a fifth level cannot be supported.

Consumers in certain terrestrial ecosystems may be relatively efficient in the exploitation of primary producers, providing that a number of species feed on different parts of the available plant population. In the east African savanna, for instance, grazing species feed on most portions (stems, leaves, flower and seed heads) of the grass plants and other herbs. In Tanzania the total grazing population of hoofed animals in the Tarangire Game Reserve is said to reach 28 percent efficiency in their herbivore role of converting vegetation. In the Queen Elizabeth National Park of Uganda, the efficiency of exploitation of available plant material by herbivores is said to be as high as 60 percent. These values suggest the possible extent to which stable and natural grasslands can be utilized by mixed populations of grazing mammals, adapted by long periods of evolutionary adjustment to the prevailing conditions. Many African ecologists believe that it will be energetically more efficient to encourage maintenance and periodic cropping of native game populations in East Africa for a protein supply among the populace, rather than maintain cattle or other presently domesticated animals.

Productivity and associated concepts

The energy accumulated by the plants and stored in the form of chemical bonds is called **primary production,** since it is the first and most basic form of energy stored (Figure 3–6). The rate at which it accumulates is known as the primary productivity of the specific plant population or community under study. This plant growth rate can be measured in units of energy, in calories; alternatively, it can be measured as dry organic matter in units of weight, in grams. Thus one frequently measures the gram dry weight of producers, consumers, and decomposers in a community.

The rate of primary production is dependent on the efficiency of photosynthesis in the particular plants of the community. As we have seen, photosynthesis is not very efficient and normally about 1 to 3 percent of the annual amount of radiation received from the sun is converted into food energy. Generally, the most productive ecosystems include the coral reefs, coastal estuaries (mouths of rivers opening into the ocean), and intensely cultivated agricultural fields. Deserts and parts of the open ocean may be very low in primary productivity, varying from 0.5 to 1.0 grams of biomass per square meter per day. The maximum rate that any extensive ecosystem can reach and hold is probably no higher than about 25 grams per square meter per day, as seen in certain agricultural crops. Ranging between these extremes are forests, grassland, and routinely farmed cropland, with production in these areas ranging from 1 gram to 20 grams per square meter per day.

Production in consumer organisms entails additional drains on available energy. A considerable portion of the plant material ingested by a herbivore may pass through its body without assimilation. The bulk of the energy actually consumed by an animal goes into maintenance of its body structure, with smaller portions going into growth, reproduction, or nearly immediate elimination and passage from the body. Maintenance energy also must be used in the capture of new food as well as its digestion. The replacement of protoplasm and maintenance of body temperature and muscular work in the animal's daily routine are additional demands on maintenance energy. Any energy left over after maintenance goes into production, including new tissue or new individuals. Efficiencies in higher-level consumers are sometimes rather high because carnivores may be able to use proteins already constructed and present in their animal food and hence directly consumed by them, whereas herbivores must completely reconstruct proteins present in their plant food to utilize them in the synthesis of animal proteins.

FIGURE 3–6

Primary production for human benefit: a haying operation on a U.S. farm.

Summary

The flow of energy through an ecosystem is unidirectional and is ultimately dependent on sunlight. Energy may be changed from one form to another (the first law of thermodynamics). In a series of energy changes, some useful energy is converted into heat energy and dissipated at each step (the second law of thermodynamics). The food chain involves a series of transfers of food energy through producer, consumer, and ultimately decomposer species. Food chains are often interlocked in the form of food webs in communities. The trophic structure of a community may be also expressed graphically in the form of an Eltonian pyramid of numbers, a pyramid of biomass, or a pyramid of energy. The ecological efficiency of each trophic level, and its productivity, may be measured in calories. Maintenance activities consume the bulk of energy available to an animal, whereas plants shunt much of their energy intake into an increase in productivity.

4 The flow of materials: biogeochemical cycles

Solar energy fixed by photosynthesis is the driving force that forms the basis of ecosystem productivity, but it is only one requirement for the organisms in an ecosystem. The interactions involving the flow of materials are as essential as the flow of energy in maintaining these living components. Thus processes in the ecosystem continually produce new biological material and decompose old material. If natural cycles did not exist to recirculate the chemical elements in these materials, entire ecosystems would be short-lived indeed.

Populations of organisms have evolved specific ways to handle the relative quantities and kinds of materials available in their particular environments. The materials necessary for their survival and growth are called **nutrients.** They include both energy-containing organic compounds such as sugars and inorganic materials such as water, nitrates, phosphates, carbon dioxide, and elemental ions such as copper and iron.

The flow of these materials through the ecosystem also involves the chemistry of the earth, or *geochemistry*, for all chemical elements that make up living things come from the earth. Of course, not all the elements found in rocks are used by living things. Some elements such as silicon are abundant in the crust of the earth, yet are rarely used. Only a few common gases in the atmosphere are taken up in significant quantities by living systems.

The cycling of materials from the earth through living systems and back to the earth is called **biogeochemical cycling.** The process of decomposition releases nutrients from dead bodies and allows the chemicals to enter the general nutrient pool of the soil or the atmosphere. Let us see now how these materials involved in living systems are cycled in typical ecosystems so that they will not be used up.

The types of biogeochemical cycles

Chemical elements tend to circulate in the biosphere in characteristic paths from environment to organism and back to the environment. These are known as biogeochemical cycles (*bio* = life and *geochemistry* = the study of chemical exchanges between different parts of the earth). The 40 or so elements essential to living organisms are known as nutrients and the pathways of exchange between living and nonliving ecosystem components are nutrient cycles. Nutrients and climate form the two major abiotic (nonbiological) parts of the ecosystem that affect organisms.

Macronutrients are present in fair abundance and are necessary in rather large amounts in organisms. These include such elements as carbon (C), oxygen (O), hydrogen (H), nitrogen (N), phosphorus (P), potassium (K), calcium (Ca), magnesium (Mg), and sulfur (S). Micronutrients are also necessary in organisms but only in smaller, trace amounts. These trace elements include iron (Fe), copper (Cu), zinc (Zn), manganese (Mn), molybdenum (Mo), boron (B), chlorine (Cl), and iodine (I), among others.

The overall pathways of biogeochemical cycles on our planet are shown in Figure 4–1. Cycles of carbon, oxygen, hydrogen, soil minerals, water, and other materials pass among living organisms of the earth's biotic communities through soil, water, and air interfaces. This somewhat complicated diagram summarizes two important points. First, materials cycle whereas energy does not and, second, consumers are not necessary for the maintenance of the cycle; it is possible for materials to cycle directly from producers to decomposers. Thus the 40 or so elements known to be required by living organisms tend to be used over and over again, or to be *recycled*. They circulate in the biosphere in characteristic more or less circular paths from environment to organism and back to the environment again and involve elements changing from an inorganic form to an organic molecule and back again.

Biogeochemical cycles have two basic components: the *reservoir*

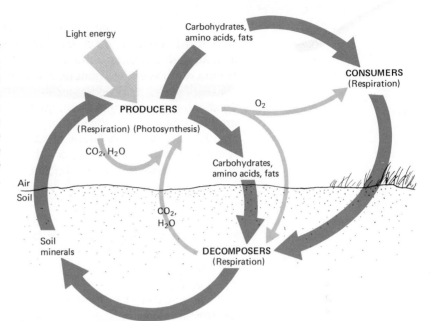

FIGURE 4-1

The overall pathways of biogeochemical cycles on earth. The narrow arrows show cycles of carbon, oxygen, and hydrogen; the broad arrows indicate cycles of soil minerals and the passage of energy through the living organisms of the earth's biotic communities.

pool, a large, generally nonbiological component of the ecosystem with relatively slow turnover rates of materials—that is, the time necessary for an element to leave a pool after having entered it may be relatively long; and one or more *exchange pools,* or *cycling pools.* These smaller, more active compartments are constantly exchanging their contents with the environment and organisms. Cycles in which the element is returned to the environment as rapidly as it is removed by living organisms are said to be more *perfect* cycles than those in which part of the material is locked up in inaccessible chemical forms or geological formations for extended times. Hence the rate and "perfection" of the cycle depend on which portion of the environment acts as a reservoir pool for the element: the atmosphere or geological formations.

The importance of biogeochemical cycles to life is twofold. (1) They help retain necessary nutrients in usable forms for the living components of an ecosystem. Animals and plants in the community develop adaptive features to aid in nutrient retention. In the tropical rainforest, for instance, many trees and shrubs have widespread shallow root systems that absorb minerals and other components from decaying matter in the leaf litter nearly instantly and hence prevent any buildup of nutrients in the soil where they could be lost by water erosion. In the tropical rainforest, almost the entire body of nutrients is retained in living plant matter. Likewise, in many temperate forests, elaborate mycorrhizal (symbiotic) associations have developed between soil fungi and the roots of temperate trees such as oaks and hickories, which aid the larger plants in absorbing nutrients rapidly and efficiently from the surrounding soil. (2) They help maintain the steady-state of ecosystems. If nutrient amounts constantly decrease or vary in quantity in the ecosystem, organisms are likely to adapt to the change but no stability will develop. Constantly decreasing amounts of nutrients prohibit development of the steady-state ecosystem since there is no constant nutrient source from the outside as there is for energy in the form of sunlight.

Structurally and functionally, biogeochemical cycles are of two basic types: *gaseous* and *sedimentary*. This classification is based on the primary reservoir pool for the element concerned—the atmosphere and hydrosphere or soils and rocks. As we shall see, the distinction is sometimes rather arbitrary.

The gaseous cycles

Gaseous cycles are said to be more nearly "perfect" than other biogeochemical cycles because the elements circulated do not become inaccessible to organisms over long periods. The reservoir pool is in the atmosphere or hydrosphere (oceans and ground water). The gaseous cycles involve four elements: carbon, nitrogen, oxygen, and hydrogen. These elements are moved about in the biosphere and environment in tremendous amounts, with the earth's atmosphere serving as the main inorganic storage reservoir. While these four elements with gaseous cycles are only 10 percent of the 40 essential elements found in living organisms, they constitute about 97.2 percent of the bulk of protoplasm, and they cycle more easily than any other elements. In these stable cycles, elements are not locked up for long periods of time. In fact, recycling takes place within a matter of hours or, at most, days. The reservoir pool is large and yet quick adjustments can occur by buffering actions. Let us now look at these cycles in more detail.

The *carbon cycle* involves a basic constituent of all organic compounds. Next to water, carbon is the most important material found for biological systems. Hence carbon, in the form of food, moves through the ecosystem with the flow of energy since energy transfers involve the construction, storage, and degradation and decomposition of carbohydrates and other carbon-containing molecules. Carbon dioxide (CO_2) in the atmosphere and dissolved in water forms the principal inorganic reservoir for elemental carbon (Figure 4–2). The photosynthetic reactions of green plants incorporate carbon dioxide from the air with water from the soil and through a series of

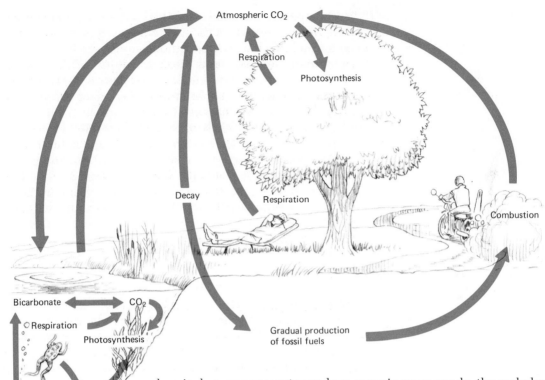

Atmospheric CO₂

Respiration

Photosynthesis

Decay

Respiration

Combustion

Bicarbonate

CO₂

Respiration

Photosynthesis

Gradual production
of fossil fuels

Carbonates;
i.e., limestones

Decay
organisms

FIGURE 4–2

The carbon cycle.

chemical rearrangements produce organic compounds, the carbohydrates, which include carbon atoms as the basic constituent of the molecular backbone. Plants are able to convert these carbohydrates in turn into the synthesis of proteins, fats, and polysaccharides. These more complex carbon compounds may be eaten by herbivorous animals and be digested and resynthesized into new carbon-containing compounds; likewise, predators eating the herbivores will manufacture additional carbon compounds of the basic units originally photosynthesized by plants.

When the green plant dies, or any of the later stages in the carbon cycles such as a herbivore or predator dies, the carbon-containing organic compounds are broken down by the actions of the decomposer organisms such as fungi or bacteria. Likewise, animal wastes are acted on by fungi and bacteria in the decay process. Considerable carbon is released in the form of carbon dioxide as a waste product from the respiration of decomposers, and carbon in this form is immediately released back to the atmospheric reservoir. Should the green plant be eaten by a herbivorous or omnivorous animal, the carbohydrates and other carbon-containing molecules from the plant will be broken down in the animal's digestive system and body cells and released to the atmosphere eventually as CO_2.

Most carbon is eventually breathed out to the atmospheric reservoir in the form of carbon dioxide, whether this be from respiration in animal cells or the decomposition activities of fungi and bacteria. (The latter, as we have noted, break down the cellular organic molecules on the death of the animal, releasing carbon dioxide to the atmosphere.) During the Carboniferous period of the Paleozoic era (Chapter 14), the superabundant carbon present in the plants and animals of the worldwide tropical swamp forests was preserved in the form of fossil fuels such as peat, coal, oil, and natural gas. Carbon was also preserved from past geologic eras in rock formations such as limestone, which resulted from huge aggregations of marine organisms. Burning of the temporary fuel reservoirs by human beings and natural weathering of limestone return carbon to the atmosphere.

Figure 4–2 shows that each part of the carbon cycle is vital to the continued circulation of this element. Aquatic and terrestrial plants take up CO_2 from the air or water for photosynthesis and make carbohydrates from carbon. Consumers and decomposers get carbohydrates from plants and animals. Should the green plants be destroyed, carbon would be unable to move from the inorganic atmospheric reservoir of carbon dioxide to be incorporated into organic carbon compounds in animal protoplasm. If the decomposers in a biotic community were eliminated, organic matter from dead plants and animals would accumulate rapidly, and the vital carbon atoms would be locked up in nondecaying bodies, destroying the cycle. Consumers, decomposers, and producer plants all return respired CO_2 to the atmosphere. Diffusion and physical stirring of water and air allow exchange of CO_2 between the ocean (the aquatic reservoir pool) and the atmosphere (the exchange pool). Surplus CO_2 is transformed to carbonates (HCO_3^-), then to bicarbonates (H_2CO_3), and is stored in the reservoir pool.

The *nitrogen cycle* is an even more complex but more or less perfect gaseous cycle (Figure 4–3). This cycle illustrates the key role played by microorganisms in many of these biogeochemical cycles. About 78 percent of the atmosphere is nitrogen gas (N_2), the largest gaseous reservoir of any element. In its gaseous form nitrogen is useless to most organisms. However, a number of bacteria and blue-green algae possess the capacity to fix atmospheric nitrogen into solid forms; that is, to convert inorganic diatomic molecules (N_2) into forms such as nitrates that are immediately usable by plants. These kinds of microorganisms include the symbiotic bacteria living in the nodules of tissue on the roots of legume plants (members of the pea family, Leguminosae) and certain free-living bacteria, as well as blue-green algae and the purple bacterium, *Rhodospirillum*, and other photosynthetic bacteria. Bacteria and algae growing on leaves in humid tropical forests also may fix appreciable quantities of atmospheric nitrogen, as may epiphytes growing on the tree branches. Researchers have found that these fixed nitrogen forms may then be used in part by the host trees. There is also recent evidence that some of the higher plants such as tropical grasses may be able to fix nitrogen. Blue-green algae symbiotic with fungi in certain lichens are able to fix nitrogen, as well as photosynthetically produce carbon compounds for the fungal associates.

The numerically most significant nitrogen-fixing bacteria, however, are those that live free in the soil or in intimate associations with legume plants, which live in little pealike nodules on the roots of clover, alfalfa, or other legumes. The bacteria fix gaseous nitrogen from the air into solid form by first forming NH_3. Oxidation of NH_3 to the form of nitrates (NO_3) occurs in the roots and shoots of higher plants and also can be carried out by a combination of *Nitrosomonas* and *Nitrobacter* bacteria. For the biosphere as a whole, it has been

FIGURE 4–3

The nitrogen cycle on land and sea, with the atmosphere serving as principal reservoir and permitting rapid recycling.

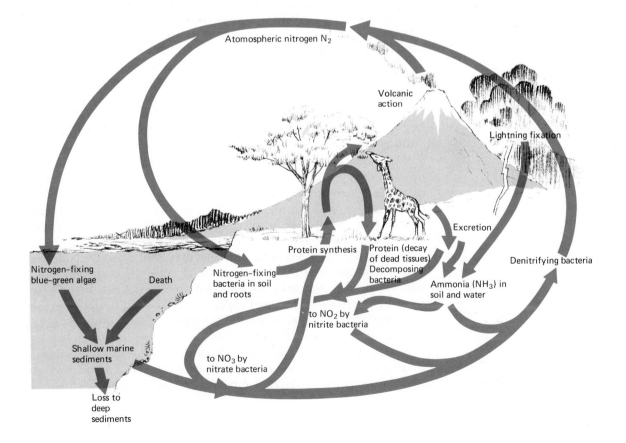

Atmospheric nitrogen N_2

Volcanic action

Lightning fixation

Excretion

Denitrifying bacteria

Protein synthesis

Protein (decay of dead tissues) Decomposing bacteria

Ammonia (NH_3) in soil and water

Nitrogen-fixing blue-green algae

Death

Nitrogen-fixing bacteria in soil and roots

to NO_2 by nitrite bacteria

to NO_3 by nitrate bacteria

Shallow marine sediments

Loss to deep sediments

estimated that 1 to 6 pounds of nitrogen per acre per year are fixed from the atmosphere. Most of this fixation is thought to be due to nitrogen-fixing bacteria, with only a small portion resulting from electrification in lightning storms or in photochemical fixation in the higher parts of the atmosphere. In very fertile areas with heavy densities of legumes and nitrogen-fixing bacteria, up to 200 pounds of nitrogen per acre may be fixed.

In water or in moist soils, such as oriental rice paddies, blue-green algae often perform the essential function of nitrogen-fixing. Increased yields result from adding algal cultures to these rice fields. Such seeding with blue-green algae (which occur naturally in rice paddies anyway) may be very important in maintaining fertility under intensive cropping. Crop rotation in terrestrial fields, especially the alternation of species of legumes one year with the crop plant the next year or two, builds useful nitrogen concentrations in the soil while maintaining soil structure and avoiding runoff pollution such as can occur with heavy nitrate fertilization.

In the tissues of the plants whose roots absorb the nitrates, nitrogen atoms are used principally to synthesize amino acids, which are built into polypeptides and proteins. The plant proteins may then be broken down and build into new protein molecules if the plant material has passed through the digestive system and body cells of herbivores (and later carnivores). Besides playing a role in the synthesis of amino acids, nitrogen is a necessary element for the synthesis of RNA and DNA, the hereditary materials.

When the plant or animal dies, decomposing bacteria and fungi cause the body to decay so that the nitrogen-containing amino acids or nucleic acids are broken down, releasing ammonia gas (NH_3). Nitrite bacteria can convert this poisonous ammonia into simple nitrite (NO_2) molecules. Still other bacteria in the soil, the nitrate bacteria, can add a third oxygen atom to nitrites and produce nitrates. This last molecule is the form of nitrogen most readily used by green plants, and the cycle is thus complete with a biologically usable form of nitrogen again.

As shown in Figure 4–3, the nitrogen may be removed from the nitrates in the soil by the action of denitrifying bacteria, which return it to the atmospheric reservoir. It can be released again from the air by either nitrogen-fixing bacteria or electrification by lightning. In the latter case, the energy of the lightning passing through atmospheric gases during electrical storms combines nitrogen and oxygen into nitrates, which precipitate onto the soil from the air.

Thus when large ecosystems or the biosphere as a whole are considered, the nitrogen cycle is relatively perfect and nitrogen is con-

PLATE 1
CHAPARRAL

The chaparral biome is patchily distributed in lower temperate latitudes around the world, occurring in regions with long dry summers and a brief, wetter winter. Areas bordering the Mediterranean Sea and foothill or lower mountain slopes in the American Southwest provide these conditions for the brushy chaparral vegetation, such as shown in this scene in the San Jacinto Mountains of southern California. Ribbonwood grows in the foreground. Most chaparral shrubs bear thick, resinous leaves that resist drought and desiccation but also are highly inflammable. The name chaparral is derived from the Indian word for scrub oak, chapparo, a common plant in this area. Animal life is relatively rich, including the Virginia opossum, California quail, Anna's hummingbird, and nocturnal ringtail cat.

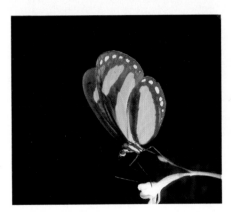

PLATE 2
TROPICAL RAINFOREST

Sweeping across much of the equatorial region of the globe lies the tropical rainforest biome and a slightly more seasonal forest, the tropical
deciduous or monsoon forest. The rainforests of Asia, Africa, South
America, and the Pacific offer the richest diversity of life in all the world's
biomes. As much as 200 or more inches of rain may fall each year, and
the equatorial sun provides ideal rapid growing conditions for a wealth
of plant life. Giant mahogany and other tropical hardwoods thrust their
crowns up through the forest canopy. Inside the cathedrallike interior fly
great numbers of butterflies, such as this glassy-winged ithomiine sipping
at a dimunitive flower on the forest floor in Ecuador, and brilliantly colored frogs, such as this poisonous Dendrobates frog from Costa Rica.
Scarlet macaws and more than a dozen gorgeous but raucous parrot species are typical denizens of the rainforest throughout Central and South
America.

PLATE 3

TROPICAL SAVANNA

Savannas are tropical grasslands with wet-dry annual cycles (rather than a warm-cold cycle as in their temperate counterparts). Many have scattered thorn trees, especially of the legume genus Acacia. Where rainfall is higher, the trees become more dense and a tropical deciduous forest replaces the open grassland. Parts of Australia are characterized by savanna vegetation; the yellow blooms of a wattle tree (Acacia) loom above the highly seasonal grass growth. Red kangaroos and spiny anteaters (Tachyglossus aculertus) are typical Australian inhabitants. The best-known tropical savannas are in East Africa, where green grass and acacias as well as fat impala antelope cover the plains in the rainy season. In the longer dry season, the landscape turns brown and trees are leafless. Baobab trees are found in the low savanna of Senegal in West Africa. Thomson's gazelles leap across the Amboseli plains in southern Kenya.

PLATE 4

TEMPERATE DECIDUOUS FOREST

Springtime in the Ozarks: a time when the trees of the temperate deciduous forest in the eastern United States blossom forth in vivid splendor. This familiar forest biome once covered most of the eastern half of our own continent, Europe, Japan, the southern part of Australia, and even the temperate tip of South America. Civilization has taken its toll of the biome in the northern zones, but scattered patches of the original forest as well as larger extents of second growth still remain. Typical animals include the red fox (Vulpes fulva), the whitetail, or Virginia, deer (Odocoileus virginianus), and the raccoon (Procyon lotor). The American mountain ash, or dogberry (Sorbus americana), flowers brighten the understory. At lower right, we see a forest interior and a close-up view of the trunk of a shagbark hickory (Carys ovata).

PLATE 5

TAIGA AND TUNDRA

The northern coniferous forest or taiga meets the treeless tundra at approximately the Arctic Circle, north of which even evergreen conifers are unable to grow. In the taiga, a bull moose feeds on willows in a boggy meadow. A spruce forest in northern Utah (below) shows the narrow growth form necessary for trees to survive the severe winter snow cover of taiga areas. A mixture of taiga conifers and alders spreads below Mt. Drum (13,002 foot volcano) in the Wrangell Mountains of Alaska. In the tundra, an Alaskan brown bear fishes for salmon. Tundra and bogs interface with scattered patches of taiga on Monahan Flats (below, far right), with the Alaska Range in the distance. Caribou (right) on the tundra near Nome, Alaska, are equivalents of reindeer in the Old World tundra regions.

PLATE 6

DESERT

The arid deserts of the American Southwest are actually richer in plant and animal species than most of the deserts of the northern and southern ends of Africa, the Middle East, Asia, and Australia, not to mention the narrow coastal deserts of western South America. In the large photograph on these two pages, the Sonoran Desert in southern Arizona is characterized by treelike saguaro cacti, several species of prickly cholla cacti, and ocotillo, a tall spindly-stemmed tree with small green leaves and bright red flowers that appear after spring rains. At upper left, a cactus wren nest provides covered shelter from the hot desert sun for the brooding adult and growing young. At far left, California poppies and yucca trees bloom in the Mojave Desert in the springtime. A kangaroo rat, a red racer snake, and a tiger whiptail lizard are typical inhabitants of the American deserts.

PLATE 7
GRASSLAND

The grassland prairies and sand hills of Nebraska have been all but obliterated by the farmer's plow in the last century, but one may still find traces of the formerly vast expanses of the native grassland of the Great Plains. Trees are restricted to river banks and other well-watered areas. The rich seasonal growth of grasses originally supported uncounted millions of grazing herbivores, such as the American Bison (Bison bison).

tinually entering the air by the action of denitrifying bacteria. Nitrogen is continually returning to the biotic cycle through the action of nitrogen-fixing bacteria or blue-green algae and through the action of lightning, and the air becomes the greatest reservoir and safety valve of the system. Loss from the cycle occurs at the nitrate stage, when erosion and the leaching of soils by precipitation combine to wash nitrates into rivers and eventually into the ocean basins. From the shallow sediments near shore, the nitrates may be brought back rather rapidly to terrestrial ecosystems by marine birds and fishes, which return to the land through migration, normal movements, or human exploitation. These consumers feed on plankton organisms that utilize nitrates from the shallow inshore sediments. Commercial fishing brings large amounts of such nitrates to shore in the form of canned and fresh fish, as does the collection of bird guano (deposits of excrement) from offshore islands for use in terrestrial agriculture.

If the nitrates in the shallow marine sediments are not absorbed by eel grass, phytoplankton, or other marine plants, they are gradually lost to deeper marine sediments by erosion and subsidence of that section of the ocean floor. In the deep ocean sediments, this nitrogen may be lost from circulation for several million years, to reenter the nitrogen cycle only upon later geologic uplift of the sedimentary beds or through volcanic action. Underseas volcanoes may erupt through these sedimentary beds and release considerable nitrogen via volcanic gases.

The sedimentary cycles

The reservoir pool of the sedimentary cycles is in the earth's crust. These pathways are said to be imperfect cycles because the elements tend to end up in sedimentary rock from which recycling is slow. In the preceding section we saw that carbon, nitrogen, oxygen, and hydrogen, constituting about 97.2 percent of the bulk of protoplasm, cycle quite easily because the principal reservoir in such elemental cycles is the gaseous form in the atmosphere. The remainder of the approximately 36 essential elements, constituting 2.8 percent of plant and animal tissues, tend to literally go downhill in terrestrial ecosystems. While a clear proportion of the element does cycle, once the bulk of the element has been removed from the terrestrial ecosystem through erosion or other means of downhill transport, the molecules or atoms have no immediate way of returning and hence their cycles tend to extend across long periods of geologic time. Compared with, say, the nitrogen cycle, these sedimentary cycles seem relatively simple in their transfers and components.

The *phosphorus cycle* is an example of a typical sedimentary cycle (Figure 4–4). Phosphorus is a necessary material in the nucleic acids DNA and RNA, bone, teeth, and organic molecules important in energy relationships in the body (ATP and similar compounds). The principal reservoir for the cycle is phosphate rock formed in past geologic ages, although deposits of excrement by fish-eating sea birds and fossil bone deposits contribute substantial phosphate in certain areas of the world. Erosion by rainfall dissolves phosphate out of these reservoir sources, forming a phosphorus pool in the soil of ecosystems. However, much phosphate escapes via runoff into streams into the sea before it is assimilated by plants.

Animals obtain phosphorus by eating plants. On death or through normal excretion of waste products from the body, they return phosphorus to the dissolved phosphorus pool. As noted, much phosphorus is lost in the dissolved state by downhill transport into the shallow marine sediment. In the past, huge colonies of sea birds have played a major role in returning a substantial amount of this phosphorus to land by feeding on fish and other organisms near shore. Migratory fish such as salmon and steelhead, which carry phosphorus in their bones far inland on their return to fresh-water spawning grounds, may also bring substantial amounts of phosphorus back to land in some areas. The ultimate source of phosphorus in the food of these marine birds and fish is the plankton, which have incorporated phosphorus into their bodies from the

FIGURE 4–4

The phosphorus cycle, with sedimentary rock serving as primary reservoir and permitting only slow recycling.

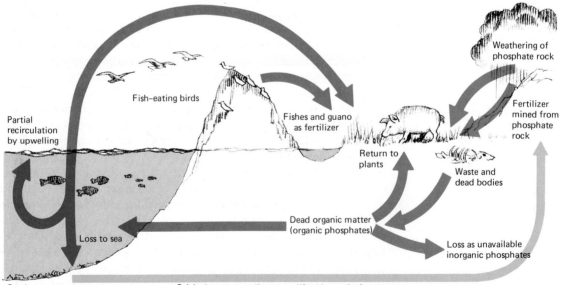

Weathering of phosphate rock

Fish-eating birds

Fishes and guano as fertilizer

Fertilizer mined from phosphate rock

Partial recirculation by upwelling

Return to plants

Waste and dead bodies

Loss to sea

Dead organic matter (organic phosphates)

Loss as unavailable inorganic phosphates

Semipermanent loss to shallow and deep sediments

Original source: sediments uplifted by geologic processes

shallow coastal sediments in estuarine waters, where their early development normally takes place. However, underwater currents and geological subsidence carry a majority of phosphate compounds to deeper marine sediments and from this area they may be brought back to terrestrial reservoirs only through major geological upheavals tens of thousands or even millions of years in the future. In these deep ocean depths the lack of sunlight for photosynthesis, combined with the low temperatures and high pressure, prevents the extensive growth of deep ocean plankton. Thus phosphorus on the deep ocean floor is simply not recycled before being covered by additional sediment, and hence it is lost to the living world. Human beings, unfortunately, accelerate the rate of loss of phosphorus, and thus help to make the phosphorus cycle less perfect, by using phosphate fertilizers. Approximately 1 to 2 million tons of phosphate rock is mined each year in states such as Florida, much of which is washed away and lost after distribution on agricultural fields. Only about 60,000 tons of elementary phosphorus per year is returned by human harvesting of marine fish.

The effect of human beings on biogeochemical cycles

Until recent times, human activities have had little effect on the massive worldwide cycling of elements in the biogeochemical cycles. Agricultural clearing and extensive deforestation might lower the rate of carbon or nitrogen cycling in local areas, or the cycling rate might be improved by planting legume crops and rotating them with more commercially valuable crops such as corn or cotton. However, the use of fossil fuels in the last half century has had major effects on biogeochemical cycles far beyond any earlier effort.

The carbon cycle has been altered in major ways and at accelerating rates by the combustion of coal and oil. Normally, the amount of CO_2 in the atmosphere is in equilibrium with dissolved CO_2 and fresh and salt water. Movement of the gas by diffusion from the air reservoir into the aquatic reservoirs equals the rate of diffusion of CO_2 back into the air from the water. Carbohydrate compounds are manufactured in photosynthesis from CO_2 in the air or water through the interaction of sunlight and chlorophyll molecules. Here the amount of CO_2 is again in equilibrium with the amount tied up in protoplasm and the amount in the gaseous reservoir state. However, since the industrial revolution began, we have been progressively adding more carbon dioxide molecules, dust, and other solid particles to the atmosphere, and in fact it is currently estimated that we are adding some 6 to 9 billion tons of carbon per year to the atmosphere through the combustion of fossil fuel. Not very surprisingly, this influx is apparently changing the composition of the atmosphere in a significant way.

In 1860 chemical procedures were sufficiently advanced that the composition of the atmosphere could be accurately measured. At that date, approximately 283 parts per million (ppm) of the air were CO_2. By 1900, the concentration had risen to 290 ppm. In 1960, however, the carbon dioxide content of the air was 330 ppm. In a century, carbon dioxide concentration has increased an average of 47 molecules of every million total molecules in the air. At our present rate of combustion of fossil fuels, the amount of carbon dioxide in the atmosphere should rise 2.3 ppm per year. The measured rise of about 0.7 ppm, however, indicates that about two-thirds of the carbon dioxide released from fossil fuels is being dissolved in the sea, or possibly is increasing the total carbon biomass of land vegetation.

The effect of this increased amount of CO_2 is not merely to increase the rate of photosynthesis in green plants, a result that ought to produce slightly higher oxygen concentrations and hence

FIGURE 4–5

The "greenhouse effect." Carbon dioxide and water vapor in the earth's atmosphere are translucent to the sun's electromagnetic radiation in visible wavelengths of light. These wavelengths heat the earth's surface and infrared wavelengths ("heat") are radiated back into the atmosphere. Carbon dioxide and water vapor, however, acting like glass in a greenhouse roof, do not allow these to pass. Hence the temperature of the atmosphere rises from the trapped infrared radiation.

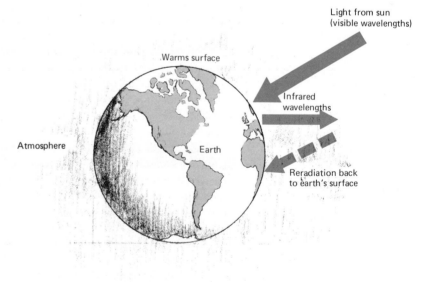

FIGURE 4–6

Air pollution from a modern industrial factory can lead to changes in local and even regional weather patterns.

be of some human interest but not critical concern. The importance of this change in atmospheric composition is that the amount of CO_2 in the air is a critical factor in maintaining the earth's temperature. Carbon dioxide along with water vapor in the atmosphere is translucent to the sun's radiant energy in the wavelengths of visible light, but it is not translucent to the longer infrared wavelengths. Of the radiant energy directed toward the earth from the sun, about 29 percent on the average is reflected back into space. About 20 percent is absorbed by the ozone layer in the outer boundary of the stratosphere (at about 30 miles above the surface of the earth). The remaining 51 percent travels through the atmosphere and is absorbed at the earth's surface. When this solar energy in the form of light rays warms the earth's surface each day, much of the light energy is reradiated as heat (infrared energy) (see Figure 4–5). However, CO_2 and water droplets trap these infrared wavelengths in the atmosphere and impede the loss of this heat energy from the earth's surface just as a blanket impedes loss of heat from the body of a person. The increase in heat produces the so-called greenhouse effect on the earth's temperature. In a greenhouse, light energy passes through the glass roof, is absorbed by the plants and soil inside the structure, and is reradiated as heat energy. This energy cannot pass through the glass as readily as the higher-energy visible light waves, and hence the temperature in the greenhouse rises. In the last 100 years there has been a 1°C rise in the average temperature across the earth (at least until 1945). This rise correlates exactly with the increasing amount of CO_2 in the atmosphere. However, since 1945 the rise seems to have been halted or even reversed, perhaps by the increased amount of particulate matter in the atmosphere. It has been suggested that these pollutant particles are now screening out more of the sun's radiation before it ever reaches the earth (Figure 4–6).

In the industrialized regions of the globe, especially the Northern Hemisphere areas of the eastern United States and western Europe, we are overwhelming natural processes involved in the sulfur cycle (Figure 4–7). The major atmospheric burden is being contributed through combustion of fossil fuels that contain high amounts of sulfur. The removal processes are slow enough (several days at least) so that the increased sulfur concentration in source areas is spread for hundreds of thousands of kilometers downwind. The results on crops and wild plants are clearly detrimental in areas of heaviest concentration of pollution, such as southern California. Regional and global air pollution caused by sulfur compounds in atmospheric circulation is thus becoming a serious health hazard and a significant example of problems generated by human effects on the normal geochemical balance of nature.

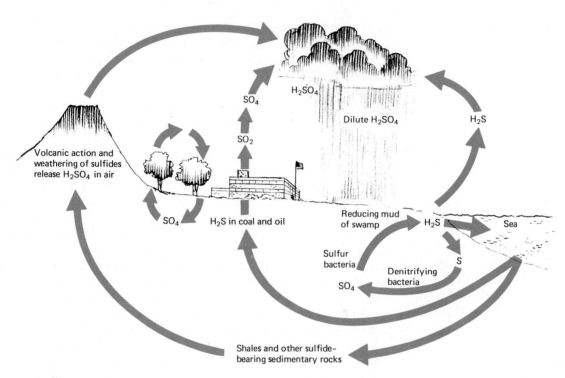

FIGURE 4-7

The sulfur cycle. When rocks are weathered, part of their products are oxidized to sulfates, which are cycled in ecosystems. In reducing muds (e.g., in swamps), sulfur is cycled between sulfur bacteria, which reduce sulfates to sulfides, and other bacteria, like some denitrifying species, which oxidize sulfides. H_2S returned to the atmosphere is spontaneously oxidized and delivered back to earth or sea by rain. Sulfides incorporated in fossil fuels and sedimentary rocks are eventually oxidized by human combustion or crustal movements of the earth and weathering.

Excess nitrates in water bodies such as fresh-water ponds, lakes, and rivers often cause algal blooms. The increased availability of nutrients temporarily stimulates the growth of these unicellular green plants, such that they rapidly use up the available dissolved oxygen in the water and choke out all other life, finally dying themselves. The decomposition of their bodies by decay bacteria continues to utilize the remaining available oxygen, and the body of fresh water becomes stagnant and so polluted that other life forms cannot survive.

Nutrient cycling and the role of limiting factors

Certain elements and other materials are essential for the growth of organisms. If these particular materials are lacking in the soil, it will be impossible for a plant to thrive or even persist in a particular environment. Animals that depend on the nutrient will be unable to secure new sources. Those materials in shortest supply are said to be limiting factors: that is, elements or other materials that are present in amounts close to the critical minimum needed for survival. Justus Liebig's *law of the minimum* expresses the idea that under steady state conditions (when the inflow balances the outflow of energy in materials), the organisms will be able to persist and be limited only by those elements that most closely approach the critical *minimum* needed for growth. When Liebig expressed this proposition in 1840, it was a result of his studies on the effects of various factors on the growth of plants. Crop yield was often limited not by nutrients that were found in large quantities, such as water, but by some mineral such as iron that would be needed only in minute quantities yet be relatively scarce in the environment. Thus he said that "growth of a plant is dependent on the amount of food stuff which is presented to it in minimum quantity."

Victor E. Shelford in 1913 modified the concept of the limiting effect of the minimum quantities by his *law of tolerance*, which states that ecological factors can also limit plant and animal existence at certain *maximum* levels. Thus there exists a precise range between maximum and minimum limits in which a particular species can exist. The physical requirements of a species dictate in general the distribution of the organism. If the individuals in that species have wide ranges of tolerance for all factors, they will likely be quite widely distributed. If, however, that species has a wide range of tolerance for one factor, and a narrow range for another, the minimum tolerance factor will dictate the organism's distribution and growth. Limiting factors may involve temperature, water, salinity, food, or other physical conditions and may involve various elements, as we have already stated. Thus in general the existence or success of an organism in a particular area depends on a complex of conditions. Any condition that approaches or exceeds the limits of tolerance is said to be a limiting condition or a limiting factor. The type of biogeochemical cycle that transports these materials may play an influential role in limiting the growth of species populations in the typical ecosystem.

Summary

Biogeochemical cycles involve the circulation of chemical elements and other materials through the ecosystem. Materials necessary for the survival and growth of organisms are called nutrients. Macronutrients are required in large amounts, whereas micronutrients are required in only trace amounts and often act as limiting factors for organisms. The two general types of biogeochemical cycles are the gaseous (perfect) and the sedimentary (imperfect) cycles, so-named after the location of their primary inorganic reservoir. Only four elements—carbon, nitrogen, oxygen, and hydrogen—have gaseous cycles; the other 36 or so elements (e.g., phosphorus and sulfur) essential to life have sedimentary cycles. Human beings are affecting biogeochemical cycles and the earth's climate through massive combustion of fossil fuels.

✗5   Environmental degradation and pollution

We have seen that the earth's biosphere is intimately related to the cycling of chemical elements. Natural processes in the atmosphere and soil have always governed the well-being of life on earth. In many ways, we have made our environment more hospitable for self-centered reasons through air conditioning and heating by consumption of fossil fuels and other manipulations of environmental factors, but we have also altered the biosphere to make it less suitable for life in general. In particular, overpopulation and industrialization have contributed to a general deterioration of the quality of our environment. In this chapter we shall be looking at natural and human changes induced in the environment and especially at the various forms of pollution that adversely affect human health. Some of the pollutants affect us directly; others affect us indirectly through their alteration of the complex environmental systems on which all human existence ultimately depends.

Natural and human alteration of the landscape

As long as the human species has been on the earth, natural processes have altered the landscape in substantial ways. Flooding, erosion, fire, landslides, and coastal wave action have all played a

FIGURE 5-1

The effects of overgrazing in the Middle East.

part in altering landscapes through time. The addition of *Homo sapiens* to the biota of the earth at first did little to affect these natural processes and events. However, in the last several centuries, the extensive natural forest and grassland acreage cleared for agriculture has been exposed to flooding and other so-called natural events.

Across the Middle East, over several thousand years forest and shrub land has been gradually converted into harsh desert through overgrazing by flocks of goats and sheep and other human activities (Figure 5–1). Yet recent satellite pictures of the 3,000-mile sweep of sub-Sahara savanna stretching from Senegal into Sudan—called the African Sahel—revealed an unexpected hexagonal island of green in the middle of the vast arid landscape. Closer inspection showed this island to be a quarter-million-acre modern ranch, fenced off with barbed wire from the surrounding desert. Within the ranch, other fences divided grazing land into five sectors, with cattle grazing a single sector at a time. The ranch has been in operation only since 1969, yet the introduction of rotational grazing has made the difference between pasture and desert. Thus there is a chance to overcome the effects of drought and overgrazing in even such long-battered lands as the African Sahel.

Ocean wave action can destroy the coastal environment wherever we have interrupted the normal environmental safeguards that prevent beach and coastline erosion. Plantings of beach grasses or the construction of offshore breakwaters of rocks or junked cars help to alleviate these problems, just as the construction of levies and networks of dams can prevent or largely alleviate the effects of damaging fresh-water floods. Let us look in more detail at several of the factors, natural and of human origin, that change, pollute, or otherwise threaten natural communities.

Fire

When plant remains accumulate in a terrestrial ecosystem or a dry season allows plants to become combustible at some point in the year, fire may occur as a natural phenomenon. Lightning or human acts will set off uncontrolled fires through such ecosystems. Historically, great fire disasters in the nineteenth and early twentieth centuries caused the American public to consider fire in natural communities as inherently undesirable. However, in recent years, it has been recognized that where fire has been a natural part of the established forest environment, periodic fires are frequently necessary for effective forest management. Thus in the southeastern United States, fire is required for successful reproduction in longleaf pine forests (Figure 5–2). These tall, graceful pines grow along the coastal plain from Virginia south to Florida and west

FIGURE 5–2

The growth stages of the longleaf pine in relation to periodic ground fires in southern U.S. forests.

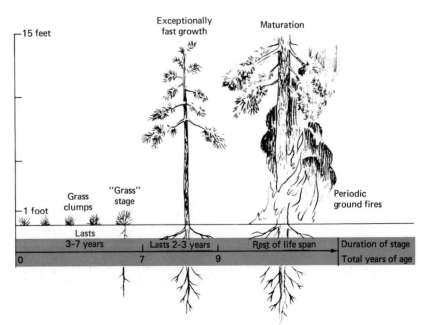

to Texas. In their natural state, longleaf pines are well spaced, and there is a light cover of grass on the forest floor. Grass fires that sweep through such forests eliminate much less valuable hardwoods or other species of pine while not affecting the longleaf pines. In the life history of this tree, for the first three to seven years after sprouting, the young longleaf pine consists of a long taproot bearing a thick terminal bud above ground, which is protected by long, tufted needles. Grass fires sweep by too rapidly to penetrate and harm such a bud. During the next two to three years, the pine grows exceptionally fast, reaching 10 to 15 feet in height and thus raising its more sensitive branch growth to a level above the vertical range of the normal grass fire. The development of thick, fire-resistant bark protects the young tree from further ground fires.

Thus as long as periodic ground fires sweep through the forests, the fire-susceptible hardwoods are kept out and the young longleaf pines can grow without much competition. Clearly, fire is necessary to maintain the economically desired climax community of longleaf pines. If periodic ground fires are not permitted in such forest areas, there will soon be no reproduction of the pines at all. Today this is recognized by forest managers, and the use of periodic control ground fires is a common practice. Likewise, in southern California and Arizona, the chaparral plant community is recognized as a fire-maintained climax (Chapter 6). Periodic fires are essential for maintaining the basic ecological character of the chaparral community — a shrub climax adapted to a dry Mediterranean climate — and thus prevent disastrously hot fires from occurring after long periods of growth in the absence of fire (Figure 5–3).

FIGURE 5–3

A forest fire in the chaparral of Angeles National Forest in southern California sweeps rapidly up the steep mountain slopes.

The clearing of land

A major, historically significant means of landscape alteration has been the clearing of land for cultivation. In recent decades, we have reached essentially the limit of suitable land available for productive cultivation. The chief remaining biotic communities seriously threatened by agricultural expansion are the *lowland coastal deserts* found on most continents and the *tropical rainforests* of the Amazon basin, Southeast Asia, and Africa south of the Sahara. Countries such as Israel are showing that the desalinization of sea water offers one practical way to surpass the chief limiting factor to productivity of the deserts, namely, the lack of fresh water. Tropical rainforests, on the other hand, present more of an ecological problem. When the forest is cleared in most areas, *laterization* of the soil occurs. Despite

FIGURE 5-4

Land-clearing in a tropical American rainforest; slash-and-burn cultivation of bananas by native Indians.

the lush appearance of tropical rainforests, the soil is usually thin and poor (Figure 5–4). When this soil is exposed to direct heavy rainfall by the clearing of a forest, silica and other soil minerals are leached downward, and organic material is oxidized. The thin topsoil is soon eroded by the rain, and a layer of aluminum and iron oxides is exposed, which in the air can form a hard, impermeable red crust called *laterite*. Once formed, this crust becomes essentially permanent, and the kinds of vegetation that can grow there, including crop varieties, are very few.

It has also been found that clearing large rainforest areas may raise temperatures, alter precipitation patterns, and in general change the basic climate of a region. Scientists working in northern South America have reported dramatic decreases in rainfall — as much as 21 to 24 percent from earlier normal rainfall amounts — over large areas of the Amazon basin rainforest in Colombia during the past 26 years. A correlation exists between these changes and the felling of vast tracks of rainforests in Colombia and neighboring countries. As one of the scientists working on the problem has warned: "Should the decrease be widely spread throughout the continent, the consequences could seriously retard the development of the tropical

Latin American countries. . . ." In Ecuador, since 1973 it has been found that the smoke pollutants emitted by burning of waste gas and oil in the oil fields of the Amazon basin area are apparently increasing the amount of rainfall. In the Columbia basin area of Washington state, several meteorologists have found that there has been a significant increase in precipitation, resulting apparently from irrigation water applied to some 300 square miles of formerly near-desert land in the center of the basin. During the 12 years following the initiation of irrigation, July and August rainfall in the basin increased significantly.

Urban clearing

The worldwide phenomenon of urbanization appears to be far more threatening to natural communities today than is the further expansion of agricultural clearing. The continued migration to cities, which increases the urban population, and the high birth rates, which is increasing the total population, swells the growth of urban areas. Continually greater demands on the available land for housing around the urban fringe are made by this increase in urban population, particularly with the trend toward spacious suburban residential development. In the United States alone, it is estimated that urban land area will have increased from 21 million acres in 1960 to 45 million acres in 2000. Formerly productive agricultural land used as "truck gardens" near cities is rapidly turned into housing as the growth in its value on the tax rolls makes it impossible for the farmer owners to continue raising crops at a profit. Natural wooded areas and bottom lands that formerly provided recreational offerings to city populations are also turned into housing developments, no longer available to fishermen, hunters, hikers, and picnickers looking for a pleasant place to spend an afternoon (Figure 5–5).

The increase in urbanized land has taken undeveloped wild areas not only from surrounding terrestrial communities, but also through filling in shallow coastal areas. Land developers may purchase shallow areas of a bay or estuary and dredge up fill material from adjoining bay bottom or bring in solid waste and soil fill from adjacent land areas. At Tampa Bay in Florida, these dredge-and-fill operations have filled in 15 to 20 percent of the original water area, and in San Francisco Bay on the West Coast some 33 percent of the original bay has been filled. Such filling destroys principal spawning areas for a host of marine fish and invertebrates, thus decreasing the flow of marine products to consumers as well as upsetting the

FIGURE 5-5

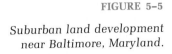

Suburban land development near Baltimore, Maryland.

ecology of the ocean areas. The destruction of bay habitat also destroys wintering places for migratory waterfowl and may even change the local weather pattern. Approximately 1 million waterfowl winter on San Francisco Bay, the most important stopping place for migratory waterfowl on the great Pacific flyway. Since some 70 percent of the bay is less than 12 feet deep and therefore easily fillable, these birds could potentially lose their entire wintering and feeding grounds. If the bay is filled in, leading to a diminution of the cool summer fogs formerly characteristic of San Francisco, meteorologists predict a sharp increase in summer temperatures and a drastic increase in smog.

These communities in California and Florida and other coastal areas are not ignoring the ecological problems of filling in bays. Many have established so-called bulkhead lines, where no commercial filling and developing is allowed beyond a certain water mark, usually at the mean high tide mark. Many areas require ecological surveys of bay areas before permits are granted for dredging, filling, or sale, and some counties have placed moratoriums on the sale and dredging of submerged land. The widespread public concern over the quality of the environment has been instrumental in bringing about these important restrictions.

Herbicides and warfare

The Vietnam war, especially in the late 1960s through 1972, brought home the potentially devastating effects that the widespread use of herbicides and other defoliant chemicals may have on the natural environment. Defoliant chemicals in considerable quantity are sprayed along roadsides and railroad tracks in the United States to kill weeds and other unwanted plant life, but nowhere in the world were herbicides used in such quantities as in Vietnam. Some 13 to 14 million pounds of the defoliants 2,4-D and 2,4,5-T (dichlorophenoxyacetic and trichlorophenoxyacetic acids), all of these chemicals produced in the United States during 1967 and 1968, were sprayed on the Vietnamese forests along with other herbicides by the U.S. military (Figure 5–6). The purpose of these herbicides was to kill all vegetation from the ground up, in order to render enemy units more visible in the thick, natural forests. Some of the herbicides were also used to destroy food crops used by enemy soldiers. Agent Orange, a mixture of the n-butyl esters of 2,4-D and 2,4,5-T, was the most commonly used defoliant for general forest and brush clearing because it attacks a wide variety of broad-leaved plants. The C-123 sprayer aircraft could carry 1,000 gallons of the defoliant and spray its load in slightly over four minutes, flying just above the treetops at about 125 miles an hour. In several weeks, the sprayed forest would become leafless, and many or all of the trees might die, depending on their susceptibility to the poison. Repeatedly sprayed areas would lose nearly all flora and fauna. Forest fires raged through the dead woods (living rainforests do not ordinarily burn), further exposing the soil to erosion and laterization. Near Saigon, a single forest fire

FIGURE 5–6

The effects of chemical herbicides on mangrove forests in Vietnam (1970). To the left is an aerial view of an unsprayed forest 60 miles from Saigon. To the right is a nearby forest area that was sprayed with herbicides more than three years previously.

FIGURE 5–7

Craterization from B-52 bombing raids in South Vietnam.

destroyed nearly 100,000 acres of woods that had been heavily treated with herbicides. These short-term changes might be relatively minor compared with the long-term effects such as destabilization of coastal lines and waterways. Mangrove swamps that had been heavily sprayed did not regenerate over a five-year period. Since mangrove trees help to stabilize coastlines and rivers with their extensive root systems, there is some fear that erosion may badly damage many of these coastal waterway areas. Ecologists have also noted that dense bamboo stands (which prevent forest reproduction) developed in the less heavily sprayed areas of rainforests. An increase in the number of birth defects among babies of Vietnamese mothers who lived in sprayed areas was also indicated by several studies.

More normal forms of warfare cause other types of landscape alteration, of course. In Indochina, great numbers of bombs dropped on both forest and agricultural areas have resulted in craterization of the landscape on an immense scale, perhaps the largest excavation project in human history. The total cratered area throughout Indochina was estimated in the early 1970s to exceed the area of the state of Connecticut by 5,000 square miles. Since these craters filled with water and remained wet even in the dry season, they provided breeding sites for mosquitoes and ponds for aquatic snails that serve as intermediate hosts for parasitic worms that later infest human beings. While no measurements of the total ecological and agricultural impact of these bomb craters have been made, at least the short-term effects are likely to have been considerable (Figure 5–7). In both wartime and peacetime activities, construction projects for airfields, highways, dams, and other projects often alter many square miles of terrain that would not normally be affected.

The clear-cutting of forests

One of the most ecologically destructive methods of logging is clear-cutting. This economical and efficient (essentially 100 percent yield) procedure involves cutting all the trees in an area rather than cutting selectively, where the yield from mature trees is about 60 percent. Unfortunately, the U.S. Forest Service has increasingly authorized clear-cutting in our 182 million acres of national forests (Figure 5–8). Under a long-standing policy formalized by 1960 federal law, the 154 national forests are supposed to be maintained for multiple use, including timber, recreation, grazing, wildlife preservation, and watershed protection. Unfortunately, as a result of political pressure in 1971 from major western lumber companies dissatisfied with the cutting rate on their own privately held lands, President Nixon ordered the Forest Service to give top priority to private logging in the public national forests.

FIGURE 5–8

Clear-cutting in a Washington forest.

Aside from the loss to recreation, grazing, and wildlife protection, perhaps the most dangerous result of clear-cutting is watershed degradation. A *watershed* is an area whose rainfall collects and drains downhill into a particular stream or set of streams. The extent and character of the hillside cover of plant growth determine how fast the runoff is and how much sediment is picked up by the water en route to the stream. Clear-cutting has been shown to cause a 7,000-fold increase in stream sedimentation. Such a concentration of sediment particles can ruin the breeding grounds of fresh-water fish. In the late winter, when heavy snow melt or rainfall on clear-cut slopes leads to rapid runoff, devastating downstream floods can result. In general, these adverse affects of clear-cutting are a far greater economic cost to the whole human population of a region than is the expense incurred by lumber companies restricted to selective logging.

Public works projects

Giant construction projects, ostensibly undertaken to "control" the environment and bend it to immediate human needs, often have deleterious side effects that make the overall scheme an ecological as well as economic loss. Any newspaper-reading member of the American public knows that great number of projects in the past several decades have been proposed and sometimes carried through despite substantial environmental objections. Thus great coal-burning power plants in the American Southwest are currently polluting vast sections of Indian reservation land. Power and irrigation dams have been repeatedly proposed for the sections of the Colorado River inside the Grand Canyon of Arizona. Only the vigilant outcries of millions of conservation-minded supporters of the Grand Canyon National Park's environmental integrity have protected this remaining wild section of the Colorado River.

One mammoth World War II project successfully stopped only by presidential intervention was the U.S. Army Corps of Engineers' Cross-Florida Barge Canal. Originally this project proposed to cut a canal across the upper, narrow part of the Florida peninsula, with the intent of allowing barge and other coastal maritime traffic between the Atlantic Ocean and the Gulf of Mexico to avoid the threat of German submarines operating off southern Florida and the Caribbean. The war ended before the project could be initiated, but in the late 1960s the plan was revived and construction was actually begun by the Corps of Engineers. The Corps also undertook an elaborate propaganda campaign, citing supposed economic benefits to be derived from cheaper shipping costs. Environment-oriented econo-mists and conservationists soon discovered that the grossly favor-able estimates used by the Corps were false. Not only was the canal shown to be a potential disaster economically, but its proponents also failed to justify its construction in the face of the ecological damage it would have done. The proposed route would have created lengthy, biologically sterile channels in place of ecologically rich rivers that currently provide considerable recreational benefit to fishermen and boaters. It also would have destroyed more than 60 miles of natural communities adjacent to the canal, five major fresh-water springs, and several large unspoiled lakes. This Army Corps of Engineers project was stopped by President Nixon in 1972. Never-theless, development boosters in Jacksonville and other communi-ties of northern Florida still narrow-sightedly desire the further de-velopment of this project, and only an aroused and ecologically informed public can prevent a small body of development-oriented, dam-building proponents from achieving such plans.

The construction of dams has often served to control environ-mental problems such as annual flooding, but in many cases it has also brought new serious environmental problems such as water evaporating from reservoirs in desert areas and silt rapidly filling up the reservoir basins behind the dams. One of the greatest dams in the world, the Aswan High Dam, was built with Russian aid in central Egypt to harness the Nile River and vitalize the stagnant Egyptian economy. Its construction has irrevocably changed the environment of the Nile River valley and the lives of the inhabitants. The results have not all been positive, and there is growing debate about whether the economic benefits of Aswan are outweighed by the ecological damage that it has caused.

Environmentalist critics of the dam point out that it has obstructed the natural flow of silt that formerly enriched the soil of farms during the annual flooding along the Nile. Thus it has been necessary for Egypt to increase the use of imported chemical fertilizers on farms downstream from the dam. The elimination of the silt flow has also increased the rate of erosion along the Mediterranean coast adjoining the Nile delta. In addition, the absence of the organic matter in the silt in the waters at the river's mouth has deprived sardines and shrimp of an adequate food supply. As a result, Egypt's sardine in-dustry, which used to produce 8,000 tons of the fish annually, has all but disappeared. The flood control provided by the dam has posed other problems. Rodent populations along the Nile's banks, which were formerly held in check by the cyclical floods, are now increas-ing and creating problems for residents. In towns bordering the river, sewage systems that were once regularly flushed out by the flooding and subsequent receding of the river have become badly clogged. A further serious criticism has resulted from one of the primary purposes of the dam: flood control. Drainage is now insufficient in more than 1.2 million acres of the nation's richest farmland below the dam. Much of that land has become increasingly saline, reducing agricultural productivity by as much as 50 percent. Schistosomiasis and other human parasitic diseases that have aquatic snails as intermediate hosts have increased in the region.

On the positive side, plants powered by electricity generated at Aswan should eventually make the country self-sufficient in fertilizer production. The loss of the sardine industry may be counterbalanced by the new fishing industry on Lake Nasser, which covers 2,000 square miles behind the dam. Fishermen are now taking river bass, Nile catfish, and carp from the lake, and government experts estimate that annual catches will eventually rise to as high as 60,000 tons. Drainage canals will be built to solve the inadequate drainage of farmland. Nevertheless, the demands of the growth rate of Egypt's population (at a rate of 1 million per year) have already outstripped the potential capacity of the dam for improving Egypt's economy and ecology.

In North America, the current largest construction project is the great Alaskan pipeline being constructed from Prudhoe Bay, on the North Slope of Alaska, some 798 miles south to the port of Valdez (Figure 5–9). After years of delay, construction work finally began surging ahead in June 1975.

FIGURE 5–9

The construction of the trans-Alaskan pipeline between Valdez and Glen Allen in the Alaskan wilderness.

The spongy oil-soaked strata nearly two miles beneath the tundra at Prudhoe Bay contains an estimated 9.6 billion barrels of oil, by far the largest deposits in the United States. Initially, the pipeline will carry 1.2 million barrels per day, an amount equal to one-fifth of the nation's current oil imports. The project is not without its environmental hazards. En route, the pipeline must cross the Brooks Range, climbing 4,800 feet in elevation, go on to span 34 major rivers and streams, and finally pass through the Alaska Range at 3,500 feet before descending to the deep-water, all-year port at Valdez. Under pressure from environmentalists, the Alyeska pipeline consortium has modified the design of the pipeline to decrease ecological threats. One major problem is the temperature of the oil, which comes up out of the ground at close to 180°F and will flow through the pipe at about 140°F. The heat could damage the tundra vegetation as well as disrupt the underlying permafrost, the frozen subsoil that covers much of Alaska. Where conditions warrant it, the pipeline will be well insulated and buried 3 to 12 feet below the ground, but for 382 miles the pipeline will be elevated 4 feet or more above the surface to keep the heat away from unstable permafrost.

To avoid disrupting the migration of caribou every spring, the pipeline will be buried for about 7 miles where it crosses the route of one herd of 450,000 caribou in the Copper River valley. To ensure that the permafrost stays frozen in this particularly sensitive area, the pipeline will be refrigerated by surrounding conduits filled with brine. The greatest fear of the environmentalists as well as the pipeline consortium is that a section of the pipeline may rupture, spraying a hot brown fog of oil across the landscape. To make the danger more acute, the line crosses three major earthquake zones. In these areas the pipe will run above ground on special mounts that will allow the 80-foot lengths of welded steel to whip 20 feet from side to side and 5 feet up and down without breaking. The Alyeska consortium believes that this arrangement will allow the pipeline to ride out earthquakes that register up to 8.5 on the Richter scale. If the pressure in the pipeline drops by 1 percent or more—a warning sign of a rupture—electronic sensors will detect the change, and technicians in Valdez can close down the affected segment of line in seven minutes. Alyeska believes that it can confine any major leak to a maximum of 50,000 barrels, and the group of cooperating oil companies has emergency plans to dam, bury, and soak up the spillage. Probably a greater danger is that the pipeline is leading to uncontrolled development of the state. The Alyeska-built service road that runs north from Fairbanks to Prudhoe Bay will eventually become a means of opening the fragile Alaska wilderness to people, including

FIGURE 5–10

Smog over Los Angeles, California.

their garbage, sewage, and guns. Bus companies have already applied to drive tourists to the North Slope on this road, and the U.S. Bureau of Land Management has plans to turn five of the present construction camps into vacation centers. Alaska presents a tremendous opportunity for both development of its enormous resources and preservation of great areas of wilderness for the future enjoyment of humankind. It remains to be seen how well Alaskans will handle both the problems and opportunities inherent in these resources.

Air pollution

Air pollution is perhaps the most familiar form of pollution to most of us. Gaseous wastes and particulate matter such as soot and dust are the principal components of air pollution, but a surprising variety of substances is introduced into our atmosphere, especially around developed urban areas and manufacturing sites. Smog has been observed over oceans, over the North Pole, and even over remote jungle areas of South America. Air pollution kills people, rots nylon stockings and windshield wiper blades, corrodes paint and steel, blackens skies, and destroys crops and forest trees (Figure 5–10).

The U.S. Public Health Service estimates that iron and steel mills, pulp and paper mills, petroleum refineries, smelters, and chemical plants contribute annually to the atmosphere 2 million tons of carbon monoxide, 9 million tons of sulfur oxide, 3 million tons of nitrogen oxide, nearly 1 million tons of hydrocarbons, and 3 million tons of particulate matter. Carbon monoxide, a colorless and odorless gas, combines more readily with hemoglobin in the red blood cells than oxygen gas does, and thus it prevents the hemoglobin from carrying oxygen. Death from oxygen deficiency occurs if a person is exposed even to low concentrations of this gas. Sulfur oxides (especially sulfur dioxide), nitrogen oxides, and hydrocarbons are the principal components of "smog." By the time we consider the fuel burned for heating offices, houses, and apartments, as well as trash burning and industrial wastes, we are including such pollutants as ozone and other oxidants, arsenic, asbestos, beryllium, fluorides, lead, organic and inorganic pesticides, and sulfides. As noted, since the advent of the human species as a technological animal, the chief sources of air pollutants have been processing industries and power-generating plants that burn fossil fuels, as well as the ubiquitous internal combustion engine. In nature, air pollution rises principally from forest fires and volcanic eruptions, but because of the infrequency with which these events happen and the short time they last, natural pollutants released into the atmosphere are soon dissipated by air movements.

The principal known effects of air pollution are, of course, on human health. Carbon monoxide occupying the hemoglobin molecules causes a reduced oxygen supply to the cells and thus the heart must work harder, as must the other components of respiratory and oxygen-transporting mechanisms (lungs, diaphragm, and other muscles). These effects may produce a critical strain in people with heart and lung diseases who live in atmospheres containing high amounts of carbon monoxide. Chronic respiratory diseases such as bronchitis and emphysema are caused by oxides of sulfur, which harshly irritate the respiratory passages and thus cause coughing and choking. The effects of air pollutants are thought to be a major cause of the abnormal death tolls that have occurred during smog disasters in London, Los Angeles, and other cities.

The effects of air pollution are not restricted to human disease. Many species of flowers, shrubs, and trees are unable to thrive or even survive in cities such as Los Angeles. Orchids, bromeliads, camellias, gladioli, and other horticulturally important flowers are damaged by smog. Agricultural crops and mountain pine forests as far as 100 miles away from the Los Angeles civic center are being damaged or killed. In late 1970, an aerial survey of the Angeles National Forest, which borders Los Angeles, showed that 261,000

Ponderosa pines had been damaged by smog. Not only are the pine community food webs and other biological relationships being altered, recreational and lumbering activities possible in such areas will undoubtedly change with the drastic rate of loss of the dominant tree species.

Smoking in closed areas such as classrooms, conference rooms, airplane cabins, or cars cause high levels of particulate pollutants and dangerous gases such as carbon monoxide. In a January 1972 report to the Congress, the U.S. Surgeon General announced that the health hazard from such pollutants is not limited to smokers. Nonsmokers in a smoke-filled room may be exposed to harmful levels of carbon monoxide that are especially dangerous to persons already suffering from chronic coronary and bronchopulmonary diseases. Carbon monoxide in such concentrations (20 to 80 ppm) causes loss of vision and hearing, decreased muscular coordination, and hampers an auto driver's ability to respond to light and to judge distance. Higher concentrations of carbon monoxide cause unconsciousness and even death by further interfering with the transport of oxygen in the blood. Of course, the smoker himself is subjected to far greater pollutant levels than is the nonsmoker. Vincent Schaefer stated in the journal *BioScience* that "in the process of smoking, the individual insults his lungs with a concentration of at least 10 million smoke particles per cubic centimeter. This is a concentration that is 10 to 100 times greater than is encountered in a very badly polluted urban area like Los Angeles or New York City."

Growing urban populations clearly increase air pollution in cities, especially those that are affected by temperature inversions. Normally, the temperature of the atmosphere decreases steadily with increased altitude above the ground; but during an inversion a layer of warm air overlying the cooler air below severely limits vertical mixing of the atmosphere, and pollutants accumulate in the cooler air trap near the earth's surface. Because of the wind patterns in the eastern Pacific and the circle of mountains surrounding the Los Angeles basin, this valley is an ideal place for the formation of inversions, usually at about 2,000 feet above the floor of the basin. Since inversions occur there about 7 days out of every 16, smog collects in abundance.

The great amount of sunshine in the Los Angeles area also contributes to its air pollution problems. Sunlight acts on a mixture of nitrogen oxides, oxygen, and hydrocarbons to produce what is known as photochemical smog. The exhaust from over 3 million cars pumped into the atmosphere of the Los Angeles basin daily, in addition to the wastes discharged by oil refineries and other industries, causes the air that the residents of Los Angeles breathe to contain far more than the usual mixture of noxious and poisonous gases. Doctors advise tens of thousands of residents each year to move out of Los Angeles for smog-related health reasons.

Water pollution

Around the world, water pollution probably offers more direct threats to human health than any other form of pollution (Figure 5–11). We seldom stop to consider the routine filtration and chemical treatment of urban water supplies today in developed countries, but without this prophylaxis, disease organisms of many varieties — such as those that cause typhoid, paratyphoid, cholera, and various dysenteries — would be widespread pollutants among our urban populations. Even today, infectious hepatitis is spreading alarmingly in the United States because chlorination is not as effective as it should be against many viruses, including those that cause hepatitis.

Other aquatic pollutant materials include almost every variety of contaminant that is poured into our water supplies by industry and agriculture. Detergents, acids, ammonia, lead, nitrates, and more exotic chemicals cause danger of poisoning and often get into ground water where purification is almost impossible. Current agricultural practices using heavy fertilization result in a flow of a heavy load of nitrates into our water supply. In themselves, nitrates are not especially dangerous, but certain bacteria present in the digestive tract will convert nitrates into highly toxic nitrites. Farm animals and human infants are particularly likely to include in their digestive tracts the types of bacteria that are able to convert nitrates to nitrites. A disease can result that terminates in suffocation. Nitrate water pollution has become a severe public health hazard in the Central Valley of California, where doctors often recommend that infants be given purified bottled water rather than tap water piped from wells.

Water pollution with sewage is of widespread concern, especially in communities along large rivers in the central United States. With increasing populations, increasingly more elaborate and expensive treatments are required to keep the water safe for human use and to maintain desirable fish and shellfish populations in the river. Storm drains often carry pollutants washed off paved streets and other drainage areas directly into a water supply, with no filtering or other treatment to which sewage is normally subjected. A major source of such pollution in the northern states is the rock salt used for de-icing streets during the winter. Thus, for example, during the winter of

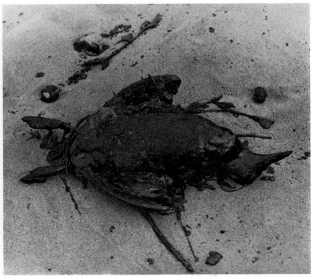

A   B

FIGURE 5–11

(A) Water pollution often results in massive fish kills. (B) Oil spills affect water birds as well as aquatic marine or fresh-water life, as shown by this oil-coated bird on a beach near San Francisco.

1969–1970, some 77,000 metric tons of salt were used for de-icing roads in the densely populated Irondequoit Bay drainage basin near Rochester, New York. Approximately 32,000 metric tons of salt went into the bay, but the rest remained stored in the soil and ground water of the basin. The chloride concentration in this small bay (1 by 3.7 miles, with a maximum depth of 75 feet) has increased at least fivefold during the past two decades. The increase has been greatest at the bottom of the bay, which has prevented complete vertical mixing of the bay's waters. The resulting lack of oxygen in the deeper waters could cause considerable reduction of the aquatic life in such areas. The changing concentration of dissolved chloride ions could upset the internal salt balance of fresh-water fish in the bay and the adjacent south side of Lake Ontario. Also, during parts of the year the chloride levels exceed the U.S. Public Health Service recommended limit for human consumption. Since chlorine compounds are known to cause genetic mutations, an increase in chlorine in water or food consumed by human beings must be reviewed with concern. Finally, the fraction of de-icing salt that is accumulating in the ground water may pose serious future problems for well-water supplies in Rochester and other areas with similarly extensive de-icing programs.

The ocean itself is not immune to pollution. The explorer Thor Heyerdahl reported after completing his trip across the South Atlantic on the papyrus raft Ra II that the ocean was always polluted with human garbage, especially floating globs of oil, as far as he could see.

While oil for a long time has been a noticeable pollutant of the seas (largely from natural leakages out of deposits on the ocean floor), only in recent years has oil become a major pollutant of international concern in all oceans. Accidents involving tankers and barges, illegal washing of tanker bilges, and accidents in handling oil cargos have created problems that have alerted people around the world to the threats of oil pollution. When released in water, oil spreads in a thin film over the surface, and it contains hundreds of substances that react with the environment. The lighter fractions are evaporated or absorbed by particulate matter and sink to the bottom. Some become dissolved in sea water, but much is oxidized by bacteria, yeast, and molds. The large, branched-chain hydrocarbons in particular persist for a long time as tarry chunks floating on the surface and lying on the bottom of the sea.

The economic and esthetic damage of oil pollution is often of more public concern than is the ecological damage. Recreational beaches and waters are ruined by oil slicks, and boats as well as seashore property are often damaged. The most devastating ecological damage is done to marine life in the intertidal zones. It may take 2 to 10 years for the area to fully recover from a major oil spill. Sandy beaches become permeated with oil. The surface oil film brought ashore is deposited on the top and eventually pressed into the sand by the waves. There it kills the inhabitants of sandy beaches, becomes mixed with subterranean currents, and is carried out again to the lower depths of the sea. Detergents used in cleaning up some oil spills on coastal rocks may create even more ecological damage by their extreme toxicity to marine plankton and to the intertidal species of animals.

Radioactive wastes are another water-polluting by-product of our modern technology. Research laboratories and hospitals use radioisotopes extensively for tracing the metabolic pathway of materials through organisms and for a variety of diagnostic purposes in human beings (localizing tumors, determining defects in normal circulatory anatomy, and determining blood volume). These studies may release considerable amounts of waste radioisotopes. Solutions containing these materials are merely flushed down the drain, thus entering the general sewer system and ending up eventually in some body of water. Nuclear power plants, now being built in many areas, form a significant potential source of radioactivity in the event of accidents to their water-cooling systems or the nuclear reactor itself. Though potentially hazardous in the future, radioactive wastes from all sources still seem minor today, compared with the other forms of pollution freely released into aquatic environments.

One of the most common and controversial forms of water pollution results from **eutrophication.** This process consists of the promotion of plant and animal growth in an aquatic ecosystem by adding substantial amounts of nutrients to the water. When eutrophication is accelerated by human activities, such as adding nitrates by stockyard runoff, the water pollution is rapid. The small, single-celled algae and other aquatic plants have an abnormal rate of increase under these advantageous nutrient conditions and thus use up most of the free oxygen present in an overnourished body of water. Larger animals such as shellfish and fish subsequently die from lack of oxygen. Runoff from stockyards, heavily fertilized farmland, and outflow from sewage plants with incomplete sewage treatment are the principal sources of excessive supplementary nutrients.

Laundry detergents that contain phosphates also contribute to eutrophication and growth. Thus environmentalists have worked to ban the growth-promoting phosphates from detergents. Some environmental engineers and the major soap manufacturers claim that since most commercial organic products such as hand soaps and cleansers contain phosphates, the legislated removal of phosphates from detergents alone is largely useless. Unfortunately, some of the chemicals used as substitutes in nonphosphate detergents present a health hazard to children in the home and to adult users who may be sensitive to them. Thus the proper ecological balance in the inclusion or exclusion of phosphates in detergents is yet to be struck, although their pollution-promoting tendency through eutrophication is well documented.

Solid waste pollution

FIGURE 5–12

Junked cars in the Florida Everglades represent an all-too-common form of solid waste pollution.

Solid waste pollution is a problem whose magnitude is just beginning to be appreciated. Solid wastes in open dumps or inadequate fills can accumulate and create extremely serious disposal problems. If they are burned, they contribute to air pollution. Water percolating through them pollutes ground-water supplies, and they serve as breeding grounds for annoying and disease-bearing organisms such as flies, rats, and cockroaches. Cans, bottles, jars, plastic bottle caps, and other packaging materials, as well as automobiles, trash, and garbage, amount to hundreds of millions of tons a year of refuse (Figure 5–12).

Current methods for dealing with the disposal of solid waste material are utterly inadequate. In the United States, disposal is commonly by one of three methods: incineration, land-filling, or reclamation. *Incineration,* or the burning of solid material, is now considered undesirable because of its major contribution to air pollution. It is being increasingly restricted to sparsely populated areas. *Land-filling* disposes of solid wastes by dumping them into a pit, natural canyon, marsh, estuary, or other uninhabited site. As the depression is filled, the waste material is covered with soil. Because of its low cost, lack of processing trouble, and avoidance of air pollution, this is the most common type of garbage and waste disposal system for towns and cities. Unfortunately, it generates subsidiary problems, such as water pollution, dust pollution, and problems with nonbiodegradable materials (those not quickly broken down by microorganisms), which do not compact easily in the fill.

Reclamation is the environmentally advantageous way to avoid solid waste pollution and is rapidly gaining favor as new technological methods are developed to make reclamation competitive economically with land-fill. Reclamation is simply the reclaiming and recycling of waste materials. Recyclable steel cans and aluminum cans are being collected by companies and resmelted. Paper is being recycled into new paper products. Glass containers can be directly recycled through reuse; other imaginative ways of recycling glass include using granular ground-up glass bottles in insulating materials, road surfaces, and building walls.

Pesticides and other chemical pollutants

Some of the most dramatic pollution problems in recent years have been concerned with pesticides. The term **pesticide** refers to any chemical substance used for the regulation of population growth in a species regarded as a "pest." Thus pesticides include *herbicides*, which are used for the control of herbaceous plants; *insecticides*, which are used for insect control; and *biocides*, which are general pesticides used against all living organisms. Most pesticides have been used principally to kill plant or animal pests, to stop the spread of disease, and to protect crops. Unfortunately, the use of certain indestructible pesticides and their subsequent accumulation in the ecosystems of the world, along with increasing pest resistance to them and their potential danger to human beings and organisms in general, have caused pesticides to be considered a major pollution problem.

Until World War II, naturally occurring compounds such as nicotine (extracted from tobacco) and lead arsenate were used as pesticides, but with the advance of chemical knowledge chemists began to synthesize pesticide compounds not derived from natural sources. Among these were the chlorinated hydrocarbons, which include such insecticides as DDT (dichlorodiphenyltrichloroethane) and the DDT derivatives: DDE and DDD; dieldrin; chlordane; and the herbicides 2,4-D and 2,4,5-T. These pesticides were developed during World War II and, unlike naturally derived pesticides, they could not be broken down easily by biological action. Thus these compounds tended to persist for many years in an unchanged state. DDT could be sprayed on the walls of houses and other living areas, and malaria-carrying mosquitoes would die even if they landed on the surface months after spraying. Hence DDT made a great contribution toward conquering malaria throughout the world and in preventing typhus epidemics (spread by body lice bites) during World War II as well. Its usefulness for increasing production in agricultural areas seemed to promise untold benefits to farmers.

Unfortunately in 1946, it was noticed that several fruitfly strains had become resistant to DDT and within a few years dozens of additional insect species were found to be resistant to the lethal effect of DDT. This increasing insect resistance has become a classic example of rapid evolution observed in progress. Essentially, the widespread use of DDT generated strong natural selection *for* genetic strains in an insect species that contained genes producing detoxifying enzymes (see Chapters 23 and 24) and *against* those genetic strains susceptible to DDT. There was also strong selection for pesticide-avoidance behavior, where genetic strains with particular behavioral patterns, such as those that kept an insect away from a pesticide, would tend to survive and reproduce more than those that were fully exposed to the chemical. Thus in several tropical species of *Aedes* and *Culex* mosquitoes, the populations in the 1940s and earlier primarily bit sleeping human beings inside houses. Now they will bite only outside houses, indicating that genotypes producing this behavioral pattern have survived to replace those that formerly entered houses, lit on a DDT-sprayed wall, and died.

Additional unfortunate attributes of persistent pesticides were soon discovered. In addition to creating strong selection for pesticide-resistant strains of insects, these pesticides were found to remain in the ecosystem for a long time and be easily spread. They were carried by water movement into streams and into the oceans. In 1967, DDT was discovered in the body tissues of penguins and other wildlife in Antarctica, indicating that DDT was dispersing generally throughout the world, even into areas where no spraying had ever been done. The persistent nature of molecules of DDT and its chlorinated hydrocarbon relatives, and their affinity for body fat, caused

A   B

FIGURE 5–13

(A) Spraying an insecticide on corn from a low-flying helicopter helps to reduce aerial drift of the pesticide to nonagricultural areas. (B) Pesticide effects on wildlife were first detected in birds that were top predators in their respective food chains, such as ospreys.

them to accumulate in the upper levels of natural food chains in all parts of the world (Figure 5–13).

In the typical food-pyramid relationship, animals at each trophic level eat large numbers of organisms from the lower levels. If these prey have any DDT molecules in their tissues, they will be incorporated in the body tissues of the animals at the next level, and so on to the top carnivores. Hence, birds of prey such as the bald eagle or osprey, being at the top of their respective food chains, should receive the highest concentrations of DDT. As it turned out, this is exactly what happened. Eagle, osprey, and pelican populations began declining in the 1960s, with corresponding increases in DDT concentrations in their bodies, and it was soon discovered that the reproductive success of these species had suddenly become almost zero.

The proximate cause of reproductive failure was thin, fragile egg shells that easily broke under the incubating parent's weight (Figure 5–13B). In subsequent laboratory experiments, DDT was found to interfere with hormonal balance, lowering the calcium ion content in the female bird's blood, which in turn lowered the calcium available for egg-shell production. Since the initial discovery that DDT causes egg-shell breakage in ospreys and bald eagles, similar devastating losses have been reported in populations of robins, brown pelicans, and other sea birds. Other chlorinated hydrocarbon pesticides have been found to cause similar effects on these birds.

These chemical pollutants have not been restricted to birds, of course; mammals as diverse as bats and sea lions have also shown dramatic reduction in reproductive capabilities in recent years. Colonies of insect-eating bats in agricultural areas have been decimated by acquiring a lethal load of chlorinated hydrocarbons from the pest insects they eat. Of course, human beings — being the top predator in many food chains — have not escaped the flow of DDT throughout the world's ecosystems. Even Eskimos have been found to have 3.0 parts of DDT per million parts of tissue in their body fat; and in areas such as Israel and Hungary, with high use of chlorinated hydrocarbon insecticides in farming, DDT concentrations in human tissue are much higher (19.2 ppm and 12.4 ppm, respectively). In the U.S. population, 5 to 20 ppm is the common range of DDT concentration. Unfortunately, DDT reaches its highest concentration in mother's milk, and in many cases such milk has actually been shown to be unfit for human consumption. The DDT molecule has also been shown to interfere with sex hormone balance and cause infertility. Thus the pollution problems inherent in the use of persistent pesticides has led to the prohibition of their use in many highly developed countries of the world, although unfortunately they are still being exported for use by unsuspecting peasant farmers in the less developed countries. Substitution for these persistent pesticides in the United States is being made by the use of organic phosphate compounds, such as parathion and malathion, or other biodegradable chemicals that may be combined with biological control methods.

Biological control

In biological control, the pest organism is suppressed by another organism or by utilizing some feature of its own biology or physiology to destroy it. Today, the principal biological control agents used are parasites, predators, and microbial diseases. Natural parasites such as parasitic wasps may be cultured en masse in the laboratory and released as needed in crop areas. Predators such as ladybug beetles may be gathered and sold to control scale insects and aphids. The most effective control is offered by bacteria and viruses that tend to be species-specific, killing only one particular pest, and do not accumulate in the environment like chemicals, merely increasing temporarily the already-existing but low-level microbial population. Thus far, insects have apparently failed to evolve resistance to these useful strains of microbial disease. The most effective biological agent in the U.S. Department of Agriculture Japanese beetle control program in the eastern United States is the milky viral disease that attacks the beetle grub underground. Other biological control agencies include resistant varieties of crop plants that are being developed by plant geneticists. Here, researchers select plants with variant genotypes that produce effective *chemical* defenses such as poisonous alkaloids or *mechanical* protection such as thicker bark.

Male sterilization has proven to be a successful biological-control technique to eradicate or control the screwworm fly in many cattle-raising areas, and the method has also been used with several injurious fruitfly species. Male flies raised in large numbers in the laboratory are sterilized by radiation and released into wild populations weekly. If a sterile male mates with a virgin female, she will lay her eggs without a further mating and they will fail to develop. The screwworm fly has been eradicated from cattle-raising areas on Caribbean Islands and in Florida by such mass-culture-and-release methods. Organic chemicals such as sex attractants may be synthesized and used to attract and destroy large numbers of adults. Specific insecticides that upset hormonal balance within an insect when absorbed through its cuticle and cause reproductive failure or death are also useful in that they do not affect other species in the same area or predators that might eat these insects.

Thus while biological control tends to have greater costs and often more complex treatment procedures than chemical spray control, the ecological results are beneficial. Many of the nonpersistent biodegradable pesticides may be safely used up by bacterial action. Often the most acceptable plan for pest control—one that avoids the limitations of pollution-producing chemicals and the occasional problems with biological control used alone—is an *integrated control* program, using several control methods simultaneously, in conjunction with the thorough study of the vulnerable periods in the pest's life cycle. The success of such integrated control programs has been shown many times over in agricultural areas that have previously met disaster with the use of chlorinated hydrocarbons alone.

Summary

Environmental alteration may occur by natural processes such as flooding, erosion, fire, landslides, and coastal wave action. Periodic fires and the control of population sizes of certain animals (whether native or introduced, as goats in the Saharan region) are necessary for the maintenance of some climax communities. Agricultural clearing and urban expansion have been alterations of major importance. The clear-cutting of national forests is advocated by lumbering interests to increase profits, but the ecological results may end up being a net loss economically to the region. Likewise, giant construction projects such as the Cross-Florida Barge Canal and the Alaskan pipeline have the potential for harming large areas of our remaining natural environment. Pollution comes about as a result of the introduction of an undesirable change in the environment caused by gaseous, aqueous, or solid waste materials. Pesticides, including persistent insecticides and herbicides, have caused major pollution problems in recent years. Biological control and integrated control programs offer alternative ways to handle pest problems without the ecological consequences of the use of chlorinated hydrocarbon chemical sprays.

PART TWO

Life at the community level

Biological communities
range from very simple to quite complex assemblages of species, and it is at this level
that we see some of the most fascinating processes and relationships in the natural
world. In Chapter 6, we explore the general ecology of communities: their organization, the differences between aquatic and terrestrial communities, the organic
productivity of communities, and the inevitable changes that take place in all communities. In Chapter 7, we discuss in detail the intimate relationships between different
species, including competition, symbiotic interactions, and predation. Finally, in
Chapter 8, we look at the nature and causes of species diversity in a community: why
there are so many species of plants and animals in a rainforest and relatively few in
the northern coniferous forest or Arctic tundra.

6 The ecology of communities

The living part of the ecosystem is the *community*, an assemblage of populations of various species of plants and animals living and interacting in a particular environment. Thus a community forms a large-scale living system with its own distinctive composition, development, structure, functions, energetics, and biogeochemical cycling relationships. Ecologists often are able to characterize communities in sufficient detail that we can recognize the same community type when we encounter it in several parts of the world. A description of the kinds of species present as well as the numbers of individuals is particularly useful in categorizing communities by their structure and composition. With these data in mind, the ecologist can look for the factors that influence the development of that community.

Distribution and organization within a community

A community is a complex aggregation of many species distributed horizontally, vertically, and temporally. All of these variables affect the structure and dynamic functioning of the community.

Stratification is the organization of a community on a vertical scale; that is, different species occur at different heights above the ground or at different depths below the water surface. As a result, the living organisms in an aquatic or terrestrial community frequently seem to be arranged in layers. Stratification results primarily from the variation in light intensity along the vertical gradient. The absorption of sunlight by the structural components of the topmost organisms in the community (e.g., leaves of forest trees or algal plants near a lake surface) decreases the intensity of light reaching the lower stories. Other environmental variables such as humidity, wind exposure, and temperature affect the vertical distribution of both plants and animals.

Stratification becomes quite complex in tropical rainforests (Figure 6–1) where a four- or five-storied arrangement may be observed among the growth forms in the community. The canopy-layer trees usually include species that control and characterize the community because of their abundance and size, called the *dominant* individuals. Their thick foliage and position at the top of the forest canopy in the maximum amount of available sunlight exert energetic dominance over the other species in the area. In the subordinate strata, represented by the younger individuals of the giant canopy species and the mature trees of other smaller species, the trees are adapted to exist at the lower light intensity that results from the absorption of sunlight by the canopy foliage. Less than 10 percent of sunlight reaching the upper canopy may penetrate through these top two

FIGURE 6–1

Stratification in a tropical rainforest of South America, as seen in vertical transect.

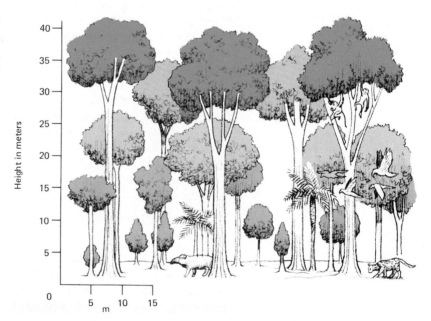

layers of trees and reach the shrub species in the third layer of vegetation. Only 1 to 5 percent of the incident sunlight in many forests
reaches the herb layer below the shrubs, and the mosses on the forest
floor may receive only a fraction of 1 percent of incident sunlight in
dense forests. Lower strata in forests normally offer little direct competition to mature canopy trees, although, of course, the seedlings of
the latter tree species must be able to meet the competition of lesser
species at the time of reproduction since all must start in the same
restrictive environment of the forest floor.

In addition to dominance, there is *dependence* among the many
species of inconspicuous organisms within the community. Most of
the mosses and ferns as well as some of the higher vascular plants
require the special environmental conditions provided by the larger
seed plants, such as shade and moisture. If the dominant vegetation
is removed, such dependent organisms soon disappear. Epiphytes
that grow on the trunks, branches, and even on the leaves of the
larger seed plants occur in the forests of the temperate zones as well
as tropical and subtropical areas. In temperate zones these dependent "air plants" are usually mosses, liverworts, or lichens and are
usually restricted to moister communities. In the subtropical and
tropical forests, epiphytes become quite conspicuous in their size,
diversity, and abundance.

Fungi and bacteria, particularly the decay fungi living in the soil
or litter of the forest floor, likewise are important components of the
community because of their activities in decomposing organic
matter. Still other fungi are host-specific, living in a symbiotic association with the roots of vascular plants.

The stratification of animals in a community is usually exhibited
in their physical distribution among foraging areas in the vegetation—for instance, in high treetops versus lower branches. It may
also be reflected in the level of nest sites in the forest or even nesting
positions on vertical rocky cliffs. Since animals distribute themselves largely on the basis of the characteristics of vegetation, particularly herbivores but also carnivores eating those particular herbivores, in a community animal stratification is apt to be related to
plant stratification because of food requirements.

The organization of a community on a horizontal scale is influenced by the distribution and spacing of each of its component
species, a topic that we will consider at length in Chapter 9. For the
present, let us mention that the individuals of a population can be
distributed in horizontal space in a community in four basic ways. In
a *random distribution*, the individuals are scattered on the community floor without reference to the presence of another individual of
the same species. In a *clumped*, or *contagious, distribution*, small

FIGURE 6-2

A lone cheetah begins its rapid run after prospective prey on the Kenyan plains.

groups of individuals are scattered through the area of the community. Between these groups will be spaces lacking individuals of a particular species. In a *regular,* or *negatively contagious, distribution,* the individuals are more evenly spaced than in a random distribution. Finally, individuals are clumped into *colonies* and the colony units are regularly distributed throughout the community area.

The reasons for this varying expression of clumping and spacing of clumps, or lack of aggregation, will be considered later. For now, suffice it to say that one species in the community is usually dependent on another in some way. For instance, insects depend on host plants, and they will be clumped in distribution if their host species forms clumps. The plant may exhibit a clumped distribution because its nutrient or moisture requirements are present only in certain areas. Among animals, generalized feeders will not be as restricted as host-specific feeders in possible dispersion patterns. Solitary hunting species such as cheetahs or leopards will space themselves out in hunting territories (Figure 6–2). Social hunting species such as African lions, on the other hand, will maintain family groupings for maximum hunting success.

Temporal orderliness in biotic communities is exhibited by different species carrying on similar functions at different times, whether in daily or seasonal cycles. Diurnal hawks and nocturnal owls may both hunt small rodents, but their periods of activity do

FIGURE 6-3

Daily periodicity in the vertical movement of marine plankton. The figure shows the cycle of up-and-down migration of three typical species of animal plankton; doubtless this movement is related to the similar daily migration of the plant plankton on which they feed. It seems likely that ultimately the migration of the plants (mostly diatoms) is aimed at attaining optimal light intensity for photosynthesis.

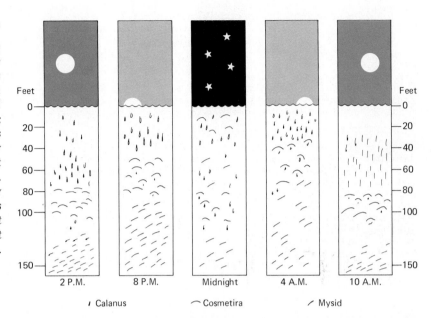

not overlap and the species of prey they catch are also active at different times. Because of the daily, monthly, and yearly cycles of light, temperature, and other environmental factors such as tides along ocean shores, species tend to adjust their behavior and physiological activity to the most favorable times. Thus within natural communities we find rhythms of function that correspond to rhythms in the environment.

In the daily *cycle*, an organism's principal activities are limited to certain hours, often to either night or daylight hours. Thus many plankton animals in fresh-water and marine habitats migrate up and down in daily cycles (Figure 6–3). As light increases in the morning, the plankton normally sink or swim downward. As light decreases in the later afternoon and evening, the plankton rise to spend the night in water near the surface. Because plankton also respond to winter and summer water temperatures in fresh-water lakes, the seasonal replacement of predominant species of algae may occur. Yellow-green algae are predominant among winter plankton. As water temperatures become warmer in the late spring and early summer, the yellow-green algae are replaced by desmids and other green algae. At the highest summer temperatures, blue-green algae may predominate but as water temperatures cool, dominance shifts back toward green and yellow-green algae. Each species has a spore stage which enables it to survive the season that is unfavorable for normal activity.

The coral reef community

Coral reefs form the most complex and yet the most glamorous, exotically beautiful community type in the ocean world. Home aquarists and businesses alike find that a salt-water aquarium with reef fishes in it is a great attraction to visitors, so much so that tropical fish importers have a difficult time keeping up with the demand for reef fishes. Their brilliant colors and fascinating behavior bring a bit of tropical splendor to the lives of those of us who do not reside in the island paradises of the Pacific Ocean and other areas where coral reefs abound. Yet, though it is an extraordinary one, the world of coral is very little understood.

It *is* known by biologists that if the tiny coral animals are to develop and maintain themselves, several environmental conditions are required simultaneously. The water must be warm, with temperatures never dropping below 50°F, and it must be clear, remaining free of sediment, sand, slime, and mud throughout the year. Coral, therefore, is found only in tropical and subtropical waters, in a belt around the globe extending no more than 32° north and 27° south of the equator. Local conditions within that area may prevent the corals from growing even though the water is warm. No corals occur, for example, off the coasts of Brazil, India, or West Africa because of the mud that is carried into the sea by the great rivers of those regions (especially during the tropical rainy season).

Corals make up the chief constituent of coral reefs, though most of the interior portion of the reef mass is dead coral deposits. These reefs are of three kinds. The *fringing reef* forms a fringe along the shoreline of a mainland or island area, lying a short distance from it and separated from it by shallow water. The *barrier reef* lies a greater distance from the shore and is separated from it by a wider and sometimes deeper channel. The Great Barrier Reef, lying off the northeastern coast of Australia, is the world's most dramatic example of this kind of reef. The third type of reef, the *coral atoll*, is a more or less ring-shaped reef that encloses a central lagoon.

It is indeed a paradox that hypotheses concerning the origin of these little islets have been controversial for more than a hundred years. As a result of his voyage around the world on the *Beagle*, Charles Darwin proposed the best and most widely accepted theory for the origin of most coral atolls. He noted that the seaward slope of a coral atoll may rise abruptly from the ocean depths for several thousands of feet, though the corals that construct it cannot exist at depths greater than about 180 feet. The reason is that corals live in a symbiotic relationship with zooxanthellae—algae that they carry in their tissues, which, in addition to acting as a dietary supplement through their active photosynthesis with light reaching them from the surface, actually help to rid the coral animal of certain ammonia and phosphorus wastes. Beyond that depth, there is not sufficient light for these algae to achieve photosynthesis. Thus Darwin reasoned that such islands, low in stature and reaching only that height above sea level to which waves cast debris and pile sand, must be based on some sort of substantial, rocky foundation, the top of which must have at one time been within 180 feet of the surface. He further hypothesized that it was improbable for all rocky banks around the world to have been uplifted to precisely that height, for in the vast areas of the oceans no land masses would rise higher than atolls. Also, no exposed mountain chains on continents show such uniformity in the altitude of their peaks. Darwin reasoned that if such rocky foundations had not been uplifted, then they must have subsided. He suggested that the sinking of mountains or other land masses below the surface of the sea provided the foundation on which all structural types of reefs (atolls, barrier reefs, and fringing reefs) are developed.

The living corals continue to grow upward and outward as the foundation sinks, because it is toward the open sea that environmental conditions (oxygen, light, food availability, absence of toxic waste materials, etc.) are most favorable. Hence the periphery continues to be maintained at the upper limit of coral growth, at the low tide level. With atolls, the outward growth of the reef and the continued subsidence of the central island will increase the size of the lagoon, until the last of the central portion of the noncoral island sinks below the surface and the true ringlike atoll is born.

Coral, despite the environmental demands that it makes, exists over a huge area of the earth—about 80 million square miles, or approximately 25 times the area of the United States and 20 times that of Europe. So far as we know, coral first appeared on earth more than 400 million years ago. In some eras, when the seas everywhere were shallow and warm, corals stretched as far north as Greenland. The reefs that we find today often have been the products of millions of years of growth. In 1952, scientists drilled two holes to depths of 4,222 and 4,630 feet on Eniwetok Atoll; both reached volcanic bedrock at this depth. The researchers discovered that the oldest sediments, just above this volcanic bedrock, consisted only of reef-building coral skeletons. Thus it seems clear that although now they are more than 4,000 feet below sea level, the organisms lived and started building this atoll reef in shallow water while they were alive about 50 million years ago.

All together, there are approximately 2,500 species of corals and a great diversity of forms. Some are tiny and cuplike, and others are branched. The formations they achieve are often as fragile as they are diverse. The "branches" of certain corals would collapse if exposed to gravity in the open air; in the sea, however, they are sustained by the water. The variety of colors—blue, pink, red, yellow, purple, and golden brown—are the result of both pigments contained in the tissues of the coral animals and of the symbiotic algae.

The coral massifs of the tropical seas are home to numerous species of fish remarkable for their form and color. These marine fishes are found only among coral formations where, since they are quite sedentary animals, they spend their whole lives. Consequently, their swimming habits are adapted to the nature of the environment here—to the channels, mazes, and crags of the reefs. They are better called reef fish than coral fish, since some of them, in fact, live in rocky environments similar to coral reefs but where the water temperatures prohibit the growth of corals. Of the 30,000 species of fishes recognized today, the species living among coral reefs are undoubtedly the most striking. They come in red, yellow, blue, orange, and green; in fact, every color of the spectrum and almost every color combination can be found. Moreover, these vivid colors are often displayed in bold patterns formed by stripes, dots, lozenges, and so on. Even within a single species, there can be a great variety of color and pattern, for the individuals of many species change their appearance with age, sex, or the season.

Most coral fishes are flat and round, having more or less the configuration of a disk. They are able to swim in almost any position—which allows them an extraordinary degree of mobility when it is necessary for them to flee from predators or to hide in the minute grottoes of a coral formation. Moreover, their comparatively large tails and short, compact fins make it possible for them to spin around quickly and to "stop on a dime." Both the vivid colors and these behavioral attributes seem to be linked to territory, providing a possible explanation to the riddle of the incredibly diverse colors of reef life forms that Darwin and hundreds of other biologists have wondered about. After spending years in observing life on the coral reef, the behaviorist Konrad Lorenz concluded that coral fishes have their colors in order to be able to recognize each other as members of the same species. Every area of the reef is a territory that belongs to a single member of a species, and that individual fish defends its territory with all its strength against any other member of the same species. Only fishes with other colors and markings do not constitute a threat to its territory, and therefore only fishes of different species are not attacked and driven out of the area.

Besides the corals and fishes, there are many other life forms that multiply and live and die in the world of coral reefs. The sea urchins and the giant clams, the starfish and sea worms, the lobsters and the marine mollusks, and many of the other inhabitants of the world of coral who have no equivalent on land make up an important section of the coral reef community. Their role is only beginning to be studied.

Because of the seasonality of optimal conditions for breeding and reproduction in many biotic communities, each species will utilize particular times of the year to reproduce and have its maximum growth. In animals, this usually occurs when the maximum amount of food is available for rearing the young. Thus insect-feeding bird species will tend to be most abundant at the times that insects are experiencing rapid population growth. In plants, reproduction is normally synchronized with periods when light conditions and water availability are optimal and pollinator and dispersal agents (for the seeds) are available.

Aquatic and terrestrial communities

The biosphere contains both terrestrial and aquatic communities in its thin film across the surface of the earth. With both types of communities, the spatial, temporal, and abundance patterns of various species are to a large extent determined by physical factors; therefore, we should begin with a discussion of these factors.

In aquatic communities, temperature, oxygen content, salinity, and pressure are critical factors determining the distribution of populations. Temperature is generally more buffered in an aquatic community than in a terrestrial community, as water is a good heat reservoir. With increasing temperature, however, less oxygen in the dissolved form can be held by the water. Hence one factor influences another. Fresh-water rains falling on land will eventually drain into rivers that empty into the ocean, and even heavy rainfall at sea will dilute the waters of the ocean in that area and thus decrease salinity. Pressure changes dramatically as we move from surface waters to deep ocean waters. If a section of the ocean heats up or cools down because of a change in the flow patterns of ocean currents, a shift in the distribution of different marine species will result. Increasing temperatures in the ocean off California may cause tuna to move further northward, while a cold snap in a tropical ocean area may cause massive fish kills.

On land, such physical factors as soil type and topography interact and determine the distribution of land plants. Other important influences are temperature, solar radiation, and associated climatic factors, especially precipitation. These physical factors combine with the geographic distribution of food plants to determine the distribution of plant-eating animals. The arrangement of temperature zones on the earth's surface explains much of the distribution of organisms with specifically restricted tolerance limits.

The productivity of communities

In Chapter 3 we looked at one of the measures of the output of community dynamics — productivity in biological communities. We also examined the means of energy transfer and learned *primary productivity* is the rate at which energy is bound in the form of organic material through photosynthesis. Productivity is most often expressed as energy in the form of calories per square centimeter per year (cal/cm²/yr) or as dry organic matter in grams per square meter per year (g/m²/yr). This, then, is a rate measurement and is to be distinguished from the amount of organic matter present at a given time per unit of the earth's surface, which is a *standing crop*, or *biomass*. It is usually expressed as dry grams per square meter (g/m²) or kilograms per square meter (kg/m²). *Gross primary productivity* is the total amount of energy bound by the community, that is, the total bulk of organic matter created by green plants per unit of surface. The energy remaining after respiration and stored as organic matter in the plant is *net primary production*, or plant growth.

Thus net primary productivity provides a means for comparing various communities and their efficiency for transforming the energy of sunlight into organic matter. The net primary productivity of major community types around the world varies considerably. Desert and tundra areas produce less than 200 g/m²/yr and forested woodlands apparently produce about 400 to 1,000 g/m²/yr, whereas mature and stable temperate zone forests in favorable environments have net productions of 1,200 to 1,500 g/m²/yr. Compared with these figures for natural communities, agricultural land ranges in net primary productivity from 100 to 4,000 g/m²/yr. Most temperate-zone cereal crops have production productivity in the same ranges as many natural grasslands and shrublands, whereas hybrid corn and other crops under intensive cultivation can exceed 1,000 g/m²/yr, and some marshlands and tropical crops exceed 3,000 g/m²/yr. Often the seasonal distribution of precipitation, the frequency of fire, and the soil characteristics influence productivity of the major community types as much as the absolute quantity of precipitation.

Succession

Since environmental changes are constantly occurring along with changes in the age and composition of a plant community, no plant community remains completely stable. The vegetational changes that occur in such a community are termed **succession** (Figure 6–4) and the various intermediate stages that appear during this process are known as *seral stages*. The *pioneer* species are the first to arrive and start the succession; the *climax* species represent those best suited for the prevailing climatic and soil conditions. The overall collection of seral stages is called a *sere*. These replacements of species occur because the earlier stages modify the local environment. Thus when mosses and lichens grow on a bare rock surface, they start to break down the rock into smaller particles. Organic matter (e.g., from dead mosses) becomes mixed in with the inorganic minerals, providing a new substrate suitable for the growth of ferns. The ferns in turn change the soil still further, making it unsuitable for their own perpetuation but ideal for the germination and growth

FIGURE 6–4

Succession may occur in several ways and at several levels. (A) Forest community succession, from annual weeds to mature forest; note the increasing vertical stratification as one seral stage advances to another. (B) Reduction of a fallen log to soil: succession of decomposers in a pine forest microcommunity.

A

Canopy				
Lower canopy trees				
Understory trees				
Tall Shrub				
Understory Low Shrub				
Herbaceous layer				
Ground layer				

Annual weed · Grass stage · Shrub stage · Young forest stage · Mature forest community

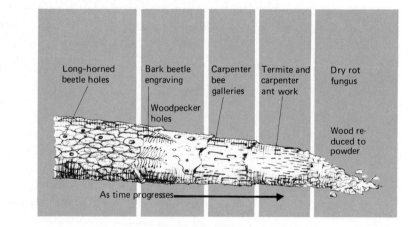

B

Long-horned beetle holes · Bark beetle engraving · Woodpecker holes · Carpenter bee galleries · Termite and carpenter ant work · Dry rot fungus · Wood reduced to powder

As time progresses →

FIGURE 6-5

Succession in a northeastern U.S. forest: (A) Lichens grow on bare rock; (B) ferns replace lichens as first soil is formed; (C) pines grow when ferns have created a better soil; (D) a hardwood deciduous forest (beech and maple) replaces the pines and becomes the climax community.

of seed plants such as columbines and quaking aspens, firs, or spruces. Eventually, the former rock has completely disappeared and a magnificent climax forest of Colorado spruces welcomes the mountain visitor (Figure 6–5).

Once a climax community has been established, succession ends and only the replacement of existing mature individuals continues. A climax community is not dying, contrary to lumber company advertisements that promote lumbering and harvesting by clear-cutting national forests. Rather, it is maintaining itself in the form of a biotic community best suited to the surrounding environmental conditions. Thus a community is normally characterized by *constant change*, which results from mature individuals dying, other younger individuals replacing them, and new species entering the area. New

species replace those of the early seral stages because of changing conditions in the soil, temperature, humidity, humus deposition, available water, and other environmental variables. Sometimes these changes are radical and rapid, as when disease or a wind storm strikes many individuals of one of the important dominant species. At other times, changes may be so slow as to be scarcely discernible over a period of years, as, for instance, in the growth of plant communities started on bare rock cliffs.

Primary succession is initiated on a bare area of new soil where no vegetation has grown previously. It may occur on newly exposed soil along the edges of beaches, along oceans, on glacial moraine exposed by receding ice, on new oceanic islands formed by volcanic eruption or coral reefs, in a landslide area or other area of extreme erosion, and on newly exposed rocks from volcanoes or earth movements. A tremendous diversity of pioneer communities is possible because of the diversity of habitats on which succession may begin in terrestrial areas. However, the influences of vegetation usually tend to make the initial habitat more favorable to other plants. This occurs by a reduction of environmental extremes, which is always reflected in improved moisture and soil conditions for a broader range of species. Thus dry, or xeric, habitats become moister and hydric, or moist ones become drier, as succession progresses.

Secondary succession results when a normal succession of seral stages or a climax community is disrupted by fire, cultivation, lumbering, windstorms, or any other similar disturbance that destroys the principal species of an established community. Often the successional stages that develop are similar to the later ones found in a primary succession for that area. However, if ecological changes had been severe, a different grouping of plants may develop.

When uninterrupted, the end of a series of seral stages in a succession is a **climax community.** The climax community may continue indefinitely because individuals that are lost for any reason are replaced by their own progeny. The energy flow through the ecosystem, as represented by the functioning of interacting individuals existing in a particular environmental state, is now at the peak of productivity. This climax community continues in a steady-state condition of dynamic equilibrium until environmental conditions change for some reason. A long-established theory in ecology postulates that a climatic region has only one potential climax, which is the most mesophytic community that the climate can support; that is, a community with average or intermediate environmental conditions, particularly with regard to moisture relationships. Given sufficient time, and an accompanying stability of climate and land surfaces, succession will proceed to such terminal, relatively mesophytic communities over much of the area within the limits of the general prevailing climate. Because of their high degree of similarity, these communities will be obviously related in growth form and the recurrence of the same genera (and often species) as dominants. Thus the regional climax formation is reflected as a *biome* unit of one general community type.

If in succession a stage immediately preceding the climax is long-persisting for any reason, it is termed a *subclimax*. Thus regular disturbance such as fire may hold succession almost indefinitely in its subfinal state. In the eastern United States, where hardwood climax communities represent the culmination of temperate deciduous forest succession, most pine forests have become effectively subclimax because of the relatively slow elimination of pine in the progression toward hardwood dominance. In the coastal plains of Georgia and northern Florida, subclimax pine forests are maintained indefinitely by the constantly recurring fires that keep down hardwoods but allow the resistant pines to grow.

If disturbance progresses to the point that the true climax community becomes modified or largely replaced by new species, the result is an apparent climax called *disclimax*. Thus the prickly pear cactus (*Opuntia*) has formed a disclimax over wide areas in Australia where it was introduced and rapidly became a dominant in desert areas. The ravages of chestnut blight fungus in the eastern deciduous forests caused oak-chestnut climax in the Appalachians in southern New England to become an oak disclimax.

Major communities and ecotones

A boundary between two major types of plant communities is called an **ecotone** (Figure 6–6). Ecotones tend to be areas of greater diversity than either parental community alone because plants and animals of both the neighboring communities invade the intermediate area and increase species richness. Thus along the wooded border of an old field, we will tend to find more bird species than if we go into the pure temperate deciduous woods or out into the center of the open grassy field. In an ecotone of this type, insectivorous birds can take advantage of the proximity of the woods for nesting and evening roosting and hunt in the open grassy meadow during the day for their food.

Major communities are those that together with their physical hab-

FIGURE 6-6

An ecotone in the Black Hills of South Dakota, where the animals and plants of a prairie community mix with the characteristic organisms of a pine forest community.

Pronghorn antelope   Grass and sunflowers   Pine trees   Deer

itat form more or less complete self-sustaining ecosystems. In such systems solar energy is the principal or sometimes the only needed input from external areas. Such communities are relatively independent of adjoining communities because of their sufficient size and level of organization. *Minor communities,* on the other hand, are more or less dependent on neighboring aggregations of organisms for an input of energy. The temporary rainpools in a desert canyon or on a trail in the tropical rainforest represent minor communities with comparatively short but intense periods of biological activities. Local frogs and toads come to these bodies of water and lay their eggs, as do many dragonflies and damselflies. These organisms are taking advantage of the very few days of existence of these minor aquatic communities to enhance their own chances for reproductive success by using pools free from normal predators (e.g., fish) found in more permanent aquatic bodies.

The significance of climax communities

As a general rule, terrestrial climax communities and their aquatic counterparts tend to be more stable and have more diverse paths of energy flow in food webs than do intermediate seral communities.

Hence they are generally the most productive of the series of natural communities in a sere. The climax community is composed of the plants and animals best suited by evolutionary adaptation to the prevailing climatic and soil conditions of the region. If a climax community is destroyed by human activities or natural disaster, hundreds of years are necessary for succession to create another climax community of the same type.

If, as in many tropical rainforest areas in Central and South America, the adjoining portions of the climax community have already been destroyed, there are no genetic reservoirs of the constituent species to replenish the area when the last stand of tropical rainforest is cleared. Several species of birds have become extinct on Barro Colorado Island in the Canal Zone in Panama during the last 50 years because the island made by the damming of Gutun Lake is apparently insufficiently large to provide the proper space in which these rainforest species can exist. With the increasing destruction of the surrounding tropical rainforest on the margins of the lake, soon there will be no opportunity for future colonization of the island by these species. Indeed, there will be no refuge anywhere for these rainforest species. Thus it becomes particularly important for us to be fully cognizant of the risk involved in any major environmental alteration, as we have seen in Chapter 5. When a climax formation is destroyed or seriously altered, it may prove impossible to recapture the genetic diversity of the biological species that once were there.

Summary

The community is an assemblage of populations of the species of plants and animals living in a particular environment. Distribution and organization within a community may be shown in vertical stratification, horizontal spacing, and temporal cycles. Species that control and characterize a community because of their abundance and size are called the dominant individuals. Some inconspicuous species are dependent and require the special environmental conditions provided by the larger, dominant plants in the community. Horizontally, individuals of a species may be distributed in random, clumped (or contagious), regular (or negatively contagious), and colonial dispersions. Many physical factors such as temperature, oxygen, salinity, pressure, soil type, topography, solar radiation, and precipitation influence the distribution of organisms in aquatic and terrestrial communities. Primary productivity of major community types varies considerably. No plant community remains completely stable; succession and replacement are constantly occurring. Vegetation in a community passes from the pioneer stage through a series of seral stages to the climax stage. The regional climax formation is represented in a biome. Ecotones are boundaries between two major types of plant communities and may have a greater diversity of plant and animal species than either parental community alone.

7 Relationships between species

Throughout all communities, organisms are constantly interacting with one another. Pathways of energy flow involve one organism eating another. The choice of a place to live involves mutual adjustment and ecological consideration between incoming species and species already present in the area. The dynamics of population growth is intimately affected by the presence of predators or organisms that serve as prey, enhancing the survival or causing the demise of the subject species. Thus on both a short-term and a long-term basis, these types of interactions result in ecological and evolutionary changes in one or more of the species involved.

There are three basic types of interspecific relations in a biotic community: competition, symbiotic interactions, and predation. In evaluating the various ecological and evolutionary adjustments that have been made in these three types of interactions, we develop a better appreciation of how the organization and dynamics of the populations of the individual component species come to constitute a functioning biotic community.

FIGURE 7–1

Competitive interactions between male Galápagos marine iguanas for territorial space and females involve aggressive posture and head-butting, but rarely result in actual injury or death.

Competition

Competition in a community represents a struggle between organisms for food, space, mates, or any other limited resource (Figure 7–1). Competition for such resources available only in restricted quantities is more acute among members of the same species than among the various species of an established community because organisms belonging to the same species will more nearly share exactly the same needs. Thus in *intraspecific* competition, the struggle is between individuals of the same species. This type of competitive interaction is common to all species whose numbers are increasing in a limited environment. In Chapter 10 we will treat it as one of the central factors regulating population growth.

In *interspecific* competition, this struggle for particular requisites occurs between the populations of two species and as such it represents an important factor during the maturation and evolution of a community. Competitive interactions among numerous species in a community act as a selective force in encouraging their diversification into different habitats, foods, times of activity, and species characters to prevent hybridization or other forms of wasted energy.

The basis for such struggles between species has been expressed as *Gause's principle*, which states that two species cannot occupy the same niche simultaneously in a given community. That is, animals or plant species that are completely overlapping competitors cannot long coexist. If two species that do occupy the same niche come together in space and time, as a rule there are three possible outcomes.

Extinction

If one species is more successful at monopolizing available resources such as food or space, its competitor will become extinct on a local basis. The adaptive advantage of one species need be only slightly greater than that of another. It may be able to reproduce more young in a given amount of time, have a slightly better means of defense, or better utilize the available food. It may also have greater tolerance for limited environmental resources such as oxygen.

G. F. Gause formulated his ecological principle after studying competition between two species of *Paramecium*, tiny one-celled aquatic fresh-water animals in the phylum Protozoa. These can be grown in test tubes in the laboratory where the *Paramecium* will compete for food (bacterial cells), oxygen, and other requirements. They reproduce by splitting into two new animals by binary fission as often as several times a day.

When grown alone in two containers in the laboratory, the numbers of *Paramecium caudatum* increase more slowly than *P. aurelia* (Figure 7–2). Eventually each species reaches the capacity of that environment to support a particular population size, and growth levels off to a size characteristic for that species in that given amount of space. However, if the two species are grown together in the same containers, *P. aurelia* increases more slowly than it did alone, and *P. caudatum* grows poorly and eventually becomes extinct after about 16 or 18 days (Figure 7–3).

In comparing Figures 7–2 and 7–3, we note that each species experiences normal growth rate for the first several days after being placed together in the same container, as long as food does not become limiting in relation to the size of the population. The population of *P. caudatum* declined to extinction eventually because of competition from *P. aurelia*. Although the *P. aurelia* grew more slowly than when cultured alone, it gradually attained the normal population size even when food became limited and competition

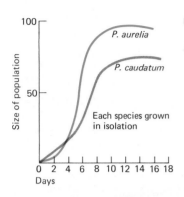

FIGURE 7–2

The growth of each of two species of closely related Paramecium *when grown alone.* Paramecium aurelia *and* P. caudatum *exhibit normal sigmoid growth in controlled cultures with a constant food supply.*

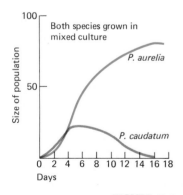

FIGURE 7–3

The effect of competition between Paramecium aurelia *and* P. caudatum *when grown together; the latter species is eliminated.*

was acute. *P. aurelia* was the more successful competitor because it was adapted to a more rapid growth rate. Hence it could trap a greater percentage of the limited available food in each generation, and it eliminated its competitor.

Competitive exclusion

A second possible outcome of interspecific competition occurs when the slight selective advantage of one of the competitors is overcome through the other species moving into other portions of the habitat. As we have seen in the preceding section, with even the slightest selective advantage between two species, no equilibrium is achieved, for one of the competitors will finally exclude the other from the community; and if there is no additional space to move to, extinction of the less satisfactorily adapted species will result.

In competitive exclusion, two species are in competition for the same niche or a critically restricted aspect of the habitat; because of the ability of the species to move into adjacent but sufficiently distinct portions of the habitat, they both continue to survive. An experimentally verified case of competitive exclusion is seen in the vertical distribution of two species of barnacles along the coast of Scotland. *Balanus* barnacles normally live on coastal rocks below the mean high-water mark of the lowest tide each lunar month. Thus these barnacles are always covered with water for part of each 24-hour period. *Chthamalus* barnacles live on rocks in the splash zone above the mean high neap tide mark (Figure 7–4).

This ecological distribution suggested competitive exclusion as an explanation, but experiments seemed imperative to eliminate the possibility that the species are restricted in their distribution not by competition but by inability to adapt to other parts of a shoreline habitat. The American biologist Joseph H. Connell removed *Chthamalus* barnacles from splash zone rocks and found that the *Balanus* species would not move into that area to live. However, when he removed *Balanus* from underwater rocks. *Chthamalus* did move down and survive below the water line. Thus competitive exclusion is involved in limiting the distribution of *Chthamalus* wherever both of these species occur together. The young *Balanus* simply grow more rapidly than the young *Chthamalus* in the lower zone and vigorously overgrow or pry off their competitors.

Character displacement

An evolutionary solution to the problem of competition involves minimizing competitive overlap in feeding or other requirements.

Tide levels

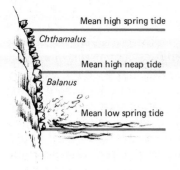

Mean high spring tide

Chthamalus

Mean high neap tide

Balanus

Mean low spring tide

FIGURE 7–4

The exclusion effect of competition between two species of barnacles. Although there is a broad area in which the larvae of both species settle, competition eliminates most of the overlap by the time the adult stage is reached; Chthamalus *is largely restricted to the zone above the level of the mean high neap tide (neap tides are the lowest tides of a lunar month) and* Balanus *to the zone below this level.*

Geospiza fuliginosa

G. fortis

FIGURE 7–5

Character displacement: Bill sizes in isolated island populations of Geospiza fuliginosa *and* G. fortis, *two of Darwin's finches, are roughly similar (see color). Where the two species compete, however, G. fuliginosa's beak is smaller and G. fortis's is larger (see beaks drawn in black).*

Thus where two potentially competing species occur **sympatrically** (together), they tend to have greater differences in their feeding adaptations (such as bill size in birds) than in areas where they do not coexist. Each species has evolved characteristics that lessen competition with other species for food resources and hence promote increased reproductive success.

This divergence of characters in areas of geographic overlap of related species is called **character displacement.** In Darwin's finches on the Galápagos Islands, for instance, bill sizes will differ distinctly between related sympatric species of *Geospiza* ground finches that live on the same island. The Medium Ground Finch has a somewhat larger and stouter bill, capable of handling bigger seeds, than the Small Ground Finch, which specializes in small plant seeds. These differences minimize the competition that would result if both species were similar enough to feed on the same prey item. On islands where these bird species occur **allopatrically** (alone), the beak size of each will often be similar or will overlap completely, representing an adaptation for generalized feeding on a variety of food items. This adaptive consequence of the influence of competition in areas of sympatry is shown in Figure 7–5.

Smaller niches

A modification of the idea of competitive exclusion combined with character displacement is that decreasing niche size in two sympatric species will lower the level of potential or actual competition. Instead of being generalized seed eaters or insect eaters, each bird spe-

A   B

FIGURE 7–6

Minimizing competition between closely related species: In the Galápagos Islands (A) the Red-footed Booby nests in trees and (B) the closely related Blue-footed Booby nests on the ground.

cializes in a different area of habitat or stakes out separate hunting grounds. In an alternate solution among some species, different locations may be chosen for breeding areas by each species, such as the tops of bushes versus ground level. Thus among the Galápagos sea birds, the Red-footed Booby nests several feet off the ground in shrubs, whereas the Blue-footed Booby nests on the bare rocky soil of these same islands (Figure 7–6). Thus we see that the two species adopt smaller niches in areas where they both occur in common, and the result is a minimization of competition.

Symbiotic interactions

Often species are able to not only minimize competition but to actually turn a sympatric living association into a way of life that provides benefit to one or both of the species. The word **symbiosis** refers to a long-term interspecific relationship in which these species live together in more or less intimate association. This is not a *social* system, such as we would find in a band of monkeys or other single-species population, but an *ecological* association involving some transfer of energy or other adaptive benefit. Symbioses have been classified into three general types: *commensalism, mutualism,* and *parasitism.*

FIGURE 7-7

A cattle egret (Bulbucus ibis) in symbiotic association with an African buffalo (Syncerus caffer) in Masai Amboseli National Park, Kenya.

Commensalism

Commensalism refers to a relationship in which one organism benefits and the other is neither harmed nor benefited. Cockroaches inhabit many American homes and they benefit from living in close proximity to human beings, who are rarely harmed or benefited by having them in their homes. While esthetically displeasing commensals to most of us, they rarely eat enough to affect our energy intake. However, in certain ghetto areas and in tropical countries, cockroaches may significantly affect human food supplies.

In the American tropics and in the subtropical parts of the United States today, Old World cattle egrets have become well established during the last several decades (Figure 7-7). These egrets follow cattle and eat the insects that the cows scare up while grazing. The cattle do not seem to be benefited or harmed by this association. However, in the African tropics where the cattle egret is native, it probably serves as a warning sentry for the large grazing mammals, alerting them to possible predator danger by sudden flight. In the New World, the egrets have carried their commensal relationship one step further and today they may be seen following tractors that are plowing fields and turning up insects and worms. Clearly this commensal behavior is of benefit to the egret but of no effect whatsoever to the tractor (and indirectly human beings).

A more intimate commensal-host relationship is shown in epi-

A   B

FIGURE 7–8

Commensalism: (A) Spanish moss, an epiphytic higher plant in the pineapple family, Bromeliaceae, hangs from oak trees throughout much of the South; (B) a more typical bromeliad species grows out from a cypress trunk in the Florida everglades.

phytic plants, such as arboreal orchids and bromeliads, which live on trees in the tropics and subtropics (Figure 7–8). These so-called air plants gain support from the tree in that they have a physical substrate on which to grow and are elevated to favorable light levels for their growth. The host is normally neither benefited nor harmed unless the weight and wind resistance of the epiphytes become so great that the structure of the tree is weakened and a wind storm brings it down.

In Central and South America, ant birds have a commensal relationship with army ant columns. These birds fly just above the head of an army ant column and eat the jungle insects that the ants scare up. Also perched on the foliage surrounding the army ant column are tachinid flies, which swoop down on any fleeing grasshoppers and parasitize these large jungle insects with their own eggs.

Mutualism

In **mutualism** both organisms involved in the symbiotic relationship will benefit. In all climates, most flowering tree and shrub species have fungi called mycorrhiza associated with their roots. Some forms cannot exist apart from their tree partner (thus having obligate relationships), whereas others may be free-living or casually associated with the tree (thus having facultative relationships). These mycorrhizae help absorb nutrients from the soil and provide nitrogen compounds to the trees through intimate associations with their roots.

In lichens, one-celled photosynthetic algae and chlorophyll-lack-

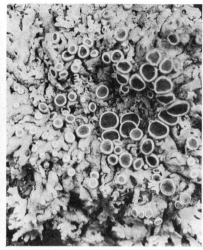

A   B   C

FIGURE 7–9

Mutualism: lichens are composed of blue-green algae and fungi living in intimate symbiosis. (A) A fructicose lichen (Teloschistes species) on a tree trunk; (B) a rock lichen (Parmelia species) with fruiting cups from which spores are released; (C) a lyngwort lichen (Lubaria pulmonaria) on a tree.

ing fungi exist together in an intimate mutualistic association (Figure 7–9). The algae, through photosynthesis, provide sugars as food for both species while the threadlike bodies of the fungus collect and hold moisture and mineral nutrients that both symbiotic partners can use.

Pollination by flowering plants and insects is perhaps the most common example of mutualism. Bees and butterflies depend on flowers for food in the form of nectar and pollen. The flowering plants depend on bees, insects, or other pollinators to carry their male reproductive cells (sperm contained in the pollen grains) to the female flowering parts of other flowers of the same species (Figure 7–10; plate 9).

Domesticated plants and animals usually enjoy mutualistic relationships with human beings since the latter care for them and ensure their continued reproduction (without which these plants and animals would soon die in competition with wild species) while depending on them for their own existence. During the course of domestication, artificial selection by human beings of these plants and animals has eliminated many of the adaptive traits useful for survival in the wild state, in exchange for traits yielding higher food production. The domestication of grain crops and other plants and animals in a mutualistic association with human beings allowed wandering prehistoric people to break their continuous cycle of hunting and gathering about 8,000 to 10,000 years ago and take up agriculture. Stable communities and increased population resulted. Thus these rather recently established mutualisms have made possible the large cities and advanced civilization we have today.

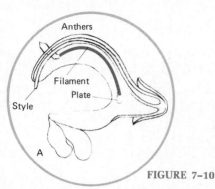

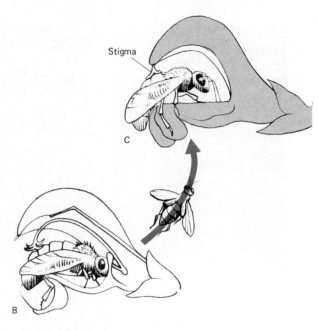

FIGURE 7–10

Pollination in the sage plant. (A) The arrow points to
a plate formed by fusion of the stamen bases of two
pollen-bearing anthers. (B) When the bee enters a
young flower and presses against this plate to get at
the nectar, the anthers are tipped forward and
deposit the pollen on the back of the bee. The female
pistil flower parts are still high in the flower and do
not touch the bee. (C) The bee visits another, older
flower. The stigma part of the pistil has grown
downward with its lobes open, and the bee's back
brushes this stigma, allowing the female stigma to
receive the pollen.

Mutualism tends to involve a long series of *coevolutionary*
changes (essentially simultaneous evolutionary adjustments in each
of several species). It would be difficult for one of the partners to
alter its biology without a concurrent change in its symbiont. Thus
symbiotic combinations that do not change concurrently will be
selected against, or at least the species will fail to continue to maintain their symbiotic relationship.

In interactions between plants and insects, for example, each
partner in mutualism will evolve traits to maintain the symbiosis.
Thus, many ants live in mutualistic relationships with plants; the
plants offer them food and a place to live in return for protection by
the ants against herbivorous insects and clinging vines or competing
grass around the base of the plant (Figure 7–11). The extreme development of symbiosis by coevolution leads to relationships in which
two species are completely dependent on each other. Such an example is shown by the unusual obligate relationship that exists between a tiny white moth (*Tegeticula yuccasella*) and the Spanish
bayonette plant (*Yucca*). By day the moths rest within the white
flowers. Soon after dusk the female yucca moth climbs up the
stamen of a flower where she begins scraping pollen from the anther
and shaping it into little pellets. When several pellets have been
made, she flies with them to another flower where she lays her eggs
within the flower ovary and then climbs up the pistil and packs her

FIGURE 7–11

A Pseudomyrmex ant feeds at
one of the leaf nectaries on an
Acacia tree in Costa Rica. The
ant protects the plant against
herbivores and encroaching
vines.

Mutualism between the yucca and the yucca moth Pronuba. *Only about 1 cm in length, the yucca moth carries a packet of pollen to a newly opened yucca flower and pushes it into the stigmatic chamber of the pistil, ensuring fertilization of the flower and hence development of the fruit parts, on some of which the moth's larvae will later feed.*

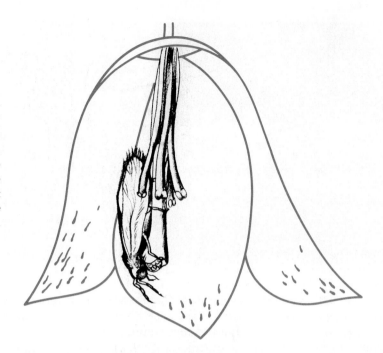

pollen pellets into a deep depression in the stigma (top of pistil; see Figure 7–12). Here the pollen grains can germinate and send the sperm cells into the stigma and down the style column to the ovary where they can fertilize the eggs. The young moth caterpillars hatch within the yucca ovary, but only a few of the developing seeds are destroyed by them. The larvae eventually pupate in the ground and metamorphose into adults at the time of the next flowering of the Spanish bayonette host.

Parasitism

In **parasitism,** one species (the *parasite*) is benefited while the other species (the *host*) is harmed by the association (Figure 7–13). Parasitism, like predation, is frequently an important factor in controlling natural populations, but despite the similarity of definition and functioning, there are significant differences between parasitism and predation. Parasites tend to be much smaller than their host, living as they do on a portion of the host's energy intake. Predators are almost always larger than their prey and tend to eat all their prey. Parasites tend to live on, in, or near the host, whereas predators associate closely with the prey but only when actually feeding on them. Parasites will kill their hosts only very slowly or not at all, while predators kill their prey quickly and consume them nearly immediately.

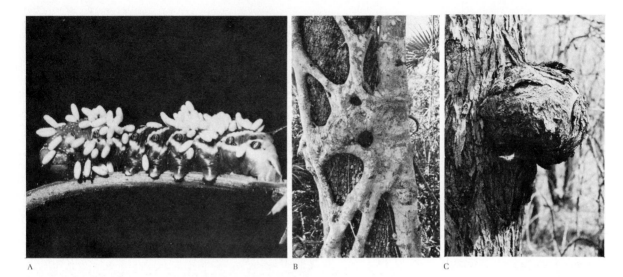

A   B   C

FIGURE 7–13

Parasitism: (A) Braconid wasp cocoons on a sphinx moth caterpillar; (B) a strangler fig (Ficus aurea) on a tropical hardwood tree; (C) a tumorous gall on an elm tree.

The most efficient parasite is one whose presence is hardly noticeable to the host. Should it kill its host before parasitic reproduction, the parasite would die without leaving any offspring. The species with this habit would rapidly become extinct (plate 10).

Predation

Predation involves one organism capturing and consuming another organism wholly or in part. The two components of such an interaction are the predator and the prey species. The **predator** is a free-living organism that lives on other living organisms, usually of a second species. Intraspecific predation, more commonly called *cannibalism*, is almost always associated with conditions of overcrowding in a population. Normal predation involves a *prey* organism eaten by a predator; the definition of prey can include plants, so that both carnivores and herbivores are predators. The spectrum of predator-prey interactions has involved many unusual ecological and evolutionary adjustments by each species in the relationship.

Predator types

Carnivorous plants (Figure 7–14) are true predators in the sense that they feed on living prey such as insects and digest part or most of the animal's proteins. The Venus flytraps (*Dionaea muscipula*) capture living prey in a leaf that closes like a steel trap. Three trigger hairs at the center of each of the bright red inner surfaces of each trap lobe will cause the trap to snap shut in about 0.3 seconds when touched by a crawling insect attracted by the color. During the next week, digestive enzymes secreted by the plant break down proteins into their constituent amino acids that are absorbed by special glands on the leaf's surface.

The sundews (*Drosera* species) are found in many boggy parts of the United States and have sticky secretions in the form of droplets that stand on the tips of thick bright red hairs above the paddle-shaped end of a leaf. When unwary insects land to investigate the bright color and glistening droplets, they become firmly stuck and the sensitive hairs turn inward, folding the prey into the center of the leaf plate. The plant cells of the trap then secrete the enzymes that break up the bodies of the prey. The chemical molecules formerly composing the prey are digested and absorbed into the cells lining the trap.

Pitcher plants use the pitfall method to capture their prey. An unwary insect, such as an ant, descends into the deep vertical trap cavity and finds hundreds of downward-pointing hairs blocking its return. It eventually falls into the fluid in the bottom of the trap, dies, and its proteins are digested by both the plant's secreted enzymes and bacterial action. Carnivorous plants, however, are relatively rare in the plant kingdom and do not form a major class of predators in most natural communities.

Animal predators are found in all major groups of animals, and predation helps regulate the population size in prey populations of

FIGURE 7–14

Some carnivorous plants found in North America: (A) the Venus flytrap (Dionaea muscipula) with a trapped prey; (B) a sundew (Drosera species), with an ant captured by the sticky hairs on the highly modified leaf; (C) pitcher plants (Sarracenia minor) in northern Florida.

A

B

C

A

B

C

FIGURE 7–15

Animal predators: (A) a frog using its tongue to capture a fly; (B) the giant Komodo Dragon, a monitor lizard in the East Indies, which may reach 11 feet in length, weigh 300 pounds, and bring down a medium-sized deer for prey; (C) a voracious spider with a grasshopper as prey.

both plants and animals (Figure 7–15). In a predator-prey relationship that has existed for a long time, coevolution minimizes or produces adjustments through natural selection whereby both the predator and prey populations exist in more-or-less natural balance. The prey is not eaten into extinction so that the predator species dies of starvation; instead, predator-prey relationships continue over long periods in a generally dynamic, balanced relationship between predator and prey. Adjustments in such factors as the birth rates and dispersal ability of each species minimize the disturbance of the system and essentially continue the status quo.

A number of kinds of animal coloration have evolved through natural selection in response to predator-prey interactions. Many predators have color lines leading forward from the eye that apparently function as aiming sights; they may also reduce glare in bright open habitats such as the grassland plains (Figure 7–16). Vertebrate predators have evolved sharp claws, caninelike teeth, rapid behavioral reactions, and other prey-capturing adaptations.

Protective coloration

Protective coloration is an all-pervasive evolutionary result of predation pressure throughout the world's biomass. The prey species comes to resemble some object in its environment through natural

FIGURE 7-16

Dark-colored eye lines serve as sighting and aiming devices for predators in catching prey. (A) Eye line of the yellow-throated vireo (Vireo flavifrons). (B) Long-billed curlew (Numenius americanus), showing direction of eye line forward of center of pupil to bill tip. (C) Head of the arboreal vine snake (Oxybelis aeneus), showing eye line and groove for aiming.

selection, thereby deceiving potential predators. Individuals that resemble nonedible objects such as bark or stones tend to be overlooked by predators. Likewise, species that resemble distasteful or dangerous animals in the area will tend to be avoided by predators. Thus protective coloration involves profound ecological as well as evolutionary implications, for many more prey species can be packed into a community if most are protectively colored in some way, reducing the impact of predation on the individual species.

In *cryptic coloration,* the prey resembles or mimics inanimate objects or background such as sticks, barks, rocks, bird droppings, thorns, cobwebs, dew drops, raindrops, and old leaves. Insects provide an abundance of examples of this type of protective coloration, and whole groups of insects may even be named after the type of coloration and structural adaptation shown, such as walking sticks, leaf katydids, and stick caterpillars of geometrid moths (plate 9).

In *mimetic* relationships, the potential prey species resembles another species in its environment that is toxic or dangerous to predators. A number of types of **mimicry** have been recognized by biologists. In *Batesian mimicry* the prey species resembles a toxic, venomous, or unpalatable species found in the same region, yet the prey is relatively harmless or even tasty. The potential prey in a Batesian mimicry complex becomes the *mimic,* and the species dangerous to potential predators is the *model* (plate 8).

Behavior is as important as color pattern and structure in the mimicry relationship; otherwise the deception provided by similar external structure, coloration, and pattern does not convince the predator that it is seeing an unsuitable prey species. Thus many harmless soldier flies resemble wasps and bees in their habitat areas, even mimicking the sounds by attaining a similar wing-beat frequency to produce the similar buzz. The monarch butterfly (*Danaus plexippus*) is common throughout North America and other areas and is distasteful to birds, lizards, and other predators. The adults are large, slow-flying butterflies with bright, showy red-orange coloration bordered with black. They are poisonous because of the diet of their caterpillars; in feeding on the leaves of the milkweed food plants, the larva incorporates into its body poisonous compounds called cardiac glycosides. When the larva pupates and the body tissues reorganize inside the pupa to form the adult, these compounds are secreted or sequestered into certain areas and then built into the tissue of the adult monarch. Predators that have previous experience with this species will avoid adults in the community. After one or two learning experiences, apparently many birds remember an unfortunate experience with the monarch for as long as several years. In the United States, the monarch is the model in a Batesian mimicry complex with the viceroy butterfly, a mimic that is edible but resembles *Danaus* and flies just like the monarch. Thus although it is edible (because its larva feeds on willow and other similar unpoisonous plants), potential predators such as birds will shy away from it, mistaking the viceroy adult for a monarch.

Sometimes a number of distasteful species share a common color pattern that warns predators that all are distasteful. This is called *Mullerian mimicry*. The potential prey species obtain a significant survival advantage because the predators in the area will learn to associate distastefulness with that pattern without trying all the species. Thus the predator has to try only one species of yellow-and-black-banded wasps in most temperate zone areas to know that it will be severely stung by an insect that looks like the species it attempted to capture as prey.

In *warning coloration*, the bright colors and markings of the individual prey species simply advertise its noxious qualities to any potential predators. Thus many brightly colored caterpillars, especially those bearing red or yellow patches, simply warn predators of their genuinely noxious qualities even though a mimicry complex may not be built up around them. Such species serve as potential models for future development of mimicry complexes, of course (plate 10).

Another interesting evolutionary result of predation pressure is found in *aggressive mimicry*, where the predator deceives the prey by its coloration and behavior. In the southeastern United States the

FIGURE 7-17

Aggressive mimicry. A female Photinus firefly has successfully attracted and is eating a Photurus male by imitating the flash pattern of a Photurus female.

snapping turtle will lie at the bottom of a muddy river and open its mouth wide, wiggling its whitish worm-shaped tongue. Curious fish coming along think the tongue is a worm, and they enter the snapping turtle's mouth to grab the tasty-looking "worm." The snapping turtle quickly closes its mouth, having deceived the prey by its coloration and behavior. Ambush bugs that live in sunflowers look like part of the yellow and orange center, and it is easy for them to attack unsuspecting bees, flies, and butterflies coming for nectar. Certain female fireflies imitate the mating flashes of the females of other species and capture and eat the males of the species that come in response to the mimic flash pattern (Figure 7–17). Thus the interaction between predators and prey may generate many fascinating developments in behavioral repertoires as well as physical features.

Summary

Organisms constantly interact with one another. Competition between numerous sympatric species in a community causes such evolutionary and ecological changes as extinction, competitive exclusion, character displacement, and smaller niches. Symbiotic interactions are ecological associations involving some transfer of energy or other adaptive benefit. In commensalism, one organism benefits while the other species is unaffected. In mutualism, both species benefit, and the association commonly involves a long series of coevolutionary changes. In parasitism, the parasite species benefits while the host species is harmed. Predatory interactions involve a free-living predator species that captures and consumes a prey organism wholly or in part. Both prey and predator show an astonishing development of physical and behavioral adaptations as a result of natural selection in these interactions, including protective coloration, Batesian and Mullerian mimicry, warning coloration, and aggressive mimicry.

8   Species diversity

One of the most striking biological outcomes of interactions of different species and environmental factors in a community is reflected in the concept of species diversity: that is, the number of species and their relative abundance in a given area. This range of species diversity is reflected in local transects as we move from a dry field into a moist hardwood forest and on a global scale as we move from the tundra communities with only several hundred species to the diverse tropical rainforests with tens of thousands of plants and animal species per square mile (Figure 8–1). How can we explain these differences among communities?

Niche structure and species diversity

Investigations of species diversity usually must proceed hand in hand with the study of niches in a particular portion of a community. The assemblage of insects inhabiting a particular foliage layer in a forest may be taken as a representative sample of species diversity in that area, and a similar sampling study elsewhere can provide comparative data. Thus biologists have studied desert lizards in parts of southern Africa, western Australia, and southwestern North

FIGURE 8-1

The tropical rainforest, as in this scene from Rhodesia, has the highest diversity of plant and animal species of all the biomes on earth.

America in an effort to understand how diversity is generated; others have searched for the same kinds of information by comparing bird communities on tropical Caribbean islands and in mainland forests. The common aim is to analyze niche structure of community species diversity, partitioning the species along the lines of their spatial, temporal, and trophic activities (Figures 8-2, 8-3, and 8-4). Species regularly replace one another along each of these three niche dimensions. For example, one warbler species may hunt for food near ground level while another forages 10 feet above the ground in the

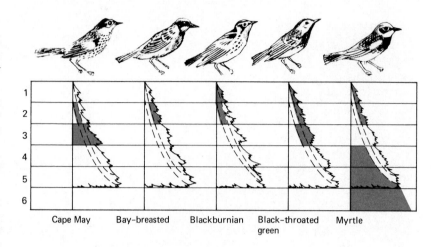

Niche separation in five kinds of eastern U.S. warblers that eat the same kinds of caterpillars (bud worms) on the same spruce trees. Robert MacArthur found that the birds hunt in different parts of the trees, so that each species utilizes a nearly private crop. The colored areas show where each kind spent more than 50 percent of its time during his observations. Different methods of hunting further separate them.

same forest. Populations of different species are often active at different times in the same area. Different species of predators living in the same area tend to eat prey of different sizes and basic types, with the larger species taking larger prey as a rule. Diversity is generated in the community by the number of kinds of organisms that can fit in along each niche dimension and yet still be separated from one another to some substantial degree.

The problem of explaining the causes of these local and latitudinal relationships and patterns in species diversity has long occupied the attention of biologists, particularly evolutionists and ecologists. The hypotheses that they have suggested for the determination of diversity are often interdependent and several may act in concert or in sequence in any given case. Overall, they fall into two primary groups: those that advance the *ultimate reasons* (the ultimate determination of diversity by evolutionary and ecological parameters) for the ecological diversity of floras and faunas and those that explain diversity in terms of *proximate reasons* (immediate causes).

Cormorant

Shag

Closely related, sympatric cormorants in Britain. The cormorant nests high on cliffs or on broad ledges, fishes out to sea, and eats a mixed diet excluding sand eels and sprats. The shag nests low on cliffs or on shallow ledges, fishes in shallow estuaries, and eats mostly sand eels and sprats. These close relatives, though very similar in appearance, obtain their living in quite different ways and thus do not compete. Their ancestors were presumably separated by character displacement, and competitive exclusion eventually resulted in genetically isolated species.

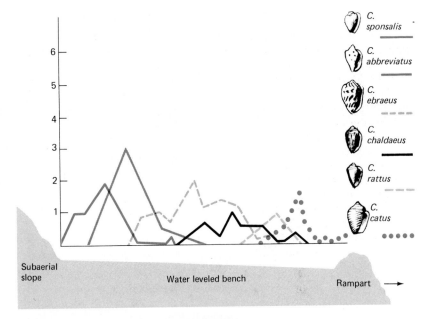

FIGURE 8–4

These closely related sympatric cone shells hunt their prey with poisoned darts along the Hawaiian shore. Each carnivorous snare species lives in a slightly different strip of shore and catches a unique assortment of prey animals, thus minimizing competition and resulting in a greater diversity of cone shell species.

Hypothetical ultimate mechanisms affecting species diversity

Ultimate reasons for the ecological diversity of biotas may be found among the historic and dynamic parameters of the biogeography of faunas and floras. Although we will be exploring geographic distribution in Chapter 17, let us here look at the arguments advanced that point to increased opportunity for tropical and subtropical regions to increase the diversity of their biotas and, thereby, the diversity of their constituent communities.

Evolutionary time

The evolutionary time hypothesis claims that species diversity increases with the age of an undisturbed community. In the equatorial tropics, the land and shallow marine areas receive more and retain more of the sun's energy than subpolar areas, and therefore they are able to sustain and generate more organic material if left undisturbed by climatic and geologic changes. The animal faunas respond quantitatively and qualitatively to the increase in plant biomass and species diversity. During the past 10 million years, the tropical and subtropical belts have been affected less drastically by climatic changes (especially Pleistocene glaciations) than have the temperate and boreal habitats. Even now, with one-tenth of the globe's land area still covered by glaciers and perpetual snow, nearly one-third of the land is in the tropical belt and a full one-third is subtropical in climate. The tropical vegetation belts are split into major subdivisions by both longitudinal and latitudinal barriers, especially sea barriers, which adds the property of spatial heterogeneity for speciation opportunities in the tropics. To the north and south, Pleistocene glaciations decimated the biota, and in the recently deglaciated areas the diversity of the fauna and flora is correspondingly very low. In this evolutionary time theory, then, the "rich" biotas of tropical areas represent the more "normal" diversity situation in a relatively undisturbed environment, whereas conditions of low faunal and floral diversity in temperate areas are thought to be exceptional and recent. Species impoverishment results from lack of time to completely occupy the available environment, though the few species present may actually utilize nearly fully the available resources through expansion of their niche.

Ecological time

According to the ecological time theory, a number of species may already exist elsewhere that could fill particular positions in the environment under consideration, but these species have not yet had opportunity or time enough to disperse into the relatively newly opened habitat space. In other words, there has been a shortage of time available for arrival of organisms in the area from neighboring zones, rather than a shortage of time for speciation and evolutionary adaptation of new organisms at the site itself. Because of the excellent dispersal ability of most organisms, this concept of "ecological time" may be of relatively little importance in most communities. For island communities, however, biologists have shown that species composition for most animal and plant groups depends on the island size, the distance from "source" areas (other islands or the mainland), and the time available for the colonization of the island.

Climate stability and predictability

Organisms living in environments with a seasonally changing and hence unstable climate must possess broad tolerance limits to cope with the wide range of environmental conditions they may encounter; plants and animals living in stable environments with more constant climates can be more finely specialized and have finer niches. Thus in tropical areas, species use a smaller fraction of available resources and the habitat will support more species.

If the changing climate is quite predictable (e.g., winter rains and summer thunderstorms on the Mohave Desert of California), organisms can evolve some specialization based on the periodic appearance of resources. Thus certain lycaenid butterflies emerge from diapausing pupae each spring following the early rains in the Mohave and lay their eggs on their spring-blooming host plants; other species have specialized to emerge and lay their eggs on the fall-blooming buckwheat plants that follow the summer rains. With either temporal predictability or annual stability in climate, then, it is possible to pack more species into the same area.

Spatial heterogeneity

The theory of spatial heterogeneity combines ultimate and proximate factors promoting species diversity. A habitat that has a complex structure will provide a greater variety of different microhabitats for species to occupy than simple habitats. Thus a rocky canyon cutting across a desert plain and containing jumbled boulders and exposed banks and sand bars will have more species than the homogeneous open and flat plain. The spatial component of niche diversity is higher and functions as an ultimate factor promoting increased species richness.

To animals, a complex forest of many plant species offers more spatial heterogeneity than a grassland. The difference in plant diversity was likely generated by ultimate factors, but its contribution to diversity among the animals that move in is proximate since the plant community developed in response to spatial heterogeneity, climate, or other ultimate components.

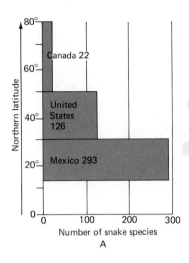

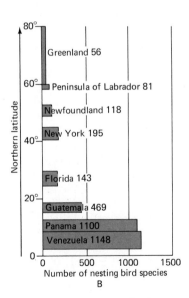

FIGURE 8–5

The relative diversity of species of two major animal groups, from the arctic to the equator.

Hypothetical proximate mechanisms affecting species diversity

Many ecological theories have been offered as proximate reasons for the diversity of species in the tropical zones and the lack of diversity in the temperate zones (Figure 8–5). Besides the premise of more ecological niches in the tropics, the niches may be more intensely exploited, or divided, by more than one kind of organism. Some biologists have suggested that competition, rather than physical environmental hardships, limits tropical population sizes and places a selective premium on developing subtle differences in niche exploitation. Others have emphasized the diversity of predators and parasites in the tropics and their selective effects on prey. Some of the hypotheses that promote diversity by proximate mechanisms follow.

A theory and a test concerning island species diversity

Until about ten years ago, ecologists tended to regard islands as "impoverished" in terms of numbers of species. This conclusion seemed logical both because of the obvious problems that plant and animal species have in colonizing islands and because islands typically support fewer species than comparable areas of mainland habitat. However, in 1967 Robert MacArthur of the University of Pennsylvania and Edward Wilson of Harvard University suggested that islands might actually be supporting as many species as possible. In other words, the species diversity on a particular island is normally in *equilibrium*. By this theory, as the number of species on an island increases, the rate of immigration of *new* species to the island should *decrease*. Eventually, the immigration rate should drop to zero as the species density of the island reaches the total number of species in the source areas surrounding the particular island system. Moreover, the extinction rate of species already present on the island should *increase* as the number of species on the island increases and interspecific competition becomes more intense. When the rate of immigration equals the rate of extinction, existing species become extinct at the same rate that new species are invading. Thus, species density or diversity reaches a dynamic equilibrium, although (because of the continual turnover of species), the actual species composition of the island can be changing.

An interesting test of this equilibrium theory of species diversity was carried out by Wilson and David Simberloff in the Florida Keys. After the arthropod faunas of several very tiny mangrove islets were carefully censused, fumigation tents were put over the islets and all arthropods exterminated with methyl bromide. The investigators then monitored the process of arthropod recolonization over a two-year period. To the surprise of most ecologists, they found that within 200 days the numbers of species of insects, spiders, and other arthropods had stabilized. Turnover rates remained quite high, but the total numbers of species on the islands remained relatively constant over a period of nearly two years, providing excellent evidence that the islands had indeed reached an equilibrium.

Competition

Biologists supporting the competition hypothesis feel that natural selection in the temperate zones is controlled mainly by catastrophic exigencies of the physical environment, whereas biological competition becomes a more important evolutionary factor in the tropics. In these diverse equatorial communities, such as tropical rainforests, populations are thought to be close to their equilibrium sizes at nearly all times, causing considerable intraspecific and interspecific competition for the available resources. With strong selection for competitive superiority, in the tropics each species has narrower food and habitat requirements, making small niches; thus more species can coexist in the overall habitat space. Organisms living in less diverse temperate and polar communities, on the other hand, are thought to be less stable in population size and often well below the carrying capacity. With this lack of saturation of portions of the community and the vagaries of harsh weather, selection for rapid reproduction and broad tolerance limits is strong and selection for competitive ability is weak. Hence, larger but fewer niches are available and less diversity evolves.

Predation

Some ecologists have claimed that there are more predators and parasites in the tropics and that these diminish the population sizes of individual prey species enough to lower the level of competition among them. The reduced level of competition by this selective or random removal of prey allows the addition and coexistence of new intermediate prey types, which in turn can support new predators in the system, and so forth, to some unspecified maximal limit of diversity (reached through trophic energy relationships). If the predators are absent, several similar prey species *cannot* coexist locally and some will be eliminated by competitive exclusion. This hypothesis suggests that competition among prey organisms may be *less* intense in the tropics than in temperate areas and does support the next hypothesis, both being alternatives to the competition hypothesis.

Rarefaction

The rarefaction hypothesis states that organisms are continuously removed from a community by density-independent means (cold snaps, predators feeding preferentially on more abundant prey types regardless of actual numerical densities, etc.; see Chapter 10). It suggests that wherever communities are not fully saturated with individuals, competition is reduced and coexistence is possible without resorting to competitive exclusion, simply because each species is relatively rare. The rarefaction and competition hypotheses are thus mutually exclusive mechanisms.

Productivity

The productivity hypothesis basically says that more productive habitats (where food is dense, as in the tropical regions where temperature, rainfall, and solar radiation allow maximum photosynthesis) will support more animal species than do less productive habitats. Each species can be more selective in what it eats and confine its diet to a limited range of vegetation or prey items. In habitats with limited food, foraging animals must be dieting generalists and remain few in number; most resources are simply too sparse to support specialized species.

Overall conclusions

As yet there is little concrete proof to support any of these considerations to the exclusion of the others. Many of the mechanisms may often act together to determine the diversity of a given community, and their relative influence doubtless varies from area to area. Many additional studies must be carried out before we have a clear concept of the relative role and operation of factors controlling the diversity of species. Yet the subject is of great intrinsic interest, and it serves to tie together some of the ideas we have been following through the biosphere and biotic communities in preceding chapters, as well as to prepare us for moving into a consideration of life on the population level of organization.

Summary

A diversity of species occupies any biotic community, and the comparative richness of species numbers varies in both local communities and major biomes. Particularly notable is the great species diversity in the tropical zones as compared with temperate and subpolar regions. Niche structure as defined in terms of spatial, temporal, and trophic dimensions is intimately involved with species diversity in a community. Ultimate evolutionary and ecological reasons for biotic diversity may include the amount of evolutionary time available for the generation of new species, the amount of ecological time providing dispersal opportunities for already-existing species to invade a new habitat, the degree of climatic stability and climatic predictability present in a particular environment, and the spatial heterogeneity of the available habitat. Proximate mechanisms for generating species diversity may include competitive interactions, predation and parasitism pressures on population size, the rarefaction of species by density-independent population regulation below saturation levels, and the relative productivity of different habitats. It is quite likely that several of these mechanisms often act together to determine the diversity of a given community and that their relative influence varies from area to area, probably even between different groups of animals. In the next section some of these controlling phenomena will be examined in more depth at the population level of organization.

PART THREE

Biology at the population level

Moving from assemblages of diverse species at the community level, we now look at the structure and biological characteristics of groupings of individuals of the same species: populations. The organization of populations (Chapter 9) involves more than the mere spatial dispersion of the constituent individuals; it includes the movement patterns and behavior of the species in mating and territorial matters. A consideration of factors involved in the growth and regulation of populations (Chapter 10) leads us to consider the present problem of explosive population growth in human beings and how to control it (Chapter 11). The genetics of populations (Chapter 12) brings us back to the evolutionary theme as we see the effects of recombination of hereditary factors at the population level. Chapter 13, on the mechanisms of evolutionary change, includes an intensive look at the validity of Darwin's concept of natural selection, the phenomena involved in adaptation to environmental change, and the processes of hybridization and speciation in populations. The history of life on earth, as revealed through the fossil record and judicious scientific detective work, paints a dramatic picture of perhaps half a billion species of organisms that have inhabited our planet (Chapter 14). The final and most significant event of this history was the origin and evolution of the human species (Chapter 15), which took over the earth in only several million years and replaced the dominant groups that had ruled in earlier eons of geologic time. We then survey briefly the diversity of present life found in the major plant and animal groups (Chapter 16). The subject of biogeography next occupies our attention, as we look for the historical, evolutionary, and ecological reasons for the broad and specific patterns of distribution found among plants and animals (Chapter 17). Our consideration of biology at the population level culminates with the basic principles of behavior (Chapter 18) and social behavior (Chapter 19), as we explore the ways that animals respond to each other and to environmental stimuli.

9 The organization of populations

A population is a group of organisms of the same species living in a particular space. Communities are composed of many species populations that vary greatly in size and character, depending on the species of organism involved and the limits of the space they occupy. Populations are like the individual organisms that constitute them in that they have a life cycle, possess a definite structure, and function in an orderly manner, growing and dying as a sort of superorganism. Additionally, many of the principles of ecology that we have considered in the preceding chapters on communities and ecosystems apply to the population level of organization. Thus populations have a spatial orderliness (which we shall consider in this chapter), a temporal organization, and a metabolic orderliness. The particular pattern of flow of energy through the population results from its position in the general ecological community, as does the flow of materials. The population is affected by environmental changes induced by human beings, and populations of many plant and animal species have become extinct through environmental degradation and pollution of human origin. Interspecific competition, symbiotic interactions, and predation all intimately affect the status of a given population. Thus in this chapter and the next several chapters, we shall be considering biology at the population level—a highly significant level of organization in the biological world. It is at the population level that evolution occurs and that animal and plant species exhibit density, growth, and regulation, as well as fluctuations and regular oscillations in population size. Here, too, the genetic character of the individuals in the population intimately influence the survival and future evolutionary history of the species, as we shall see in Chapter 12.

At first sight, it might appear that the properties of a population are simply an aggregate of the characteristics of the individuals in the group. However, when considering the organization of populations, we see that a population can show characteristics that an individual cannot, such as evolution, dispersion (distribution), dispersal (moving from one place to another), and density (degree of crowding). It is also possible for a population to show growth that can be measured in several ways not possible at the individual level — for example, the population's natality (birth rate), mortality (death rate), age distribution (the fraction of the population present in each age class), **biotic potential** (the maximum possible growth rate under ideal conditions), and growth form (the manner and rate of population growth). Each of these properties is not characteristic of the individuals in the group but is unique to the aggregation, and collectively they influence the ecological status as well as evolutionary future of a population. When we understand the normal role of these population attributes, it becomes easier to understand how environmental pollution and other changes brought by human beings can radically affect natural populations of organisms. In many ways, then, the population is an extremely important structural unit in biology.

The spatial arrangement of populations

The spatial arrangement of the individuals composing a population is called the population's **dispersion,** which is simply the internal distribution pattern or placement of individuals within a particular population. A great many factors influence this distributional pattern. The individual organism can live and reproduce only within certain physical limits in its environment; this set of conditions has been called its *tolerance range*. Since the tolerances of different individuals will vary to some extent, a range of tolerance of a local population becomes somewhat larger than any one of the individuals in it. Populations are not limited solely by their tolerance ranges, though it is clear that a population cannot exist under conditions outside its tolerance range. These conditions may include purely

Uniform

Random

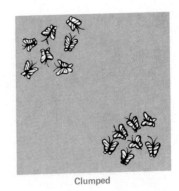

Clumped

FIGURE 9–1

Uniform, random, and clumped dispersion of organisms.

physical phenomena, such as the climate or the effect of local rocks on soil suitability, or the presence of other kinds of organisms, such as a large pine tree shading a population of small shade-intolerant plants. The direct effects of other kinds of organisms such as herbivores, predators, and parasites, and interference by directly competing species, may also set the boundaries of a population's distribution.

Populations, of course, are not static structural units but are generally constantly increasing or decreasing in size. A population's parameters of dispersion may be determined by the slow or fast rate of dispersal of that species. From another perspective, species that are more tolerant of environmental extremes of all kinds will tend to increase their population dispersion more rapidly than those that have narrow limits of tolerance.

The physical placement of individuals within the population may be distributed according to several general patterns. These types of spacing often function as a regulatory mechanism for population size since normal reproduction, dispersal, and other behavior of the individuals is clearly influenced by the degree of crowding in most species. Three basic ways in which individuals in a population may be dispersed—that is, spaced relative to each other—are in uniform, random, or clumped patterns (Figure 9–1).

Uniform dispersion

In *uniform dispersion*, the individuals are regularly spaced throughout the area occupied by the population. This kind of distributional pattern is fairly frequently encountered in nature, especially among invertebrates. Uniform dispersion is characteristic of situations where competition between individuals for resources is severe, or where positive antagonism promotes even spacing. Among the ver-

FIGURE 9–2

Gannets nesting in a dense colony on Bonaventure Island, Quebec.

tebrates this type of distribution occurs in nesting colonies of sea birds on islands where nesting space is divided among the parents at approximately the distance of the length of the mother bird's neck-to-bill-tip reach (Figure 9–2). Among many desert and other arid-zone plants, intense competition for the little annual precipitation means that shrubs that are well spaced in a reasonably uniform dispersion pattern will survive, whereas those seedlings that sprout too close to existing shrubs will not obtain sufficient water to grow to maturity. Additionally, the roots of many desert and chaparral species secrete antibiotics that inhibit the growth of seeds or neighboring plants within the radius of their root systems. This biochemical warfare results in uniform spacing patterns similar to those caused by competition for water.

Random dispersion

In *random dispersion* individuals are scattered over an area without any regularity or any affinity for each other. This type of spacing is relatively rare in nature, and it generally occurs in uniform environments where resources are spread evenly throughout the area of the population's distribution. Additionally, the species' behavior must

FIGURE 9–3

Life on the floor of the deep ocean trenches must cope with total darkness, cold, and great pressure. Here in Romanche Trench sea anemones, sea urchins, and other small creatures that look like small shrimp exist at a depth of 24,360 feet below the ocean surface.

include neither a tendency to aggregate because of social attractions nor a tendency for individuals to repel one another. In nature we almost always find a nonrandom dispersion of resources, whether these be minerals of a particular type, food organisms, or even sunlight. Thus organisms are almost never random in distribution. Organisms living at great depths in the ocean below the level of sunlight penetration, and hence below the area where plants can grow, show essentially random dispersion across the ocean floor (Figure 9–3).

Clumped dispersion

By far the most common type of spacing is some form of close aggregation or groups of individuals, that is, a *clumped dispersion* pattern. This kind of spacing involves an irregular, nonrandom distribution of individuals throughout the population. Even mating pairs and their newborn offspring surrounding them may be regarded as clumped; and since in plants many seeds fall close to the parent, populations of most living things generally tend to show a clumped distribution. Besides clumping for reproductive reasons, animals aggregate in herds, swarms, or colonies for greater ease in feeding and sometimes also in order to enhance defense against predation. Clumping distribution may vary in intensity throughout the year and even be present only at certain seasons or at certain times in the organism's lifetime. Clumps or aggregates of individuals may themselves be scattered uniformly or at random throughout the area of distribution, depending on the distribution of resources or degree of social behavioral interactions between the subgroupings within the population.

The reasons for animal and plant aggregation as a common feature of dispersion arrangements are not difficult to see. Local habitat differences within the population's area of distribution provide optimal and suboptimal habitats for individual survival, and hence more individuals tend to survive in the areas with the best combination of environmental factors. Weather conditions in certain seasons in both the temperate zone and tropical areas may promote aggregation. A rainstorm in the tropical rainforest will provide temporary rainpools even in an area where it rains almost every day, and these pools attract scores of species of frogs and toads for breeding. During the dry season in southern Florida, populations of most wildlife species will aggregate around "gator holes," depressions in swampy areas that have been made by alligators, where water will stand long after the open flat marshlands have dried up. Many mammals and birds have become adapted to feed and function best as groups and hence social attraction promoting clumping is adaptively advantageous to them (Figure 9–4). The activities of reproduction require at least one male

FIGURE 9–4

Social clumping: a pair of Swallow-tailed Gulls in the Galápagos Islands.

and female cohort to aggregate, of course, and many birds and mammals are additionally stimulated by group reproductive behavior during a particular season of the year.

Social distance

The concept of social distance is also important in understanding the spacing of individuals in populations. A minimal *individual distance* between members of a population always exists even in the most clumped aggregations. Animals seem to have a particular minimal distance that when penetrated will cause nervousness and a tendency to move away. *Flight distance* represents the distance at which animals are likely to flee. Sable antelopes in Africa and pronghorn antelopes on the American Great Plains are able to detect potential danger at a considerable distance because of their excellent eyesight and thus have a great flight distance. Rhinoceroses, on the other hand, have relatively poor vision and their flight distance is not great. Wildlife biologists also use another term in discussing spacing, *charge distance*—the distance at which animals are likely to charge when approached. Sable and pronghorn antelopes are not aggressive and thus their charge distance may be only several yards. On the other hand, rhinoceroses have an aggressive temperament and their charge distance may be considerable—several dozen yards. Despite this extensive charge distance, rhinoceroses are usually alerted to the presence of an intruder by their keen sense of smell rather than by their poor eyesight.

One may well ask why all animals do not aggregate if advantages exist for clumping behavior. It is clear that aggregation often increases the survival of individuals in the group. Because of the protection that the group offers against being attacked by predators and the frequently beneficial modification by the group of the local habitat or microclimate, the survival rate of the grouped individuals is enhanced. Thus sufficient heat is generated within a beehive to allow the colony to survive cold night temperatures in the late fall that would kill any solitary bees resting outside the hive. The principal negative aspects of aggregation are that intraspecific competition is increased for needed nutrients, food, and space. For these and other reasons, many species remain solitary in normal behavior.

Dispersal and migratory behavior

The actual act of moving from one area to another is termed *dispersal* behavior, and it may involve movement on both a local and a long-range scale. The simplest local type of dispersal movement is exemplified by protozoans, simple single-celled animals that live in ponds and other bodies of water. In kind of a random process analogous to the Brownian movement of molecules, these tiny animals disperse uniformly through the medium, or at least wherever food and light are uniformly available. A population that is increasing in size, area, and density will often relieve the increased crowding by dispersing juvenile or adult individuals out to the margins of the area as they move into adjacent habitats. This normal individual movement on a local scale establishes the patterns of distribution of organisms within a population that we discussed in the preceding section. We may also look at the problem of population dispersal in a wider sense, that is, the movement of individuals into or out of the population area.

When one-way movement into an established population area or an uninhabited area occurs from neighboring areas, we say that **immigration** is taking place. If the area is undercrowded in relation to available resources present, the immigrants are normally successful in establishing themselves. **Emigration** is one-way movement out of a particular population. It is commonly found in situations where overcrowding is resulting from excessive reproduction or environmental stresses such as drought. Thus vast swarms of so-called migratory locusts move across the plains of Africa during periods of overcrowding in their places of origin.

Migration, on the other hand, involves the periodic departure and return of individuals to and from a population area. Each year hundreds of species of birds migrate across the North American continent to the far Arctic slopes to breed and raise their young. After the short Arctic spring and summer, the adults and young fly south thousands of miles to the southern United States or even South America for the winter. Monarch butterflies fly as far as 2,000 miles from northern Canada to southern California and Mexico, as well as some of the Gulf states, and pass the winter in a semidormant state hanging by the thousands in trees in certain sheltered coastal areas. The green turtle, *Chelonia mydas,* is found in oceans around the world, and most populations undergo remarkable migratory journeys (Figure 9–5). A particular population will lay its eggs on a single beach of a remote oceanic island or coastal region. An adult female resulting from an egg hatched on that beach will return to it in each breeding season that it is capable of laying eggs, approximately every two to three years. During the many intervening months, she feeds with the other adults in shallow sea areas along coasts where turtle grass grows. Thus the populations that lay their eggs on remote Ascension Island in the middle of the South Atlantic Ocean will swim 1,200 to 1,400 miles from the mainland coasts of South

FIGURE 9–5

A map showing recovery of tags throughout the Caribbean Sea from 2,000 adult female green turtles (Chelonia mydas) marked at Tortuguero, Costa Rica. The arrows indicate only geographic spread from the tagging locality at the sole nesting beach for these turtles and do not show exact routes of migration.

America to this tiny six-by-seven-mile oceanic island at two-or-three-year intervals, where each female lays about a hundred or so eggs in a pit dug in the sand and then leaves the beach, not to return for several more years. When the eggs hatch about 60 days after laying, the tiny hatchling turtles rapidly navigate across the beach to the sea and swim off in the equatorial current which they apparently follow downstream to the bulge of Brazil. There they grow up on a diet of turtle grass in the rich offshore feeding grounds and repeat the cycle of return with the adults to Ascension Island when they mature some years later. The remarkable studies of Professor Archie Carr at the University of Florida have shown that over 80 percent of these nesting turtles will "home" to the exact island beach from which they had emerged years earlier.

Mating systems and population organization

As we have seen in the discussion of clumped dispersion, sexual reproduction and the breeding activities of animal populations may affect the organization and distribution of the individuals involved. The term *mating systems* refers to a complex of factors that brings males and females together and affects gene exchange in the population. Mating systems—their varieties and evolution—have been particularly well studied in the higher vertebrates, and we shall consider their effect on population organization in these animals.

Mating requires individuals of the two sexes to become synchronized in activities and to become physically close to one another. The types of behavioral mechanisms that bring about mating in animal populations may be collectively referred to as mating systems.

The loosest structural form of mating systems is found in **promiscuity,** in which males and females have complete freedom to mate with any individual of their species that they encounter. Promiscuous mating systems are relatively frequent in nature and may be found even among the highest primates such as chimpanzee bands; however, there is usually some form of discrimination by the females that prevents truly random gene mixture. The purest form of promiscuity is probably reached in wind-pollinated land plants and certain marine invertebrates such as oysters, which release their germ cells to be dispersed randomly by wind or ocean current. Hence the organization of such populations of promiscuous species involves individuals that are generally scattered more or less at random, at least as far as the mating system is concerned.

When true pair bonds are involved, two or more individuals will stay together in a close relationship. In **monogamy** one female and one male share a pair bond and both parents typically care for the young (Figure 9–6). This situation is rather rare among mammals since the male plays a relatively minor role in nurturing the offspring, but in birds monogamy tends to be the predominant mating pattern where the only activity at which males are not equally adept as females is egg-laying. In fact, about 91 percent of all bird species are monogamous. Many terrestrial carnivorous mammals are also monogamous; here the male can aid in prey capture.

In **polygamy,** one individual has simultaneous pair bonds with more than one member of the other sex: in **polygyny,** one male has pair bonds with two or more females at the same time; and in **polyandry,** one female may mate regularly with several males. Probably because of the relative reproductive roles of males and females,

PLATE 8

MIMICRY

In Batesian mimicry, a distasteful or dangerous model species is copied by an edible or harmless mimic species. The top two pictures show a Batesian mimicry complex in the Eastern United States: the monarch (Danaus plexippus) at left is the distasteful model and the viceroy (Limenitis archippus) is the edible mimic. In Mullerian mimicry, both species are distasteful and share the same color pattern and behavior. Here, an ithomiine butterfly (Hypothyris mamercus, above left) and a heliconiine butterfly (Heliconius numata, above right) have converged in color pattern even though they are members of separate families; both are distasteful to birds and other tropical forest predators. Mimicry reaches bizarre extremes in a black katydid (left) that precisely mimics a large black wasp species living in the same wet forest area of Ecuador.

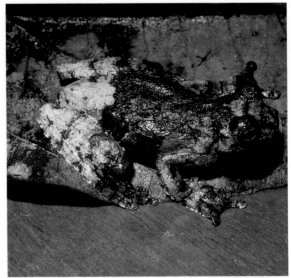

FIGURE 9-7

Territorial behavior in relation to the entire home range area of a typical male songbird during the breeding season.

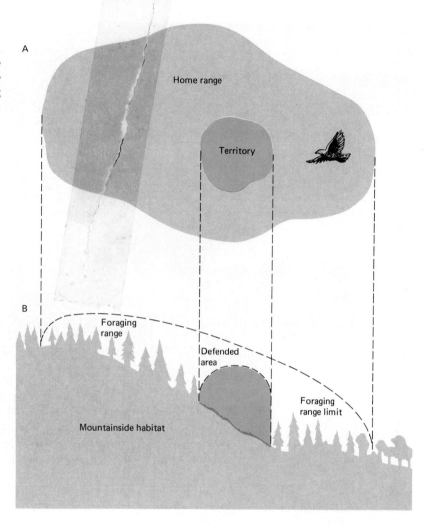

Territorial arrangements

Most vertebrate animals and the more advanced invertebrates commonly spend their whole lives in a definite area. This area of normal activities is called a *home range*. Here the animals search for food, seek mates, rear the young, and carry on their normal activities. If all or a part of this area is actively defended, the defended portion is called a **territory** (Figure 9-7). The role of the territory in spacing a population becomes clear when we realize that an individual (or a

FIGURE 9-6

A pair of African lions in Ngorongoro Crater, Tanzania, early in the mating sequence.

the latter state is quite rare among all animal groups, although it appears to occur in some of the button quails, rails, and certain other tropical birds.

It is clear that even the most highly evolved vertebrates such as the primates and grazing herbivores exhibit every possible social structure, from solitary conditions to very casual associations and tightly knit groups, and mating systems vary from promiscuity to regular polygamy and strict monogamy. In both primate and lower groups, males of many species are incorporated within a group throughout their lives and play a significant role in the upbringing of the young. Thus it is hard to draw any conclusions about the evolutionary course of mating systems throughout all phyla; however, it is clear that they influence the organization of populations through their effect on the distribution of individuals.

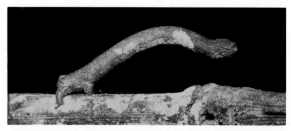

PLATE 9
CRYPTIC COLORATION

Cryptic coloration conceals even fairly large mammals as well as countless insects from the sharp eyes of potential predators. A young margay cat (top, far left) is concealed by its spotted coat on the leaf litter of the forest floor. A Sargassum filefish (Monacanthus hispidus) and a Sargassum crab (Portunus sayi) are superbly camouflaged among the highly dissected branches of Sargassum weed (far left, bottom). A black, brown, and white hylid frog is readily hidden by remaining motionless on matching leaf litter in the Ecuadorian rainforest. A similarly colored Ecuadorian snake is likewise concealed on the tropical forest leaf litter. A green leaf mantid matches living leaves quite well, while a brown-colored species blends into the environment as it perches on a dead leaf. A stick caterpillar of the geometrid moth family has irregular blotches of coloration that look like lichens growing on a natural twig. An adult dobsonfly conceals its outline by resting on the bark of an alder trunk.

PLATE 10
WARNING COLORATION

In frightening coloration, an organism attempts to scare off a predator before it can be eaten. Thus the American io moth (Automeris io) uses large false eyespots, exposed on its hindwings when the forewings are projected forward, to scare off birds that might disturb it in daytime resting places (top left). In warning coloration, the animal's colors or patterns deliberately advertise its presence and the organism depends on its distasteful or dangerous qualities to ward off enemies, like the famous poison-arrow frogs (Dendobates) of South America (above). A coral snake uses vivid red, yellow, and black colors to warn potential predators, and many harmless snakes (lower left) in the tropics use some combination of these colors to mimic a coral snake species and gain some measure of protection. Some harmless animals use a color pattern and structure to deflect predator attacks from vital areas. Thus the hairstreak butterfly (middle left) bears a false head complete with hairlike "antennae" at the rear edge of its wings; the real head is largely protected between the heavily veined, leading edges of the forewings.

pair or family group) will defend such an area against intrusion by any other individual. This naturally results in a spacing out of the population, frequently in direct relation to the available resources (especially food supply) of the environment. In other words, a territory is not merely dependent on the total amount of physical space required for each territorial individual. Territoriality was first observed in nesting birds and described as part of social behavior theory in the early part of this century. Today it is known that a great many frogs, lizards, crocodiles, alligators, birds, and mammals defend territories and are dispersed in accordance with the territorial requirements of that particular species. We shall consider territoriality at greater length in Chapter 19, when we look at its role in social behavior. For now let us note that the advantages of the dispersion resulting from territorial isolation are threefold. First, competition between individuals is reduced because of the more or less equal spacing throughout the population area. Second, energy that would be spent in antagonistic fighting, if all individuals were close together, is conserved during critical periods such as the mating season. Aggressive displays may be common at this time between individuals holding neighboring territories, but actual fighting is nonexistent unless an individual's territory is physically invaded. Finally, because of the kind of spacing that is provided by territoriality, overcrowding and exhaustion of the food supply is prevented.

The human population structure

From the fossil record and a comparison of population sizes in surviving societies of primitive peoples, anthropologists and biologists believe that the earliest human beings existed in small groups of hunters and food-gatherers. Living in small bands of 20 to 40 members, such groups wandered across the plains and lower mountains from season to season, gathering roots, tubers, grains, and opportunistically obtaining meat from animals encountered by the hunting parties. We can reasonably assume that these groups of hunters and food-gatherers were spaced out in accordance with available resources.

It is particularly instructive to look at the aborigines of western and central Australia, although they currently occupy only a relic area of that continent and must be regarded as adjusted to survive under more severe desert environmental conditions than formerly

FIGURE 9-8

Among Australian aborigines, like these in the Finke River area of the Northern Territory, the "old men" of the tribes are the ruling elders, masters of ceremonial rituals, and the teachers through which the sacred laws and myths concerning kinship, marriage, hunting, and other social activities are passed down from generation to generation.

(Figure 9–8). At the time of its discovery by Western man, the continent of Australia is believed to have contained approximately 300,000 people of an apparently early Caucasoid stock. Among these aborigines, the smallest population unit is the *horde*, a family clan of some 40 related individuals. Mating normally occurs by bringing females into the horde from other similar groups. Thus the effective breeding unit for Australian aborigines is the *tribe*, defined as a group of hordes in a common territory that share a common dialect and a common range of personal mobility. Although its size varies across the continent, the tribe averages about 500 people and therefore contains about a dozen hordes. The territorial limits of each tribe are determined by geographical boundaries such as mountain ranges, canyons, rivers, and ecological boundaries formed by changes in plant communities, climate, and the occurrence of surface water. One authority, J. B. Birdsell, has demonstrated that the *size* of the tribe is correlated with the amount of annual rainfall in the area and that the size of the tribal *area* occupied is also correlated with the amount of annual rainfall over the area. Thus in early primitive hunting-gathering societies, population structure was intimately adjusted to the ability of the environment to support the population.

With the coming of the agricultural revolution, about 8,000 years ago, people began to settle in permanent communities and grow their own food crops. The production of surplus food from crops such as grains enabled small towns to be built up around these agricultural centers, where craftsmen, priests, and other classes could develop without the need to grow their own families' food. They could be dependent on the excess food grown by the farmer class. Human population growth in such areas continued to accelerate as long as the agricultural base was sufficient to maintain the population of that area.

At the time of the industrial and scientific revolution, beginning about 1600 A.D., advances in sanitary procedures and medical practice greatly lowered the human death rate, as we shall see in Chapter 11, and the growth of cities and other urban areas increased even more rapidly. Today it is clear that the general human population dispersion is concentrated in vast urban centers around most of the world and that these urban centers are dependent on agricultural production of areas located often at considerable distances from the cities. Starting during the winter of 1973, the costs of fossil fuels jumped greatly. It was no longer possible for many of the poorer nations to conveniently transport food from one area to another at a cost compatible with their resources. The cost of fuel for mechanized agricultural equipment rose prohibitively, and nitrogen-based fertilizers that were essential for cultivating the "miracle strains" of grains could no longer be obtained at reasonable prices. Thus in nations such as India, Pakistan, and Bangladesh, the threat of mass famine became acute in even the remote but productive agricultural areas, let alone the large urban cities.

From this brief overview, it can be seen that the human species placed itself in a position outside of the normal controls that highly evolved dispersion patterns tend to exert on population growth. The territorial systems of various birds and mammals, a limitation of breeding and rearing space, the availability of nesting sites, and even the establishment of peck orders (Chapter 19) tend to restrict the right to feed and to breed within a species to those individuals who can successfully cope with competition inside the population. The result of *dispersion according to resources* is that the available food, opportunities for breeding, and space for the rearing of young are distributed over the habitat occupied by a species, and crowding is prevented. Individuals unable to find space in the original population area must emigrate to new and unoccupied sites where new competitive situations can be explored, tested, and possibly a new population started. Thus the social organization of most species has evolved to their present status through the forces of natural selection, and this dispersion assists in balancing the population with the extent of its available resources.

Primitive peoples matched this condition for survival by adjusting their group size to that which could be supported by the environment. Today, our burgeoning population threatens to exceed the supporting capacity of the environment taken on a world basis, let alone a local one. Consequently, the specter is raised of impending future disaster for the human species unless the demands of this species in terms of numbers, needs, and aspirations are related in a balanced manner to the finite resources available to it on a sustained basis. The right to feed and the right to breed are important to the individual but also to the species as a whole, and cultural controls must supersede the individual's complete freedom to breed or the entire species will lack the freedom to feed! Viewed in another way, in order to achieve a stabilized social system within its population dispersion throughout the world, the human species must obtain a balance between its numbers, dispersion, and available resources, thus keeping the total population within environmental capacities.

At the beginning of the agricultural revolution, about 8,000 to 10,000 years ago when people began settling into cities, the habitable portions of the globe were occupied by about 10 million human beings. It has been argued that this was a stable number that reflected the sustaining capacity of the environment of a population that gained its livelihood in hunting and food gathering and that had at its command a limited degree of mobility and a limited source of external energy in the form of fire and stone tools. These early human groups were, as we have noted, small in number and probably never far from the threat of starvation. The control of additional energy through the use of agricultural methods and the buildup of surplus storable foods led to an expansion of the population in limited areas of the temperate zones. Thus the earth became a more productive

FIGURE 9–9

The growth of the world's human population from 1800 to 2000 A.D.

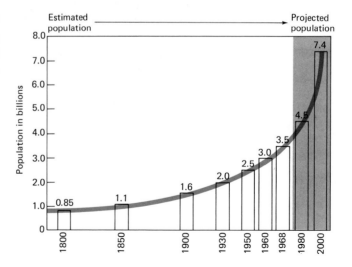

planet both in terms of people and food. By the beginning of the Christian era, the earth's inhabitants had reached a total estimated to be between 250 and 350 million. Today the population of the earth is approaching 4 billion and by 2000 (Figure 9–9), at its present growth, the population will double to more than 7 billion, 10 billion by 2018, and 20 billion by 2068. By 3068, there would be, at the present rate of growth, 358 billion people or 11 persons per acre uniformly spread over the 33 billion acres of the land surface. Of course, the spatial organization of the human population would have long disappeared in the chaos of cultural and social collapse.

Summary

The organization of a population includes dispersion, or the spatial arrangement of individuals within the aggregation. The tolerance range and minimal individual distance between members affects the overall dispersion, which is generally reflected in uniform, random, or clumped patterns of distribution. The most common kind of plant and animal spacing patterns is some form of close aggregation, or clumped dispersion. This results from local habitat differences, reproductive activities, survival advantages, and social needs. Dispersal behavior involves immigration, emigration, and true migration activities. The mating system of a species, whether involving promiscuity or true pair bonds as in monogamy and polygamy, influences the distribution of individuals. Territorial behavior also spaces individuals but normally in relation to available food resources. Human population structure and dispersion have changed radically in the past 10,000 years as people moved from a nomadic hunting-gathering existence in small groups to a sedentary, agricultural and urban mode of life. Thus the present human dispersion pattern has no real relation to resources, but is commonly dependent on modern technology and transportation to supply human needs. The potential instability in this arrangement is reflected in our present fossil fuel crisis.

10   The growth and regulation of populations

When Darwin was writing *On the Origin of Species*, he calculated that potentially a single pair of elephants could have over 19 million living descendants after 750 years. Elephants have a gestation period of well over a year and produce only a small number of offspring; yet even with this low potential for reproduction, a great number of elephants could be produced if there were no population limitation. The American entomologist L. O. Howard studied the common housefly some years ago and determined that the female produces an average of 120 eggs at a time and that about half of these eggs develop into females. He also observed that there are normally seven generations per year. Based on these data, then, it would be possible for one female to produce 120 individuals for the first generation, of which 60 would be females, each capable of producing 120 eggs. The number of offspring in the second generation would be 7,200, and the third generation would comprise 432,000 flies. If we assume that after reproduction all individuals of the reproducing generation die — that is, they survive only one generation — we can still see that the first fertile female would be responsible for ultimately producing over 5 trillion live flies in the seventh generation.

With this potential for reproduction, it is little wonder that population growth, especially of household and agricultural pests, is of such great concern to us. Many factors in nature act against a potential geometric rate of increase, and in this chapter we shall be looking at the two opposing forces operating in the growth and development of a population: namely, the ability to reproduce at a given rate and the influential factors of the physical and biological environment in which an organism exists. With these two opposing sets of variable forces, we can easily see the complexities inherent in the problems of population growth and regulation.

Measuring population growth

Population growth itself is the increase in the number of individuals constituting an aggregation. It does not necessarily result from an excess of births over deaths but may be caused by increased survivorship, movement into the population of new organisms, or other factors. If there is no increase in emigration or other types of removal processes, growth of a population causes an increase in **density,** which is simply the size of the population within a particular unit of space.

Population density may be measured and expressed as the number of individuals per unit area or volume. The *crude density* is the number (or even total biomass) per unit of total space, including all environments within the population's described boundaries. Sometimes ecologists wish to measure the number of biomass of organisms or a species per unit of habitat space, that is, the area actually suitable for species to inhabit. This *ecological,* or *specific, density* describes the number of individuals in the area or volume actually available to be colonized by the population.

At any particular time, the population may be changing in size and hence a measurement of density may not mean very much, especially in organisms that have short generations and very rapid changes in population size. Thus rather than obtaining an absolute estimate or measurement of population size, ecologists frequently find an *index of relative abundance* to be a more useful statistical reference. This is particularly true for agricultural entomologists who may be attempting to measure the severity of infestation of a crop field or orchard. Thus an entomologist in the Imperial Valley of California might estimate that there are 5,000 *Colias* sulfur butterflies per acre, a relatively meaningless estimate if the alfalfa plants on which the butterflies' larvae are feeding are of different ages, sizes, and densities in the several fields composing that acre area. A figure indicating the average number of larvae per mature plant versus average number of larvae per juvenile or immature plant would be more meaningful, especially if the two fields constituting this acre contain plants more widely spaced in one than in the other. The number of larvae per plant provides an index of the severity of the impact that these herbivores will have on leaf production and eventually flowering.

Regardless of the measurement of density employed, positive or negative population growth is generally occurring, for populations are almost never static in size. *Demography* is the study of these changes in population size. In the restricted sense, demography means a study of fluctuations in the size of human populations, but it may be applied to the study of the population dynamics of many organisms. Thus in a demographic sense, populations acquire additional numbers either by **immigration,** the movement of members into the population from other populations, or by **natality,** the occurrence of births. Likewise, such a population may lose members either by **emigration,** the departure of some individuals, or by **mortality,** the death of members. We have noted in the preceding chapter how immigration and emigration affect population structure. Thus for the present we shall assume that these factors either balance each other or do not occur to a significant extent, and we shall here examine the standard demographic ways of expressing natality and mortality and their effect on population growth. These determine, in partnership with survivorship, the pattern of increase or decrease for a population.

Natality

The rate of birth, or **natality,** is simply the increase in the population due only to the normal rate of reproduction. New individuals may be continually introduced by either asexual or sexual reproduction, but in general the maximum possible natality is rarely reached because conditions ideal for maximum reproduction of new individuals do not exist in natural situations. Under such ideal conditions there would be no ecological limitations to reproduction, only such physiological limitations as the maximum number of eggs the female could produce per unit of time. (The ability to reproduce at a maximum rate is referred to as **biotic potential,** a subject that we shall consider later in this chapter.) Instead, we will use natality to refer to the observed population increase resulting from reproduction under a particular set of ecological conditions. Unlike the maximum possible natality for a species adapted to a certain environment, this actual birth rate is not a constant but will vary with the size and composition of the population as well as physical environmental factors.

We normally express natality in mathematical notation as $\Delta N/\Delta t$, or number of new individuals produced in a population, divided by change in time. The natality rate for a flock of sheep producing 120 new lambs per year would be approximately 0.3 lambs per day (120 lambs divided by 365-1/4 days). For human demographic purposes, the birth rate, or natality, is normally expressed as the total number of births in a year divided by the total population at midpoint of that year (on July 1), multiplied by 1,000. Thus during 1971, there were approximately 3,769,000 live births in the United States. In the middle of 1971, the country's population was estimated to be 207,100,000; therefore the birth rate for the year in 1971 was 3,769,000/207,100,000 = 0.0182. As this is calculated to be the rate per person in the population, we multiply 0.0182 by 1,000 and find that the United States birth rate was 18.2 that year.

If we look at birth rates throughout the world, we find that natality figures seem to be largely concentrated at two points: between 40 and 50 live births per thousand head of the population, or between 14 and 20–24 live births per thousand members of the population. This is not a random distribution of numbers but is attributable to the factor of population-control policy. Some nations have adopted population control methods on an extensive scale and have reduced their population rates to below 20; other less developed nations have little or no population control and hence birth rates stay naturally up in the high 40s and 50s. We shall explore this demographic problem at greater length in Chapter 11.

Mortality

The death rate, or **mortality,** of a population is calculated in a similar manner to natality and simply refers to the number of individuals dying per unit of time. The lowest possible mortality would equal the population loss under ideal or nonlimiting conditions; that is, it is evident that even under the best of environmental conditions, individuals will die of old age at a point determined by their physiological longevity. As cells stop dividing and body systems wear out, the individuals of a particular species will die, a maximum age that is more or less characteristic of each species. The mortality rate we actually observe in nature is the rate of loss of individuals under given environmental conditions where predators, accidents, competition, and other factors share in causing deaths. This actual mortality is what we normally consider the mortality of a natural population.

The death rate for most species is expressed as the number of organisms dying per unit of time, such as 54 rabbits dying per year in the population inhabiting a particular field. Demographers calculate human mortality from the total number of deaths that occur during the year divided by the total population at the midpoint of that year multiplied by 1,000. Thus in 1970, for example, the death rate for the United States was 9.6 people per thousand individuals in the total population and in Belgium 12.8 deaths per thousand. West Africa had among the highest rates that year: 26 deaths per thousand in Guinea and Cameroons, 25 deaths per thousand in Nigeria. The death rate has been lowered in the advanced industrial nations, of course, by large-scale introduction of DDT and other pesticides to control insect vectors of epidemic diseases and the discovery of medical drugs and immunizations that protect the populace against previously fatal diseases. Thus India, Pakistan, Chile, and Ceylon have had a remarkable reduction in death rates since the end of World War II because of these innovations, and we may expect similar reductions in the remaining areas of the underdeveloped world as the full range of modern medical technology reaches them.

Survivorship

The percentage of individuals living to various ages represents the **survivorship** in a population. This situation is usually expressed graphically in a survivorship curve. The surviving percentage per thousand born is plotted from the maximum of 1,000 alive at time of birth through a series of intervals to the time of the death of the last individual, representing the maximum life span for the species. Examples of survivorship curves representing four types of natural situations are given in Figure 10–1.

These survivorship curves indicate the number of survivors out of an initial population of a given size after a specified time interval. We assume that in constructing such a curve all the individuals in the population at the beginning of the observation period were of similar age, and we also assume that all members of the population

FIGURE 10–1

Four different types of survivorship curves, depicting the number of individuals still living at various times after birth (or emergence from pupae in the case of the fruitflies). See text for explanation.

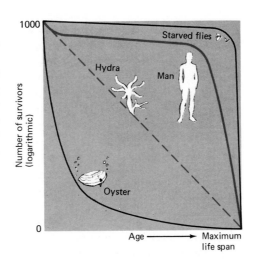

originally were born with the same capacity for survival and that environmental effects on birth rate are ignored.

In the first type of survivorship curve, as illustrated by a population of starving adult fruitflies in a laboratory bottle (curve A), most of the *Drosophila* survive for almost equal lengths of time, living out the potential life span under conditions of no food. The curve therefore shows no real drop in survivorship until very close to the point when mortality becomes universal, causing a precipitous drop in survivorship near the point of maximum life span. This gives a right-angle curve to survivorship patterns of this general type. Normally, environmental factors kill some fraction of a population early in life and hence a strongly angled curve of this type is rarely encountered.

The survivorship curve for the human population of an industrialized society is represented in curve B, in which there is an initial drop in survivors because of deaths at birth and early childhood, followed by a period of good survival until the death of individuals in old age. Mortality losses among human infants and juveniles have been very greatly reduced in such societies, as have losses among older adults under 65 years of age. Thus if the organisms, whether human or large animal, survive infancy with its relatively high death rate, they have a good chance to live out almost the entire maximum life span for the species. Since the actual length or physiologically possible duration of human life has not been substantially increased by modern medicine, this survivorship pattern tends to concentrate the ages at which people die into their late sixties, seventies, and early eighties.

Occasionally, we find a form of survivorship curve in which mortality occurs at a constant rate throughout the life span of individuals in the population (curve C). Thus in a population of fledged birds of equal age, mortality at any age after fledging (acquisition of flight feathers) is about equally probable. A laboratory population of the fresh-water coelenterate *Hydra* also shows a survivorship curve in which the mortality rate at all ages is constant.

Perhaps the most extreme early changes in survivorship are shown by many natural invertebrate populations, such as that illustrated for oysters (curve D) and for most plant populations. Here most individuals die early in life in either embryonic or infant and juvenile stages; the members of the population who survive these early phases of risk subsequently enjoy an extremely low mortality. For example, large numbers of oyster larvae die soon after hatching because of the hazards they encounter as they are moved by the currents through their marine environment. These tiny larvae are particularly vulnerable to predation and are very likely to land in places where they cannot survive when they settle on the bottom to begin life as sessile adults. Thus the survivorship curve for the population of these marine bivalves drops rapidly at first; however, the few larvae that detach to a satisfactory substrate will survive well and live for many years, barring capture by an oyster fisherman.

Within any of these survivorship curves, differences arise for a number of reasons, primarily the extent of environmental pressures on each age group. If all environmental restraints were removed, we would expect a typical survivorship of 100 percent of the juveniles until the maximum life span had been reached. This would result in the maximum possible rate of increase for the population since all individuals in each generation would survive to reproductive age. As we have noted, this concept is called the biotic potential of a population; while it is rarely if ever reached in nature, it allows us to estimate the extent of operating ecological restraints.

Biotic potential

We can calculate the biotic potential of a population by subtracting the minimum mortality from the maximum natality as measured under restrained conditions in the laboratory or outside enclosures where there is no lack of food, no predation, parasitism, or competition for space. Of course, since these growth-restraining factors are present in natural situations, we actually find and normally measure in nature only the *realized* rate of increase. The difference between the biotic potential as measured under ideal conditions in the laboratory, or from estimating total egg production possible in females, and the realized rate of increase is in a measure of the *environmental resistance* present, that is, all the limiting factors in the environment that act on this particular population. As a result of this interaction between the biotic potential and environmental resistance, populations tend to have characteristic patterns of increase or population growth form, which we can now look at in detail.

Population growth form

In order to understand the interaction of biotic potential with environmental resistance, we can best start by seeing what happens when a population increases at a rate approaching its biotic potential. From 1879 to 1881 a total of 435 striped bass were released in San Francisco Bay after being moved from the Boston area. Striped bass were not native to this Pacific ocean environment, and their natural enemies and other limiting factors were apparently absent. By 1899, more than 1 million pounds of striped bass were being harvested per year from the San Francisco Bay. To take a simple laboratory example, consider the yeast organism. If we put a single yeast cell into a laboratory test tube, under ideal conditions and in only 10 hours we will have over 500 yeast cells. These examples of explosive population increase represent one extreme of population growth form, the *exponential growth curve.*

Exponential growth curves appear in species or situations where population density is permitted to increase rapidly, that is, in exponential or compound interest fashion (Figure 10–2). Here a population continues to increase at an accelerating rate. Finally, instead of leveling off, the rate precipitously decreases to zero by a large die-off of the population. In such instances, the growth curve has a J-shape. The more individuals that are added to the population, the faster it increases because all of those that are added also breed and hence increase the total growth rate of the population. The process is analogous to a savings account in a bank where interest is compounded daily, the interest being added to the principal balance and thus providing a larger base on which to calculate the interest for the next time period. Unlike accounts in most banks, however, this exponential J-shaped growth rate may stop abruptly as environmental resistance becomes effective more or less suddenly. The population then suffers a crash in numbers regardless of population density.

These J-shaped curves can be observed in the growth of annual

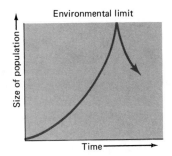

FIGURE 10–2

The J-shaped, or exponential, growth curve of a population. After an initial establishment phase, the population increases exponentially until the environmental limit causes a population crash.

FIGURE 10–3

The S-shaped, or sigmoid, growth curve of a population. The indicated points along the curve are (1) the establishment of positive acceleration phase; (2) the logarithmic increase phase; (3) the inflection point, where the growth rate begins to slow down; (4) the deceleration phase; (5) the maximum population size or carrying capacity.

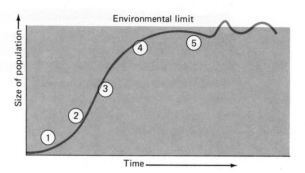

plants in fields and along roadsides. They increase rapidly through the spring and summer, but then die off suddenly as the photoperiod drops below a critical level or as frost arrives. Likewise, insects with short life spans such as plant thrips or microscopic algae in ponds will grow very rapidly during a period of favorable weather and then crash rapidly when the weather changes in the autumn. Such populations must survive the unfavorable periods in a dormant or resting stage, maintaining an equilibrium level that is far below their maximum annual population size.

A second major pattern of population growth is the sigmoid growth curve, or S-shaped growth form, where growth starts slowly, accelerates rapidly in exponential form and then decelerates and continues thereafter at a more or less constant level. The growth form of such a population is definitely S-shaped, or sigmoidal (Figure 10–3), for several reasons. In terms of absolute numbers, the initial growth period is slow, but it is followed by a period of rapid increase and then a subsequent slowing down at the upper level. The period of increase contains an inflection point in the logistic curve of the positive acceleration phase, where the population growth no longer continues to accelerate but begins to decelerate. This deceleration phase is a slow-down of population growth caused by a gradual increase of environmental resistance present in the ecosystem. The growth rate decelerates more and more gradually until it approaches the zero point.

At the zero point, there is no net change in the population size. It is at equilibrium with its environment; that is, it has reached the upper level beyond which no major increase will normally occur. This is the population size at which the population limits itself, often called the **carrying capacity** of the environment. New individuals are certainly still being added to the population and others are being continually lost through death, but in the equilibrium period these two forces, natality and mortality, are in balance. The population increased when they were out of balance in favor of natality and decreased when mortality exceeded natality. The fact that a more or less equilibrum level was reached and maintained below the environmental limit of the area indicates an important way in which populations are regulated. Let us look now at the ecological components of population regulation.

Population regulation

As we have seen in considering the sigmoid growth form, many populations, especially of larger organisms with long life cycles and lower biotic potentials, tend to more or less stabilize themselves with respect to their environmental carrying capacity after an initial phase of population growth fills the available area. The equilibrium level implicit in this logistic or sigmoid curve does not imply absolute constancy but rather a dynamic state of fluctuation around an approximate mean. The limiting factors, which we have loosely called environmental resistance, vary in effect from year to year and season to season. Thus in some populations the amount of fluctuation in size is minimal after the initial stages of establishment, as in the classic case of the introduction of sheep onto the island of Tasmania in 1819 (Figure 10–4). In many, perhaps most, other populations the amount of departure from the "equilibrium" level is considerably greater than with the Tasmanian sheep, and it may be quite irregular in nature. This situation has caused a number of population ecologists to state that no true equilibrium really exists in natural populations, but merely a constant adjustment to prevailing natural conditions. In still other populations, there is a quite regular oscillation around the mean. This state of affairs particularly is exhibited in certain predator-prey interactions, which we shall consider later in this chapter.

FIGURE 10–4

The logistic growth of the sheep population of Tasmania and the maintenance of that population in a state of semiequilibrium. The dots represent averages for five-year periods.

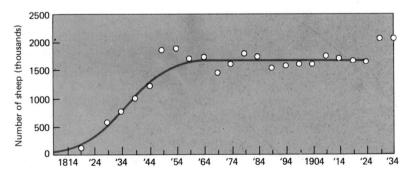

The factors affecting population growth are commonly grouped into two categories: **density-independent** factors and **density-dependent** factors. A population is said to be density-independent (in terms of the total population) if its effect is constant regardless of the number of individuals in the population. A density-dependent factor is one whose effect on the population varies with the density. Let us now look at some of these factors and their role in controlling population size and distribution.

Density-independent population regulation

Density-independent factors exert the same effect on the population regardless of the number of individuals present. Thus they tend to be largely abiotic in nature and include the influence of climatic factors and mineral deposits as well as other physical environmental factors. *Climatic changes* such as seasonal *photoperiod* variations (that is, the daily amount of sunlight at different times of the year), rainfall, and temperature are almost always density-independent in their effect.

In the populations of thrips insects (*Thrips imaginis*) living on roses in South Australia, J. Davidson and H. G. Andrewartha found that the number of adult thrips on roses is low at the beginning of the spring (September in the Southern Hemisphere) and increases dramatically as late spring approaches, reaching a maximum size about December 1. The thrips then go into a precipitous decline, producing an essentially J-shaped growth curve, and are very scarce until the next August, when a minor peak of abundance occurs during the southern winter prior to the great spring expansion. This cyclic abundance, observed over many years of study (Figure 10–5), occurs because in southern Australia flowers are abundant during the springtime in rose gardens but not during the long, dry summer or in the growing period of the cool winter. With so many flowers suddenly appearing about December, the abundant food and available living space provide an essentially unlimited environment for expansion of the thrips populations. However, they never have a chance to become excessively abundant on individual flowers because the long, dry summer occurs first, and the desiccating climate kills the thrips that attempt to emigrate from the drying blooms.

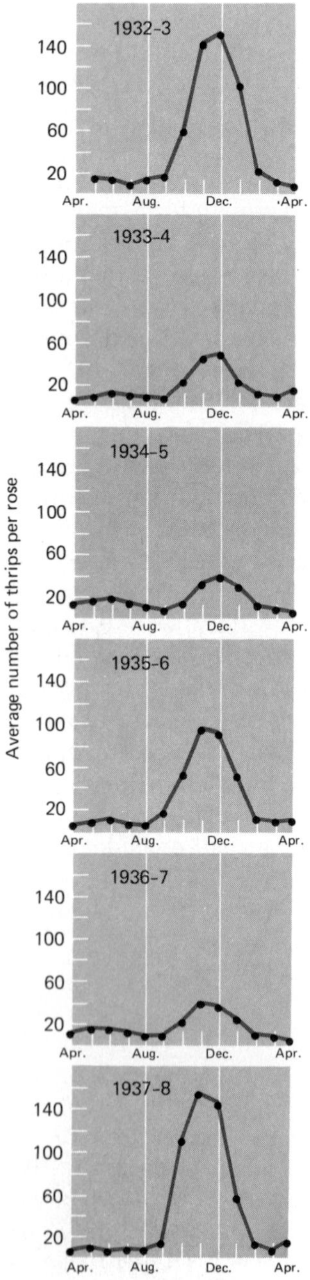

FIGURE 10–5

Seasonal changes and cyclical abundance in a population of Australian thrips insects living on roses.

There is no absolute shortage of food, only a seasonal change in its accessibility. Thus despite the regular seasonal changes in abundance of thrips each year, as shown in Figure 10–5, the climate is operating on the individual thrips insects regardless of the size of the population. They always die as the summer begins, and birth rate and survival rate are very low during midsummer to winter. Eggs laid in tissues of the flowers are relatively safe from desiccation during the summer, as are those nymphs and adults fortunate enough to reside in living plants. The species survives at low population levels but does not increase again until food becomes readily accessible because of climatic change.

Thus Davidson, Andrewartha, and their coworker L. C. Birch suggest that abundance and distribution of thrips are related to the seasonal change in availability as well as accessibility of food and that this latter factor is dependent on the influence of weather, particularly rainfall, and the growing period for the roses. From this work, Andrewartha and Birch have extended the principle of weather regulation to a number of other natural populations.

Generally, climatic factors are most important in governing the population growth of small organisms, such as insects or aquatic plankton, that have short life cycles and high biotic potentials. Weather and other physical environmental factors such as soil or water temperature are obviously quite important in determining the length of favorable periods for annual growth of such organisms. These species lack stability in their population size, which generally changes rapidly through the season. In many plant species, population size is affected by the level of sodium, phosphorus, or other nutrients in the soil. Independent of the number of plants growing in the ecosystem, the growth and distribution of the plants depends purely and simply on the presence or absence of the necessary mineral in that area. Of course, we should recognize that low nutrient levels, even in an area where the mineral occurred, would tend to also restrict population growth through physiological responses resulting from pure crowding, a density-dependent reaction.

It is also possible for density-independent effects to regulate populations of large vertebrates. Heavy rains, drought, human destruction of habitat, or even pesticide use can greatly affect nesting and hatching success as well as survival of young game birds, rabbits, and other birds and mammals. The drying of the Everglades in southern Florida affects populations of fish regardless of their size, because during droughts the pools dry up completely independent of the density of the resident populations. Muskrats are continually subject to the vagaries of water level fluctuation and are forced to migrate overland when drought dries up marshes or ponds and their burrow or lodge entrances are exposed. Flooding during periods of heavy rainfall, of course, can also result in the same disruption of normal activities. When beavers and muskrats move overland, foxes, coyotes, and minks can usually attack and kill at least the young adults. When human beings drain prairie potholes, or destroy extensive areas of natural habitat through clear-cut logging or widespread agricultural cultivation, native animal populations whether of squirrels or migratory waterfowl will be seriously affected, regardless of their density. The former widespread use of persistent pesticides during the 1960s decreased songbird populations greatly in sprayed areas. Insectivorous birds were picking up concentrated DDT from earthworms and other invertebrate food sources that had fed on leaves sprayed by plane or from the ground to control Dutch elm disease. The degree of mortality caused by the poison was independent of the size of the robin or other songbird populations.

Density-dependent factors

Factors that operate by exerting effects varying proportionately to the size of the population tend to be those involving interspecific biotic relationships. The huge numbers of intraspecific and interspecific interactions in nature commonly contribute to and, indeed, often generate biological regulation of population size, and they seem to be responsible in many cases for the steady-state population sizes seen in the terminal portion of the sigmoid growth curve. These factors operate throughout population growth, not solely at one point in time, and they begin to act well below the carrying capacity. Indeed, their effect is first noticed when they cause the inflection point in the sigmoid growth curve. Their effect intensifies as the upper population limit is approached. Should the population size exceed the equilibrium density or carrying capacity of the environment, density-dependent factors exert an even stronger effect and cause a greater rate of loss, which operates as a brake on population growth until the equilibrium density is reached or reached again. If population size drops below the equilibrium density, the density-dependent factors normally ease their effect and the population is permitted to increase. Because of the time lag between cause and effect in growth and reproduction, the population size tends to oscillate either regularly or irregularly about the equilibrium density. Overpopulation in most animal and plant species is believed to be prevented in the primary sense by density-dependent factors.

Competition between individuals is commonly an important regulating process in natural populations. Individuals of the same species compete for a resource that is in limited supply, and under such conditions a population can obviously accommodate only a certain maximum density. The population members may compete for food and space or even basic favorable physical factors such as temperature and exposure to sun or weather. An increasing mortality rate will result if the population exceeds the equilibrium density. Competition accelerates as the population increases. Since the younger animals are usually most affected, the addition of new reproductive animals to the population is curtailed relatively rapidly, and the birth rate will drop back to a level that yields an equilibrium density.

Parasitism and *predation* also control population density, sometimes with devastating results. In 1904 the sac fungus *Endothia parasitica* was accidentally introduced to the United States from China. In its native land, this fungus is parasitic on the bark of the oriental chestnut and is kept in check by a variety of natural mechanisms. However, when the fungus reached the United States and transferred to the beautiful towering American chestnut, there were no such control mechanisms available. By the late 1940s the chestnut, which had been a dominant tree in the forests of the Appalachian region of eastern North America, was virtually extinct. Its population had been reduced to an extraordinarily low level by an abnormal parasite.

Under normal conditions, the parasite and host are adjusted by way of adaptation and selection to exist at a population level safely above the extinction level. Both host and parasite populations continue to exist, although the parasite may have had at one time the potential for completely exterminating its host. Long-existing parasite host relationships exist in a delicate balance, yet if a parasite is introduced to a new association, as with the American chestnut, the parasite tends to devastate the new host rather than regulate the host in a balanced equilibrium sense. In populations of animals and plants, parasites and pathogens (disease-producing organisms) may spread more easily from host to host in high-density populations than in low densities and hence reduce the density of the host through mortality. Predators can find more prey individuals in populations of high density; and as long as they are not swamped by extremely high numbers of potential prey, they operate as a very efficient density-dependent factor.

Some species of animals have regular *emigration* as a density-dependent regulatory mechanism. At high population levels, some of the population leave to move into unoccupied territory. Other density-dependent processes include *physiological* and *psychological* control mechanisms. In many mammals the social stresses caused by crowding have been shown to act on the individual through the endocrine system, particularly the pituitary and adrenal glands. Increased population density leads to the inhibition of sexual maturation, lowered sexual activity, and inadequate milk production in nursing females. Stress caused by crowding will also cause spontaneous abortion or absorption of tiny fetuses in the uterine walls of pregnant females, thus curtailing the number of offspring produced per litter.

In general, species controlled by density-dependent factors tend to be larger organisms, such as birds, mammals, trees, or shrubs. They tend to have long life cycles that usually last more than a single year. The rate of production of new offspring is lower than for those species regulated primarily by density-independent factors, and thus they tend to have lower biotic potentials and more stable population sizes. Yet it should be clear that no plant or animal population maintains a perfectly constant density over time and that these various regulating factors that we have been examining are not perfect. It should also be apparent that density-dependent and density-independent factors interact to control population growth and size for most species.

Population fluctuations

Fluctuations in numbers of a population may be correlated with seasonal or annual changes and physical limiting factors such as temperature and rainfall, as well as various long-term relationships with parasites, predators, and other competing species. In both tropical and temperate areas, population growth is keyed to favorable periods in the annual cycle. Animals seek shelter for a dormant stage, or a portion of the population dies during periods of environmental stress. In the temperate zones, populations of many plants and animals go into a state of **hibernation,** in which an animal greatly reduces its metabolic processes and remains quiescent during the winter months. Often this hibernation takes place in burrows or dens deep underground, where the air temperature changes and humidity fluctuations are minimized. In desert areas, during seasonally hot times, animals frequently escape the heat by **estivation,** a state of reduced metabolism and behavioral activity similar to hibernation, but carried out to survive the hot period rather than the cold period. In many tropical areas of the world, rainfall is seasonally distributed and is the key factor governing seasonal changes in population structure and functioning. Temperature changes are relatively minor, indeed often insignificant, through the course of a year. Many organisms pass the tropical dry season in a state of **diapause,** or dormancy, hidden underground or in trees or other shelter away from the desiccating temperatures and low humidities characteristic of this time of year. The first rains of the wet season give the cues for breaking of diapause and these organisms become active again. Annual plants spend the dry season in the tropics or the cold season in the temperate zone in the form of dormant seeds. Leaf production and leaf drop are then correlated with the alternation of wet season and dry season, or warm season and cold season, in both tropical and temperate forests.

There are longer-term oscillations in density, however, that are not obviously related to seasonal or annual change, and there may be long, regular cycles of abundance with many years between peaks and depressions. The best-known examples come from the Arctic and occur among the small mammals. Thus many northern mice, lemmings, voles, and their predators, such as the snowy owl and the arctic fox, exhibit a three- or four-year cycle of abundance. Lemmings become extremely abundant every three or four years in the northern tundra areas and then crash to very low population levels within a single season. The populations of their predators likewise increase and then crash, following the sudden decline of the lemming population. Arctic foxes that are resident in these areas will starve and lose some members of their population because of increased mortality following a lemming population crash, but the snowy owls respond to the decline in lemming numbers by migrating far to the south in search of food, even to the southern United States.

An even longer-term oscillation is the 10-year cycle of the snowshoe hare and the Canadian lynx, a large relative of the more familiar bobcat. From the early 1800's, the Hudson Bay Company in Canada has kept records of the number of pelts of fur-bearing mammals taken by trappers and purchased by the company each year. Various ecologists have plotted these data (Figure 10-6), and it is apparent

FIGURE 10–6

The changes in population numbers of the snowshoe hare and the lynx in the Canadian subarctic, as indicated by the number of pelts received by the Hudson Bay Company from 1845 to 1935. This is the classic case of long-term cyclic oscillation in population density.

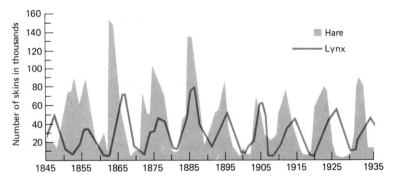

that the lynx reaches a population peak every 9 to 10 years (average of 9.6 years). These peaks of abundance are often followed by crashes or rapid declines in numbers, and the lynx becomes very scarce for at least the next several years. When the numbers of pelts of the snowshoe hare are plotted into the same time period over the past 250 years, it is found that the hare usually follows the same general cycle, with its peak abundance usually preceding that of the lynx by a year or more. Since the Canadian lynx is largely dependent on the snowshoe hare for food in these Arctic areas, it seems obvious that the cycle of the predators is related to that of the prey. However, the ultimate factors causing either the 3-year cycles in small mammals or the 10-year cycle in the hare and lynx are not clearly understood as yet. They have been linked to the cycle of sunspots in the sun, the amounts of pathogens building up in the rabbit populations, nutrient cycles in the soil, changes in food quality and quantity, and physiological and psychological stresses, as well as to the cycles of other animals in the Arctic.

Summary

The growth and regulation of natural populations is a subject of great practical interest since most species have the capacity to increase at a tremendous rate. The density of the population is established by the relative balance between natality and mortality, and also between rates of immigration and emigration. Because of environmental resistance, few species ever reach their biotic potential, the physiologically maximum rate of increase that is possible under ideal conditions. Each species has a distinctive survivorship pattern and growth form, depending on the extent of ecological restraints operating on that population. These restraints include density-independent factors, such as climatic changes and availability of soil nutrients, and density-dependent factors, such as competition, parasites, predators, migratory behavior, and physiological and psychological control mechanisms. Population size often seems to fluctuate about a certain mean level, held near or below the carrying capacity of the environment by density-dependent factors. Periodic long-term oscillations occur in the population sizes of some animals, particularly small mammals and their predators in the Arctic, for reasons that are as yet not completely understood.

11 Human population growth

The present explosive growth of the human population on earth has its historical roots in our distant past. If we look only at Figure 11–1, it seems that the world population of the human species has followed a typical growth curve. Such a curve, however, obscures the fact that the growth has not been uniform either in time or space across the earth. In a temporal sense, the rate of growth has accelerated each time that we have made a major cultural advance or gained control over a new source of energy. There have been four principal turning points in human population growth.

Turning points in growth rate

In earliest human history, some 1 to 3 million years ago, people lived on the open plains of Africa in small family groups. Their primitive primate ancestors are believed to have been African forest species that live alone or in small, closely knit groups. With major changes in late Cenozoic climate, causing a retreat of the forest and the expansion of open savanna plains, these early primates developed **bipedalism,** that is, walking on the hind limbs alone. This novel evolutionary development freed the forelimbs for other purposes

FIGURE 11–1

Human population growth for the past half-million years. If the Old Stone Age were in proper scale, its base line would extend about 18 feet to the left.

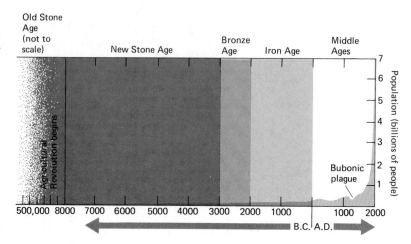

such as tool making and weapon using. The advent of tool making in these early peoples, the utilization of fire for cooking and warmth, and the development of the use of caves and other reasonably permanent shelters enhanced the human survival rate. These new cultural features entered the repertoire of survival equipment over 600,000 years ago, when the early Australopithecines developed into Peking Man, Java Man, and similar *Homo erectus* forms. This major *cultural revolution* set off the first real surge in world population growth as human beings spread across the Eurasian continent from Africa.

The second notable increase in the rate of human population growth was started by the *agricultural revolution* about 8,000 to 10,000 years ago; this led to settled communities in which people could rely on surpluses of storable foods. At this time in several major regions of the world, human groups changed from a primitive hunting-and-gathering existence that involved every member of the nomadic group in food acquisition to an agricultural society where only a part of the group was needed to raise enough food to feed the whole group. Population growth increased because the sedentary mode of existence—raising crops and livestock—presented fewer hazards to the average family than the former migratory wandering of life. Also, farming activities tended to require more hands to support the agricultural work, and farmers were motivated to increase their family size for practical reasons as well as because of increased survivorship. Food, in the form of grains, could be stored for times of scarcity and drought. The death rate consequently decreased, and the live birth rate increased.

The third and fourth identifiable major turning points in human history are sometimes combined. They are the *scientific revolution*, which began in the late sixteenth and early seventeenth centuries, and the *industrial revolution*, which began in the eighteenth century in western Europe. Medical knowledge and public health measures began lowering the death rate significantly, and this trend has continued to the present day in underdeveloped nations as scientific advances have reached them. Growth was further accelerated by social changes associated with industrialization. Both stages are based on an increased understanding of nature — the exploitation of fossil and nuclear fuels, conversion of energy into electrical energy that could be widely dispersed, and an accompanying technology that enhanced the productivity of the individual as well as that of the environment. Thus the environmental carrying capacity was enlarged by stepping up the yield of food per unit of cultivated ground. Technological advances in housing and transportation of food have made it possible for human beings to inhabit even remote polar or desert areas today, where previously population growth was limited by the lack of a sufficient natural supply of food or suitable living conditions.

Regulating factors for human population growth

Population sizes in early hunting-gathering human groups would have always been density-dependent, that is, limited by the availability of food resources. These resources in turn would be determined by environmental factors such as climate, soil conditions, topography, and the group's degree of effectiveness in exploiting the food resources. If we look at modern societies, we see that females are potentially capable of reproduction between the ages of 13 and 55, and males can reproduce generally from 13 years of age on for the rest of their lives. Child bearing can occur at intervals as close as 11 or 12 months, and the occurrence of single births is by far the most common circumstance (about 97 percent). In the human population as a whole today, under conditions of excellent medical care and diminution of normal environmental hazards, each female is potentially capable of producing a maximum of about 42 babies, nearly twice the largest recorded actual number (24).

We can assume from this picture that various factors must have intervened in preagricultural life to reduce drastically the number of births per female. One of the chief factors is that the age of death was probably about 30 years throughout most of human history, permitting the female to live only less than half her potential reproductive life (Table 11–1). Also, many primitive groups even today nurse the young as long as two and one-half to three years; if such long nursing periods were usual in early people, the onset of renewed fertility would have been delayed, and hence the frequency of gestation. Finally, recently weaned children probably had a very high incidence of infant mortality, especially during times of seasonal extremes such as the long tropical dry season in Africa and the cold winter of the northern temperate zones. The numbers of offspring borne by individual females in some contemporary groups of food gatherers and hunters such as in central Australia, Greenland, Indonesia, and the southern tip of South America indicate that three to six children is about the outside range for normal family size. On the basis of such comparisons, the average number of children borne by a female in Pleistocene human populations was probably five. In modern living populations where medicine has not been yet extensively introduced, infant mortality prevents about half the offspring born of a female from surviving to reproductive age.

Thus it seems likely that early groups of our human species were not subject to rapid increases in population growth. In fact, the balance between natality and mortality was probably critically adjusted to local conditions in order to keep the group at its ecologically optimal density. No general increase in population growth could occur with this mode of existence. Clearly, population growth would be density-dependent, and carrying capacity along with population density would increase stepwise with each development leading to the improved efficiency for extracting a living from the environment.

TABLE 11–1

Estimated average life span of human populations.

Population	Years
Neanderthal	29.4
Upper Paleolithic	32.4
Mesolithic	31.5
Neolithic Anatolia	38.2
Austrian Bronze Age	38
Classic Greece	35
Classic Rome	32
England, 1276	48
England, 1376–1400	38
United States, 1900–1902	61.5
United States, 1950	70

Source: After E. S. Deevey, Jr., "The Probability of Death." Copyright 1950 by Scientific American, Inc. All rights reserved.

It is quite possible that the establishment of the earliest human settlements, for instance, could have been independent of the development of agriculture. Groups could have successfully settled wherever continuously productive food webs were available, as, for example, on the shore of a lake or sea. Eventually, some of these early shore settlements adopted agricultural practices. With this development came the domestication of farm animals and the further selection of cultivated plants that were of economic importance. At this time, it was no longer necessary for people to wander as nomads over the face of the earth. By changing the level in the various food webs from that of a secondary or a tertiary consumer to that of a *primary* consumer, human beings increased the carrying capacity of the habitat, by 10 times in one case and 100 times in the second case, for our total species population. (In Chapter 3 we noted that normally energy transfer to the next highest trophic level is at most 10 percent efficient.) Of particular importance to the enhanced survival of agricultural groups was the ability to survive periodic drought or cold weather and thus avoid the normal annual juvenile losses found in hunting-gathering groups at this time.

The increased urbanization associated with regional development of agriculture and the consequential increases in carrying capacity of that habitat for human beings led to the foundation of cities. The well-known ancient city of Jericho has been excavated as far back as about 6,000 years. Several other cities in this part of the Middle East may be 2,000 or even 3,000 years older than Jericho. The effects of urbanization on rate of population growth were, of course, not always positive because crowding of people made it possible for any disease or parasite to spread more rapidly than it could among scattered nomadic tribes. The early urban societies may have overcome density-dependent regulation by their agricultural development of food resources only to have a further density-dependent factor develop strength in the form of epidemic diseases and related restrictions on population increase.

Thus it was not until the scientific and industrial revolutions began that population growth took off at its present rate of increase (Figure 11–1). In many parts of the world, the scientific and medical advances that originated in Europe centuries ago did not arrive until World War II. From 1940 to 1945 the Allied nations introduced on a large scale public health practices such as malarial control and sanitation of public water supply, and the death rate in such countries as India and Pakistan fell. The high birth rates characteristic of these unindustrialized nations, however, continued unabated. Because of the gap between the birth rate and the death rate on a worldwide scale, the population growth rate in all countries has accelerated to the point where it is estimated that the world population will jump from our present 3.9 billion to 7.4 billion by 2000. The trend of the world's population growth since the beginning of the Christian era is shown in Figure 11–1.

Population growth around the world

The story of human population growth, as we have seen, is not primarily a story of changes in the birth rate, which has remained more or less constant through history, but of changes in the death rate. Each major cultural advance resulted in the lowering of the death rate. The advent of tool making and the use of permanent living sites such as caves about 600,000 years ago improved human survival prospects. The development of culture was responsible for a great increase in human brain size, from about 500 cubic centimeters in the Australopithecines about 2 million years ago to about 1,350 cubic centimeters with the origin of *Homo sapiens* about 200,000 years ago, when the growth of brain size leveled off. Early human beings added to the store of cultural information by developing, teaching, and learning techniques of social organization as well as group and individual survival. These features gave a selective advantage to individuals with the large brain capacity necessary to take full advantage of the novel cultural developments. Such advances caused the slight decline in the average death rate, which persisted until the agricultural revolution, producing, however, an estimated average annual rate of population increase of only 0.002 percent.

By 8000 B.C., the world population was at about 5 million, having increased from perhaps 125,000 individual Australopithecines in Africa 2,000,000 years ago. Human groups of the time of the agricultural revolution had spread from Africa to occupy the entire planet. They probably first entered the Western Hemisphere about 30,000 B.C., migrating across the Bering Strait during the period that a land bridge was exposed because large quantities of water were frozen in the northern glaciers. During this period from perhaps 30,000 B.C., to 8,000 B.C., increased hunting and gathering efficiency, especially with the use of finely crafted projectile points on arrows and spears, may have led to the extinction of many large mammals, such as great ground sloths, sabertooth tigers, and woolly mammoths. Then, as we have seen, the agricultural revolution helped to raise the general standard of life, which began to lose some of its hazards so that human life expectancy began to creep upward from its primitive level of perhaps 25 to 30 years.

Modern trends in population growth

During the Middle Ages in Europe, the breakdown of the feudal system gradually destroyed the huge estates of the landed aristocracy. Simultaneous immigrations to the New World also left new land available in Europe for remaining residents. In 1500 the ratio of people to available land was about 27 per square mile; the addition of the vast, virtually unpopulated frontiers of the New World produced a ratio for Europe and the Western Hemisphere combined of less than 5 per square mile. Thus the land shortage in Europe was in part alleviated at the same time that new income was pouring in from the exploited natural wealth of the Western Hemisphere. With scientific and industrial innovation, it is little wonder that a population boom began in Europe between 1650 and 1750.

In Asia, a similar boom began in the century following 1650, with a population increase of some 50 to 75 percent. The new agricultural policies of the Manchu emperors, who took over China after the collapse of the Ming dynasty in 1644, doubtless led to a depression of death rates. India, being in a period of economic and political instability at this time (after the collapse of the Mogul empire in 1707), was racked by war and famine and probably had little population growth.

Between 1650 and 1700, then, the world population grew at a rate of about 0.3 percent per year. This growth rate increased to approximately 0.5 percent between 1700 and 1850. The industrial revolution, as well as improved sanitation and the introduction of small pox vaccination toward the end of this period, allowed the population of Europe to double, even in the face of substantial immigration into the New World where the population jumped from some 12 million to about 60 million in the same period.

Between 1650 and 1850, growth in Asia was slower than in Europe, amounting, as we have noted, to an increase of about 50 percent. The factors that favored rapid increase in Europe's population were to appear much later in Asia.

The population size of Africa until the middle of the nineteenth century has remained obscure. It is generally accepted that between 1650 and the time of European colonization, the population remained more or less constant at about 100 million people. With the influx of European technology and medicine into Africa, death rates started to drop and the population increased perhaps 20 to 40 percent between 1850 and 1900. By 1950 the population had doubled to 200 million.

Between 1850 and 1950 the average growth rate of the world population was about 0.8 percent per year. In absolute numbers the population increased during that century from slightly more than 1

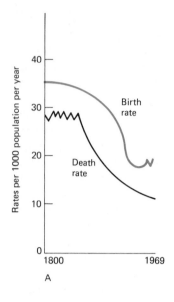

A

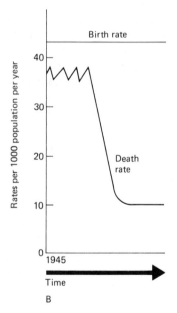

B

billion to almost 2.5 billion. Populations more than doubled during this period in Europe and Africa, while the population did not quite double in Asia. In Latin America the population multiplied about fivefold and in North America it increased more than sixfold. A continuing decline of the death rate from 1850 to 1900 came as a result of the industrial revolution and continuing advances in agriculture and medicine. The lot of Western people had been substantially improved by the reduction of chances for crop failures and famine, especially through mechanized land and sea transport of food to areas afflicted by famine and improved access to more remote resources. As the industrial revolution progressed, another significant trend appeared. Birth rates in Western countries began to decline, following by about a century the drop in death rate.

This diminution of the birth rate paralleling the lowering of the death rate has been called the **demographic transition.** It is thought that this decline in human fertility among Western populations was caused principally by the change in social attitudes toward children. With the progression from an agrarian society to an urban-industrial society, children became less of an economic asset and more of an economic burden. Children were useful workers on the farm, but in the city the father was the principal wage earner. Older children worked for very low wages and young children prevented the mother from working. Under the crowded urban conditions and the socioeconomic pressures of industrial work, a small family became the most desirable state, and urban families were motivated to curb their birth rate by whatever means they could. Generally, this trend occurred well in advance of knowledge of the usefulness of mechanical contraceptives, let alone chemical contraceptives.

The underdeveloped countries of the twentieth century have had uniformly higher birth rates than those of preindustrial Europe, and these have continued without any decrease. In the meantime, these countries have experienced sharply decreasing death rates, the result of the introduction of modern health measures (Figure 11-2). Instead of a gradual invention and introduction over a period of centuries, as in Europe, the purification of water supplies, vaccination campaigns, and malarial spraying campaigns were initiated almost simultaneously in a few years' time following World War II. Thus today, the demographic situation in these underdeveloped countries has been rapidly brought to a point that superficially resembles

FIGURE 11-2

(A) The historical occurrence of the demographic transition in western Europe during the last two centuries. (B) The birth and death rates in underdeveloped countries in the twentieth century up to the present.

western Europe's population parameters from about 1700 to 1800. There is considerable question, however, as to whether there will be a demographic transition in these less developed countries. That is, it seems extremely unlikely that the urbanization process will foster a natural transition of motivation toward smaller families. The reasons for this are rather complex.

At the time of the demographic transition in western Europe, the population of the continental area was small enough that industrialization and urbanization affected the social values of nearly everyone in those countries. In India and other less developed areas today, the populations are already so large (nearly 600 million in India alone) that the majority of people will simply not be affected by the current level of socioeconomic progress before food crisis becomes severe. Thus a number of population biologists and agricultural scientists have predicted that the death rate in these countries will greatly increase again in the late 1970s and early 1980s without an earlier drop in the birth rate. While scores or perhaps hundreds of millions of people die, the birth rate among the remaining populace will decrease because of malnutrition and a desire not to bring children into a starving society. Michael M. Sligh has suggested that this will cause a **demographic transposition** effect, in that the relative positions of the birth rate and death rate will be transposed at least for a time (Figure 11–3). While this demographic transposition will have tragic immediate effects on the populations of many less developed countries, it may require tragedy on such an immense scale to cause these countries to fully adopt absolute population control and zero population growth as national dogma.

In general, then, it seems unlikely at this point that worldwide population growth will be curtailed by the same changes in economic and social values that led to the rise of small families as a norm in Europe and North America. Far more likely, it seems that the possibilities of *war, famine,* and *pestilence,* as proposed by Thomas Malthus nearly two centuries ago, will continue to serve as ultimate population control factors on the human population. These biological constraints, including epidemic diseases and massive famines, will continue to sweep the world or parts of it wherever overcrowding becomes a major factor. Voluntary limitation on population size would be far preferable to these disasters, of course. While there are a number of methods for population control currently available, many of them are promoted as family planning methods that can be used for timing conception and contraception in marriage. Even in the underdeveloped nations, the average married couple wants a family averaging 3.5 children in size. In the United States, substantial subgroups of the population still desire a

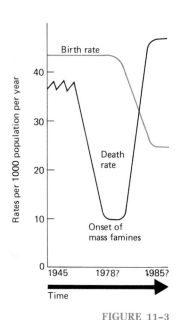

FIGURE 11–3

Predicted demographic transposition in the underdeveloped countries in the 1978–1985 period.

family of 3.5 to 4.0 children, 1.4 to 2.0 children more than enough to replace the parents. Thus there is little question that as long as the family is "free" to plan its size, there will be increased population growth and that as death-control measures increase in influence in less developed nations, there will be an even greater increase in world population growth.

Despite hopeful propaganda, the so-called Green Revolution and increased production from the sea do not offer substantial solutions to the world population dilemma. Many of the advantages of the new miracle strains of grain delivered by the Green Revolution have been due to intensive fertilization and mechanized agricultural practices, which may soon be impractical in large areas of the world because of the diminishing amount of fossil fuel available and its rapidly spiraling cost. The sea is gradually being poisoned by heavy metals, oils, and pesticides, and is probably being fished to its maximum capacity at present. The intensive cultivation of algae might offer a solution. Even so, it has been carefully estimated that massive algal cropping could feed perhaps only double the present world population, should people decide that they would like to survive principally on algal flour.

The earth's carrying capacity in terms of food, energy sources, and industrial energy has reached the point of saturation. Some experts argue that this point has been exceeded already by the human population as a whole and that despite that great aspirations of the developing nations, there is little possibility — indeed a total lack of probability — that everyone on earth in a population of the present size will be able to enjoy the benefits that the highly developed nations have enjoyed over the past 50 years or so. We must become reconciled to the fact that the world's population will actually have to decrease a substantial amount and that wise use of our remaining resources will have to be carried out, if we are even to *hope* to maintain a high standard of living analogous to that enjoyed in the most developed areas of the world at present. Unless population growth is curbed, then, it is clear that a greater and greater proportion of the world will live in more and more misery, as Malthus gloomily predicted, and the growing number of ecological catastrophes around the world, including mass famines and epidemics, will affect even the highly developed nations. Fortunately there are means of population control available besides voluntary abstinence from sex, and these products of the scientific-medical revolution have considerable merit for worldwide implementation and success as control measures. Let us look at some of these methods and programs now and see what hope they offer.

Population control measures

In primitive societies, population growth was controlled by voluntary as well as involuntary means. Excess individuals were eliminated from the group involuntarily by famine, disease, or warfare. Since a hunting and gathering group requires a large territory to furnish what it needs during the course of a year, a low population density in the form of widely spaced groups is necessary. Should a group exceed the carrying capacity of its area, individuals can survive by voluntarily emigrating from the group to a new area and establishing additional groups. Since there would probably be strong selection against individuals in groups that did not have some form of population control, it is not surprising that people themselves developed cultural traits that would control reproduction in their society.

Many existing primitive groups, and presumably our ancestors as well, have extensive marriage and reproductive taboo systems. Frequently, minimum and maximum ages are set by custom for motherhood or fatherhood, and required child spacing is not uncommon. Reproductive contact may be prohibited or restricted until a new child reaches a certain age. Widows are often forbidden to remarry for a period. Consecration of virginity is common to almost all modern human groups as well as many primitive groups. In some societies, celibacy is advocated as a high calling. The Old Testament contains obvious references to the practice of withdrawal, or coitus interruptus.

Ancient Egyptians used crude barriers to the womb made by stuffing leaves or cloth into the cervical canal. Ancient Greeks practiced population control by discouraging heterosexual marriage and encouraging homosexual relationships. The condom, or sheath, dates back at least to the Middle Ages when devices made of linen, fish skins, or sheep intestines were used. Douching, a practice of flushing out the vagina with water or a solution immediately after intercourse, has a similarly long history in Europe. The trend toward late marriage and the widespread practice of withdrawal are believed to be responsible for the reduction of European birth rates that followed the industrial revolution, prior to the introduction of modern contraceptive methods.

Among the many conventional modern methods and devices for birth control are mechanical, chemical, and hormonal **contraceptives.** All of them are intended to prevent the meeting of the sperm and ovum or preventing the newly developing egg from becoming attached in the womb.

Mechanical contraceptives

Mechanical contraceptives include the *condom*, one of the most popular and effective means of birth control. Usually made of very thin rubber, this sheath fits tightly over the penis during intercourse and retains the semen after ejaculation. Its advantages lie in simplicity of use, low cost, and wide availability in the United States as well as most other developed nations. In the mid 1970s it was estimated that 25 percent of all married couples in the United States use this method of contraception. The failure rate is quite low, especially if the couple has been instructed in its proper use. No fitting or prescription by a doctor is required, unlike other mechanical contraceptive devices.

The *diaphragm* is essentially a rubber cup with a rubber-clad rim of spring steel that is inserted in the vagina before intercourse to cover the cervical entrance to the womb or uterus where it acts as a barrier to sperm during intercourse and for several hours afterward. Before insertion the edges of the diaphragm are coated with a spermicidal agent to prevent sperm from slipping underneath the rim. It is removed several hours after intercourse. Available by prescription in the United States, the diaphragm has to be fitted by a doctor, although the device itself is of nominal cost. Because of the relatively complicated routine of fitting and periodic insertion, the diaphragm has not been utilized in mass population control programs in the less developed countries.

The *cervical cap*, like the diaphragm, bars entrance of sperm into the uterus. It fits tightly over the cervix and may be left in place for three weeks at a time, with removal necessary only near and during the menstrual period. When properly fitted, the cap is extremely effective. The principal disadvantage is the necessity of placing it correctly.

The *intrauterine device* (or IUD) is a plastic or metal object that is inserted into the uterine cavity and left there to prevent contraception as long as desired. IUDs are designed in different shapes, from loops and rings to spirals and bows. They probably act by dislodging the fertilized egg or by preventing its implantation in the uterine wall. Similar foreign bodies in the uterus act to prevent pregnancy in domesticated animals such as camels. It has also been suggested that the presence of the foreign object in the uterus hastens the muscular movement of the oviducts and speeds the ovum to the uterus, preventing it from being fertilized in the tubes. The failure rate of the IUD is among the lowest for all contraceptives. About one or two women out of 100 using an IUD will conceive. The cost is minimal, and the primary advantage is that once in place, the IUD can be forgotten. Pills do not have to be taken daily and no contraceptive materials have to be inserted before intercourse. The IUD is probably most suitable for women over 30 who have completed their families, for they are least likely to have a birth control failure, to expel the IUD, or to have problems with side effects. Younger women generally find a hormonal contraceptive to be superior.

Chemical contraceptives

Chemical contraceptives are frequently used in combination with the mechanical contraceptives. These include sprays, foams, tablets, suppositories, creams, and jellies designed to kill or immobilize sperm before conception occurs. They are placed in the upper vagina with special applicators. Although these agents tend to be less effective than the mechanical devices discussed above, they are easier to use and are obtainable without prescription or fitting by a doctor. They are effective about 60 percent of the time, hence not very reliable for population control.

Hormonal contraceptives

Hormonal contraceptives, which include pills or injections to prevent ovulation through chemically simulating the conditions of early pregnancy in the body, have proven to be the most effective ones in present use. The pill is composed of the steroid female hormone, estrogen, and a synthetic chemical, progestin, similar to the natural hormone progesterone that is produced by the ovaries. These hormones are administered either sequentially or in combined form for three weeks, or 20 or 21 days of the 28-day menstrual cycle. The pill method is simple, inexpensive, and virtually completely effective if taken according to instructions. There may be a few undesirable side effects, but most of them wear off within a few months or can be eliminated by adjusting the dosage or changing brands. Many of these side effects resemble the symptoms of pregnancy, such as tenderness and swelling of the breasts, weight gain and retention of fluid, nausea, headaches, depression, nervousness, and irritability. Only about one in four or five women taking the pill experience undesirable side effects, however. Male steroid hormonal pills are as yet unsatisfactory because of the severe side effects affecting secondary sexual characteristics and general mental state. For most women it appears that the benefits of taking the pill outweigh the risks, which can be minimized by the close supervision of a good physician.

Natural contraceptive methods

Among the so-called natural contraceptive methods, the *rhythm method,* or periodic abstention, is the only method of birth control approved by the Roman Catholic Church. Since the timing of ovulation (release of the mature egg by the ovary into the oviduct where it can be fertilized) usually occurs 16 days before the menstrual period, one can abstain from intercourse from several days before ovulation until about a day after ovulation, and thus avoid fertilization of the egg. Unfortunately for the success of this method, the timing of ovulation varies among women and even from month to month in the same woman. In fact, one out of six women have such an irregular cycle that the method will not work even haphazardly for her. Carefully kept records of previous menstrual and temperature cycles must be used to predict the time of ovulation. In general, the period of abstention amounts to a considerable fraction of the month, which tends to have a detrimental effect on the conjugal relationship between most couples, especially since the fertile period may be a time when the woman is relatively more receptive to sexual relations.

The rhythm method has a notoriously high failure rate, both because of the irregularity of many women's cycles and because of the occasional lapse of discipline on the part of the participants. Other "natural" contraceptive methods include the ancient coitus interruptus, douching, and, of course, abstention from intercourse.

Unfortunately, contraceptive methods are generally promoted on the premise that family planning is the ultimate goal. Yet when a couple decides to have more children than merely their replacement rate of two, the use of contraceptives proves primarily a population growth depressant, *not* a population control. The basic problem still remains of achieving an international motivation for small families of at most two children since effective contraceptive methods are already available to stop world population growth.

Abortion and sterilization

Should contraceptive methods not be employed or fail in their intent, a woman who conceives may prevent the birth of an unwanted child by induced or therapeutic **abortion:** expulsion of the fetus from the uterine cavity. During the first 12 weeks of fetal development, abortion can be accomplished by several methods without much chance of injury to the mother. Beyond about 16 weeks, however, the risk of death of the mother by hemorrhage increases considerably. The most commonly used medical method for abortion is dilation (of the cervix) and curettage (scraping) of the uterus. A physician working under sterile conditions can do a D and C with a young fetus with less risk than delivering a full-term pregnancy. Cruder forms of abortion have long been practiced in every area of the world, and abortion represents the single greatest factor contributing to the death rate among pregnant women. The early abortion laws in Europe, Canada, and the United States were passed to protect women from its danger under primitive medical conditions. Recently there has been a trend toward legalizing abortions in the developed nations, based on the premise that the decision of a mother to have a child should be left to the individual, not society.

Sterilization is being advocated by many groups at present. Most of the time, sterilization is completely effective as a birth control method and has no side effects on sex life or normal physiological functioning. Most sterilizations are obtained by couples who have completed their families and wish to avoid the continued use of contraceptives. In the male, the operation involves tying off or cutting the vas deferens tubes leading from the testes to the urethra tube in the penis. This prevents the ejaculate from containing sperm cells. This operation, done through the scrotal sac wall, is called a *vasectomy* and takes about 15 minutes under local anaesthetic in a physician's office or field clinic.

An analogous operation, called a *salpingectomy*, can be carried out in the female, but special abdominal surgery under general anaesthesia is required unless the operation on the Fallopian tubes is done when they are accessible following the birth of a baby. With either the sterilized male or female, sperm or eggs continue to be produced regularly but eventually break down and are reabsorbed by the body or picked up and engulfed by the phagocytes (specialized white blood cells from the circulatory system). A good surgeon has at least a 50 percent chance of reversing either a vasectomy or salpingectomy operation, should the recipient wish to become capable of reproduction again.

Major operations resulting in sterilization include the *hysterectomy*, in which the uterus itself is removed. This, of course, would rarely be done for birth control purposes alone, but is rather commonly done for women experiencing menopause. More severe sterilization operations include *castration* (removal of the testes) of the male and *oophorectomy* (removal of the ovaries) in the female. Unfortunately, these operations result in detrimental hormone imbalance because the sex organs are endocrine glands that produce hormones controlling the secondary sexual characteristics of an individual. Without these hormones circulating in the blood, the body may experience serious physiological problems. The operations mentioned earlier, vasectomy and salpingectomy, only prevent fertilization; they *do not* prevent hormonal production or upset any other normal bodily activity.

In India and other countries, male sterilization has proven a small but useful part of population control programs. Western cultures are gradually increasing the number of male sterilizations as ignorance is dispelled over the exact nature of the male operation and unfounded fears that a man's sex life will end are swept away. Vasectomies in particular seem to be an effective and easy population control measure to advocate among married men who already have one or two children and who, along with their spouse, do not wish a larger family.

Prospects for the future

From the preceding discussion, it is clear that a continued increase in human population will have ever-more-detrimental effects on natural communities, including the continued existence of other species and even our own biological and psychological uniqueness. It seems impossible to plan in a rational way for the housing and feeding of the additional billions of human beings soon to join our standing crop of *Homo sapiens*. Instead, we must work toward finding the quality of life that most of us would like to have and make possible its attainment by limiting population size across the face of the earth. Only by a reduction from our present world population size can we even hope to allocate resources fairly among all the nations of the world, so that each person may live a life full of happiness and reasonably free of despair. Without population control, or even if we continue at our present level of ineffective attempts at control, it seems clear that the less developed nations will shortly be in serious trouble, and not much further into the future the presently "highly developed" nations will join them. Our tenure on earth may soon reach an untimely end if we do not learn to face this ultimate problem of population control.

Summary

Human population growth has been accelerated each time that *Homo sapiens* has made a major cultural advance or gained control over a new source of energy. The cultural revolution about 600,000 B.C., with the utilization of fire for cooking and warmth and the development of advanced tools, weapons, and reasonably permanent shelters, set off the first significant surge in world population growth. The agricultural revolution about 8,000 to 10,000 years ago led to settled communities subject to fewer environmental hazards for their continued existence. The scientific revolution, beginning about 1600 A.D., and the industrial revolution, beginning in the eighteenth century in western Europe, brought in greatly lowered death rates because of medical advances and public health measures and economic factors associated with the industrialization of society. Population growth has especially surged upward, then, since about 1650, although specific areas of the world have joined this latest increase in growth rate at different times. In a technical sense, population control measures are already available to halt this worldwide growth, including mechanical, chemical, and hormonal contraceptives, sterilization, and abortion; but the primary problem is not technological, but psychological: to convince people and governments around the world that zero population growth, with a replacement rate of not more than two children per couple, is an absolute necessity for continued survival.

12 The genetics of populations

When we look at human populations, or indeed populations of any plant or animal species, we see tremendous variation among the individual members. This variation provides the raw material for *evolution*, that is, *change in characteristics of populations*. With this chapter, we begin our consideration of the processes of this evolutionary change and how it affects the characteristics of organisms. In 1859, when Darwin published *On the Origin of Species*, the basic framework was laid for the understanding of organic changes through evolution. He proposed that the raw material of variability in plant and animal populations was shaped and molded by the action of natural selection — that is, the differential survival and reproduction of certain varieties at the expense of other varieties. The weakest point in the argument that Darwin and his followers presented was that of genetics. If he had known that at the time an obscure Austrian monk was laying down the basic principles of genetics with experiments in his monastery garden, Darwin might have been able to explain why rare genetic traits persist in a population and are not simply blended out of existence. However, the significance of Gregor Mendel's experiments was not recognized by his contemporaries, and for the next half century evolutionists proceeded independently of any secure knowledge of genetic principles in their concepts of how populations changed over time. It was not until the 1920s that geneticists and naturalists melded their conceptual schemes into the so-called synthetic theory of evolution, which includes the ideas of population genetics and natural selection. In this chapter, then, we shall look at how the genetically controlled attributes of a population arise and are established, recombined, or lost from a particular group of organisms.

Variation and the units of heredity

In Darwin's day, most biologists thought that the units of inheritance were present in each cell of the body as "pangenes" and that at the time of reproduction these heritable units came together and passed with a stimulating rush, noted as orgasm, into the fertilized egg that would create the new organism. Thus in this view, the characteristics of both parents were always blended in the offspring. Darwin observed that heredity was normally stable during the process of reproduction; in other words, like begets like. His basic postulate was that variations occur at every generation in the population and that at least some of these variations are heritable. However, in every species the number of individuals that survive to reproductive age is very small compared with the number produced. Therefore, the determination of which individuals will survive and which will perish comes from their *fitness* in general, not by chance processes, and this fitness in turn determines the contribution that each individual will make in the next generation. Unfortunately, blending inheritance could not answer the question of why rare mutations or changes in appearance persist from generation to generation in a population, and even increase over time. One would suppose under blending inheritance that the effects of such change would be swamped and lost.

Part of the problem in 1859 and later years was that Darwin had no knowledge of the physical basis of the exact transmission of genetic information. Mendel, who in 1865 published the first principles governing the inheritance of **phenotypic** characteristics (that is, the various physical traits that characterize an organism, such as color, shape, and size), did not understand the physical basis for transmission of factors either, although he correctly postulated the existence of a cellular mechanism that physically separated particulate genes (hereditary factors) during cell division. For Mendel, the units of inheritance were invisible but discrete factors that governed single traits. In the early twentieth century, geneticists who rediscovered

Mendel's principles described these factors in terms of their most obvious phenotypic effects, such as *red eye* versus *white eye* and *long wing* versus *short wing*. The factor that controlled a particular character came to be called a **gene.** The alternate states of expression of a particular trait, such as eye color, came to be known as **alleles** of a single gene. A particular allele would produce a particular character state, for example, red eye for one allele versus white eye produced by a second allele of the gene for eye color. With careful observations of the process of **meiosis,** that is, cell division in which a dividing reproductive cell endows each of its daughter sperm or egg cells with half of its chromosome complement, several geneticists discovered that the behavior of certain genes followed exactly the inheritance of particular chromosomes.

Through this kind of evidence and other lines of study, it was soon found that genes were located on the chromosomes and that a complete set of chromosomes was located in each cell in the body. Each set of chromosomes actually contained genetic material from both parents that originally made the fertilized egg that developed into the body of the individual. The pairs of chromosomes occurring in the cell nucleus were a combination of parental contributions, each member of a pair having come from one of the two parents. Thus this discrete physical basis for the genes offered an explanation of why traits could be preserved generation after generation without being lost in a blending effect. A characteristic that was **recessive** in effect (not expressed when in combination with its **dominant** allele on the other member of a chromosome pair) could be retained in the population through individuals mating at random until at last it came to combine in a fertilized egg with another chromosome bearing the same recessive allele. Now the new offspring carrying both recessive alleles would be able to express the characteristic determined by that allele.

Eventually, the physical site of each gene on a chromosome became known as a **locus,** and it was discovered that any particular locus might have one, two, three, four, or many more alleles. Naturally, any one individual could have at most two different alleles of that gene because of having just two loci available on the two members of a chromosome pair. However, within a population composed of many individuals, the variability can be quite great. A number of populations of alfalfa sulfur butterflies, for example, have been found to have as many as 12 or 14 alleles for a single protein enzyme locus on a particular chromosome. How does this variability come about, and why should a population maintain such a high level of variability or at least have such a high level of variability at any particular time?

Mutations: the ultimate source of variability

In our changing concept of the gene during the past century, geneticists have moved from viewing the gene as a factor that governs single traits to a physical locus on a chromosome that could affect more than one phenotypic characteristic, a phenomenon called **pleiotropy.** At the same time it has been realized that this phenotypic effect of the gene, whether on one or many characteristics, is usually exerted through a single chemical product of the gene, normally an enzyme molecule. About 40 years ago, it was found that X-rays and other sources of radiation could **mutate,** or change the genetic material at particular locus sites along the chromosome. Thus it was possible to induce changes in the gene that controlled, for example, eye color. New eye-color mutations could be produced that were also found in nature, and some were induced that were not found in nature.

With the discovery of the role of DNA as the repository of genetic information, the machinery of inheritance became clear. The structure of DNA makes possible great precision in the copying of the genetic information, and the biochemical mechanisms by which the genetic information is stored in the DNA molecule shapes the development, structure, and functions of an organism through its control of chemical products. Thus it was found that radiation or even chemicals such as chlorine gas could alter the structure of the DNA molecule and create slightly different chemical products from that gene. In other words, a new allelic form of that gene had been created and this mutation now produced an altered gene product, often culminating in an externally visible trait such as a different eye color.

Thus mutations are the ultimate source of variability in a population, and with mutations occurring at a rate of approximately 1 out of every 10,000 individuals carrying a particular gene, it is not very surprising that variations appear fairly regularly among organisms, especially species that have a high rate of reproduction, large population sizes, and have been around for long periods of time. Darwin recorded many instances of mutants among animals and plants, particularly in domesticated strains where the unusual trait had been selected by human intervention and allowed to be passed on to succeeding generations. Thus a species as familiar as the common house dog, *Canis familiaris,* has hundreds of different varieties controlled by mutations of various genes having regulatory and structural effects. Under the conditions of domestication, many mutations that might have been disadvantageous in nature, such as short legs in sheep, would become established under the care of animal husbandmen.

The constant occurrence of mutations may be understood more clearly when we realize that cosmic rays and other background radiations are incessantly bombarding the **genome** (the total collection of genes) of every species, and thus changes occur even without human inducement. The average spontaneous mutation rate for a given gene locus has been estimated to be one or two new mutations per 100,000 genes per generation. Thus within every 100,000 sperm cells, for instance, one or two may be expected to carry a new mutation for a *particular* gene. Since there are probably at least a minimum of 10,000 gene loci in a *Drosophila* fruitfly, for example, the total mutation rate will be $1/100,000 \times 10,000$, or 1 out of 10. In other words, 1 out of every 10 sperm cells will probably carry at least one new mutation among *all* the genes they are carrying. Important mutations occur also by chromosomal rearrangements that alter patterns of gene activity during development.

The vast majority of mutations are detrimental and hence do not survive unless selected. The reasons for this disparity between frequencies of harmful and beneficial mutations are apparent from the fact that the system is already probably well balanced and running well, and any change, as in the works of a finely crafted Swiss watch, will likely be deleterious. The type of mutation produced by X-rays, chemicals, ultraviolet light, or cosmic gamma radiation is completely at random in the sense that a directed change is virtually impossible to produce. There is no evidence that organisms are able to elicit specific genetic mutations in their genomes when required; they must await the rare random mutation that will confer a beneficial effect and thus enhance the survival of its carrier, which can pass it along to the offspring of any matings. We would like to know, then, how recombination of these novel units occurs, because if the mutation originates in only one individual, it is important to understand how it is transmitted throughout the population.

The process of recombination

During sexual reproduction the male and female exchange gametes; that is, the sperm and egg leave their independent existence and unite in the form of a fertilized egg. The egg now contains one complete set of hereditary units from the male parent and one complete set from the female parent. Should one of these parents have a mutation in place of the normal allelic form of a particular gene, the new zygote will undergo development and form a new individual that will mature, still carrying this mutation. Since large numbers of sperm are produced by the male parent and several score or more eggs are produced by even the least fecund female, there is a good chance that some of the siblings will also be carrying the mutation, and subsequent matings between offspring in the population may allow two mutated alleles to come together in the same individual. The effect of this **inbreeding** is to establish the rare variant trait in a number of individuals. Naturally, this can occur more easily in a small population than in a large population where the sibling offspring would be less likely to meet. On the other hand, if the mutation is not deleterious in itself but is recessive in effect, it may persist in some individuals for many generations until, by chance, two individuals carrying the same recessive mutant allele will meet and mate and combine these two mutants in a single offspring.

Generally, most plant and animal species are **outbreeding** in nature and avoid self-fertilization or fertilization with close relatives. Thus many primitive human societies have strong prohibitions against incest between nuclear family members and even marriages between first cousins; these cultural prohibitions probably arose as a result of observing deleterious effects in continually inbred families. The vigorous characteristics of hybrid corn and other crop plants are derived by crossing two different varieties to produce the seeds for each planting. Hybrids are less likely to be carrying two deleterious recessive alleles at any particular locus. It is also possible that the greater vigor results from merely having two unlike alleles of a locus together in the same cell. At any rate, the advantages of outbreeding are considerable for the offspring.

Thus recombination of chromosomes occurs through sexual reproduction in the normal course of events, and mutant alleles are spread throughout the population unless they are deleterious in effect; selection removes individuals carrying those alleles from the population, usually before they are able to breed. On the population level, then, if individuals are mating freely throughout the area, we say that **panmixis**—free recombination—is taking place.

The general collection of genes in the population is termed the **gene pool** (Figure 12–1). We can also refer to the frequency of particular alleles in the population as **gene frequency.** Thus, for instance, in a population of *Drosophila*, we may say that the gene frequency for the red-eye allele is 0.60 and for the white-eye allele 0.40. This means that 60 percent of the alleles in the population for eye color are red-eye alleles, whereas 40 percent are white-eye alleles. We know from common observation that the genotype of a species stubbornly resists change because of the selection for the optimal

Gene pool:

= A = 0.6   = a = 0.4

FIGURE 12–1

A population gene pool, where 60 percent of the butterflies are homozygous for the white-winged gene and 40 percent are homozygous for the spotted-winged gene.

"average" phenotype, and therefore recombination on the population level generally does not affect the overall appearance of a species. The average organism is a product of tens of thousands of years of evolution and is generally well adjusted to its environment. Thus unless the environment changes, wide variations in structure or function are almost always doomed to failure.

However, the continued existence of genetic variation in a population does prepare it for possible future climatic changes or other new evolutionary pressures. This fact sets up an essential contradiction between our feeling that natural selection is eliminating undesirable variation from the population and the continued existence of genetic variation that seems to be essential for long-term adjustments to change. In essence, the problem is solved by the way variations are stored in a gene pool and their maintenance. If the variation is neutral in effect, it will remain in the population in the same ratio over an indefinite period of time as long as the population is large and there is no selection against that particular genotype, no mutation increasing its frequency, and no unequal migration into or out of the population. The Hardy-Weinberg principle, named for the first discoverers of this general concept, is a fundamental description of events in the genetics of natural populations.

The Hardy-Weinberg principle and equilibrium conditions

To begin, let us consider a population as an assemblage of genes existing through time rather than as a collection of individuals reproducing themselves generation after generation. An assemblage of genes can be described in terms of the frequencies of their various alleles. These gene frequencies, or proportions of the different alleles of a gene present in a population, are obtained by counting the total number of organisms with various genotypes in the population and estimating the relative frequencies of alleles of all. It should be recalled, of course, that we are usually dealing with **diploid** organisms, that is, sexually reproducing organisms and chromosomes from both male and female parents, and therefore there can be two allelic forms of a gene for any one locus. If both alleles at this chromosome locus on the pair of chromosomes are the same, the organism is said to be **homozygous.** If they differ, the organism is said to be **heterozygous.**

As an example of how to determine gene frequencies, let us consider a population of roses in a garden. In this particular rose species, red flower color, R, is incompletely dominant over white, r, with the heterozygous condition being pink, a distinct phenotype from either homozygote. Our population of 100 individuals contains 70RR (red), 20 Rr (pink), and 10 rr (white) individuals. Thus in this population there are a total of 200 genes with respect to the gene locus controlling flower color. Of these, 140 plus 20, or 0.8 of the total (160/200), are R genes, and 20 plus 20, or 0.2 of the total (40/200), are r genes.

The same gene frequencies could also be calculated from the frequencies of the three different genotypes — 0.70 RR, 0.20 Rr, 0.10 rr — according to the following formulas:

Frequency of $R =$ **frequency** $RR + 1/2$ **frequency** Rr (i.e., $0.7 + 0.1 = 0.8$)

and

Frequency of $r =$ **frequency** $rr + 1/2$ **frequency** Rr (i.e., $0.1 + 0.1 = 0.2$)

Thus any time that we wish to calculate the gene frequencies based on the samples of individuals themselves, we can calculate the frequency of one allele, say A (the dominant allele frequency is by convention designated as gene frequency p), as the number of AA homozygotes plus one-half the number of Aa heterozygotes. The frequency of the recessive allelic form a (designated gene frequency q) is defined as one-half the number of Aa heterozygotes plus the number of aa homozygotes. Since A and a are the only alleles at this locus, the sum of their frequencies equals 1.00, or $p + q = 1.00$. That is, the frequency of allele A plus the frequency of allele a equals 100 percent of the alleles for that gene in the population, since they are the only two alleles at that locus. Now we can see how Hardy and Weinberg developed their description of the behavior of such allele frequencies in the total gene pool of the population.

The gene pool can be considered as a gametic pool, that is, a pool of sperm and eggs from which samples are drawn at random to form the zygotes of the next generation. Thus the genetic relationship between an entire parental generation and the subsequent generation of offspring is very similar to the genetic relationship between a particular parent and its own offspring. So now we can ask the specific question: How will the frequencies of genes in the new generation depend on their frequencies in the old generation? We could almost ask: How are these gene frequencies "inherited"?

Some of the early geneticists thought, for instance, that with *dominant* alleles present, no matter which frequency one started out with, the result would be the stable equilibrium frequency of three dominant individuals to one recessive since this was the Mendelian segregation pattern for these genes (Chapter 23). In 1908, Hardy in

England and Weinberg in Germany disproved this concept, showing that gene frequencies are not dependent on the presence or lack of dominance but may remain essentially unchanged or be *conserved* from one generation to the next, under certain conditions. Let us look now in detail at this phenomenon of the conservation of gene frequency.

The principle disclosed by Hardy and Weinberg may be simply illustrated by an example from human genetics. In human beings, the difference between those who *can* taste and those who *cannot* taste a certain chemical, *phenylthiocarbamide* (PTC), resides in a single gene locus that can have two alleles, *T* and *t*. The allele for tasting, *T*, is dominant over *t* so that the heterozygotes *Tt* are tasters as well as the homozygotes *TT*, and the only nontasters are *tt*.

If we were to choose an initial population composed of an arbitrary number of each genotype, we would ask: What will be the frequency of those genes after many generations? Let us, for example, place on an isolated south Atlantic island a group of teenage students fleeing in a plane from a potential atomic war in 1990. They have the genotypic ratio among them of 0.40 *TT*, 0.40 *Tt*, and 0.20 *tt*. The gene frequencies in this initial population are therefore $0.40 + 0.20 = 0.60 \, T = p$ and $0.20 + 0.20 = 0.40 \, t = q$. Let us *also* assume that the number of individuals in the population is large, perhaps about 200; that tasting or nontasting ability for this particular chemical has no effect on survival rate, fertility, or attraction between the sexes; and that there is an equal number of males and females.

As these teenagers mature in isolation in this tropical paradise, away from the world at war, they will undoubtedly choose their mates at random from those of the opposite sex regardless of their tasting abilities. That is, while some may have preferences in hair color, they probably will not have even the slightest interest in testing whether the prospective mate has the ability to taste or not to taste PTC. Matings between any two genotypes, then, can be predicted solely on the basis of the relative frequencies of these genotypes in the population.

The types of random mating combinations and their relative frequencies in a population such as we have described are predicted in Table 12–1. As shown in this table, nine different types of matings can occur, of which three matings are reciprocals of each other (e.g., $TT \times tt = tt \times TT$). In all, therefore, there are six different mating combinations between these genotypes; these will produce offspring in the next generation in the following ratios (Table 12–2): 0.36 *TT*, 0.48 *Tt*, and 0.16 *tt*.

Now we note that although the frequencies of genotypes — that is, zygotic frequencies — have been altered by random mating, the *gene* frequencies have not changed. For T, the gene frequency is equal to $0.36 + 1/2 \ (0.48) = 0.60$, and the frequency of t is $0.16 + 1/2 \ (0.48) = 0.40$, exactly the same as before. Under these conditions, no matter what the initial frequencies of the three genotypes, the gene frequencies of the next generation will be the same as those of the parental generation. For example, if the founding population of this island had contained 0.25 TT, 0.70 Tt, and 0.05 tt, the gene frequency for T would be $0.25 + 1/2 \ (0.70) = 0.60$, and for t it would be $0.05 + 1/2 \ (0.70) = 0.40$. However, despite the new frequencies of genotypes, the offspring are again produced in the ratio of 0.36 TT, 0.48 Tt, and 0.16 tt, or a gene frequency of 0.60 T to 0.40 t.

In algebraic terms, p equals the frequency of T and q equals the frequency of its allele t. If no other alleles exist, $p + q = 1$. The equilibrium frequencies of the genotypes are given by the binomial expansion of this expression squared. That is, $(p + q)^2 = p^2 + 2pq + q^2$. In this algebraic expression, p^2 is the equilibrium frequency of the TT genotype, $2pq$ is the frequency of the Tt genotype, and q^2 is the frequency of the tt genotype.

Two important conclusions follow from this discussion. First, under conditions of random mating in a large population where all genotypes have an equal chance of survival to the age of reproduc-

TABLE 12–1

Types of random-mating combinations and their relative frequencies in a population of 0.40 TT, 0.40 Tt, and 0.20 tt genotypes.

	Males		
	TT (0.40)	Tt (0.40)	tt (0.20)
Females			
TT (0.40)	0.16	0.16	0.08
Tt (0.40)	0.16	0.16	
tt (0.20)	0.08	0.08	

TABLE 12–2

Relative frequencies of the different kinds of offspring produced by the matings shown in Table 12–1.

Parents		Offspring		
Type of mating	Frequency	TT	Tt	tt
TT × TT	0.16	0.16		
TT × Tt	0.32	0.16	0.16	
TT × tt	0.16		0.16	
Tt × Tt	0.16	0.04	0.08	0.04
Tt × tt	0.16		0.08	0.08
tt × tt	0.04			0.04
		0.36	0.48	0.16

tion, the gene frequencies of a particular generation depend on the gene frequencies of the previous generation and not on the genotype frequencies. Second, the frequencies of different genotypes produced through random mating depend only on the gene frequencies in the potential mates. This allows us to express the conservation of gene frequencies in the form of a principle or law.

The Hardy-Weinberg equilibrium principle describes the behavior of genes in large populations as follows: *In the absence of outside forces—namely, selection, differential migration, and differential mutation of genes—and with random mating between all genotypes, the initial gene frequencies in a population will be maintained from generation to generation. Also, after the first generation, the genotype frequencies will remain stable, that is, at an equilibrium point.* We can represent genetic events in a randomly mating population of this sort in the following way:

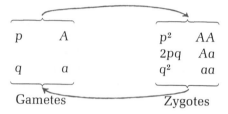

$$p \quad A \qquad p^2 \quad AA$$
$$\qquad \qquad 2pq \quad Aa$$
$$q \quad a \qquad q^2 \quad aa$$

Gametes   Zygotes

Undisturbed, the cycle shown above will exist indefinitely. Each generation produces gametes that carry the alleles A and a in proportions p and q. These in turn unite at random to give rise to zygotes AA, Aa, and aa in the proportions p^2, $2pq$, and q^2.

It must be remembered, of course, that this Hardy-Weinberg model we have outlined is essentially an artificial situation since our results are accurate only when the conditions of random mating and large population size are met and differential mutation or differential selection are omitted. The Hardy-Weinberg principle states that in the absence of forces that change gene populations, the relative frequencies of each gene allele will tend to remain constant from generation to generation. Needless to say, these forces are rarely absent in natural populations. But the Hardy-Weinberg principle does provide a yardstick against which to measure change from *expected* proportions of genotypes and gene frequencies in a population, and hence the evolutionary biologist or geneticist can look for the factors that are causing the population to shift away from the expected neutral distribution of gene frequency. We will be examining natural selection at length in the next chapter, but let us now look briefly at the factor of population size and its effect on the genetics of population.

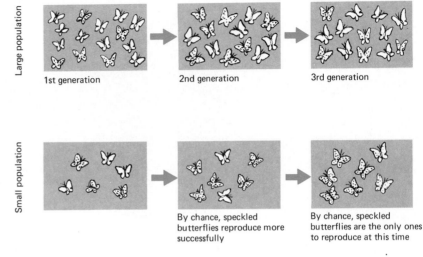

FIGURE 12–2

An example of genetic drift. In the large population, chance events are canceled out by the large numbers and random mating, and on the average the same relative number of genotypes persists from generation to generation. In the small population, butterflies of one genotype may, by chance, hatch late or otherwise not reproduce as successfully, and within several generations only one type remains.

Large population

1st generation   2nd generation   3rd generation

Small population

By chance, speckled butterflies reproduce more successfully

By chance, speckled butterflies are the only ones to reproduce at this time

Small populations and gene frequencies

We carefully specified that the population must be reasonably large for the Hardy-Weinberg principle to work. In a large population, there is greater chance for a random mixture of gametes through sexual reproduction by males and females of different phenotypes. In a small population, sometimes a mutation or rare allele is carried by only one or two individuals. If those individuals are accidentally lost from the population or do not manage to reproduce at one generation, the allele that they are carrying is lost from the population's gene pool. The evolutionary effects of this element of chance in small populations have been described in several concepts of population genetics.

In **genetic drift,** a small population may lose a rare allele or have a common allele established as 100 percent representative of a particular gene because of random chance affecting the adults able to reproduce in a particular generation (Figure 12–2). Sampling variations due to chance alone have a greater role to play than selection or mate choice in determining which genes are present among the sample of gametes that will form the progeny constituting the following generation. The smaller the population size, the greater is the chance that the gene frequencies in the successfully fusing gametes

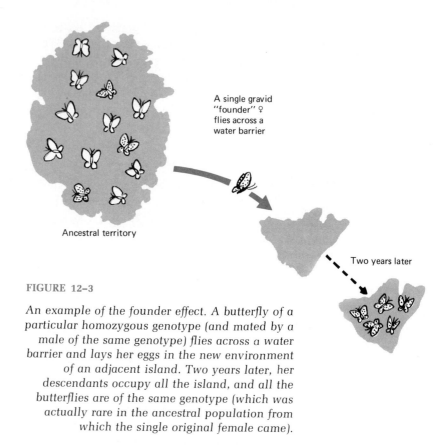

A single gravid "founder" ♀ flies across a water barrier

Ancestral territory

Two years later

FIGURE 12–3

An example of the founder effect. A butterfly of a particular homozygous genotype (and mated by a male of the same genotype) flies across a water barrier and lays her eggs in the new environment of an adjacent island. Two years later, her descendants occupy all the island, and all the butterflies are of the same genotype (which was actually rare in the ancestral population from which the single original female came).

will be different from those at large in the parental population. In large populations, random changes in gene frequency are insignificant enough to be ignored.

Another way in which gene frequencies can be changed in small populations is called the **founder effect** (Figure 12–3). Sometimes new populations are started by only one or several fertile individuals arriving in a new area. The sample of alleles of the various genes in their genotype provides the founding gene frequencies for that population. If a particular rare allele is not present in those founding adults, it will not be present in the new population until another mutational event or a new carrier immigrant arrives.

From the expectations expressed in the Hardy-Weinberg principle, it is also important, of course, to have random breeding, or panmixis, in which an individual has an equal probability of mating with any other member of the opposite sex regardless of its genotype. Otherwise the free rearrangement and recombination of genetic units required in the Hardy-Weinberg expression would not be possible.

Clearly, it is also important that the immigration of foreign individuals into the population or the emigration of members out of the population does not change the frequencies of the alleles in the gametes being produced. This would alter the genetic constitution of the parental population if the migrants carried gene frequencies different from those of the original population.

Chromosomal variability on the population level

The collective phenotype is the total sum of all of the morphological, physiological, ecological, and behavioral attributes of an individual during its lifetime. Since the genotype—that is, the total assortment of genes received in the zygote—determines the phenotype of the individual within a particular environment, it is perhaps not surprising that various chromosomal alterations can vastly affect this phenotype, especially since the genes are located in a particular order on the chromosomes. Therefore, major types of chromosomal alterations in structure have important effects on the genetics of populations as the chromosomes are shifted about through reassortment and gametic exchange. We shall take up the physical nature of these chromosomal alterations in Chapter 24, after discussing the details of the character and structure of the genetic material along the chromosomes.

Human population genetics

In a sense, human beings are an accumulation of genetic variability ranging through some billion or more years of history of organic life on the earth. The basic cellular attributes found in the present human body plan can be traced through phyla in the fossil record to the dawn of life, as we shall see in the next two chapters. At the time of their emergence from a primate stock and for many hundreds of thousands of years afterward, human beings were shaped physically and behaviorally by the forces of natural selection operating on the heritable diversity within their gene pool. Thus early human history is probably quite similar to that of any other species of plant or animal. Individuals adjusted to their environment or they ceased to exist. The many different human populations existing across the face of the earth indicate that the human species became genetically highly diversified and adaptable. Across an extremely broad range of diversified habitats, human beings maintained a high reproductive rate and in biological terms their evolutionary success in inhabiting the entire world has been unprecedented among organisms.

The invention of culture in the last 500,000 to 600,000 years of human history has meant the additions of such aspects as refined tool making, development of agriculture, scientific and technical equipment and resources (including most recently the highly significant invention of analog and digital computers), and the use of symbolic thought and language. Through these cultural additions contributing to evolutionary progress, the human species has been profoundly altered to a large extent and has learned to control its environment and way of life. No other species has been placed in the position of being able to essentially control its own evolution. What effect has this had, and will it have in the future, on the population gene pool of the human species? It is certainly clear that the human gene pool is being less and less altered by selection as modern medicine and other cultural advances bring escape from the normal operation of natural selection. Is our gene pool still sufficiently adaptable that we can accommodate to every created environment?

It is clear from the fossil record, as we shall see in Chapter 15, that the human brain size and other physical characteristics have not changed significantly for the past 100,000 or more years. While we have no direct knowledge of the genes from these earlier populations and are basing these judgments on skull sizes and artifacts left by early peoples, it does seem that the genetic evolution that gave rise originally to our culture is now a thing of the past and that cultural evolution itself is the principal—perhaps the only—influence that will affect the future of humankind.

Most biologists believe that there is still considerable selection on variability going on in the human species. There are different fertility rates among the different segments of the world population. There are changing patterns of disease, famine, and war, and there are environmental circumstances of population increase and decline, all of which affect the total composition of the gene pool. Actually, the major cultural impact that has caused significant human environmental changes has taken place only during the last two or three centuries. The genotype that developed into *Homo sapiens* through various intermediate forms for some 2 million years was originally adapted for a nomadic life of hunting and gathering. For approximately the past 8,000 years we have had an agrarian lifestyle in many parts of the world, and large numbers of people have been exposed to a highly urbanized environment only in the last several centuries, particularly in the last few decades. As Carl P. Swanson has said, "Culture, in all its forms and degrees of complexity, has created an enormously more diversified environment, but no data suggest that the created environments of today favor certain geno- types in the same manner that the Arctic environment permitted the Eskimo genotype to maintain itself. Time has been too short, and the number of generations too limited, to bring about detectable change, even though such change could possibly be taking place."[1]

We can perhaps presume that human genetic variability is great enough to accomplish these changes without deleterious effects on our biological being, but we do not know for certain. We do know that natural selection has been altered greatly in modern times as ad- vances in medicine have brought infectious diseases under control and major health problems have been solved through the use of anti- biotics and improved sanitation. Millions of individuals who in past centuries would have died from disease and accident now survive to reproductive age and have many children. It is, of course, quite un- likely that such diseases so influential in past ages affected all geno- types uniformly. The better-adapted genotypes were the ones to sur- vive the influx of epidemic diseases and the other stresses of life. In modern populations, with improved sanitation and medical care, such genes would not have the same degree of selective advantage and the resistance they confer might be lost through genetic drift or through lack of continued selection. As we shall see further in Chapter 13, evolution is a *change in gene frequency,* and hence humankind in this sense is undergoing evolution even without the normal operation of "natural selection."

Modern medicine has also been able to offset or depress the effects of certain deleterious genes and bring the carriers to reproductive age. Thus the number of deleterious genes in our population is in- creasing, such as those for diabetes and hemophilia. When we remove natural selection against carriers of such genes, the frequen- cies of these genes will rise to whatever level they are being pro- duced by random mutation and then the existing gene frequencies will be conserved according to the Hardy-Weinberg principle. Con- sequently, a greater social burden will develop for society as a whole, even though each of the individuals concerned may be benefited directly by this medically promoted preservation. Overall, the genetic load of the population is clearly becoming larger, and the cost to society will increase. This trend toward reduction of early mortality in general, and our greater life span with concomitantly- lowered ages of retirement, is worrisome to many demographers, psychologists, sociologists, and population biologists. As central-

[1] *The Natural History of Man* (Englewood Cliffs, N.J.: Prentice-Hall, 1973), p. 367.

ized governments extend their authority to every aspect of daily life in countries around the globe, the possibility is raised of the human population heading toward a relatively static way of life at a minimum level of existence, devoid of individual challenge and opportunities for initiative. Some scientists hold that the average level of human intelligence worldwide is being eroded through differential reproduction of less intelligent groups. Presently, in their view, the impoverished and untrained segments of every population are contributing more heavily to the future gene pool than those more advantageously placed, as we saw in Chapter 11. Some believe that a greatly increasing population density will lead to the selective *necessity* of having a docile, conforming genotype for mere survival. Under such conditions, the things that have made us unique and enhanced our drive toward domination over the rest of the universe could well be lost.

Summary

Concepts of population genetics have provided an essential key, missed by Darwin, toward our understanding of how evolutionary changes in characteristics of populations occur. Mendel's 1865 discovery of genes governing particular traits included a correct theory as to how these genes were separated and recombined in the gametes that would produce the next generation of offspring. Early in the present century, his findings were confirmed by microscopic observations of chromosome behavior, and population geneticists and natural selectionists melded their conceptual schemes into the so-called synthetic theory of evolution. The ultimate source of new allelic forms of a gene, and hence of genetic variability in a population, comes from mutational changes induced in gene loci by radiation or other agents. These alleles (alternate forms of the same gene) can be assorted into new combinations through sexual reproduction and mating between different individuals. New mutations may become established through inbreeding of siblings in a small population; but generally outbreeding has greater selective advantages, and a new advantageous allele will become established because of its physiological advantages to the individual carrying it, rather than through inbreeding. Almost all mutations are deleterious in effect and hence eliminated by selection because the organism is generally already well adjusted to the environment. On a population level, the Hardy-Weinberg equilibrium principle states that the initial gene frequencies in that population will be maintained from generation to generation in the absence of outside forces, namely, selection favoring a particular allele, differential migration of individuals carrying a certain allele, or differential mutation rates of genes changing to certain allelic forms in preference to others. These forces are rarely absent in natural populations, but this principle does provide a reference point to measure evolutionary changes from *expected* proportions of genotypes and gene frequencies in a large population at genetic equilibrium. Gene frequencies in small populations are affected by chance events, such as in genetic drift and populations established by a few founder individuals. Human population genetics is being heavily influenced by cultural developments such as new ways of living, medical advances, and differential reproduction of less advantageously placed social groups that lack access to population control methods or lack economic and educational motivation to limit family size. The consequences of these trends may be to eliminate the genetic traits that have led *Homo sapiens* to short, extraordinarily rapid ascendancy and domination of the earth.

13   The mechanisms of evolution

Across billions of years of time, great changes have taken place in our universe. None are of as great interest to us as the changes that have led to the human species becoming the dominant organism on our planet. Until the last century, it was difficult for people to conceive of how this could have occurred other than by creation through a Supreme Deity. Not only human beings but all species of plants and animals on earth were believed to be products of separate, instantaneous creative acts, perhaps having taken place in as little as six days. Much of Western-world thinking on human origins and the rest of life stems from the Genesis account in the Bible, and strongly literal translators and interpreters in the Christian, Jewish, and Mohammedan faiths have consistently held to an exact acceptance of the Biblical creation accounts.

Yet there is considerable room in the language and context of the Genesis texts to see them as a broad and general overall account of God's creative actions in the universe, and indeed most modern scientists of these faiths see no conflict between the Bible and their viewing of evolution as a description of the Supreme Being's continuous creative activity in His universe. In other words, creation becomes an ultimate concern of God and evolution is a word describing the proximate activity that we can observe.

Thus much of the social-political turmoil and religious furor that greeted the publication of Darwin's *On the Origin of Species* has disappeared today (though somewhat curiously revived in the early 1970s in California, Michigan, and various eastern states by a group called the Creation Research Society). Indeed, the last hundred years have seen a remarkable fertile research effort into the various mechanisms of evolutionary change; and with the wealth of data available today, few question that evolutionary change does occur currently and has occurred in past eons of geologic time.

The principal questions that evolutionary biologists now address themselves to are centered on *how* evolutionary change occurs. The various theories of evolution thus center on the *process* of this change, not the *fact* of the occurrence of change, for it is believed that there is now sufficient evidence that biological change from generation to generation is a fact basic to almost all life on earth.[1] Let us look at the main types of evidences for evolutionary change that biologists have found and then consider some of the theories that have been proposed to account for the mechanisms involved in the process of evolution.

Evidences for evolution

FIGURE 13–1

Fossil evidence for evolution: (A) cycad leaves from the Petrified Forest in Arizona; (B) a fish from the Green River shales in Wyoming.

By far the strongest evidence for the past occurrence of evolution comes from the **fossil** remains of prehistoric plants and animals preserved in the earth's sedimentary rocks (Figure 13–1). As eroded particles of sand and silt were deposited layer by layer in fresh water or shallow oceanic depressions, over millions and millions of years,

[1] There are a few "living fossil" groups of organisms that have remained essentially unchanged for long periods of geologic time.

A

B

FIGURE 13–2

A comparison of several developmental stages in the vertebrates: a fish (shark), a marsupial mammal, and a human being (a placental mammal).

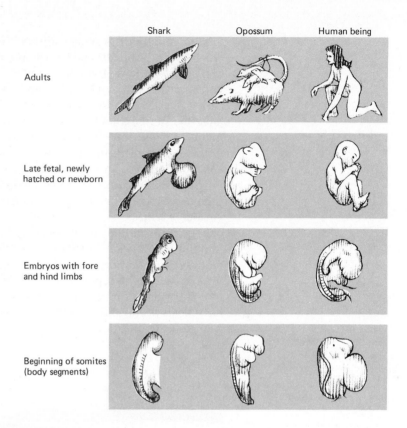

Shark   Opossum   Human being

Adults

Late fetal, newly hatched or newborn

Embryos with fore and hind limbs

Beginning of somites (body segments)

the dead bodies of organisms were occasionally entombed and the hard parts or impressions of hard parts survive today in the solid rock strata. Later uplift of these strata due to movements in the earth's crust have exposed these fossil-bearing layers in many currently terrestrial areas of our planet, where paleontologists (scientists who study ancient life) can extract them and re-create a picture of what life forms were present in past ages. Thus we have a remarkably broad picture through time of an estimated 2 billion years for the development of life from primitive single-celled, soft-bodied organisms in the early seas to the dominant, highly organized vertebrates and flowering plants of today. The gradual replacement of earlier primitive forms by more advanced forms in subsequently deposited rock strata has provided the best factual proof for Darwin's theory that evolution had occurred in past eras of geologic time, and this evidence is even stronger today than it was in Darwin's day.

Another kind of evidence supporting the concept of evolution involves the stages of development, or **ontogeny,** of the vertebrate embryo in the higher vertebrate groups (e.g., mammals). The earliest embryonic stages of the mammal show features of the more primitive vertebrate groups, and as development progresses the mammalian embryo appears to go through stages where it resembles a fish, amphibian, and reptile before it clearly becomes a mammal embryo (Figure 13–2). This recapitulation of earlier evolutionary stages during development can be explained by assuming that all vertebrates are best built in one developmental way, and hence it is not surprising that the body plan of the higher group is built or elaborated from the more primitive one during development.

A third kind of evidence for the occurrence of evolution has been the frequent observation of **subspeciation** (the formation of distinct races of a species in different geographical areas) in modern species of a great many plant and animal groups (Figure 13–3). Biologists have extrapolated from these observations the conclusion that in geographic race formation we are observing the beginning of the speciation process: that is, the splitting of a species into two or more distinct kinds that will continue to evolve in different ways until they become so far apart in characteristics that they can no longer be considered the same original species. The pattern of subspeciation in certain California salamanders (Figure 13–3) suggests this kind of evolutionary example. The development of domestic varieties, through human selection of breeding stock with particularly desirable characteristics, offers another model of how evolution can occur.

There are many more technical modern contributions to the evidence for evolution, including changes in chromosomes and other

FIGURE 13–3

The salamander Ensatina eschscholtzi *occurs in the Coast Range and in the Sierra Nevada mountains, encircling the Great Central Valley of California. From the Oregon border to Mexico, the coastal populations (E. picta to E. eschscholtzi) have a generally tan to reddish-brown, unblotched appearance. The interior Sierran populations exhibit an increasingly blotchy pattern, culminating in the black-and-yellow E. klauberi. Interbreeding occurs in areas of overlap, except where E. klauberi and E. eschscholtzi are adjacent. The lack of interbreeding between these two terminal races or "end" populations in the circular distribution of the salamander suggests to some biologists that evolutionary divergence has resulted in separate species there.*

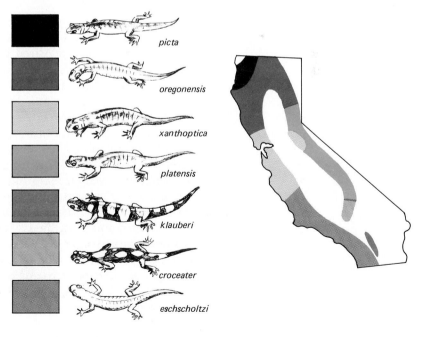

cellular constituents, endocrinology (the comparative study of hormones in animals), patterns in geographical distribution of groups, and correlations in behavioral patterns in groups of varying degrees of relatedness. In Chapter 5, for instance, we discussed the evolution of pesticide-avoidance behavior of tropical mosquitoes. But the principal "hard" evidence is the fossil record.

Three theories of evolution

When biologists say that they believe *evolution* is a fact, then, they mean that the biological world has apparently not been static and unchanging, but that over the course of geologic time there have been successive and indeed progressive (in the sense of being more advanced and complicated in structure) changes in living things on earth. These changes have been dramatic, as we shall see in the next chapter on the fossil history of life on our planet. By evolution they also mean that the frequencies of various genes in a species are often changing from one generation to the next. Here, however, we want to examine briefly the theories of evolution that *best* account for the changing process that has led to our present diversity of some 2 million living species. These theories of evolution can be grouped under three headings.

Natural selection

In 1858 Charles Darwin and Alfred Russell Wallace proposed in a paper that the phenomenon of natural selection best accounted for the progressive development of new species (Figure 13–4). *On the Origin of Species,* which contained hundreds of pages of data supporting Darwin's view, followed the next year. According to the concept of natural selection, the individuals of a species that are best adapted to their biological (competitors, predators, etc.) and physical environments will survive and hence pass their genes along to the next generation in their offspring. Thus geographically separate populations of a widely distributed species will often be exposed to differing climates or other environmental conditions, and natural selection will tend to cause the divergence of these populations as each becomes adapted to its own unique set of conditions. Also, for any particular population, the climate of the locality itself will gradually change over hundreds of years and natural selection will cause changes to appear in the former characteristics of that local

FIGURE 13–4

The English naturalists who were the codiscoverers of the theory of evolution by natural selection: (A) Alfred Russell Wallace (1823–1913) and (B) Charles Darwin (1809–1882).

A

B

population. If these changes become great enough, we say that the population has become a new species, quite different perhaps in descriptive aspects from the one that existed there some hundreds of generations previously.

Thus natural selection provides a mechanism by which differential survival of variant types in a population can occur. Those best fitted to the prevailing environmental conditions survive better and leave more offspring to form the next generation than the unsuitable potential parents. The basic idea of natural selection still guides much of evolutionary thinking. Yet while its existence seems well supported by many field and laboratory observations, there is still considerable current argument over its exact role and contribution to the speciation process, that is, the creation of a new species (or *kind*) of organism. The distinction is made by many between **macroevolution,** the appearance of new species in time, as we can observe in the fossil record, and **microevolution,** the development of new races (but *not* species) from already existing species through the gradual accumulation of genetic differences between geographically isolated populations. The latter process was and still is held by the so-called Neo-Darwinists to also account for macroevolution and, by extension back to earlier eras, higher taxonomic categories such as genera, families, orders, classes, and phyla (see Chapter 15). Since the Neo-Darwinian view is still present in almost all current textbooks, let us look at it in more detail.

The Neo-Darwinian theory of evolution

The great evolutionary biologists of this century—Julian Huxley, Ernst Mayr, George Gaylord Simpson, Alfred Sherwood Romer, G. Ledyard Stebbins, and Theodosius Dobzhansky—supported the natural selection theory of Darwin and said that the principal pathway of evolution was the evolution of new species by geographical isolation and accompanying gradual accumulation of small genetic differences. In other words, the same processes at work in microevolution were said to be responsible over longer periods of time for the evolution of new species and higher groups, that is, macroevolution. A great number of studies in the last hundred years have shown that natural selection is responsible for virtually all geographic or other genetically meaningful variations within species, the few exceptions being explained by genetic drift, the founder principle, and other formulations (Chapter 12). By simple (and logical) extension of these observations, it would seem that only the length of time required to create a new species has prevented us from observing actual speciation. Thus the attractiveness of the Neo-Darwinian view is great, and from the 1930s into the 1970s it has held sway as *the* best explanation for the evolution of new species (Figure 13–5). Only a few "radicals," considered to be on the fringe of rational biology, have questioned this interpretation. Foremost among these early questioning biologists was Richard Goldschmidt, who published in 1940 *The Material Basis of Evolution* in which he severely questioned that microevolution and macroevolution were two ends of the same spectrum. Instead, he observed that microevolution, while common in the natural world and even widespread through all major taxa, had never been actually shown to lead to the creation of a new species and hence could not be a process controlled by the same factors as macroevolution. To explain how macroevolution occurred, he invoked the idea of **saltational** ("jumping") evolution. He was severely criticized for his nonconforming views, but suddenly in 1973 these views received dramatic new support, albeit in a slightly different context. Let us look, then, at this saltational theory of evolution.

Saltational evolution

In Goldschmidt's original statement of this theory, he emphasized that "macroevolution cannot be understood on the basis of the Neo-Darwinian principle of accumulation of micromutations. This is true for the first step of macroevolution [the formation of species], and still truer when the higher categories up to phyla are concerned." Instead, Goldschmidt proposed that a macromutation was necessary

PLATE 11

THE GALAPAGOS ISLANDS: NATURAL LABORATORY
FOR ECOLOGICAL AND EVOLUTIONARY STUDIES

Since Charles Darwin visited the Galápagos Islands in 1835, this archipelago has been one of the most famous areas in the world to biologists. Located some 600 miles out in the Pacific, off the coast of Ecuador and northwestern South America, these remote islands have served as an invaluable resource of exciting data for generating hypotheses on the course and nature of organic evolution. Below, a pair of Blue-footed Boobies (Sula nebouxii) prepares to start a family on North Semour Island.

PLATE 12

The sparrowlike Darwin's Finches have provided a classic example of beak adaptation and evolution from the time of Darwin. Clockwise, we see here the Warbler Finch (Certhidea olivacea) on Tower Island, the Large Ground Finch (Geospiza magnirostris) on Tower Island, the Cactus Finch (Geospiza conirostris) on South Plaza Island, the Sharp-beaked Ground Finch (Geospiza difficilis) feeding on seeds on the ground on Tower Island, the Sharp-beaked Ground Finch feeding on ticks on marine iguanas at Punta Espinosa on Narborough Island, a brown female Small Ground Finch (Geospiza fuliginosa) in the bushes on Hood Island, and a black male Small Ground Finch against the blue sky on Hood Island. At far left, a large male land iguana (Conolophus subcristatus) surveys the cactus-dotted landscape of South Plaza Island, with the red cliffs of Santa Cruz Island in the background. The selection pressure exerted by herbivorous tortoises has caused some species of prickly pear cactus in the Galápagos to exhibit an arboreal, or treelike, growth form, such as this Opuntia species growing on Bartolome Island. Another endemic but low-growing cactus, Brachycereus nesioticus, grows in the volcanic lava cracks on Punta Espinosa.

PLATE 13

The brilliant breeding-season colors of the Hood Island race of the Galápagos marine iguana
(Amblyrhynchus cristatus) appear in the males of the
Hood population about early December and last for
several months while mating occurs and females prepare to lay their two eggs in burrows on the scarce
sandy beaches among the volcanic rocks. These
iguanas feed on marine algae and are capable of diving for up to an hour in depths of 30 to 40 feet. In the
seven insectivorous species of lava lizards (genus
Tropidurus) on the Galápagos Islands, the female
bears the brighter colors, as in this Tropidurus delan-
onis female on Hood Island (right). The giant
Galápagos tortoises were greatly reduced in numbers
by whaling and naval vessels provisioning their
crews with tortoises during the last several centuries,
but a few races still survive today, including the
Hood tortoise (lower right), with a high Spanish-
saddle-shaped carapace.

these views as incompatible with the findings of population genetics. Population geneticists theorized that a single mutation in one individual could only very rarely lead to a new population, simply because of the low probability of successful reproduction (it might or might not breed, it might or might not survive compared with its already well-adapted fellows, etc.). The only real exceptions in the Neo-Darwinian view were chromosome mutations and hybridization resulting in polyploidy (Figure 23–15), which at least in some groups of plants (particularly ferns and some flowering plant families) appear to have often led to the origin of new species. Goldschmidt argued in 1940 that these saltational changes had to occur only very infrequently to account for the evolution of new species and groups over the course of time (Figure 13–6). Nonetheless, the Neo-Darwinians carried the day for the next 30 years, not only by amassing great amounts of evidence showing microevolutionary trends in a host of species but by a series of brilliant books synthesizing their views of evolutionary theory. A whole generation of evolutionary biologists was trained under Neo-Darwinism, and a situation developed that has become an almost classic repetition in the history of science through the centuries: almost everyone both teaching and doing research in biology assumed that the Neo-Darwinians had adequately answered the problem of how to account for evolutionary change in the living world, and the orientation of their instruction and research proceeded from this presupposition.

To the great surprise of biologists, then, in August 1973 two leading exponents of the Neo-Darwinian revolution announced their dissatisfaction with this now-traditional explanation of the evolution of new species by the gradual accumulation of small genetic differences. The idea of saltational evolution as an explanation of macroevolution was revived independently by the population geneticist Hampton L. Carson, known in particular for his brilliant work on the evolution of fruitflies in the genus *Drosophila*, and Ernst Mayr, perhaps the leading evolutionist of our day and author of *Animal Species and Evolution* (1963), called the greatest landmark book in evolutionary biology since the publication of Darwin's *Origin of Species*.

Carson and Mayr, speaking at the First International Congress on Systematic and Evolutionary Biology at Boulder, Colorado, proposed that each species has a species-specific gene complex located in parts of its chromosomes (held together by inversions and other mechanisms) that contains the genes that govern the principal characters of that species. This gene complex is only occasionally broken up in sexual reproduction by unusually great (and very infrequent) relaxation of selective pressures in the environment (Carson's pri-

FIGURE 13–5

Natural selection and English moths in unpolluted and polluted woods. (A) Two adults of Biston betularia, *one typical and one melanic, resting on an unpolluted lichen-covered tree trunk. (B) Two adults of* B. betularia, *one typical and one melanic, resting on a soot-covered tree trunk near Birmingham. In less than a century, moth populations of this species living in polluted woods have become almost entirely melanic because of intensive selection by birds against moths that do not match their background.*

for this type of evolution, causing a major change in the body system of that particular organism (a "jump" to a new form in one generation—hence, "saltational evolution"), which "leads at once so far toward the new type that selection can immediately be efficacious, and which permits a large evolutionary process to take place in a time as short as, or even shorter than, is ordinarily required for the production of a subspecies." Hence, "species and the higher categories originate in single macroevolutionary steps as completely new genetic systems." Goldschmidt cited considerable developmental, genetic, morphological, and paleontological evidence as support for his views.

Simpson and the other Neo-Darwinian evolutionists attacked

FIGURE 13–6

A theoretical example of saltational evolution in birds.

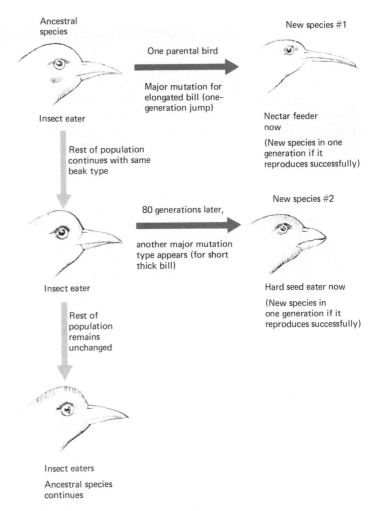

Ancestral species

Insect eater

One parental bird

Major mutation for elongated bill (one-generation jump)

Rest of population continues with same beak type

New species #1

Nectar feeder now

(New species in one generation if it reproduces successfully)

80 generations later,

another major mutation type appears (for short thick bill)

Insect eater

New species #2

Hard seed eater now

(New species in one generation if it reproduces successfully)

Rest of population remains unchanged

Insect eaters

Ancestral species continues

mary hypothesis) or by the introduction of sufficiently novel and strong selective pressures to which the species is exposed only infrequently over geologic time (Mayr's view). When this highly integrated gene complex is broken up in individuals of a particular population, evolution of new species (and, by extension, higher taxonomic categories) may result.

Under normal circumstances, however, this species-specific gene complex remains intact and is passed along from generation to generation. This accounts for the relative stability of the appearance of a species in widely separated populations (for example, an alpine butterfly species on a high summit in the southern Rocky Mountains, where a population was left by receding Pleistocene glaciation, and the same species still living north of the Arctic Circle in Alaska). Geographic variation and other forms of microevolution such as **polymorphism** (several forms present in the same population) are controlled by genes outside of this species-specific gene complex and hence can vary independently of the basic species characteristics. Thus, for instance, a mimetic *Heliconius* species can vary tremendously in wing coloration over its geographic range, resembling a different model species in each of a dozen different areas; yet its wing venation and other basic morphological characters remain constant and individuals from different populations that look totally unalike can mate and reproduce quite successfully if brought into contact in the laboratory (Figure 13–7).

It is too early to tell how this revival of saltational evolution will be received since further testing of the theory remains to be done. This section has tried to emphasize that the *theory* of evolution is still in an unsettled state and that no one has yet come up with a perfect, all-inclusive explanation of how the process of evolutionary change works. The *fact* of evolution seems well established by the fossil record and our observations of change in present-day populations; the main task faced by evolutionary biologists is to press on from this evidence to discover new insights into how evolution works; and to do this successfully, as in all fields, curiosity cannot be bounded by the biases of the past (plates 11–13).

FIGURE 13–7

The evolution of dramatically differernt subspecies in a tropical Heliconius *butterfly species.*

The process of adaptation

The thorniest problem in evolutionary biology is that of speciation. As we have seen, there is considerable difference of opinion over how this process of species formation is accomplished. Yet almost every biologist today agrees that the features of an organism — whether structural, physiological, or behavioral — exist for a reason, which is that they serve an adaptive function; that is, in some way each feature helps the individual organism (and hence the species) to survive. These *adaptations* to successful living in the environment are almost invariably found to have rather complex genetic and developmental control mechanisms. Almost any structure could be cited as an "adaptation"; otherwise, the species would not be expected to have retained it. (In fact, Paul Ehrlich and Richard Holm have suggested in *The Process of Evolution* [1963] that "adaptation" is a meaningless term for just this reason.) Nonetheless, we usually think of the more unusual survival features of organisms as a special category of adaptations, and the word has some descriptive content in this context. For instance, the differences in beak size and shape

FIGURE 13–8

A schematic representation of the relationship between bill structure and feeding habits in 10 species of Darwin's finches (Geospizinae) from Santa Cruz Island in the Galápagos.

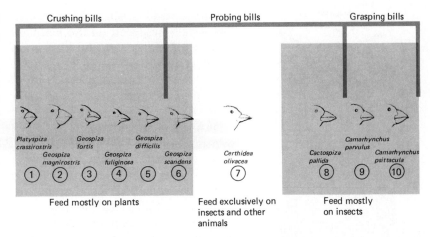

among the Darwin finches of the Galápagos (Figure 13–8) are a dramatic example of adaptations for different feeding and foraging strategies. The shapes of the carapaces of the different races of the giant Galápagos tortoise species show clear adaptation to their particular island environments.

Adaptation can be observed on all taxonomic levels, but we primarily think of it as referring to species and geographic variation within a particular species. The formation of **races** of a species occurs because geographically isolated populations diverge in adaptive characteristics in response to local environmental conditions. Thus a population of lizards living on light-colored sand dunes will gradually come to be distinctly light-colored and even sandy-textured in overall phenotype, because dark-colored individuals are more easily seen (and removed) by predators before they can breed, or even because the dark-colored individuals (in, say, the normal range of color variation in that lizard species) get overheated too easily by the desert sun. On the other hand, a population of the same lizard species that lives on an adjacent rocky plateau of black lava or other igneous rock will assume a dark adaptive coloration over time because of natural selection by predators of the more easily seen light-colored individuals in that colony.

When a species of organism first reaches a new environment through dispersal, as, say, to an oceanic island, there may be no or very few competitors present. In such cases, there is sometimes an almost explosive **adaptive radiation** of the original species into a diversity of new species, each adapted to a particular habitat and way of life. When the first female finch reached the Galápagos Islands from the mainland, or perhaps from Cocos Island off Costa Rica, it apparently found an insular environment empty of any similar land birds. Its descendants gradually assumed the roles of wood-

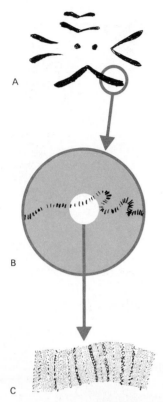

FIGURE 13-9

The chromosomes of a fruitfly, Drosophila melanogaster. (A) Chromosomes from a diploid body cell of a female; one of the four pairs of chromosomes is very small. (B) The giant salivary-gland chromosomes of a female, also in the diploid condition. (C) An enlarged drawing of a section of one of the chromosomes, showing the banding that represents particular genetic regions of the chromosome. Changes in the order of the genes, such as major deletions or inversions, can be readily detected under the microscope as shifted banding patterns.

peckers, warblers, sparrows, and other small ground and tree birds that would be found on the mainland but that were absent on the island archipelago. While this adaptive radiation was primarily evidenced in beak structure, rather than coloration or even general size, it was sufficient to result in some 13 specialist species that reach probably greater total population levels than they would if all were still like the original generalist finch.

The different adaptive forms that occur within a single population of a species is known as **adaptive polymorphism.** Each form of the species present in that population represents an adaptive advantage in some way, though one form may be best suited for conditions at a particular time of the year. For instance, in many *Drosophila* populations in the California mountains there are flies with several distinctly different chromosomal arrangements (Figure 13-9). One type will be predominant during the cool spring months; another type will be predominant later in the hot summer period. The different alleles controlling the constitution of hemoglobin on human red blood cells provides another example of an adaptive polymorphism. The person who is homozygous for normal hemoglobin has normally shaped red blood cells that can smoothly flow through capillaries, but is subject to malaria if bitten by an infected mosquito. The person who is homozygous for sickle-cell anemia has defective hemoglobin and the red blood cells are sicklelike in shape; they jam in the capillaries of the brain and elsewhere, usually causing death before the age of 12. The heterozygous person will not get sickle-cell anemia (although he or she may have some sickling of the red blood cells under conditions of low oxygen pressure, as at high elevations in the mountains); the heterozygote is also protected against the effects of malaria.

Adaptive coloration, still another dramatic outcome of the adaptation process, was discussed in detail in Chapter 7.

Isolating mechanisms

In this chapter we have been looking at ways in which genetic variants, once they arise, are preserved through their adaptive value for local conditions and are transformed into new geographic races or even new species. A reduction in interbreeding (i.e., flow of genes from one population to another) is necessary in this process. Otherwise, divergence would not occur, or if mutations did appear, they would be swamped by normal alleles and the two sets of variants would become identical again. Evolutionists have identified various types of **isolating mechanisms** that are involved in decreasing in-

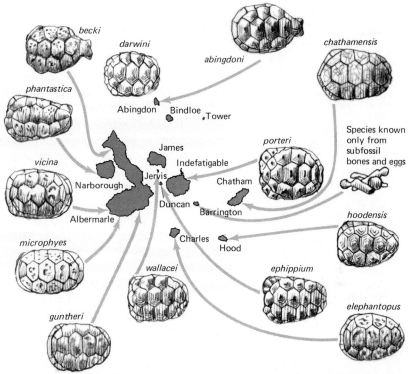

FIGURE 13–10

The distinctive races of the giant tortoise Geochelone elephantopus in the Galápagos Islands. There is a tendency toward a saddle-backed form on the northwestern islands and a dome-shaped carapace ("shell") on the central and southern islands.

FIGURE 13–11

The ecological isolation of two spiderwort species. (A) Tradescantia canaliculata normally grows on cliff tops, and T. subaspera at the foot of the cliffs. (B) Where the slopes are somewhat more gentle, the two species come together and hybridize in the intermediate habitat.

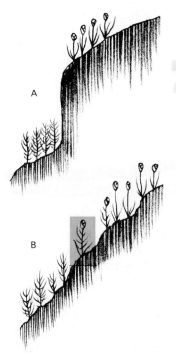

terbreeding between separate groups or organisms in the same geographic vicinity, and for the purposes of discussion we may group these under five categories.

1. *Geographic isolation.* Geographic isolation is simply *space* between populations and is usually a prerequisite for speciation—the formation of new species—in the traditional Neo-Darwinian view, as we have already noted. Also, of course, if two races or even two species are separated by a geographic barrier, they will not be able to interbreed and become one morphologically similar population again. Most of the 15 races of giant tortoises on the Galápagos Islands are distinctly different because of the geographic isolation preventing all but very infrequent interbreeding between the different populations (Figure 13–10). The Kaibab squirrel on the North Rim of the Grand Canyon is distinctly different from its close relative, the Albert squirrel, on the South Rim, and they are kept from interbreeding by the mile-deep canyon between them.

2. *Ecological (or habitat) isolation.* Ecological isolation and the three types of isolating mechanisms that follow operate between species living in the same area. In *ecological isolation*, two or more related species utilize different habitats in the same region (Figure 13–11). Different species of warblers may feed at different levels off

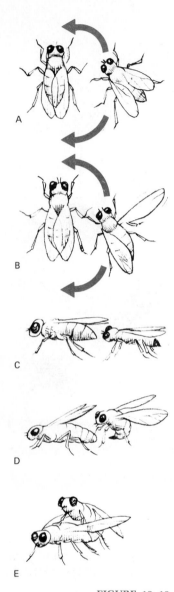

A

B

C

D

E

FIGURE 13–12

Species-specific courtship movements of the fruitfly Drosophila melanogaster *that result in behavioral isolation from other* Drosophila *species: (A) orientation, (B) vibration, (C) licking, (D) attempted copulation, (E) copulating pair.*

the ground in the same tree, and hence males and females of different species rarely come into direct contact with each other. Among the many aquatic species of turtles in Florida, some species are specialized to live in slow-moving rivers while others live in backwater sloughs or pools. Thus the habitat preferences keep the several species out of potential reproductive contact with each other.

3. *Seasonal isolation.* With seasonal isolation two species that might potentially be capable of interbreeding are prevented from doing so by having different breeding seasons. The timing of pollen production in plants is often different for related species living in the same area, or they may each have quite different pollinators (e.g., bees versus beetles) that never visit the other related plants in the region. Courtship and mating seasons in animals are often distinctly separate in related species.

4. *Behavioral (sexual) isolation.* Behavioral isolation involves factors that prevent the males and females of different species from responding to each other even if they come in contact while in reproductive condition (Figure 13–12). Elaborate *courtship displays* are part of the repertoire of many birds, such as the strutting of the gobbler turkey, the drumming and dancing of the grouse, or the spectacular feather displays of the birds of paradise. *Visual recognition* through colors or patterns often gives the instrumental cue to the individual of the opposite sex that this proposing "mate" is indeed the correct species. **Pheromones,** those ubiquitous chemical signals released by many animals (to be discussed in Chapter 19), are often involved in providing the key identification that this is the "right" species with which to court and mate. The *flashing light signals* of fireflies and *sound production* in birds, frogs, and a great many insects are behavioral isolating mechanisms that help sort out animals for correct pairings during the mating seasons.

5. *Genetic isolation.* If all other isolating mechanisms fail and two individuals of different species are about to mate or actually do mate, genetic isolating mechanisms come into play (Figure 13–13). Differences in the structure of the male and female genitalia, especially in insects, may prevent successful coupling and the insemination of the female. The sperm, even if introduced, may be unable to survive in the environment of the reproductive tract of a female of the wrong species. Anomalies in development of the embryo following fertilization will probably occur because of an unequal division of the genetic material at each mitotic division and the absence of half of the parental genes normally present from each parent individual in single-species matings. Finally, even if the hybrid develops to the adult stage, it is usually sterile because of meiotic failures (the chromosomes do not pair up or synapse properly, etc.). Thus the

FIGURE 13–13

An example of genetic isolation between two domesticated species: the Merino sheep and the milch goat.

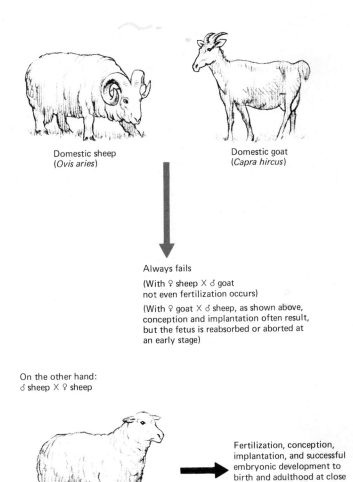

Domestic sheep
(*Ovis aries*)

Domestic goat
(*Capra hircus*)

Always fails

(With ♀ sheep × ♂ goat
not even fertilization occurs)

(With ♀ goat × ♂ sheep, as shown above,
conception and implantation often result,
but the fetus is reabsorbed or aborted at
an early stage)

On the other hand:
♂ sheep × ♀ sheep

Fertilization, conception,
implantation, and successful
embryonic development to
birth and adulthood at close
to 100% frequency under
controlled conditions

overall effect of these and the other isolating mechanisms is to prevent or at least greatly reduce the chances of diversity being decreased through hybridization.

Summary

The ultimate origins of human beings and other forms of life are unknown, although there is a wealth of evidence that evolutionary change occurs currently and has occurred in past eons of geologic time. Thus evolutionary scientists assume the fact of this change but still debate the mechanisms and processes that may bring about such change. The evidences for evolutionary changes in organisms come primarily from the fossil remains of progressively more complicated prehistoric plants and animals, the stages of embryonic development in higher vertebrates, the frequent observation of subspeciation or formation of distinct races of a modern species inhabiting different geographic areas, and the development of domestic varieties of plants and animals by selective breeding. The most widely accepted scientific theories put forth to account for the changing process of evolution include natural selection (in the original concept by Darwin and Wallace), the Neo-Darwinian theory of evolution (combining natural selection, geographic isolation, and the gradual accumulation of small genetic differences), and saltational evolution (major mutational changes toward a new type of organism). The Neo-Darwinian view dominated biological thinking for nearly 50 years (and still does), although recently new saltation-oriented theories have been advocated by evolutionists. No all-inclusive explanation is yet available concerning how the process of evolutionary change works to create new species. However, adaptive features of an organism are usually explained as part of the process of natural selection. Isolating mechanisms, such as geographic, ecological, seasonal, behavioral, and genetic isolation, help to maintain species distinctness.

14 The history of life on earth

Few biological topics arouse so much interest as the fossil history of life on earth. Our imagination and curiosity are stirred by viewing huge dinosaur bones and other evolutionary relics of the distant past, and even the ancient Greeks wondered about the animals and plants that left behind fossil remnants surprisingly similar to present-day life. Actually, as paleontologist A. Lee McAlester has pointed out, "In spite of such well-known exceptions as dinosaurs and trilobites, most of the principal kinds of life that have *ever* arisen are still living on Earth today. Far from being a strange progression of bizarre and exotic creatures, the history of life is a record of the persistent survival of familiar organisms found around us in our modern world." We shall look first at the types of evidence used to establish the historical record of organic diversity, then examine theories about the origin of life, and finally review the actual record itself down to the present day.

The geologic fossil record

The strongest evidence we have for the history of living things is the geologic record. With few exceptions fossils are restricted to sedimentary rock layers or strata, laid down in ancient lakes and ocean

FIGURE 14-1

Sedimentary rock layers exposed on an eroded coastal cliff in Mexico.

basins as sediments washed in from the surrounding land (Figure 14-1). The sediments that constitute most of these strata were deposited in shallow seas where marine animals lived in abundance. During the times of geologic emergence of these regions, plants and other land life occupied the broad lowland surface near bodies of water. Individuals of either marine or land species were occasionally buried and preserved as fossils, being embedded in the accumulating sediments and thus incorporated in the strata.

At least three-quarters of the known sedimentary strata are of marine origin, a not-surprising finding since sediment normally collects ultimately in the seas by land erosion and compared with sediment deposited on the land, sediment buried beneath the sea stands little chance of disturbance. While many of the fossil animals, mainly shellfish, resemble creatures of similar groups living today, most are at least different species. They are most radically different in the lowest and hence the oldest strata and least different in the highest and youngest layers that we find today. Because of these changes in form through time, traceable by the record of fossils left in the strata, it is possible to match up the *relative* ages of strata in different parts of the world. New forms of marine animals rapidly spread from their place of origin to other shallow areas. Hence today we can say that two strata that occur on two continents but contain similar fossils must be of similar geologic age. This inference became generally accepted by geologists and paleontologists a century before radioactive dating provided *absolute* dates for the strata (and simultaneously confirmed the inference). Because of this basic similarity, strata in different parts of the world can be matched or correlated, even though there exists no physical continuity between them today. Correlation by means of fossils has made possible a worldwide standard scale of strata in the form of a *geologic time scale* (Figure 14-2).

FIGURE 14-2 Geological time scale

Periods	Duration of interval (in millions of years)	Beginning of interval (millions of years ago)	Appearance of life forms
Cenozoic era			
Quaternary			Age of mammals. Rapid evolution of horses and numerous other mammals. Flowering plants, grains, grasses, and cereals abundant.
Pleistocene epoch	1	1	
Tertiary			
Pliocene epoch	12	13	
Miocene epoch	12	25	
Oligocene epoch	11	36	
Eocene epoch	18	58	
Paleocene epoch	4	63 ± 2	
Mesozoic era			
Cretaceous	70	135 ± 5	Deciduous trees. Dinosaurs common, but extinct by end of period.
Jurassic	45	180 ± 5	Reptiles abundant. First true mammal and first true bird (Archaeopteryx) appeared.
Triassic	50	230 ± 5	Amphibians attained maximum size. Dinosaurs first appeared during the late Triassic. Forests resembled modern evergreen forests.
Paleozoic era			
Permian	50	280 ± 10	Conifers first important. Reptiles abundant. Numerous groups of invertebrates became extinct.
Pennsylvanian (upper Carboniferous)	25	305 ± 10	First reptiles. Age of insects. Vegetation abundant.
Mississippian (lower Carboniferous)	40	345 ± 10	Age of crinoids (sea lilies). Blastoids and foraminifera abundant.
Devonian	40	405 ± 10	First amphibians and lungfishes. Age of fishes. Primitive seed ferns, scale trees, corals, brachiopods, and echinoderms common.
Silurian	20	425 ± 10	First land plants and first air-breathing animals (scorpions). Eurypterids abundant.
Ordovician	75	500 ± 15	Bryozoa, corals, and first known vertebrates (primitive fishes or ostracoderms). Graptolites common.
Cambrian	100	600 ± 20	At beginning of period life already at high stage of development. Animals already developed into many phyla: trilobites, brachiopods, algae, cytoids, worms, sponges, gastropods, and coelenterates. Graptolites.
Precambrian era			
		3,500	Soft-bodied life, poorly preserved in the fossil record.

The geologic time scale is divided into four great eras that have been subdivided into periods and epochs as well as smaller units, all reflecting the proper relative positions of the known strata in an idealized cross section of oldest to youngest sedimentary records. The periods and other units of the time scale were originally named for layers of rocks that contain characteristic fossils, but their names are also the names of the corresponding units of radioisotope-determined time. Thus the Devonian strata were deposited during the Devonian period, which lasted from about 405 to 345 million years ago, or approximately 60 million years.

Fossils are formed in sedimentary rock in several ways (Figure 14–3). (1) Preservation of hard parts of the original animal occurs by *mineral infiltration* of interstitial spaces where living tissue has decayed. This may happen by *impregnation* of the original organic material (especially porous substances such as skeletal elements of vertebrates) or by actual *replacement* of the original organic molecular material by another substance (usually silica or calcium carbonate). (2) Fossils may be formed by *impressions* on a rock surface, where all traces of the animal are gone and only an imprint (e.g., a footprint) remains. (3) Sometimes fossils are represented by *a volume of space in rocks*, where an animal trapped in a forming layer of sediment or other rock material decays; minerals infiltrate this natural "mold" and a three-dimensional *cast* of the animal is formed. (4) Fossils may be found in the form of a *dark film of carbonaceous material*, where carbon (usually in plant leaves and stems) is left on the surrounding rock material when the plant decays. The concentration of carbon into coal formations is a similar type of carbonization. (5) A rarer kind of fossil is found in *amber*; small animals, especially insects, become embedded in the pitch, or externally dripping sap, of conifers, and this material is later exposed to the right conditions to change to the mineral amber. The animal, sealed off from the outside air, remains unchanged and even the internal anatomy may be studied thousands of years later. (6) The rarest fossils are those of *actual animals* frozen in extremely cold regions, especially in the deep permafrost soil layers or glacier crevasses of the far northern Arctic or high alpine mountains. In Siberia and Alaska baby and adult mammoths have been found frozen with undigested buttercups and grasses still fresh in their stomachs and frozen flesh preserved well enough to cook and eat.

In general, then, the fossil evidence of life from about 600 million years ago to the present is comprehensive in variety and worldwide sampling and seems sufficiently well dated by radioactivity and geologic evidence to establish the geologic time scale of fossils on a reliable basis.

FIGURE 14–3

Fossils are formed in a variety of ways: (A) mineral infiltration preserved part of the skeleton and plated skin of a Trachodon dinosaur in the Cretaceous period; (B and C) three-dimensional casts of trilobites and ferns are common in certain sedimentary beds; (D) impression fossils are represented here by dinosaur tracks at Rocky Hill, Connecticut; (E) a dark carbonaceous film traces the fossilized parts of a fern frond; (F) fossilized resin preserves a tiny fly in golden amber from the Baltic Sea of the Eocene—Oligocene epochs; (G) a well-preserved baby wooly mammoth, dug out of the frozen Alaskan tundra and preserved today in a freezer at the American Museum of Natural History.

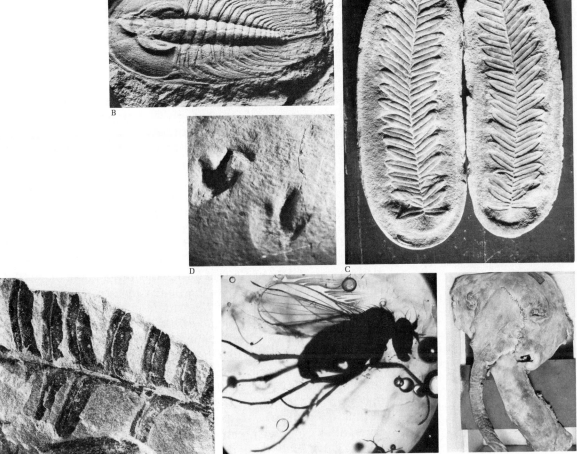

Let us look now at the major events in the history of life and the issues that are important to our understanding of the evolutionary record, as well as those that are still undecided and controversial among biologists.

The origin of life

Probably more words have been written on the origin of life by scientists with fewer facts and direct evidence at hand than on any other topic in biology. The facts we have to go on are mainly those derived by inference from *astronomical* observations of the atmospheres of nearby planets, the *spectroscopic* analysis of comets and certain rare meteorites containing carbon, and *biochemical* studies of chemical reactions under conditions alleged to be similar to those on the primitive earth. The only *direct* evidence of the nature of the earliest life on earth is a few Precambrian era fossils (earlier than the Cambrian period, which began about 600 million years ago). These fossils include single-celled and filamentlike algae and bacteria and are dated as far back as 3.1 billion years ago.

The most widely accepted astronomical theory for the origin of the earth says that the solar system formed about 4.6 billion years ago from a diffuse dust cloud; the central portion condensed to form the sun and areas in the outer cloud condensed to form the planets (Figure 14–4). In order to evolve by natural means even the simplest bacterial cell in 1 billion years or less, a host of assumptions and hypotheses about the nature of the process and early terrene conditions must be met. While a detailed discussion is beyond our needs here, it is worth reviewing the general scheme of events that has been proposed by various recent workers in the field of molecular evolution.

It is presumed, first of all, that during the formation of the earth by condensation of dust particles the lighter elements and molecules such as gases were lost from the earth's gravitational field. Thus most of the carbon and nitrogen now on the surface of the earth and present in the earth's atmosphere must have volatilized from the earth's interior because of high heat evolved from the decay of radioactive elements. Some hydrogen was present, mostly in the form of water vapor. Thus the primitive atmosphere is believed to have been *reducing* in nature, with no free molecular oxygen present until photosynthetic organisms appeared.

Operating from these premises in the 1920s and 1930s, the English biologist J. B. S. Haldane and the Russian biochemist A. I. Oparin, pioneer workers on the origin of life, put forth the following general

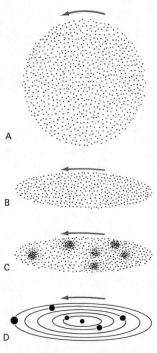

FIGURE 14–4

The hypothetical formation of the solar system from a cosmic dust cloud. (A) The spherical cloud has collapsed to a disk. (B) The material in the dust cloud is aggregating to form larger objects. (C) Aggregation is continuing and the objects are condensing to form the planets. (D) The aggregation is complete and the solar system has formed.

FIGURE 14-5

Two scientists widely known for their theoretical work on the origin of life: (A) A. I. Oparin and (B) J. B. S. Haldane.

A

B

High-voltage electric current

Spark

Water
Ammonia
Methane
Hydrogen

Condenser

Boiling water

Water containing amino acids

FIGURE 14-6

The laboratory apparatus used in Miller's experiments on the formation of amino acids from a reducing gas mixture.

hypothesis, with some modification of details by later workers (Figure 14-5). The simple reactive molecules that were to take part in further organic syntheses were formed in the atmosphere, mainly by the action of electrical lightning discharges and ultraviolet solar light. Enough energy was available to produce very large amounts of organic material in the period from perhaps 4.6 billion to 3.6 billion years ago. These new organic molecules probably included formaldehyde, hydrogen cyanide, aldehydes, nitriles, acetylenes, and amino acids—simple molecules that can be produced in a laboratory flask containing water, ammonia, methane, and hydrogen (Figure 14-6). A graduate student at the University of Chicago, S. L. Miller, first performed just such an experiment in 1953 and discovered that the passage of a high-energy electrical spark through this mixture created amino acids, the building blocks of proteins.

The next hypothetical step does not have such modern experimental evidence behind it. While some larger molecules may have been formed in the atmosphere, the synthesis of most of the more complicated organic molecules probably had to occur in aqueous solutions. Hence reactions between various organic compounds that were dissolved in surface waters supposedly led to the formation of carbohydrates and proteins, and perhaps even early amino acids. At these times, then, the waters of the oceans and inland lakes would have resembled a dilute organic soup.

Now the Oparin-Miller hypothesis assumes the formation of protein-coated droplets in concentrated tidepool areas of this "soup." Coacervate droplets (aggregates of protein with a boundary layer mainly held together by surface tension) would somehow evolve membranes that allowed the selective passage of organic compounds into the droplet ("feeding") and hence stimulated the development of metabolism, that is, energy-transfer reactions. The evolution of metabolism meant these heterotrophic "organisms" could free themselves from the necessity to feed on preformed organic molecules in the concentrated prebiotic soup. Instead, like modern bacteria, they could now synthesize all the biochemical compounds they needed from a few, very simple starting materials in the soup. The later evolution of photosynthesis made possible complete independence of organisms from the prebiotic soup; the consumption of free oxygen produced by photosynthetic algae and other plants allowed a significant improvement of energy production in both these plants and the heterotrophic animals (see Chapter 25). Up to this point, the droplets still formed in a random manner, with no continuity of structure. Life was awaiting, as Miller puts it, the formation of a substance that could be capable of catalyzing the synthesis of protein enzymes according to a *set pattern*. The nucleic acids provided this mechanism — though how their complex regulatory activity (Chapter 24) originated, few have dared to even hazard a guess.

Clearly, no biologist will ever know the complete story of the origin of life because of the lack of opportunity for us to return in geologic time for direct observation of the relevant events. The outline presented in the preceding paragraphs is merely the most generally accepted present hypothesis for the biochemical events that could have been involved in a proximate sense. A number of biologists feel that these hypotheses are compatible with a belief in a Supreme Deity; others hold to a purely mechanistic viewpoint. Neither point of view can be proved by physical evidence; thus research in this area is intellectually interesting and stimulating but without promise of a definitive answer. However, the history of life starting about 600 million years ago becomes abundantly documented by fossil evidence, and we can turn to that subject with considerably more confidence in the factual basis of our study.

The Cambrian and Ordovician seas

Most of the major groups of animals and plants originated in the sea in Precambrian times, more than 600 million years ago (Figure 14–2). In fact, life was restricted to the oceans through the Cambrian and Ordovician periods of the Paleozoic era, ranging down to 425 million years ago. It was not until the Silurian period (425 to 405 million years ago) that the first marine life came out onto the barren land surface and made the transition to a terrestrial existence.

At the beginning of the Paleozoic era, the Cambrian seas already swarmed with life. Astonishingly, all but one phylum of invertebrate animals with preservable hard parts first appear in the fossil record at or near the boundary between the Precambrian and Cambrian. The lone exception is the phylum Bryozoa (moss animals; see Chapter 15), which first appears in the sedimentary fossil beds near the middle of the Ordovician period. The major, phylum-level body plans of invertebrate life were apparently determined in one giant evolutionary radiation during late Precambrian and early Cambrian time. From all indications, the common ancestors from which they arose either lacked preservable hard parts or occurred so rarely and locally that they are not preserved in known Precambrian rocks. This first geologic era, the Precambrian, occupies about 85 percent of geologic time (Figure 14–2), and thus there was considerable opportunity for geologic activity that could have destroyed most early fossil evidence. Since then, there has been extensive evolutionary change *within* each of these phyla, but, with the exception of the Bryozoa, no appearances of new invertebrate phyla.

During the Cambrian and Ordovician periods, representatives of many primitive *classes* within each phyla filled the oceans (Figure 14–7). Later, these classes mostly became extinct and were replaced

FIGURE 14–7

(A) The Cambrian seas teemed with marine invertebrates such as the early trilobites, echinoderms, and coelenterates. (B) In the Ordovician seas (475 million years ago), a huge predatory cephalopod (Endoceras) is surrounded by smaller Endoceras, snails, trilobites, colonial corals, and other invertebrates.

A

B

by more successful classes that dominate our modern oceans. Thus the strange *trilobites* of the phylum Arthropoda (see Figure 14–3) dominated the early seas, representing 60 percent of all Cambrian fossils, whereas the major modern arthropod class of crustaceans was represented by only a few shrimplike species in Cambrian seas. It appears that the Cambrian period was a time of experimentation within the major phyla of invertebrate animals; many relatively short-lived and unsuccessful groups arose. The Ordovician period, which followed, was a time of secondary radiation and modernization that led the best-adapted of these early groups to evolve into more successful and longer-lived classes. By the late Ordovician, most of the invertebrate classes that dominate the seas today were well established, and very few classes became extinct throughout the rest of the history of life.

The conquest of the land

Life originated in the sea and most major groups of algae and invertebrate animals (and even fish among the vertebrates) are still predominantly or exclusively marine. It was not long after the great radiation of marine life in early Paleozoic times, however, that the first plants and animals began to colonize the lifeless surface of the land, for the oldest land fossils are found in Silurian rocks (425 to 405 million years ago).

The problems faced by early marine organisms in making the transition to terrestrial life were many and formidable. Animals and plants living in the sea have an inexhaustible supply of water, whereas land-dwellers must obtain water from rain, streams, soil, or the food they eat. To prevent the evaporation of the water they get, land organisms must have tough, relatively watertight coverings, such as the waxy surface coating of leaves or the horny skin of reptiles. They must also employ special membranous structures for breathing the oxygen of the atmosphere rather than absorbing oxygen from the surrounding water. Reproductive activities present still other kinds of problems. In the sea or fresh water, organisms normally release their gametes (eggs and sperm) directly into the water, where fertilization occurs on contact. On land, special adaptations are required to bring about the meeting of the delicate single-celled gametes without desiccation. Yet despite these considerable difficulties, there were distinct advantages to colonizing the land. For green plants, there were large areas of unoccupied space with abundant direct sunlight for photosynthesis. For animals, there was abundant free oxygen to breathe in the atmosphere and, after land

FIGURE 14-8

At the end of the Silurian period, swamps and coastal marshes were already occupied by primitive low-growing vascular plants. The tallest plants in the foreground are Psilophyton, and on the left side near the water are the forerunners of the present club mosses.

FIGURE 14-9

A Silurian eurypterid, an early marine ancestor of the land scorpions and spiders.

plants began spreading from the ocean margins, an almost limitless supply of food.

The first land plants most probably originated from green algae living in coastal marshes (Figure 14–8). Exposed to periodic drying, these plants would be under intense selection for genotypes capable of withstanding the desiccating effects of the atmosphere. A vascular system for transporting water and nutrients from the soil was an essential development for land plants. In water, nutrients could enter the plant by absorption over the entire body surface, but land-dwelling plants developed roots specialized for this function. Likewise, the chlorophyll-bearing leaves and supporting stems were developed for allowing photosynthesis at considerable heights above the ground, and a waxy surface cuticle prevented loss of water from the surface of these exposed plant parts.

The first animals that reached the land in Silurian times resembled our modern scorpions (Figure 14–9), members of the great phylum Arthropoda ("joint-footed" animals). While many phyla of invertebrate animals have become adapted to living in rivers and lakes, only two—the mollusks and the arthropods (e.g., sow bugs, insects and spiders)—have representatives that are fully adapted to life out of water. Even among the mollusks, only gastropods (single-shelled animals) have made the transition to become "land snails." The arthropods have been by far the most successful land invertebrates. Of these, some fully terrestrial crustaceans have evolved, but it is two other major arthropod groups—the arachnids (spiders and scorpions) and especially the insects—that are the invertebrate masters of the land.

The success of both gastropods and arthropods in invading the land was largely due to physical characteristics already present in their aquatic ancestors (Figure 14–10). Gastropods had a watertight shell into which they could withdraw for protection against drying out. Likewise, arthropods had a tough, waterproof, external skeletal covering. With their mobility, aquatic gastropods and arthropods could have left the water for brief periods and scavenged along the shores of prehistoric seas for detritus from marine life. As land plants became abundant, selection would have increased the respiratory and other capabilities of these animals to move about independently of water in search of food on land.

The oldest known land animals are Silurian scorpionlike arachnids, but both insects and arachnids apparently first became really abundant in the Carboniferous period (345 to 280 million years ago). The vast coal deposits of this period sometimes include associated shales with large fossil insect faunas, both primitive flightless forms and winged insects. One dragonflylike species had a wingspread of over two feet. Today insects make up over three-fourths of all living species of organisms.

FIGURE 14–10

Typical arthropods and a gastropod: (A) a giant desert centipede (Scolopendra heros), (B) a land snail (gastropod), (C) a box crab (Calappa species), and (D) a Chinese praying mantis (Tenodera sinensis).

The rise of the vertebrates

In the Middle Ordovician rocks are found bone fragments of primitive fish, the earliest fossil vertebrates. They arose from some sort of invertebrate ancestor, but, as is so often the case, the exact ancestral group is uncertain because of the absence of relevant intermediate fossils. On indirect evidence, several present-day soft-bodied marine animals offer a clue to the probable ancestral type for the vertebrates. These animals have a backbonelike structure starting in an early stage of their life history; unlike the backbone of the vertebrates, however, it is not divided into separate segments, or *vertebrae.* Instead it is a solid rod, called a **notochord,** made of stiff, gelatinous organic matter. A similar solid notochord is found in the early embryological development of modern vertebrates, but it is replaced by separate vertebrae in later stages. Unfortunately, these notochord-bearing animals have almost no fossil record, although three different kinds are found in modern seas. The three groups, along with the vertebrates, make up the four subphyla of the phylum Chordata —animals with backbones. The three notochord-bearing subphyla can be thought of as *invertebrate chordates* since they have a backbonelike structure but lack separate vertebrae (Figure 14–11).

In a way not yet understood, an invertebrate ancestor of the chordates—perhaps the larval stages of echinoderms such as starfish and sea urchins, which resemble closely the larvae of certain invertebrate chordates—developed a notochord. This notochord in the ancestral vertebrates and the internal skeleton that developed from it were key adaptations that paved the way for the extraordinary evolutionary success of the vertebrates. In almost all invertebrate animals

FIGURE 14–11

Invertebrate chordates: (A) sea squirts, or tunicates (Polycarpa species), and (B) an acorn worm, or hemichordate.

A

B

the supporting skeleton is external, a condition that places great limitations on the strength of the skeleton and the ultimate size and mobility of the animal. In contrast, the internal skeleton of the vertebrates served as an excellent strengthening support system, yet it could remain relatively light and flexible because it did not have to cover the entire exterior of the body. The internal bone skeleton, and the efficient muscle and nervous systems that developed with it, permitted the evolutionary development of many large and mobile vertebrate animals that are without parallel in the invertebrate world.

The age of fishes and the origin of land vertebrates

The earliest vertebrates, from which all others arose, were primitive fishes (Figure 14–12). Indeed, four of the eight classes of vertebrates are fishes (jawless fishes, armored fishes, sharks, and bony fishes), and today, as in the past, fishes far outnumber land-dwellers in numbers of both species and individuals. They remained a relatively insignificant group, however, until the end of the Silurian period,

FIGURE 14–12

Certain primitive fishes during Devonian times developed the ability both to swim and to wiggle across land seeking water during droughts. The early lobe-finned fishes, called crossopterygians, developed from the primitive bony-plated fishes, the placoderms. Some had simple lungs.

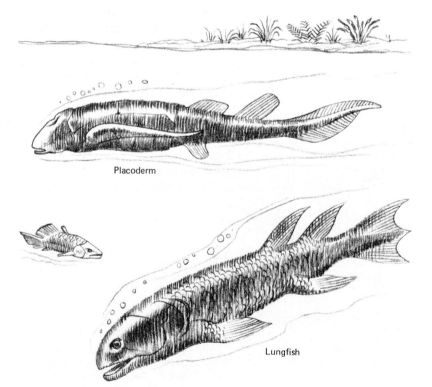

Placoderm

Lungfish

FIGURE 14-13

Lobe-finned fishes, such as this Devonian crossopterygian (Eusthenopteron), were the ancestors of land quadrupeds, the labyrinthodont amphibians.

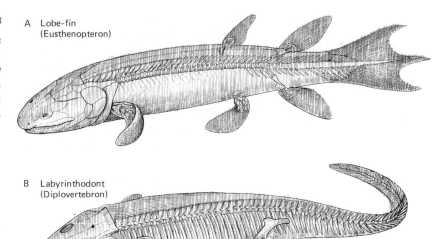

A Lobe-fin
(Eusthenopteron)

B Labyrinthodont
(Diplovertebron)

when fishes began an explosive evolutionary radiation that lasted throughout the succeeding Devonian period (405 to 345 million years ago). Since most of the evolutionary expansion and replacement of the earlier fishes took place during this period, a key time in fish evolution, the Devonian is sometimes called the Age of Fishes.

The Devonian was also the time when vertebrates first made the transition to land, for the oldest fossil amphibians are found in late Devonian rocks of Greenland and eastern Canada. In adapting to life on land, these early vertebrates faced the same problems of reproduction, water retention, and oxygen respiration that were solved in the Silurian period by plants and invertebrate animals. In addition, however, they faced a unique problem of locomotion on land. Land plants are essentially immobile, of course, and the first land-dwelling invertebrates were arthropods and snails that originated from marine ancestors that were already adapted for moving on the solid surface of the sea floor by means of legs or a muscular foot. Fishes, on the other hand, were adapted by fins and musculature to a swimming mode of life. Hence a profound modification was required for life on land.

The early lobe-finned fishes had the potential for developing a means of locomotion on land. Their bone arrangement permitted the fins to move freely at their point of attachment to the body; in addition, and also unlike the other fishes, the muscles extended *into* the fin to allow precise control of fin movements (Figure 14–13). This lobe-finned arrangement, with its complex muscles, was ideally suited for development into elongated, flexible limbs to support and move the animal on land.

Another adaptive feature of the lobe-finned fishes that fitted them for life on land was their set of auxiliary lungs. During dry seasons, they could wander between drying fresh-water streams and ponds, using their lungs to obtain oxygen directly from the air when the water became deficient in oxygen because of stagnation or evaporated from their home areas. Such lungs, which are also found in the living lobe-finned fishes (the lungfish of Africa, Australia, and South America), made possible life on land. As a means of survival from drying ponds or as a way of escaping aquatic ponds, or perhaps even because of abundant food on land, terrestrial life proved a useful advance, and the amphibians arose. Thus in the Devonian period we find a group of lobe-finned fishes, the *crossopterygians*, that gave rise to the land-dwelling amphibians. Because of the fossil skull and skeleton similarities, and the modification of the lower fins into stubby limbs for walking on land (Figure 14–13), the crossopterygian-amphibian transition is one of the best documented in the entire fossil record.

Although they flourished until the end of the Cretaceous period, crossopterygians are unknown from Cenozoic rocks and were long thought to be extinct. In 1938, however, a South African fisherman caught a living specimen off Madagascar, a so-called coelacanth, which was later named *Latimeria* (Figure 14–14). Since then, several dozen have been obtained, and research on these amazing "living

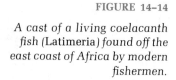

FIGURE 14–14

A cast of a living coelacanth fish (Latimeria) found off the east coast of Africa by modern fishermen.

fossils" has helped to clarify the structural features of the Paleozoic crossopterygians that gave rise to the amphibians.

The amphibians rapidly expanded and diversified during the Carboniferous period, when most of the world's land areas were tropical in climate, low in elevation, and covered by lush wet forests. The vegetation of these great swamps comes down to us today preserved largely in the form of coal. These early amphibians, called *labyrinthodonts* (for the labyrinthlike structure of their primitive teeth), gave rise to the surviving modern amphibians (frogs, toads, newts, salamanders, and their relatives) in the early to mid Mesozoic era. In the Carboniferous, however, most of them looked something like fat, stubby-nosed alligators; some were as large as 10 feet in length. Throughout the Carboniferous period the labyrinthodonts were the dominant land vertebrates, but they lacked one important solution to permanent land life: reproduction apart from water. Even present-day amphibians must return to the water to reproduce, for their delicate membrane-covered eggs quickly dry out and die if deposited in the air. It remained for the reptile descendants of the labyrinthodonts to solve the reproduction problem by developing eggs with tough outer coverings to prevent drying.

Reptiles and mammals

By late in the Carboniferous period the first reptiles had evolved from closely similar amphibian ancestors. During the succeeding Permian period (280 to 230 million years ago) the reptiles went through an explosive evolutionary radiation that began their long dominance of the land, a dominance that was to last throughout the great Mesozoic era (230 to 63 million years ago). As the numbers of reptiles expanded in the Permian period, the less efficient amphibians began to decline in importance. Labyrinthodont amphibians were extinct by the close of the Triassic period (180 million years ago); since Triassic time the principal amphibians have been frogs and salamanders, which have survived to the present day by remaining small and living in wet tropical forests and around lakes, rivers, and streams.

The transition between amphibians and their reptile descendants is somewhat obscured in the fossil record because the principal differences between the two types of vertebrates is not in skeletal features but in mode of reproduction. Reptiles overcame the problem of reproduction on land with the shelled egg, a device for allowing the embryo to grow in its own self-contained liquid environment (Figure

FIGURE 14–15

The first reptiles arose from forms such as **Seymouria**, *whose body structure displays features typical of both amphibians and primitive reptiles. About two feet long,* **Seymouria** *had sharp teeth and probably lived on any smaller animals it could catch on land or in the water. The first fossil remains of this animal were found in 1901 near Seymour, Texas.*

14–15). After development is completed, the animal breaks out of the shell as a small but otherwise fully formed and independent animal. Many early reptiles are believed to have retained the semi-aquatic habits of their amphibian ancestors, and thus it is possible that the shelled egg, which must be deposited on land, first evolved as a means of avoiding loss of the developing larval stages to fish, giant arthropods, and other aquatic predators. Nevertheless, the reptilian shelled egg also freed land vertebrates from the necessity of living near large bodies of water and ultimately permitted them to wander freely over the land surface in search of food and favorable habitats.

The most dramatic descendants of the early reptiles were the dinosaurs, which dominated Jurassic and Cretaceous landscapes during the Mesozoic era (Figure 14–16). In fact, the reptiles so thoroughly occupied the air, land, and aquatic habitats in the Mesozoic that this era has been aptly called the Age of Reptiles. The dinosaurs evolved many specialized herbivores and relatively few carnivores, which preyed on the herbivores. Herbivorous dinosaurs included the huge, long-necked *sauropods*, the largest land animals ever to evolve. They probably spent much time feeding on aquatic vegetation in swampy rivers and lakes, where their huge bodies could be buoyed up by water. The herbivorous *horned dinosaurs* developed the greatest defensive weapons. They had huge skulls with long, stout horns that must have been very efficient weapons for defense against the great carnivores of the Mesozoic. The bizarre *stegosaurs* had tiny heads and large, bony plates running down the spine; the related

FIGURE 14–16

A diversity of dinosaurs filled the rich savannas, swamps, and forests of the Mesozoic era. For example, Stegosaurus, a 20-foot-long armor-plated vegetarian, had a double row of thick, bony plates running down the back and possessed a short, thick tail armed with two pairs of long, bony spikes; Brontosaurus ("thunder-lizard"), one of the biggest land animals that has ever existed, weighed about 20 tons and reached a length of 60 feet.

The discovery of a living fossil

In 1938 a South African fishing boat, dredging deeper than usual off East London, southwest of Madagascar, brought up an unfamiliar fish—a big, deep-bodied, five-foot fish with large, bluish scales. The specimen was brought back to port in South Africa, and since it was decaying, it was skinned, stuffed, and mounted. When a zoologist from a nearby college saw the fish, he was as surprised as if a dinosaur had been resurrected before his eyes. To the astonishment of the scientific world, this unknown fish proved to be a surviving coelacanth, *Latimeria*, a lobefin fish belonging to a group supposedly extinct since the days of the Mesozoic dinosaurs.

Little of the original animal remained for scientific perusal except the skin and outer skull bones. The outbreak of World War II prevented scientists from immediately searching for more specimens. Once the war ended, however, the scientist who originally saw the first coelacanth, Professor J. L. B. Smith, circulated printed handbill descriptions among fishermen all along the East African coast. In the deep waters off the coast of the Comoro Islands, north of Madagascar, the coelacanth home was discovered. Soon a number of living specimens were caught and the physiology of some was even studied.

These present-day "living fossils" do differ in some specialized ways from their earliest ancestors. The coelacanths of the fossil record were freshwater fishes, complete with lungs for movement on land or in very shallow water. The living coelacanths have shifted not only to salt water but to the deep-sea regions, and as marine fishes for millions of years, lungs would be useless for breathing purposes. Yet the modern *Latimeria* still retain the lungs as a pair of ventral outgrowths from the throat, much in the fashion of their fossilized forebears.

FIGURE 14–17

One of the great predatory dinosaurs of the Cretaceous period, Tyrannosaurus *("tyrant-lizard") was the biggest biped dinosaur that ever existed. About 40 feet long and 17 feet high, it was carnivorous, and its great jaws and daggerlike teeth were well equipped for tearing flesh. Footprints as well as fossil skeletons have been found. One footprint was 28 inches long and over 31 inches wide; the distance between strides was 12½ feet.*

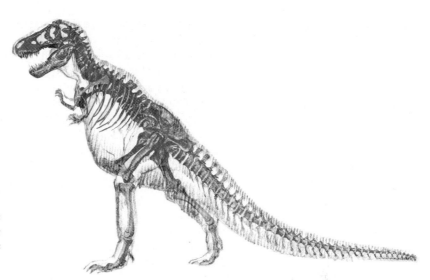

ankylosaurs were low and short-legged, with a more complete covering of bony armor. Both of these latter groups also developed peculiar weapons at the end of the tail: long spikes in the stegosaurs and a heavy, bony, clublike structure in the ankylosaurs. The great *theropods* were the ruling carnivores, and the familiar *Tyrannosaurus rex* was the most awesome flesh eater ever to evolve (Figure 14–17).

Almost unnoticed among the thousands of species of Mesozoic dinosaurs were the first true mammals. During Permian and much of Triassic time the dominant land vertebrates were the *mammallike reptiles*, an abundant and diverse group that preceded the dinosaurs and gave rise to the mammals in late Triassic time (Figure 14–18). Just as early reptile fossils are difficult to distinguish from their am-

FIGURE 14–18

Moschops, a mammallike reptile of the Permian period, was herbivorous and lived along South African rivers. More than six feet in length, this therapsid had a third eye, sensitive to light, on top of its large, wide skull.

FIGURE 14–19

A fossil Archeopteryx, *the earliest known bird.*

phibian ancestors because the two groups differed primarily in mode of reproduction, the most significant differences between reptiles and their mammalian descendants are again reproductive and physiological rather than skeletal. Most importantly, mammals maintain a constant body temperature, whereas the body temperature of reptiles is determined by solar radiation and the surrounding air temperature. This warm-blooded adaptation permitted mammals to lead a far more active and diversified life than reptiles, for they could survive in cold regions and search for food at all times of the year and during the cool of the night as well as the warmth of the day.

Though originating at about the same time as the dinosaurs, throughout the Mesozoic era of dinosaur dominance, the mammals remained small, inconspicuous shrewlike animals that are relatively uncommon fossils. The early evolutionary history of the birds is even more obscure. The Jurassic limestones of southern Germany have yielded several remarkably complete fossils of a transitional bird, *Archaeopteryx*, which possessed a reptilelike skeleton as well as wing and tail feathers (Figure 14–19). The beak had teeth (all modern birds are toothless), and without the excellent preservation of feather impressions in the fine-grained limestone these fossils proba-

FIGURE 14–20

Early Cenozoic mammals cluster at a Paleocene water hole 65 million years ago. At top right, the flesh-eating, wolverinelike Oxyaena crouches above the water as a horned, 10-foot Arsinoitherium approaches the pond. At the upper left corner, an 8-foot, root-grubbing Barylambda, a primitive forerunner of horses and cows, cautiously views the other visitors. Below, a tiny primitive opossum flattens at the sight of the carnivorous Oxyaena. In the pool, a semiaquatic Moeritherium raises its head.

bly would have been described as reptiles. Since relatively few comparable bird skeletons are known from younger rocks, we have had only glimpses of the evolutionary history of the birds. However, the extinction of the dinosaurs at the end of the Mesozoic marked the rapid rise of many kinds of mammals into the great Cenozoic evolutionary expansion of that class.

The ancestral placental mammals (those with a special structure, the *placenta*, inside the uterus of the female for the nourishment of the developing young from the mother's body fluids) were small, superficially mouselike representatives of the insectivore order. This order includes long-snouted primitive creatures such as the modern shrews, moles, and hedgehogs (Figure 14–20). In the Paleocene and Eocene epochs (63 to 36 million years ago), the insectivores went through an explosive evolutionary expansion that ultimately led to such diverse mammals as whales, bats, elephants, and human beings. Many of the early representatives of modern orders were quite small in stature and unspecialized. Thus ancestral catlike carnivores were no larger than rabbits, and horses were four-toed herbivores about the size of fox terriers (Figure 14–21). Throughout the Oligocene and Miocene epochs (36 to 13 million years ago) some members of each order tended to increase in size and to diversify into increasingly specialized types. Some orders, such as the elephants and the horses, reached their evolutionary climax earlier in Cenozoic times and are reduced in numbers and diversity today. Other orders, such as the rodents and primates, appear to have steadily increased in diversity until modern times.

Just as with the Mesozoic reptiles before them, mammals became the dominant herbivores and carnivores of the land surface; they

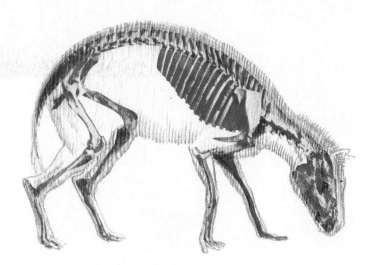

FIGURE 14–21

Eohippus, *a four-toed horse of the lower Eocene.*

FIGURE 14–22

Diversity among modern African mammals: (A) a hippopotamus feeding on land, (B) a male gereneuk feeding on an acacia tree, (C) an olive baboon female and young cleaning a root, and (D) giraffes feeding on a thorn tree.

also spread into the air with flying forms (bats) and returned to the sea as marine carnivores (seals, porpoises, and whales). The greatest diversification of land mammals has developed among the many herbivorous orders, which range in size and habit from tiny burrowing rodents to arboreal monkeys and huge grazing mammals such as elephants and rhinoceroses (Figure 14–22). Relatively few orders of placental mammals have become extinct and since mid Cenozoic time their general history has been a progressive modernization leading to the familiar mammals of today. During the Pleistocene (1 million to 10,000 years ago) and even more recent times, however, a few large species in several orders of mammals became suddenly extinct in Europe and subsequently in North and South America. Their disappearance and that of large flightless birds coincides with the appearance of early human hunting groups.

Since human origins and evolution are of such particular interest to us and of such staggering impact on the rest of the natural world, we shall take up human evolutionary history in a separate chapter.

Summary

The origins of life from nonliving systems are shrouded in the distant past and are inaccessible to us by direct observation of even fossil remains. Most theories on the origin and early development of life are derived from hypothesized environmental conditions of the early earth, but some modern experimental evidence exists to show possible intermediate chemical steps from nonliving molecules to cellular organization as we know it today. The first fossil plant organisms date to about 3.1 billion years ago, and the fossil evidence for animal life begins abruptly (and in relatively advanced form) only 600 million years ago. All phyla and most major groups except the vertebrates appear at the latter date—the start of the Cambrian period of the Paleozoic era. During the remainder of the 370 million years of the Paleozoic, invertebrate life flourished in the seas, and the first plants and scorpionlike arthropods moved onto the land. In the middle of the Paleozoic, the first vertebrates moved onto the land in the form of amphibians, derived from lobe-finned fishes. Their place was taken by the reptiles in the Permian period as this group, with their shelled eggs, radiated widely and dominated the land during the entire Mesozoic era. The mammals and birds originated from reptile groups during the Mesozoic, but did not become highly diversified until the Cenozoic era. At almost the close of the Cenozoic, human beings appeared on the scene and began to have great impact on larger mammal and bird species.

15 The origin and evolution of human beings

Late in the Cenozoic era, the origin of our own species. *Homo sapiens,* occurred—the final and most significant evolutionary event that had appeared in eons of geologic time. The story of human evolution is complex, with tantalizing new pieces of fossil evidence being found almost every year. The reinterpretation of previously discovered specimens has also clarified concepts on the origin of human beings and the history of their primate ancestors. While we have unique attributes, we are still basically animals and the product of the same evolutionary forces that have shaped other forms of animal life.

The early primates

The evolutionary development of the Hominidae, the human family of large-brained, tool-using primates, is extremely recent, having taken place within the last several million years. However, the primates, the order to which human beings belong, underwent adaptive radiation throughout the Cenozoic times. As we saw in the last chapter, most of the orders of placental mammals arose early in the

Cenozoic from relatively unmodified, shrewlike insectivore ancestors. The oldest primates in the fossil record—isolated teeth and
jaws from early Paleocene rocks—are in fact almost indistinguishable from insectivores. While the insectivores were ground-
dwelling, primarily insect-eating mammals, the primates became
adapted to a different mode of life, that of the omnivorous tree-
dweller. Only a few primates, notably human beings, are adapted for
life on the ground. Thus many of the specialized characteristics of
the early and present primates evolved as adaptations for an arboreal
mode of life. The most significant of these new adaptive changes, as
far as the human species was to be concerned, was the development
of the grasping hand with its opposable thumb, an almost universal
primate character that evolved as an organ for grasping, manipula-
ting, and exploring objects. Depth perception is important to a tree-
living animal, and the majority of the primates are unique in possessing binocular or stereoscopic vision wherein the visual fields of
the two eyes overlap. In most mammals the eyes are placed toward
the sides of the head, which allows a wide field of vision from each
eye but, because of the lack of overlap in the visual fields, does not
allow precise stereoscopic vision. Correlated with the grasping hand
and the arboreal habit, then, were a forward placement of the eyes
and a relative flattening of the face so that the eyes could stereo-
scopically view the same area. While arising in primates as adaptations for life in the trees, in human beings these developments were
to become the foundation of later evolutionary success in terrestrial
life where coordinated hand and eye movements led to such activities as hunting with hand-made weapons and tool making.

Closely associated with the development of great visual acuity and
increased dexterity of the hands was the marked expansion of the
brain; in the highest primates, such as human beings, progressive enlargement of the brain culminated in the development of higher
mental faculties.

The origins of modern primates

Thus the grasping hand and stereoscopic vision marked the early
primate fossils that appear by Eocene times, when tree-dwelling
primates about the size of large cats were common in North America
and Eurasia. These were not true monkeys, but a less advanced
group called the prosimians (Figure 15–1). After the origin of true
monkeys and apes in Oligocene time, the prosimian order decreased
dramatically in importance. They have not become extinct, however,

FIGURE 15–1

A tree shrew, modern
representative of the
prosimians.

but survive to the present day in certain large tropical islands or other isolated environments. Examples are the lemurs, found today only on the island of Madagascar, and the large-eyed tarsiers of Borneo, Sumatra, and the Philippines, as well as the lorises, a lemurlike group found in India, Africa, and Southeast Asia, but not in Madagascar.

There are three living groups of higher monkeys and apes, namely, the New World monkeys (Ceboidea), the Old World monkeys (Cercopithecoidea), and human beings and the great apes (Hominoidea) (Figure 15–2). Hominoids differ from monkeys in lacking a tail and in other fundamental adaptive features. During the Oligocene epoch, these three anthropoid superfamilies differentiated from the basic primate stock. Unfortunately, fossils from this period are extremely rare. Until very recently they were known only from half a dozen bones and teeth discovered in the early years of this century in northern Egypt. While additional fossils have been found recently, all specimens indicate that the separation of apelike hominoids from the monkeys was an early and rapid event in primate history, one that shows clear differentiation by the close of the Oligocene time.

During Miocene and early Pleistocene times (about 25 to 10 million years ago), apelike primates inhabited the Old World. The fossil record of apes is somewhat more complete than that of monkeys, and certain fossils clearly represent members of the ape family (Pongidae) such as *Pliopithecus*, a Miocene gibbonlike creature that is generally regarded as ancestral to today's gibbons. Miocene sediments of Europe and Asia, especially in India, have produced teeth and jaws of *Dryopithecus*, the oak ape, so-called because of the

FIGURE 15–2

(A) New World capuchin monkeys, (B) an Old World vervet monkey, and (C) Old World baboons.

A

B

C

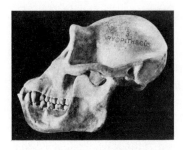

FIGURE 15-3

A reconstructed skull of Dryopithecus, *an early anthropoid primate.*

presence of oak leaves in these fossil deposits (Figure 15-3). These primitive oak apes were the early ancestors of the modern orangutan, chimpanzee, and gorilla, and indeed these generalized apes seem to be ancestral to *both* human beings—the Hominidae—and to the recent great apes—the Pongidae.

One of the Miocene apelike types close to human lineage is *Proconsul africanus,* discovered in the 1930s by the anthropologist Louis Leakey on an island in Lake Victoria, East Africa (Figure 15-4). While no complete skulls or skeletons are available, the anatomical features of the known parts of the skull and teeth are remarkably unspecialized for an ape. The generalized characteristics of these skeletal components have suggested to some investigators that *Proconsul* is not far removed from the basic common stock from which apes and human beings arose. It was primarily a tree-dweller, but apparently was capable of descending frequently to the ground and foraging. The basic transition from tree-dwelling to ground-living primates might well have first appeared in the primate lineage at this time. During the Miocene epoch, the great expanse of tropical forest in East Africa dwindled and left patches of forest separated by open grasslands and scrub. Selection would have favored development of the capacity for ground walking, which would permit arboreal forms to cross the open plains from one patch of woodland to another.

FIGURE 15—4

Proconsul, *an early apelike primate from the Miocene of East Africa.*

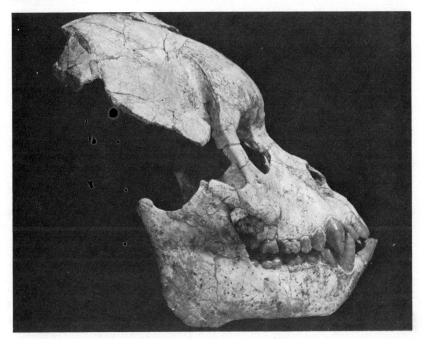

Our first known ancestors

The first truly humanlike genus of Miocene-Pliocene hominoid appears in the fossil record about 14 million years ago. Fragmentary remains of an upper jaw from the Siwalic hills of India and an upper jaw from Kenya, East Africa (found by Louis and Mary Leakey) apparently represent the point where the hominid lineage separated from the pongid assemblage. The pattern of dentition, unlike that of modern apes, is extraordinarily humanlike. These two fossil forms, named *Ramapithecus* and *Kenyapithecus*, respectively, apparently ranged widely throughout Africa and Asia in late Miocene and early Pliocene times and probably represent a direct ancestor of modern human beings (Figure 15–5).

FIGURE 15–5

An outer view of the mandible (lower jaw) of Ramapithecus.

About 2 million years ago, major climatic changes occurred that had not been present on earth since the Permian and Precambrian times. Great sheets of continental glacier ice covered large areas of the Northern Hemisphere in a series of vast expansions and contractions. The extreme climatic changes of Pleistocene times caused rapid changes of distribution in animal and plant species. Our own species, *Homo sapiens*, developed in this unique geologic and climatic setting.

The evolutionary sequence of early human forms is more clearly revealed in the fossil record of the Pleistocene than in that of the preceding periods. Because of the rapid discovery of fossils in the late 1800s and early 1900s, and the rather self-centered desire of each authority to give a new name to each fossil specimen without concern for its similarities to previously discovered specimens, the Pleistocene record seemed at first to include many species and genera. Within the past decade, however, reconsideration of these Pleistocene fossils in the light of modern biological principles has revealed a much simpler story. In essence, there are two main branches in Pliocene and Pleistocene human evolution: *Australopithecus* in the Pliocene and early Pleistocene, which then became extinct; and *Homo erectus* in the early and middle Pleistocene (with *Homo sapiens* appearing in the late Pleistocene).

The Australopithecines

The Australopithecines were first discovered in South Africa in 1924 by Ramond Dart, when a single well-preserved skull of a child was found. Since then, considerable numbers of adult and children's skulls, jaws, and limb bones have been collected in South Africa and the Olduvai gorge in Tanzania, and fragmentary remains at sites in Asia, suggesting that Australopithecines were widely distributed in the Old World tropics (Figure 15–6). This new genus *Australopithecus* was probably represented at first by two species that were generally similar but differed in size and geographic distribution. The larger and very robust kind of these African ape species, *Australopithecus boisei*, was about the size of a gorilla and was apparently a vegetarian. This massive-jawed northerner probably began specializing to feed on rough vegetable matter sometime between 6 and 10 million years ago. Its canines and incisors became smaller, while molar teeth and jawbones became increasingly large.

While *A. boisei* was developing, a more lightly built gracile species, *A. africanus*, about the size of a chimpanzee, evolved simultaneously from a mutual ancestral type. It specialized in a more omnivorous diet and hence was under less selective pressure to evolve large molars and jaws than it was to improve its meat-eating, hunting, and tool-using attributes. Like the northern *A. boisei*, it walked bipedally and dwelt in open country.

During the next several million years, as *A. africanus* spread across Africa, some paleoanthropologists suggest that two additional species of Australopithecines evolved from this successful hominid type. In the north (including East Africa), the gracile types evolved

FIGURE 15–6

A reconstruction of an early hominid in Africa: Australopithecus africanus.

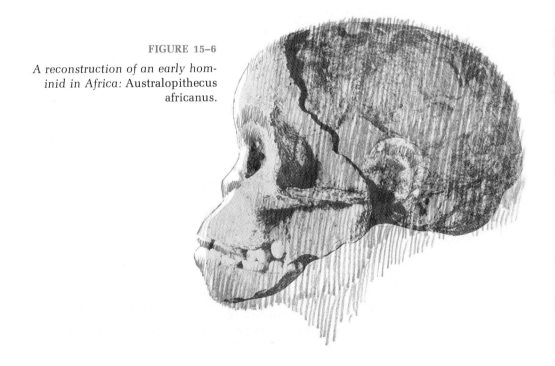

into "*Homo*" *habilis*, a tool-using hominid that was sympatric and contemporaneous with *A. boisei* until the latter disappeared about a million years ago. In the south (especially South Africa), the gracile populations of *A. africanus* did not continue on and evolve into a *habilis* type; instead, *A. africanus* became a somewhat robust type that has been named *A. robustus*. Like *A. boisei* in the north, *A. robustus* would soon die out, probably from competition with the rising populations of *Homo erectus*. Still a third line of *Australopithecus* (*A. modjokertensis*) is believed to have developed in Asia prior to the appearance of *H. erectus* there.

The Australopithecines were human in every feature except for their smaller brain case. This suggests that erect bipedal locomotion on the ground preceded the development of a large, complex brain in human evolution. The upright stance completely freed the hands from use in locomotion, which then could be used exclusively for such tasks as tool making and weapon throwing. These tasks in turn put a premium on an increased intelligence and awareness of environmental stimuli, and thus lead to selection pressures for a larger brain. Several sites with Australopithecine fossils contain crude stone tools as well as the fractured skulls of baboons and other animals, indicating that at least the later Australopithecines were able to use formed tools and make effective weapons.

Homo erectus appears

In the middle Pleistocene, a larger-brained species similar enough to the modern human species to be considered in the same genus succeeded the Australopithecines. This species, *H. erectus*, has been found in fossil deposits distributed throughout the Old World (Figure 15-7). The first fossils of this species were those of the famous Java Man, discovered at Trinil, Java, in 1894 by a young Dutch army surgeon, Eugene Dubois. Dubois had been profoundly influenced by the writings of Charles Darwin in the 1860s and 1870s and believed that he could find the origins of human beings in the Far East. After requesting an army post in Java, to search for human fossils in his spare time, he may have been as surprised as the world was when his efforts were rewarded by the actual discovery of the earliest human being there. Tropical areas are notably poor regions in which to seek fossils of any kind because of the rapid decay of dead organisms and strata erosion from the heavy rainfall, and primate bones are delicate enough to be even less likely to be preserved. Additional specimens were found in China 30 years later, which became known as Peking

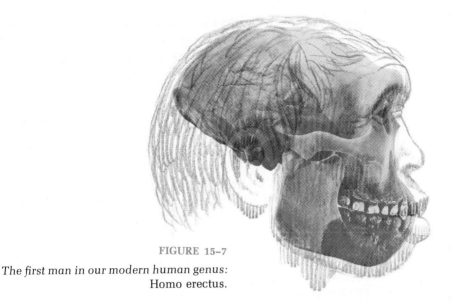

FIGURE 15-7

The first man in our modern human genus:
Homo erectus.

Man. Since that time, many additional skulls, jaws, and limb bones have been reported from Africa and Europe as well as in Asia. In fact, recent discoveries of skulls, hip bones, and other bones of human beings living some 1.5 to 3 million years ago in Africa's Great Rift Valley are providing anthropologists and biologists with valuable new information on the origin of *H. erectus.*

New findings were made in 1975 by Mary Leakey, Richard E. Leakey, director of the National Museums of Kenya and son of Louis and Mary Leakey, and Donald Carl Johanson, curator of physical anthropology at the Cleveland Museum of Natural History. In October 1975, Mary Leakey announced the discovery of the oldest reliably dated early human remains ever found. The fossil jaws and teeth of eight adults and three children were between 3.5 and 3.75 million years old. The bones were found at Laetolil in Tanzania, some 25 miles south of Olduvai Gorge, site of many famous early discoveries by Louis and Mary Leakey. The Laetolil teeth resemble the *Homo* teeth found by Johanson in Ethiopia's Afar Depression, where he also discovered more than 150 fossilized bones of two infants and from three to five adults. Potassium argon dating indicates that the Ethiopian bones are at least 3 million years old, and probably 3.5 million years old. Teeth and jaws resemble those of early *Homo,* not *Australopithecus.*

In March 1976, Richard Leakey announced that he had discovered a complete skull of *H. erectus* in 1975 at Koobi Fora in northern Kenya. Estimated by him to be 1.5 million years old, the skull is almost identical to *H. erectus* fossils found in 1927 in Choukoutien Cave near Peking, China, which were dated at only half a million years. This newly found skull represents a very important link in human evolution, for Leakey claims it confirms that *Homo* and *Australopithecus* were coexistent. Thus even the gracile *Australopithecus* forms may have been relatives rather than forebears of humankind, living at the same time as *Homo* but reaching an evolutionary dead end. The *H. erectus* skull from Koobi Fora is clearly related to a skull found in 1972 east of Lake Rudolf in the same region of Kenya. Leakey believes that these two skulls of the same type and a hip bone found in 1975 belong to the genus *Homo*. These fossils have been dated at from 2.5 to 3 million years old. Thus our picture of early human history is rapidly changing, and it is a history in which there are as yet no absolute assertions. Many authorities question Leakey's controversial interpretations. With further discoveries each year, the relationship in the evolution of *Australopithecus* and early *Homo* forms ought to present an even more fascinating picture than we have today.

Australopithecus had a cranial capacity of about 600 cubic centimeters (cc) and stood about four to five feet in height; however, *H. erectus* had a brain volume of 900 to 1,100 cc, which is about intermediate between that of *Australopithecus* and modern human beings (1,300 to 1,500 cc). This brain increase was all in the neocortex, a region of the brain involved with associative learning behavior. *H. erectus* was about five feet in average stature. Until the recent discoveries by Leakey and Johanson, specimens of *H. erectus* had been found in fossil deposits that ranged in age from about 700,000 to about 200,000 years ago. Now it appears that *Homo* may be much older. This low-browed man was a tool maker and a hunter who learned to use fire and probably had some powers of speech. He used stone tools, principally large hand axes made from pebbles of flint that were sharpened on one side by chipping. Throughout middle Pleistocene time, these tools developed in design and workmanship, indicating that human beings became more adept tool makers concomitantly with their increase in brain size.

Modern Homo sapiens takes over

The emergence of the modern species *H. sapiens* began about 500,000 years ago, and therefore our species is believed to have been contemporaneous with *H. erectus* for about 200,000 years (Figure 15–8). A large-boned race of *H. sapiens*, Neanderthal Man, appeared on the scene in Europe about 100,000 years ago and roamed over the

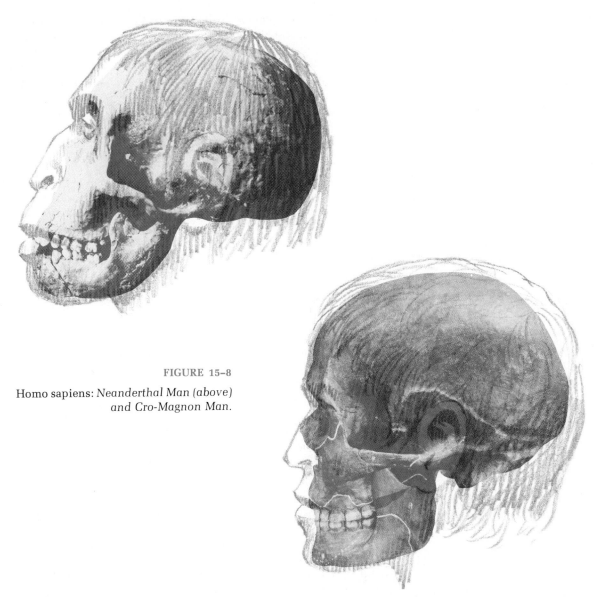

FIGURE 15–8

Homo sapiens: *Neanderthal Man (above) and Cro-Magnon Man.*

continent until about 40,000 years ago (late Pleistocene). Then he dramatically disappeared. One of the best known of fossil men, the Neanderthal Man was a cave dweller, short in stature (about five feet) but powerfully built, with prominent facial brow ridges and a large brain with an average capacity of 1,450 cc. This brain capacity is actually somewhat larger than the worldwide average (1,350 cc) in modern human beings. He was replaced by a modern type much like ourselves, grouped under the common name of Cro-Magnon.

Human fossil hunters

The history of human fossil discoveries is replete with sagas of perseverance by anthropologists and paleontologists. One thinks especially of Eugene Dubois, born in 1858 in Holland, who at the age of 29 resigned his lecturer position and a promising career at the University of Amsterdam and took off for the wilds of Sumatra, where, he claimed, he would solve the mystery of human origin. Unable to secure financial backing for his proposed expedition, he enlisted as a doctor in the Dutch East Indies Army and left on a seven-week voyage in October 1887 aboard a Royal Dutch Mail packet, with his young wife and baby daughter. After several years of hospital work and an attack of malaria, he was transferred to the drier climate of neighboring Java and placed on inactive duty. Now he was free to spend all of his time searching for fossils. The colonial government even supplied him with a crew of native convict laborers for his excavations. In 1893, Dubois announced that he had discovered the "missing link of Darwin" between the apes and human beings: *Pithecanthropus* (from the Greek words for ape and man) *erectus* (meaning upright). In 1895, he returned home with his fossils (a skullcap, a thighbone, and two teeth) and proceeded to show them at many scientific meetings. He lived until 1940, convinced that Java Man was a missing link between human beings and apes rather than human, as many other experts thought. Today his Java Man specimen is identified as belonging to the species *Homo erectus* and these bones (probably of one individual) are still considered one of the greatest fossil finds ever made.

Partway around the world, Louis and Mary Leakey, a husband-and-wife team of East African anthropologists, were to find in 1959 some fossils of human ancestors that could be more accurately called "missing links." Over a period of 28 years the Leakeys had commuted from Nairobi, Kenya (where Leakey had a museum position), to Olduvai Gorge, a dry river gulch in northern Tanzania, several hundred miles away. Early in their search it took seven days to get there from Nairobi on the primitive, rough track to the gorge. Over the years, and with the help of other experts, they recovered an enormous number of animal fossils from the various layers of sediments, along with remnants of extremely ancient pebble tools. But fossil hominid bones never could be found. Then almost by accident late in the afternoon of July 17, 1959, Mary Leakey spotted part of a hominid face, with large brown teeth, sticking out of Bed I, the lowest layer of the gorge. By a miracle of geology, this new fossil (named *Zin-janthropus boisei*, after the Charles Boise Fund, which had supplied considerable financial support to the Leakeys) could be dated. The skull lay just above a layer of gray volcanic ash, and this ash could be accurately dated at about 1.75 million years of age by the then newly discovered potassium-argon dating technique. Mary Leakey raced back to camp, got her sick husband, and both returned to the spot at a run. There, Louis said, "I turned to look at Mary, and we almost cried with sheer joy, each seized by that terrific emotion that comes rarely in life. After all our hoping and hardship and sacrifices at last we reached our goal . . . we had discovered the world's earliest known human."

Only a year later, on December 1, 1960, Louis Leakey discovered the first skull of *Homo erectus* at Olduvai's Bed II. All about the site lay more refined tools, such as flaked stone hand axes. Intensifying their search, both at Olduvai Gorge and in neighboring regions of Tanzania and Kenya, they and their son Richard Leakey began to discover some even more ancient specimens of early human beings, as related in this chapter. The remarkable book of life at Olduvai Gorge seemed to represent part of the primordial African center of human expansion and evolution, uncovered because of the incredible perseverance of a dedicated family.

For erroneous reasons, we have grown accustomed to the idea that Neanderthal Man was our direct ancestor. Actually, it now appears that we may be closer to the truth by considering him a geographical offshoot of *H. sapiens*. Classic Neanderthal Man was isolated in western Europe by glaciation, while to the south and east more modern types were developing. The fossil record shows particularly great human variability in the Middle East between about 100,000 and 30,000 years ago. It has been suggested by Theodosius Dobzhansky that the extraordinary variability in Middle East human fossil deposits represents contact zones and intergradation between geographic races. Cro-Magnon representatives of the modern *H. sapiens* can be traced back in the fossil record to about 35,000 years ago. These remains have been found in many sites in western and central Europe, but little is known of modern human beings on other continents during the time that Cro-Magnon flourished in Europe.

It is not even certain, as was once widely accepted, that *H. sapiens* originated in a single area and then spread over the world to differentiate into the four basic human races of the Australoids, Mongoloids, Negroids, and Caucasoids. Independent development of groups of early *H. sapiens* from late *H. erectus* populations may represent a more accurate picture in the view of some paleoanthropologists. The aforementioned primary racial divisions, which were probably once geographically separated to a large extent, have intermingled and intercrossed for untold thousands of years, and the distinguishing features of the basic racial groups have become increasingly blurred by the countless migrations and intermixings. Today, *H. sapiens* lives in essentially one great reproductive community.

The influence of cultural evolution on human development

We should mention that the evolution of human culture has given us a double heritage. We are the product of both biological and cultural evolution. In the foregoing account, we have stressed the direct fossil evidence for human ancestry, but throughout the Pleistocene, the human fossil skulls, bones, and teeth are extremely rare in comparison with the stone tools and other remnants of a sequence of cultural events that can, in a general way, be correlated with human physical evolution. Human exercise of reason and ability to communicate rational thoughts has led to the emergence of human culture, transcending in evolutionary importance purely biological inheritance.

Children gradually acquire customs, beliefs, and values by instruction and imitation of the adults around them. Each new generation, then, is able to draw on the rich store of past accumulated knowledge and ideas.

We can see the effects of this cultural influence in the development of tools. The earliest stage in human cultural evolution, the Old Stone Age (Paleolithic), was characterized by chipped stone tools; the New Stone Age (Neolithic) by ground and polished stone tools; and, finally, the Age of Metals by copper and bronze, and later iron, tools. Most of human Pleistocene history falls into the Paleolithic, or Old Stone Age, for Neolithic cultures began only about 10,000 years ago and were followed about 5,000 years ago by metal-using cultures. Thus the culture of *Australopithecus* and *H. erectus* as well as most of the existence of *H. sapiens* is represented by Paleolithic chipped stone tools. These show a progressive improvement in tool-making techniques throughout the Pleistocene, however, ranging from early crude hand axes to highly advanced, carefully styled projectile points and knives made by late Paleolithic *H. sapiens*. Bone was used in late Paleolithic times for making fine tools such as needles, and a highly developed art was present, as attested in the elaborately carved bone objects and the well-known cave paintings of France and Spain (Figure 15–9).

FIGURE 15–9

Neolithic axes made by early human groups in France.

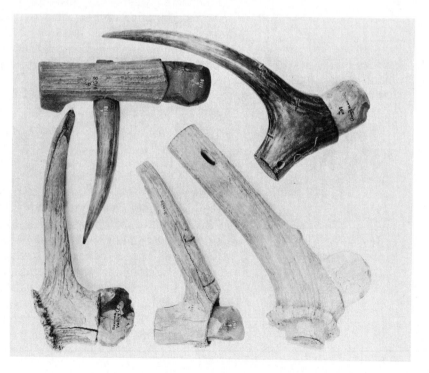

As the Neolithic peoples developed more advanced stone tools made by grinding and polishing, they also learned to make pottery for storage of food and water. And most important of all, they began to cultivate plants and domesticate animals. With this step, for the first time in several million years of Pleistocene history, human beings no longer had to depend on hunting and the gathering of wild plants but could grow their own food and remain in one place. The development of agriculture in Neolithic times may be one of the most significant events in all human history, for it allowed the development of permanent communities and permitted a division of labor: some people provided the food and others became craftsmen, scholars, priests, and tradesmen. After the agricultural revolution began, human culture evolved rapidly. At about the same time, the development of writing in Egypt and Mesopotamia led to the beginnings of recorded human history. We still have only tantalizing hints of the origin of human language, the development of clothing, and the social structure of early human communities; but because of the Pleistocene record of fossil bones and tools, we now understand more clearly human development from the Cenozoic primate ancestors — an event that must be marked as one of the most important in the history of life following its origin billions of years ago.

Summary

Human evolution is a complex story but reasonably well documented by fossil evidence and cultural accouterments. It appears that human beings originated in central Africa at least several million years ago, having descended from a long line of earlier primates. The three most recent groups of humanlike creatures have been the Australopithecines, *Homo erectus*, and our own species, *H. sapiens*. The Australopithecines were the first primates to stand upright, walk bipedally, and use tools fabricated from stones and pebbles. The later Australopithecines were contemporaneous with the oldest and most primitive true human beings of our own genus, *H. erectus*, who lived across much of Eurasia during the middle Pleistocene times, between 700,000 and 200,000 years ago, and were apparently found in Africa much earlier (to perhaps more than 3 million years ago). *H. erectus* had a somewhat larger brain capacity and was distributed, like late *Australopithecus* species, throughout the Old World tropics. *H. sapiens* emerged about 500,000 years ago and at times several geographic subspecies met, interbred, and replaced one another. Modern human beings are a product of both biological and cultural evolution, and the latter factor has the most significant influence on our present evolutionary directions.

16   The diversity of life: animals and plants

The late scientist-mathematician-philosopher Alfred North White-head once said the man who invented religion killed God and the man who invented taxonomy killed science. **Taxonomy** is the study of the classification of plants and animals, and despite Whitehead's remark it is the oldest branch of biology and has probably involved greater scientific effort by more workers up to the present century than any other field of science. Even the area of ecology is a product of only recent decades, though the word was coined in 1869.

Why should taxonomy have occupied such a prominent position in biology? The study of the diversity of life involves classifying the various characteristics of plants and animals into categories that will allow later identification of these organisms by biologists other than taxonomists. Since almost all biologists need to know which species they are working with (from molecular research to ecological studies), biology is firmly rooted in classification for practical as well as historical reasons. For the origin of taxonomy so early in the development of science, we must look to certain characteristics that human beings have always exhibited. We inhabit a world of diverse living species and seem to have an almost innate urge to classify them and give them names in order to systematically accumulate knowledge about these species. The basic principles of biological classification (more rigorously referred to as taxonomy or systematics) are really simply an extension of the principles that govern the classification of nonliving objects. From the records that have been preserved, it seems apparent that as soon as people began to be thoughtfully observant of the world in which they lived, they desired and needed a systematic scheme in which to place the objects that surrounded them. From the time of the ancient Greeks, we have records of attempts to arrive at a satisfactory classification of living things. Classification, of course, depends on the grouping of similar objects, and the ways in which different groups of people choose to group objects depends on their aim and the way they perceive them.

Classification concepts

In even the earliest schemes, people made a distinction between animals and plants. This distinction is easy to construct with the more highly organized groups but not with single-cell organisms and simple aggregates of cells, where neither structure nor habits furnish satisfactory criteria for assigning these organisms to one or the other of these major categories. Nevertheless, these attempts served at least to emphasize the fundamental unity of biological systems and to exclude other parts of the natural world such as minerals and stones (included even in some eighteenth-century biological classification systems). In essence, modern studies have confirmed the basic unity among all life in terms of molecular composition and basic physiological activities. Of course, we have seen in earlier chapters how atoms that at one time were in the ground water, in solid minerals, or in some other nonliving reservoir such as the atmosphere may later become part of a living organism and when that organism dies be returned to the inorganic reservoir. This principle of cycling in ecology allows us to see the organism as continuous with its inorganic environment and in dynamic equilibrium with it. Yet we do not need to classify nonliving components of the ecosystem with our taxonomic hierarchies of life because of this ecological relationship. There are still fundamental differences between living and nonliving matter (Chapters 14 and 20).

The general goal of modern taxonomy is to establish classification systems that reduce the multiplicity of living things to some kind of order, ideally one that expresses **phylogenetic** relationships — the establishment of the genealogical tree for the phyla, or a pedigree for all living things. The use of comparative anatomical studies and

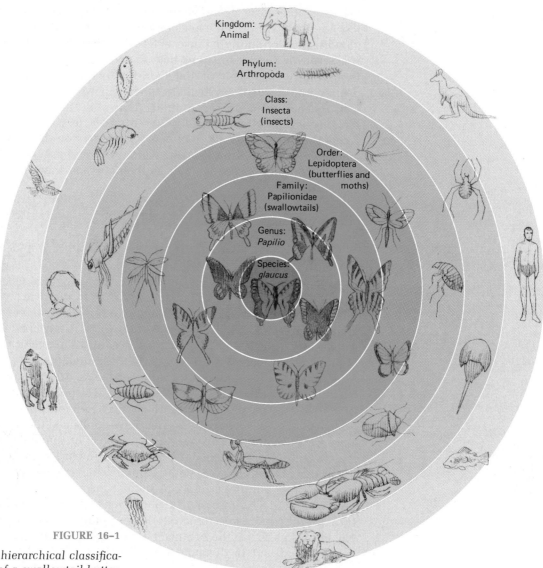

Kingdom:
Animal

Phylum:
Arthropoda

Class:
Insecta
(insects)

Order:
Lepidoptera
(butterflies and
moths)

Family:
Papilionidae
(swallowtails)

Genus:
Papilio

Species:
glaucus

FIGURE 16–1

The hierarchical classification of a swallowtail butterfly, from kingdom down to species.

comparative embryology during the last 150 years has particularly contributed toward expressing the fundamental relationships between living things and to organize them in some kind of natural order. Let us look now at the basic taxonomic concepts that have served to organize the classification schemes formerly in use and those presently used by biologists.

The basic units of classification schemes, into which biologists divide both plants and animals, are called **species** (singular and plural: species). Species is a Latin word that signifies "kind." Species can be subdivided into the minor categories of subspecies, varieties, and races, and they are collected or grouped into the major hierarchical categories of genus, family, order, class, and phylum (Figure 16–1). The early definitions of species used by taxonomists prior to the twentieth century tended to emphasize anatomical characters, referring to species as groups of more or less similar individuals that have distinctively different traits from other comparable groups. Many of the early naturalists considered a species to be a group of organisms that had been created with the distinctive anatomical characters fixed and unchangeable.

One such believer in the independent creation of each species was Carolus Linnaeus (1707–1778), a Swedish botanist and natural historian whose introduction of the principle of identification of plants and animals by a Latin genus and species name was immediately adopted by other taxonomists and is still in use throughout the world. This system of **binomial nomenclature** applies a unique pair of genus and species names to each species of organism, a combination that is repeated in no other group. The utility of this simple handle for identification and classification was immediately recognized, for in previous centuries long Latin phrases had been used to identify different kinds of organisms—phrases that could not be conveniently used to refer to a specific kind of organism in technical publications or even common identification manuals (Figure 16–2).

FIGURE 16–2

(A) Carolus Linnaeus and (B) John Ray, pioneers in taxonomic classification.

A

B

John Ray (1627?–1705), a British botanist and perhaps the greatest plant systematist of all times, had anticipated the usefulness of a simple name referring to each species and the necessity for some standards in applying the term species to them. Without such standards it would have been impossible to name and thus to accumulate information about the nearly 18,000 plant species with which he was familiar. Ray reserved the term species for morphologically distinct types that would reproduce their own kind from seed. If two or more species appeared in the offspring of a single plant, then these species were considered to be actually one. The utility of this principle is so obvious and important that it has been the cornerstone of systematic biology to the present day. It is still the only universally acceptable attribute common to all species: A species must be characterized by some heritable difference that distinguishes it from all related species.

The next great advance in the study of biological classification was Darwin's lucid discussion of organic evolution in *On the Origin of Species*. Evolutionary theory provided an explanation for the existence of groups of similar organisms. Clearly, in Darwin's view, members of a species were relatives who owed their common traits to a common ancestor. Because the characteristics of these organisms are controlled by many genes that may vary somewhat in frequency from population to population, an enormous variety of heritable changes of type is possible. Sometimes morphological differences alone are insufficient to establish the validity of one species differing from another, and genetic crosses must be made. These crosses will indicate the degree of close relationship by the extent to which hybridization is successful.

The modern concept of species, then, has tended to move toward a *genetic* definition rather than a morphological one as in Linnaean times. It defines a species as *a freely interbreeding population of possibly variant phenotypes that, unless prevented by geographic or physiological or ecological barriers, share in a common gene pool.* Different races of this species may result from mutations among the genes in this pool and subsequent isolation; however, if mating can occur between representatives of these populations in nature and result in the production of fertile offspring, these populations are considered representatives of a single species. Genic compatibility is thus implicit in this concept. If the genes were incompatible the stocks could segregate from one another further and essentially be able to create or establish a new species.

The divergence between species, then, provides some measure of the closeness of their relationships, and this divergence may be expressed in anatomical, physical, and behavioral characters, capacity for hybridization, or even in subtle differences in biochemical

FIGURE 16–3

Homology and analogy in structural features of animals. Homologous features used for the same function found in two organisms are similar because of common ancestry; thus bird and bat wings both represent evolutionary modifications of the vertebrate forelimb. Analogous features used for the same function are similar, but have no common ancestry; the butterfly wing originated as a flap of the insect exoskeleton, whereas the bird wing originated as a modification for the front pair of limbs in certain reptilian vertebrates.

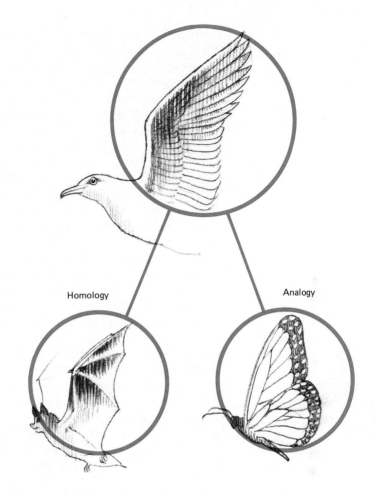

Homology

Analogy

processes and molecules. However, no single criterion can be used consistently to delimit species. Reproductive isolation — that is, the lack of interbreeding between two segregates — is relative and exists in every conceivable degree between different groups of organisms. It is often impossible to apply the genetic definition in practice. Widely divergent populations may frequently be experimentally hybridized and other, distinct species may come to closely resemble each other superficially from entirely different causes. The only logical use of the term species was probably offered in 1859 by Darwin, who states: "I look at the term species as one arbitrarily given for the sake of convenience to a set of individuals closely resembling one another. . . ."

In studying plant and animal systematics today, the bases of classificatory systems are related to recognition of **homology,** or correspondence in structure and function (Figure 16–3). We look for homologies in adult structures and establish them through study of the comparative morphology of living and extinct forms (fossils). If we are able to study the organisms in the field or in laboratory culture, homologies may also be sought in juvenile stages and in developmental and physiological processes. Thus evidence to build criteria for phylogenetic affinities and for taxonomic systems based on phylogenetic relationships comes from comparative anatomy, comparative embryology, biochemistry, physiology, ethology (behavior), and ecology. Where certain characters are recognized as fundamental and therefore rather common to all members of a group, they presumably were representative in the ancestral form from which the members of that group have all descended. Derived characters, which have particular value to the more specialized descendants, may arise from these fundamental traits through mutational changes.

As we have seen in Chapter 14, the first biological systems that have left recognizable fossil traces in the earth's crust were already well-defined algae and representatives of all the major invertebrate animal phyla. A great time and information gap exists between the stages of nonliving matter, of which the early earth and its atmosphere were composed, and the living matter that must have originated from it at some point. Likewise inaccessible for our study is the subsequent great gap between the organization of these first living systems and even the simplest single-celled systems known in the fossil record. While we can only conjecture as to the stages that must have taken place in these earliest organisms, leading to the hosts of species in the fossil record and eventually to presently existing diversity of life, we may assume that the *basic* characteristics we observe today among all organisms had to originate at those earliest times.

We shall now look at the animal groups as they exist today and their general characteristics, biology, ecology, and major taxonomic divisions. Then we shall look at the major plant groups and their general characteristics and biology.

The kinds of animals

Classification schemes for the major groups of animals vary considerably from one authoritative taxonomist to the next. Great differences of opinion differ even at the level of the kingdom, the highest classification category. Some recognize only two kingdoms, the plants and the animals; others recognize as many as five kingdoms, including the unicellular blue-green algae and bacteria, the protists (unicellular or colonial algae and protozoans), fungae, true plants, and multicellular animals. Our purpose here is to examine the major animal phyla as they exist today, and we shall not worry excessively about their arrangement into kingdoms. However, it is worth spending a little time on the differences between animals and plants, a problem we shall be keeping in mind especially as we look at the single-celled animals and plants, where the distinction often blurs.

Basic differences between animals and plants

Some of the important and widespread characteristics that distinguish plants from animals include basic types of metabolic pathways for energy extraction from the environment, ability for physical movement, kind of structure and organization, degree of irritability and responsiveness, arrangement of reproductive cycles in sexually reproducing species, and physiological devices for maintaining a constant internal environment.

(1) Most plants have photosynthetic pigments that allow them to *extract energy from sunlight*; thus they are *autotrophic* (they produce their own food). Many of the single-celled protists (green algae, diatoms, protozoans) have photosynthetic capabilities, but no organisms recognized as multicellular higher animals are photosynthetic. All of them are *heterotrophic*; that is, they derive food by feeding on other organisms. (2) Another fundamental difference between plants and animals is the degree of *mobility*. Most juvenile and adult plants are attached to the substrate and are nonmotile; their dispersal phase is restricted to specialized reproductive structures such as spores or seeds. While many animals are also attached to the substrate, such as sponges and mussels, most of these have at least one mobile phase during their development. The great majority of animal species are mobile throughout their lifetime. (3) Plants tend to have *continuous growth* throughout their life, sending out new leaves and stems from shoot tips, whereas most animals have an organized and more definite body plan involving growth at only limited (and usually early) phases of development. Individual plant cells usually possess *rigid cell walls*, whereas animal cells normally have only a *membrane*.(4) Most animals have nerves and muscles that make possible a *high degree of responsiveness* to conditions in the environment (an attribute frequently called irritability). *Special receptors* for environmental stimuli are an almost universal attribute of animals, whereas almost all plants lack special receptors. (5) Most plants have a sexual cycle involving *two distinct body forms* (the body form existing between meiosis and fertilization having only half the normal diploid

number of somatic chromosomes). In animals the development of this phase is highly unusual, and in sexually reproducing species essentially the entire life of the animal is spent as a diploid organism. (6) Most animals maintain devices to establish and regulate a *constant internal environment*, and their internal fluids in the cells tend to resemble sea water in being rich in sodium chloride (ordinary salt). Plants, on the other hand, have less evident devices of this type.

The most basic difference overall, though, between plants and animals seems to be in the way that food is acquired or synthesized. The other adaptive features of animals and plants alike appear to be more or less closely related to this basic difference, particularly in regard to the animals' mobility for obtaining food from other organisms rather than manufacturing it as plants do from raw materials brought to them by environmental phenomena such as sunlight, rainfall, and air currents.

With this background, let us look at 11 important animal phyla and some of their basic characteristics.

Phylum Protozoa

The protozoans are protists, having structural and evolutionary affinities to other simply organized organisms such as slime molds, fungae, and algae (Figure 16–4). Strictly speaking, therefore, they are not animals, but they tend to be the most animal-like protists. They are among the most successful of all organisms, with some 30,000 described living species. They are heterotrophic or parasitic, engulfing their food in the form of other living creatures. The entire body

FIGURE 16–4

The delicate structure of various species of Foraminifera, a major group of the phylum Protozoa.

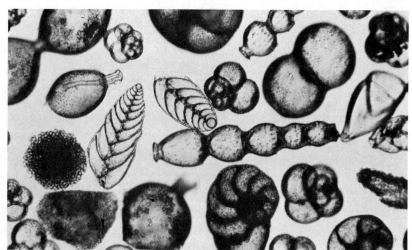

plan is composed of one cell, except for colonial groupings of cells in certain species, and most are found in aquatic situations. Conservatively, at least 100,000 species of phylum Protozoa can probably be assumed to exist at present. Some 10,000 extinct species have been described to date from geologic time periods as early as the Precambrian. As the protozoologist Theodore Jahn has remarked, the protozoans are anything but simple unicellular organisms (the definition given in many college textbooks): "This definition is outstanding for the magnitude of the erroneous impression which it gives. No protozoan is simple; some are not unicellular, and some may not, in the strict sense of the word, be considered animals. The protozoa cannot be considered simple in any sense of the word. Each individual is complete in that it contains, often within a single cell, the facilities for performing all of the body functions for which a vertebrate possesses many organ systems." As we shall see in a later chapter, the plan of cellular organization in a single cell such as an ameba or *Paramecium* is almost unbelievably complex (plate 14).

Phylum Porifera: the sponges

FIGURE 16–5

Sponges (Microciona) in the phylum Porifera.

About 5,000 species of sponges are known; all of them are aquatic and most are marine (Figure 16–5). Their structure is simple and resembles that of the Protozoa, in that the separate cells often seem to live semi-independently of the whole group. But in most sponges the cells form a few well-differentiated kinds with different functions, and some simple coordination of these incipient tissues is found in the two layers of cells composing the body. The body is permeated by pores, canals, and chambers, through which a current of water flows, bringing in food in the form of microscopic particles with the current. Fossil deposits show that sponges have been around since the dawn of life. The adults are sessile and often live in colonies; the larvae, however, are flagellated and free-swimming. About 150 species of horny sponges live in fresh water, while all of the other known species in phylum Porifera are marine, including glass, chalk, and horny sponges. These sponges often provide a home for numerous other animals such as shrimps, crabs, barnacles, and brittle stars, which take shelter in the spaces of the sponge colonies (plate 14).

Phylum Cnidaria: the coelenterates

The approximately 10,000 species of cnidarians (formerly called phylum Coelenterata, referring to the hollow body) constitute a phylum of great importance, both in the historical record and in a

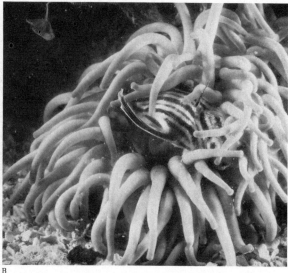

A   B

FIGURE 16–6

Marine representatives of the phylum Cnidaria: (A) a jellyfish (Chysaora quinquecirrha), showing the floating bell at top and long tentacles and mouth-area arms that capture prey, and (B) a sea anemone (Bunodosoma species) devouring a fish after capturing it with a multitude of tentacles.

modern ecological sense (Figure 16–6). They are believed to represent the most primitive living multicellular animals, and today they are possibly the most common macroscopic animals of the oceans. The corals and their relatives form the coral reefs and coral islands across the tropical and subtropical seas. They have two cell layers separated by a gelatinous substance, and the individuals may be either polyps (cylindrical sessile animals that are often colonial) or free-swimming, bell-shaped jellyfish types. The name now used for the phylum comes from the most common trait of the animals, the presence of *cnidoblasts*—stinging cells found singly or in groups around the mouth, on the tentacles, and elsewhere on the body. A cnidoblast contains a horny, stinging capsule, or *nematocyst*, which has a pointed spike attached to a coiled hollow thread. These function in trapping and holding the prey, and some can secrete a paralyzing toxin on stimulation by touching a prey or enemy. The 200 or so species of jellyfish are perhaps the most commonly seen representatives of the Cnidaria. Tide pools along the coast are filled with sea anemones, sea pins, and sea pansies (about 6,000 recent species), and the corals are familiar to anyone who has visited tropical waters (plate 15).

Phylum Platyhelminthes: the flatworms

Approximately 12,700 species of living flatworms are found today; about two-thirds of them are parasitic (Figure 16–7). The body is flattened dorsolaterally, from back (dorsal surface) to underside (ventral

FIGURE 16-7

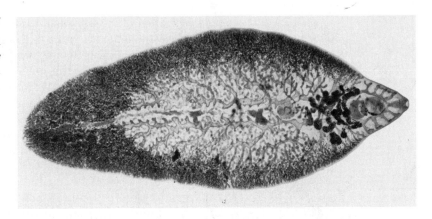

The sheep liver fluke (Fasciola hepatica), a flatworm in the phylum Platyhelminthes.

surface), and the digestive system has only a single opening, at the mouth. They lack a circulatory system and most species are hermaphroditic — both types of sex organs are located in a single worm. Some tapeworms are up to 20 yards or more in length, but in general the flatworms are small animals, ranging from microscopic sizes to lengths of two to three inches. The head region contains the chief nerve centers and sense organs; the mouth is on the underside of the body, commonly in a middle location. The free-living flatworms are undoubtedly the more primitive members of phylum Platyhelminthes and are carnivorous; the parasitic members are specialized evolutionary derivatives. Included in this phylum are the planarians (who can regenerate missing heads and tails), the tapeworms, and the flukes. The number of flukes (6,250 species) and tapeworms (3,400 species) in the gut and other organ systems of most vertebrates and invertebrates is truly astounding. Besides their usefulness in regeneration research, planarians are also well known for their ability to be trained by conditioning. The worms can be conditioned in an aquatic maze and then cut up into several pieces; after the pieces have regenerated to whole animals, each still "remembers" the training of the original worm. A chemical memory substance has been implicated in these knowledge-storage and knowledge-transfer processes. The life cycles of the parasitic members are quite complex, often involving a number of intermediate hosts before the primary host is involved.

Phylum Nemertina: the ribbon worms

FIGURE 16-8

A ribbon worm, phylum Nemertina.

The ribbon worms, or proboscis worms, include about 600 species of largely marine animals (Figure 16-8). Some are terrestrial, but very few live in fresh water. All Nemertina species have an alimentary system with a separate mouth and anus, thus being somewhat advanced over the flatworms. Also, they possess a circulatory system, and the sexes are separate. Most of the species range in length from less than an inch to under two feet, but one species of *Lineus* can be up to 100 feet long. Many of the species are brightly pigmented with red, green, brown, and other colors forming striped patterns. The terrestrial species are occasionally seen in temperate and tropical forests, climbing over moist logs and along the ground.

Phylum Aschelminthes: the rotifers and nematode worms

The diverse phylum Aschelminthes includes approximately 1,500 species of rotifers, which are nearly microscopic aquatic animals with an anterior wheel organ, and an as yet undetermined but astounding number of nematode worms (Figure 16–9).

The rotifers are mostly fresh-water animals, with a few marine species. They live as swimmers, floaters, creepers, and sessile forms, and their shapes tend to reflect their modes of life. Creepers and swimmers are elongate and roughly worm-shaped, while floaters tend to be globular and saclike. Sessile species are vase-shaped. The major distinguishing feature of all the groups is the wheel organ, an anterior wreath of beating filaments used in swimming propulsion and in creating water currents to bring in food to the mouth, which has a muscular pharynx with jaws. One of the amazing features of the rotifer class is that each member of a species is believed to be constructed from exactly the same number of embryonic cells. Each individual retains the same number of nuclei at exactly fixed locations and thus any two members of a species are structurally identical down to the cell and nucleus. By actual count, a rotifer consists of 1,000 to 2,000 cells or nuclei, and each species possesses a particular number somewhere in that range.

The roundworms, or nematodes, are cylindrical, elongated worms that often have bristles but lack any hairlike cilia on the outside of their body (flatworms and annelid worms have such cilia). A great number are parasitic species; some types reach lengths of well over three feet but most are under two inches in length. Only 12,000 named species have been described, but informed guesses by nematologists place the number of existing nematode species at about half a million. After the insects, then, roundworms are probably the most abundant of all animals. Virtually every other animal species harbors at least one type of parasitic nematode, and roundworms are also found extensively in plants. It has been said that the free-living species in fresh-water, ocean, and soil habitats, and the parasitic species in other organisms, would probably adequately outline all

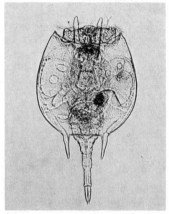

A

B

FIGURE 16–9

(A) a rotifer, or wheel animal (Platyias quadricornis), and (B) a roundworm (Nematode species) in the phylum Nematoda.

FIGURE 16–10

A top shell (Calliostoma species), gastropod representative of the phylum Mollusca.

FIGURE 16–11

Segmented worms in the phylum Annelida: (A) the marine fringed worm (Cirratulus grandis) from the intertidal sand of the west coast of Florida and (B) the common earthworm (Lumbricus terrestris).

surface, geological, and biological shapes on the face of the earth if by some mysterious means all matter other than roundworms was removed from the earth.

Phylum Mollusca: the mollusks

Phylum Mollusca includes the clams and their bivalve relatives, snails, chitons, octopuses, and squids, as well as a host of related kinds, amounting to some 128,000 described species (Figure 16–10). The soft mollusk body is covered by a thin dorsal fold, the mantle, which usually secretes a hard shell of one or more pieces to protect the body. The digestive tract is usually complete with a mouth and anus and is often placed in a U-shape. The ventral foot provides locomotion for most of the species although one class—the squids, cuttlefish, and octopuses (700–800 recent species)—have a jet-propulsion system, utilizing expulsion of water from the mantle for movement (plate 15).

Phylum Annelida: the segmented worms

With about 15,000 species, annelid worms are important members of all major environments (Figure 16–11). Here for the first time in the lower phyla we find a series of highly organized body segments. The first anterior segment typically forms the head and the last contains the anus, with the intermediate segments all alike in at least the developmental stages. Numerous Annelida species also have modified trunk segments; up to 800 trunk segments can be produced in some annelids (plate 16).

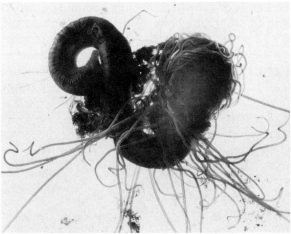

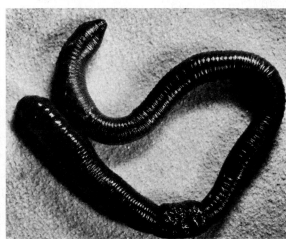

A

B

All living species possess a tubular digestive tract, complete from mouth to anal opening, and a closed circulatory system with longitudinal vessels having side branches in each segment. They occur in marine, fresh-water, and terrestrial habitats in great abundance. The earthworm class includes some 3,100 species, and the marine polychaete group contains over 5,800 known species. The leeches are familiar parasitic members in fresh-water and tropical habitats, widely known far beyond their numbers (about 300 species) because of their blood-sucking habits.

Phylum Onychophora: *Peripatus* and its relatives

Only about 73 species of the strange animals in phylum Onychophora are known, but they are the most clearly perfect "missing link" between two other animal types known in zoology. Their characteristics are intermediate between annelids and arthropods, one of the best proofs of the close relationship of these major phyla. The Onychophora are widely distributed in tropical and subtropical regions of the world. These worms, such as the well-known *Peripatus*, are about two or three inches in length and live in damp, leafy places on the ground, feeding mainly on insects (Figure 16–12). The body looks like that of a velvet-textured caterpillar, and the head, consisting of three segments (one head segment is found in the annelids and six in most arthropod heads) has a pair of annelidlike eyes and a pair of insectlike antennae. Along the body there are many pairs of claw-bearing legs and corresponding internal segments, with an absence of external segmentation. Thus annelid and arthropod traits are thoroughly mixed. The sexes are separate and most species have internal fertilization, the zygotes being retained in the female until live young are ready to be born. Some species are truly viviparous, with the mother's body contributing to the nutrition of the embryo.

FIGURE 16–12

The odd tropical Peripatus, *sometimes placed in the phylum Onychophora as a "missing link" between the Annelida and the Arthropoda.*

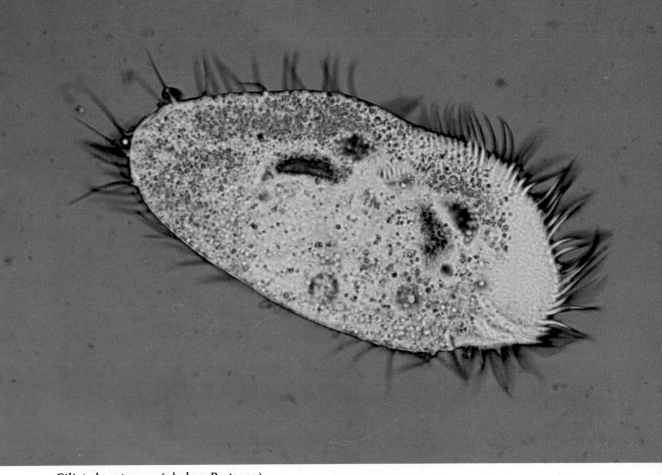

Ciliated protozoan (phylum Protozoa)

Breadcrumb sponge
(phylum Porifera)

PLATE 14

Lima file clam (phylum Mollusca)

LEFT:
Portuguese man-of-war (phylum Cnideria)

BELOW:
Sea anemone (phylum Cnideria)

ABOVE: *Scallop (phylum Mollusca)*

LEFT: *Dog welk and egg case (phylum Mollusca)*   PLATE 15

Polychaete worm (phylum Annelida)

Horseshoe crab (phylum Arthropoda)

PLATE 16

Starfish (phylum Echinodermata)

Sparrow hawk (phylum Chordata)

Differential grasshopper
(phylum Arthropoda)

Phylobates frog (phylum Chordata)

Bull elk (phylum Chordata)

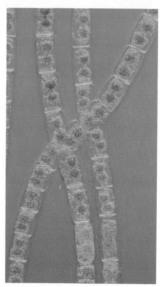

ABOVE: *A marine garden of red, green, and brown algae (divisions Thallophyta and Chlorophyta)*

FAR LEFT:
The green alga
Acetabularia *(division Chlorophyta)*

LEFT:
The green alga
Zygnema *(division Chlorophyta)*

Liverworts of the genus Marchantia
(division Bryophyta)

Mosses (division Bryophyta)

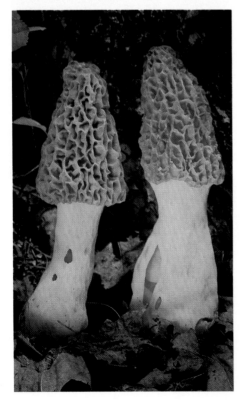

Cup fungus of the genus Galacti
(division Mycophyta)

Morel fungus (division Mycophyta) **PLATE 17**

Slime mold
(division Myxomycophyta)

Narcissus in bloom
(division Tracheophyta)

Spanish "moss" (a ctually a flow-
ering plant in the family
Bromeliaceae) and oak trees (both
in the division Tracheophyta)

PLATE 18

A   B

FIGURE 16–13   **Phylum Arthropoda: the joint-footed animals**

Two representative of the phylum Arthropoda: (A) an asparagus beetle, with eggs (class Insecta), and (B) the lobster (Homarus species, class Crustacea).

Some 1 million described species make phylum Arthropoda the largest not only among animals but all living organisms (Figure 16–13). More arthropods are known than all other kinds of living creatures, including plants. About 75 percent of the described species are insects, but according to some estimates, as many as 10 million insect species may actually exist and remain to be described. The beetles, one of the orders of insects, include some 300,000 species alone, a fact that makes this order larger than any other whole phylum of other organisms. Arthropods have been found on the deepest sea bottoms and in the air miles above the earth. They occur in all environments and ecological niches, including many not open to other forms of life. Their main competitors are other arthropods, bacteria, and vertebrates, but their essentially unrivaled success is a consequence of their unique body construction based on a series of segments and external skeletal armor. Their interior construction is remarkably similar among the arthropod classes, but the exterior chitin-covered parts vary greatly and have made the phylum very diversified.

The basic external skeleton of arthropods is divisible into head, thorax, and abdomen, with a series of paired jointed appendages, usually one per segment in the primitive forms. They have a straight, complete digestive tract and an open circulatory system (lacking a system of continuous blood vessels for conveyance to the tissues, the blood sloshes around the body by muscular movements). The central nervous system has a series of ganglia nerve centers that provide arthropods with good reflexes and rapid responses to stimuli. The diverse arthropod phylum includes horseshoe crabs, sea spiders, true spiders, scorpions, daddy-longlegs, crustaceans, sow bugs, millipedes, centipedes, and insects (plate 16).

Kingdoms and classification schemes

One would think that scientists could reach definitive decisions on taxonomic classification after two hundred years of classifying organisms almost entirely on morphological criteria. This would seem to be especially true at the kingdom level; after all, the kinds of geographic and varietal variation found within species hardly extend even to the generic level. However, the number of living kingdoms of organisms in the world has been hotly debated since Linnaeus' time.

In 1758, Linnaeus (and earlier scientists) thought that there were three kingdoms: plants, animals, and minerals. Biologists soon dropped minerals from their definition of life, but the two-kingdom scheme of animals versus plants was not totally satisfactory either. So in one commonly-used classification system, the bacteria and blue-green algae, considered part of the plant kingdom in the two-kingdom system, are grouped as a third kingdom, the Monera. This system groups in the Monera all one-celled organisms without a nuclear membrane. Other systems group all one-celled eukaryotic organisms (cells with nuclear membranes) together as protists, and put the fungi in with the plants, arriving at three kingdoms: Protista, Plantae, and Animalia.

The most recent proposed classification systems divides organisms into five kingdoms: the Monera, Protista, Fungi, Plantae, and Animalia. In this scheme, the Monera are the only organisms that have prokaryotic cells, those with no organized nuclei in a nuclear membrane. They also lack mitochondria and chloroplasts and have simple chromosomes (a single circular molecule of DNA). The Protista are also unicellular but have eukaryotic cells and complex chromosomes of DNA plus protein. They have a nuclear envelope and mitochondria, as do the Fungi, Plants, and Animals. The Fungi lack chlorophyll and have a cell wall made of chitin and other non-cellulose polysaccharides; otherwise, they share most charcteristics with at least some or all of the Plant and Animal kingdoms.

No one classification system for kingdoms has been universally accepted, but regardless of the scheme used, the Linnaean binomial system for species and the categories for general classification of taxa (genus or higher) remains unaffected.

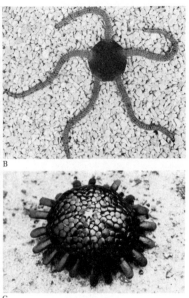

FIGURE 16–14

Three animals from the phylum Echinodermata: (A) a white sea urchin (Lytechinus variegatus), (B) a brittle star (Ophiopholis species), and (C) a plated sea urchin (Colobocentrotus species).

Phylum Echinodermata: the sea stars and their relatives

With about 5,700 species, phylum Echinodermata includes the sea stars (starfish), sea cucumbers, sea urchins, sand dollars, sea lilies, and serpent stars (brittle stars) (Figure 16–14). Unlike most of the other phyla we have been examining in the preceding sections, the members of this group are radially symmetrical, that is, formed on a circular pattern of organization rather than a bilateral one. They are all spiny-skinned animals and are exclusively marine. The bilaterally constructed larvae are quite mobile, but most of the adults are sessile; while having basically five arms or parts, they lack internal segmentation. The bilateral larvae metamorphose (change their structural organization) into sessile or sluggish adults when they settle after a period of swimming in their early life.

One of the most notable features of the adults is a tube foot system, which aides in both feeding and locomotion. Tube feet (Figure 16–14) act as suction cups, which can be used to grasp food as well as the bottom substrate. To open an oyster, for instance, a starfish grasps its prey in a multirayed embrace. Hundreds of sucking disks, or tube feet, fasten on to the oyster's two shells and begin pulling in opposite directions. The tubular disks tug in alternate fashion so that some can rest while others pull. Gradually the oyster's muscles tire and the shell opens. At this point, the starfish will evert its stomach out of its mouth, surround the soft parts of the oyster, and begin digesting it (plate 16).

Phylum Chordata: the protochordates and vertebrates

Most of the chordates are vertebrates — the familiar fishes, amphibians, reptiles, birds, and mammals (Figure 16–15). The most primitive forms in phylum Chordata of some 50,000 species, however, are the protochordates. These animals share the common traits of all chordates, namely, a rigid rodlike notochord of supporting cartilage near the dorsal (back) surface of the animal, a dorsally located central nerve cord, and gill slits or pouches in the throat or pharynx. The most primitive members of the phylum are undoubtedly the headless and unsegmented tunicates, which are marine and comprise about 2,000 species living in sand, mud, or attached to rocks. Many of the tunicates look rather like sponges superficially, but they are actually highly organized animals with tissue layers and incurrent and excurrent openings serving as the food-gathering device, drawing water in and pumping it out. In the tadpolelike larvae of the tunicate's life cycle we see the relations to many of the other chordates. The highly mobile larva has a notochord and a dorsal nerve cord (plate 16).

Another primitive group of chordates are the essentially headless and segmented cephalochordates, or lancelets. Amphioxus is another common name for these cephalochordates. These marine organisms live in shallow coastal waters of various places around the world, dwelling in burrows on sandy bottoms. Only two genera are known, which consist of about 30 species in all. The typical lancelet is pointed at both ends, slender, laterally compressed, and about two to three inches long; it possesses a notochord made of cartilage, a dorsally located nerve cord, and the other attributes of the chordate phylum. Although amphioxus evidently is quite specialized in many respects, the animal does display numerous vertebrate features in a very primitive form. Consequently, it suggests what ancient vertebrates might have been like and probably exemplifies a transitional stage in the evolution of ancestral larvalike tunicates to the first primitive vertebrates: jawless fishes (Chapter 14).

The highest animals in the phylogenetic scheme of life are the vertebrate chordates. They have segmented bodies, with a head, trunk, and tail, and a bony or cartilaginous skull enclosing the brain. The embryo of all vertebrates has a notochord, and the adult possesses either a notochord alone or a vertebral column of cartilage and bone. The subphylum Vertebrata contains eight generally recognized classes: jawless fishes, extinct armored fishes, sharks, bony fishes, amphibians, reptiles, birds, and mammals. We have looked at the characteristics of these groups in Chapter 14, when we considered the evolution of the vertebrates from the earliest fishes in the Ordovician through the appearance of human beings in the Pleistocene.

FIGURE 16–15

The Chordata: (A) a lancelet, or Amphioxus species (Cephalochordata); (B) a clown or anemone fish (Pisces); (C) a frog (Amphibia); (D) an Arizona coral king snake (Reptilia); (E) young barn owls, Tyto alba (Aves); (F) African rhinoceroses, Diceros bicornis (Mammalia).

The kinds of plants

At present there are approximately 288,800 species of plants known to be living on the earth. About one-half of the species of plants are known as higher plants and include such groups as the club mosses, ferns, and seed plants. The higher plants are separated by some authorities into 341 families, the majority of which are tropical and subtropical in distribution. The major groups of lower plants are classified in a great many ways, often as many as 12 phyla (called divisions for plants) being recognized. We shall look only at the major recognized divisions here (plates 17 and 18).

Division Cyanophyta: the blue-green algae

The undifferentiated single-celled or colonial body of blue-green algae places them among the most primitive plants known. The nucleus structure is primitive, lacking a nuclear membrane. No sexual reproduction is known to occur, and there are no motile stages. The blue-green algae are photosynthetic, and many of them are quite successful in fresh-water and marine habitats as well as damp soil and rocks. About 1,500 species are known. Their characteristic color is due chiefly to the blue pigment phycocyanin and chlorophylla, but a few species also possess a red pigment. Fossils of this group date from the Precambrian era to recent times.

FIGURE 16–16

Algae: (A) a marine brown alga (Sargassum), (B) a North Sea green alga (Cladophora), and (C) a fresh-water green alga (Spirogyra).

Division Chlorophyta: the green algae

Approximately 5,500 species of the photosynthetic green algae are known (Figure 16–16). The *chloroplasts* contain *chlorophyll* a and b pigments (to be discussed in Chapter 25) that, along with other as-

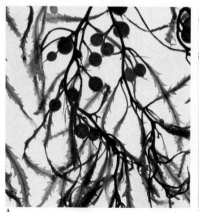

A

B

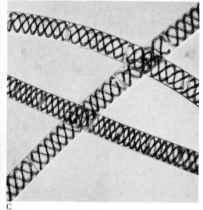

C

sociated pigments, relate these plants to the pigments found in higher plants. Food is usually stored as starch, and in the Chlorophyta both sexual and asexual reproduction occur. The plant body may be single-celled, colonial or multicellular, and both motile and nonmotile stages occur. They are common in fresh-water and marine habitats.

Division Chrysophyta: the diatoms and yellow or golden algae

More than 5,000 living Chrysophyta species, mostly diatoms, are known. They are photosynthetic and store their food as an oil or as a complex carbohydrate. This is a truly diverse division, including motile and nonmotile single-celled forms and colonial forms composed of long filaments. They live in soil and in fresh and marine waters.

Division Thallophyta: the brown and red algae

The thallophytes are multicellular fresh-water and marine seaweeds and kelps. They are photosynthetic, with the chloroplasts containing chlorophyll a and c or d, as well as brown or red pigments. There are about 4,500 species known; most of them are marine. The plant body is non-motile, usually large and complex, and a conspicuous part of the marine environment. Food is stored as a complex carbohydrate.

At least several other divisions of algae, including groups with other pigment types and flagellated algae, are commonly included in division classification schemes of the plant kingdom (or kingdoms). The appendix on classification gives further details on these relatively small groups.

Division Schizophyta: the bacteria and their relatives

The true bacteria form only one of the orders within this phylum but comprise about 1,300 out of the 1,600 known species (Figure 16–17). The viruses have also been included in this division by some authorities. Most of the species reproduce by simple fission of the cell into two daughter cells. A few have simple whiplike flagella filaments. The group is known in the fossil record from early Precambrian to recent times.

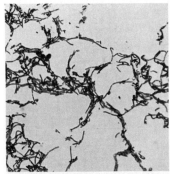

B

FIGURE 16–17

(A) Bacteria (Bacillus subtilis) and (B) slime mold (Physarum virice). The slime mold plasmodium (body) forms a flat blanket from which spore-bearing sporangia arise.

Division Myxomycophyta: the slime molds or slime fungi

While only 450 species have been described of this interesting division, Myxomycophyta, their attributes are extraordinary enough to merit special mention. The plant body of a slime mold is a naked mass of protoplasm. The nuclei seemingly float freely through a flat sheet of cytoplasm that may be several inches to even a foot or more in diameter. Slime molds travel by pseudopodia (or amebalike false feet) over rotting wood and engulf their food much like an animal. The plant body is noncellular in the sense that there are no cell walls, but, as noted above, contains scattered nuclei (Figure 16–17).

Division Mycophyta: the true fungi

Division Mycophyta includes yeasts, molds, and mushrooms, with about 36,500 species in all (Figure 16–18). The plant body is usually composed of long, multicellular filaments, called hyphae, that form a plant body from which reproductive parts grow. Chromosomes are present in these multicellular forms; evident chromosomes are lacking in the single-celled yeasts. Many of the species have elaborate reproductive structures, such as big clublike sacs or stool-shaped structures, that disperse millions of spores following sexual exchange between the hyphae.

FIGURE 16–18

True fungi: (A) a coral fungus (Clavaria species), (B) a fly mushroom (Amanita muscaria), and (C) satyr's bread fungus (Hericium erinaceum).

FIGURE 16–19

*The Bryophyta: (A) pigeon
wheat moss (Polytrichum
commune) and (B) liverworts
(Marchantia species).*

A   B

Division Bryophyta: the mosses and liverworts

The bryophytes are multicellular terrestrial or aquatic plants that
lack vascular (conductive) tissues (Figure 16–19). They are photosynthetic and contain chlorophyll a and b pigments. Most species
have multicellular sex organs that produce a parasitic spore-bearing
structure on top of the leafy sexual phase in the life cycle. About
8,000 species of liverworts and 320 species of hornworts are known;
these flattened bryophytes grow on moist soil and rocks in a leather-
like body. The mosses, with about 14,000 species, have a leafy plant
body with a radial symmetrical arrangement of leaflike structures
around a stem. Spores are released from a capsule elevated above the
sexual-phase plant body.

Division Tracheophyta: the vascular plants

The tracheophytes, or vascular plants, comprise the ferns and seed
plants (Figure 16–20). They have specialized conducting cells (vascular tissues called *xylem* and *phloem*, whose structure and functioning we will consider in a later chapter). In the most primitive
forms, the leaves are usually absent as well as the roots, leaving a
plant body composed of branched stems; there are only four living
species of that ancient group. Somewhat over 900 species of club
mosses are known and about 25 living species of the familiar horsetails (*Equisetum*), which grow along the edges of ponds and marshes.
The ferns, however, include over 9,000 species distributed in both

FIGURE 16–20

The vascular plants: (A) horsetails, (B) ferns (marginal wood fern, Dryopteris marginalis), (C) gymnosperms (lodgepole pine, Pinus contorta), and (D) an angiosperm, or flowering plant (white water lily).

temperate and tropic zones around the world. They are among the most successful colonizers of remote oceanic islands in the tropics because their light spores can be borne on the winds for great distances. The living gymnosperms have naked seeds not surrounded by ovary tissues but borne on the open surfaces of leaves or in cones; there are about 710 species existing at present. These pines, firs, spruces, and their relatives were more abundant over the earth in past ages. The most advanced class in this division of vascular plants are the angiosperms, or flowering plants; the seeds are enclosed by special structures called carpels. The leaves are typically broad rather than needlelike, as in the gymnosperms; a pollen tube specifically transfers the sperm cells to the developing eggs, rather than pollen merely drifting directly to the exposed ovule, as in pines and related gymnosperms (we will look at plant reproductive strategies in more detail in the last chapter of this book). Some 250,000 species of flowering plants are known, the largest families being the composites, such as the sunflowers, and the orchids, each family having more than 20,000 species. The angiosperms are all multicellular plants, with vascular tissues for transport of fluids and food materials. Most are photosynthetic to at least some degree, having chlorophyll a and b pigments. A few parasitic species lack chlorophyll.

Summary

With close to 2 million described species of living plants and animals, the diversity of life could overwhelm us in our biological studies were it not for the efforts of taxonomists to organize systems of classification and categorize each species by genus, family, order, class, and phylum. The term species refers to a group of organisms that shares a common set of genes and phenotypic attributes and that normally does not interbreed with another distinct species. Homology, or correspondence in structure and function, is used to establish classificatory systems. At the most basic level, animals are distinguished from plants by types of metabolic pathways for energy extraction from the environment, ability for physical movement, kind of structure and organization, degree of irritability and responsiveness, arrangement of reproductive cycles in sexually reproducing species, and physiological devices for maintaining a constant internal environment. The principal phyla of animals include the protozoans (Protozoa), the sponges (Porifera), the coelenterates (Cnidaria), the flatworms (Platyhelminthes), the ribbon worms (Nemertina), the rotifers and nematode worms (Aschelminthes), the mollusks (Mollusca), the segmented worms (Annelida), Peripatus and its relatives (Onychophora), the joint-footed animals (Arthropoda), the sea stars and their relatives (Echinodermata), and the protochordates and vertebrates (Chordata). The main divisions of plants involve many types of algae (Cyanophyta, Chlorophyta, Chrysophyta, Thallophyta), bacteria and their relatives (Schizophyta), the slime molds (Myxomycophyta), the true fungi (Mycophyta), the mosses and liverworts (Bryophyta), and the vascular plants (Tracheophyta).

17 Ecological and historical biogeography

At the beginning of this book, we looked at the worldwide distribution of major vegetation zones, or biomes, the dozen or so major community types that circumvent the globe at similar latitudes on each continent. The geographical distribution of life in these biomes is established by both ecological and historical factors.

The distribution of organisms

As we have seen (Chapter 9), in an ecological sense the distribution of any species of organism is ultimately limited by the distribution of suitable environments. Rainforest orchids do not occur in the deserts of southern California nor do the ground-dwelling cactuses of the Mohave Desert invade the wet soils of the rainforest. The limiting environmental conditions may be associated with rainfall, temperature, annual cycles of light, or, for aquatic organisms, degree of salinity in the water. The limiting conditions may also be exerted by other organisms, which constitute a major part of any species' environment. However, the distribution of a species in the world as a whole is not determined solely by ecological factors; for example,

South America

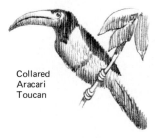

Collared
Aracari
Toucan

Africa

Gray
Hornbill

FIGURE 17–1

Ecological equivalents in families from similar environments: (above) a toucan from South America and a hornbill from Africa; (right) a cactus from the American Southwest (left) and a euphorb from Africa. An ecological biogeographer would view the deserts or tropical rainforests containing these look-alike organisms as worldwide biomes. An evolutionary biogeographer, however, would separate Africa and South America into different faunal or floral regions because these organisms are not closely related even though they look quite similar.

the cactus species of southern California are replaced by members of another family, the Euphorbiaceae, in the great deserts of Africa (Figure 17–1). This distribution is determined by purely historical factors.

The study of the geography of living things and the factors that determine their distribution is called **biogeography.** When we consider the ecologically caused resemblances and differences in communities, we are involved in *ecological* biogeography, that is, the study of the functional and adaptive associations of plants and animals in different geographic locations. *Historical* biogeography is concerned with the differences in the histories of regions. The frequently puzzling differences among regional biotas inhabiting the same general kind of ecosystem—the flora and fauna of the Sahara in Africa versus those of the deserts in the American Southwest, for instance (Figure 17–2)—are due to the fact that plants and animals arrived in each region at different times and from different sources, and their pathways of evolutionary divergence differed once they had arrived in that continental area.

Let us first review the broad principles of environmental influence on the distribution of organisms and then look in more detail at the historical reasons for the patterns we see in the distribution of life today.

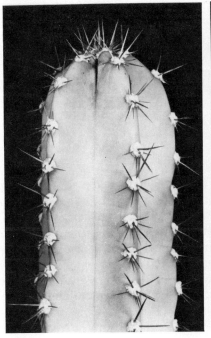

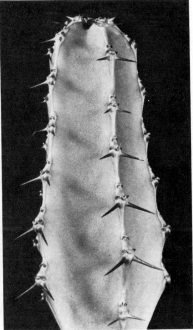

FIGURE 17-2

The desert in (A) North America (Death Valley) and (B) southwest Africa (Kalahari).

A

B

Ecological biogeography

In Chapter 2, we looked at the organization of the biosphere and the climatic and geologic processes that affect the formation of climax communities and the major biomes. The differences in environment from area to area cause different adaptations in the plants and animals inhabiting a particular environment. In order to coexist with the other biological members of a community—an intimate part of any organism's total environment—each organism must adapt its niche to accommodate, tolerate, and exploit diverse members of that community.

Plants, especially terrestrial species, tend to play a predominant

Lystrosaurus in Antarctica

The discovery of the first Antarctic fossil specimen of the Triassic reptile *Lystrosaurus* illustrates the frequent role of chance and the factor of serendipity (being prepared to interpret or exploit a fortuitous happening) in solving scientific problems. An unidentified fossil bone fragment was brought out of Antarctica in 1968 by a New Zealand geologist, and when paleontologist Edwin H. Colbert identified it as a piece of the jaw of a labyrinthodont amphibian (the first land vertebrate fossil ever to be reported from Antarctica), the U.S. National Science Foundation decided to sponsor a search for Triassic vertebrates in the Transantarctic mountains.

Thus in the following southern summer of 1969–1970, a group of about 20 scientists flew to a field camp near Beardmore Glacier, within about 400 miles of the South Pole. They arrived on a cold Saturday afternoon and spent much of their time in insulating the sides of their hut with snowblocks. The next morning — Sunday, December 4, 1969 — the expedition leader, geologist David Elliott, decided to go over to the nearest rocks, a ridge known as Coalsack Bluff. No field work had been planned there; in fact, the expedition's presence near that particular area was more by accident than by design, since it was principally chosen because it was a good place for the big ski-equipped Hercules cargo plane to land. Elliott's spur-of-the-moment trip over to the bluff was mainly for want of something better to do — serendipity reigns again.

When Elliott reached Coalsack Bluff that Sunday, the element of chance entered once more. As a result of an argument with a paleobotanist over the age of the sandstones that were exposed in the side of Coalsack Bluff, Elliott climbed up to these sandstones and began poking about on the cliff faces. He soon found two tiny fragments of what seemed to be fossil bone.

On Elliott's return to the Beardmore Camp at lunchtime, Colbert identified the samples as fossil bone and after lunch set out for the bluff with a large group of scientists. A bonanza of fossil bones were discovered. That very evening Colbert identified a piece of skull collected by James A. Jensen as belonging to the key reptile *Lystrosaurus*, and the news was flashed across the world. Within a few hours of his arrival on the southern continent, Colbert had encountered a fossil of the very animal he had hoped to find.

Here was a land-living fresh-water-feeding herbivorous reptile of ancient age, virtually identical to animals in Asia and Africa, which could not have possibly negotiated the wide seas that now separate southern land masses. In the words of Lawrence M. Gould, Professor of Geology at the University of Arizona and former Chairman of the Committee on Polar Research of the National Academy of Sciences: "No further proof was needed for the former existence of Gondwanaland. . . . This is the key index fossil of the lower Triassic in the major southern land masses of the former great southern continent of Gondwanaland. . . . Not only the most important fossil ever found in Antarctica but one of the truly great fossil finds of all time."

role in the ecological biogeography of biotas because they form the foundation of all food chains, and the character of the flora in a given place strongly influences the nature of later links in the chain and therefore of the whole biota. Plants are sensitive to precipitation patterns, solar radiation, soil type, temperature level, and other environmental variations that have north-south gradients from tropic to arctic areas. Progressing northward, the seasonal variation of temperature becomes greater, and cold winter temperatures prevent plant growth for longer and longer periods of time during the annual cycle. Radiation is not as intense, and its distribution through the year in terms of daily photoperiod becomes increasingly uneven, so that the Arctic has months of continuous daylight followed by months of continuous darkness.

Because the rainfall patterns are influenced by air movements and by topography, an east-west differentiation in available moisture also takes place across the continent. Thus these horizontal zonations of climate tend to create horizontal zonations of vegetation that influence the distribution of animals, creating zones of distinctive biotas. Similar changes occur as one ascends a mountain. Figure 17–3 shows the relationships between geographical and altitudinal climatic zones. Temperature and precipitation change at different elevations in the same manner as at different latitudes. Terrestrial communities tend to respond closely to these ecological changes, and the result is seen as different biotas inhabiting different regions.

In summary, then, the basic outlook of ecological biogeographers

FIGURE 17–3

The parallel between latitudinal (horizontal) and altitudinal (vertical) distributions of vegetation zones.

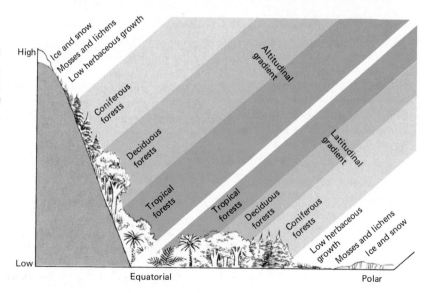

A   B   C

FIGURE 17–4

Vegetation zones at different elevations in Arizona: (A) the lower Sonora desert at about 3,200 feet elevation, with many saguaro cacti; (B) an upper Sonoran oak woodland at about 4,300 feet elevation; (C) a Canadian zone coniferous forest at over 8,000 feet elevation.

emphasizes the importance of plants, since plant distribution is especially sensitive to physical environmental factors and plays a major role in determining animal distribution. The dominant factors controlling plant distribution include temperature, solar radiation, and precipitation, which vary both horizontally and vertically across continental areas. Thus there is a parallel arrangement of major plant communities observed in both altitudinal and latitudinal transects, a biogeographic phenomenon particularly well seen in the high mountains of western North America (Figure 17–4).

Historical biogeography

In historical biogeography we are primarily concerned with the resemblances and differences of the faunas and floras of large areas, such as major continents and oceans. Ecological biogeography is more concerned with the smaller-scale distribution of plants and animals—why they live in a certain place. Much of the attention of evolutionists and other biologists during the nineteenth century was directed toward the description of regional patterns of floras and faunas that could be explained better by historical factors than ecological factors. In particular, by the beginning of the twentieth century, the basic biogeographic regions of the world were outlined by the accumulated facts concerning the overall distribution of land mammals and birds.

The world's faunal regions

These early biogeographers proposed five basic faunal regions with various subdivisions: namely, the Holarctic, Oriental, Ethiopian, Neotropical, and Australian (Figure 17–5). Some general faunal resemblance was seen to exist throughout each region, and each possessed distinctive components that differed from those of any other region, apparently because of historical reasons. This pattern of historically significant biogeographic regions does differ considerably from the patterns of biome distribution described in Chapter 2, where the emphasis was on existing vegetational zonation caused primarily by climatic and associated environmental factors. Let us now look at these historical biogeographic regions in more detail (Figure 17–6).

The *Holarctic region* encompasses the whole northern areas of Europe, Asia, and North America. Many biogeographers divide the Holarctic into two major regions, the Nearctic (New World northern areas) and the Eurasian Palearctic (Old World northern areas). These are the major temperate zones of the world, excluding only those at the southern tip of Africa, Australia, and South America, which are sometimes placed in their own restricted regional classifications. The Holarctic region shares in North America and Eurasia such animals as moose (called "elk" in Europe), timber wolves, various hares, and great stags (called "elk" in America). Different genera of deer occur on each continental area. The number of genera in common in the mammalian faunas of Eurasia and North America is large, and many times the species of the same genus are so closely alike that they may be regarded as subspecies of the same form. Thus on each continent beavers, moose, reindeer, and bears, respectively, are very closely related, while bison, lynxes, and the various deer are only a little more distinct. In Tertiary fossil strata, horses and camels are found on both continents and closely resemble one another.

The *Palearctic region*—the Old World part of the Holarctic—is composed essentially of Eurasia above the tropics and the small northern corner of Africa above the Tropic of Cancer. Thus it is

FIGURE 17–5

The faunal regions of the world. These zoogeographic regions were established because of basic similarities between bird and mammal faunas of the land areas within each region, and their relative distinctness from faunal groups in other zoogeographic regions.

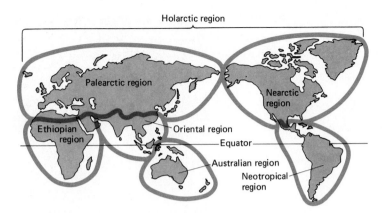

north-temperate in climate and includes that part of the Arctic inhabitable by life. Extensive deciduous forests range through the wetter parts of eastern Asia, including the temperate parts of China and Japan. Much of the interior of Asia, on the other hand, is arid and open, and this area extends through southwestern Asia to north Africa, grading into grassy steppes, especially along its northern edge. Across northern Asia and Europe lies a great stretch of coniferous forest, and to the north the Arctic tundra.

The *Nearctic region*—the New World part of the Holarctic—is composed of North America above the tropics. This region, too, is north-temperate in climate, with an Arctic fringe. Originally, the vegetative cover of much of eastern North America in the midlatitudes was deciduous or mixed forest. To the west and the middle part of the continent are extensive grasslands, and further west lies a complex of more or less arid desert and canyon country from which mountains arise, bearing strips of mixed deciduous or totally coniferous forests. To the north, as in Eurasia, stretches a giant band of coniferous forest across Canada and then a band of tundra, before the final band of perpetual ice and snow.

The *Oriental region* is essentially composed of tropical Asia, from southern China through India and Malaysia, and closely associated continental islands, including Ceylon off India, Sumatra, Java, Borneo, Formosa, and usually the Philippines. While it is mainly tropical, the Oriental region extends a few degrees north of the tropics, especially in northern India. There is much rainforest, especially in the eastern part of the region, but it is interrupted in places by more open, drier country. High mountains are also found in the Oriental region, including the eastern Himalayas. In the islands between mainland Asia and Australia, the boundary between the Oriental and Australian faunal regions is impossible to exactly delimit, and opinions as to their proper limits have varied considerably (as

FIGURE 17–6

Characteristic species of the world's faunal regions.

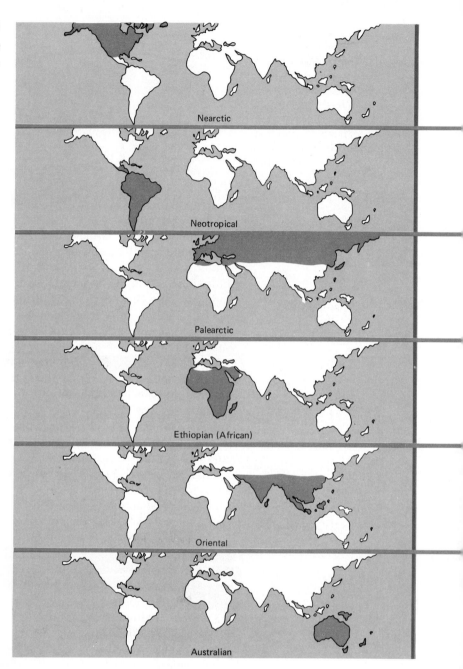

Nearctic

Neotropical

Palearctic

Ethiopian (African)

Oriental

Australian

Mountain goat

Blue jay

Musk ox

Caribou

Sloth

Tapir

Capybara

Anteater

Polecat

Hedgehog

Bison

Reindeer

Elephant

Gorilla

Zebra

Giraffe

Water buffalo

Indian elephant

Tiger

Malay tapir

Koala

Kangaroo

Flying phalanger

Wombat

they have with other regions). However, the Oriental region typically includes such mammals as Indian elephants (a different genus from the African elephant), gibbons, orangutans, tigers, and certain deer. The Orient has many pheasants and many pigeons, but only a moderate number of parrots. Some 66 families of birds are represented on land and in fresh-water habitats in the Oriental region. However, the only exclusively Oriental family is the Irenidae, the fairy bluebirds and leafbirds, with 4 genera and 14 species. Thus the Oriental bird fauna is rather generalized and actually resembles the Ethiopian bird fauna more than any adjacent region. The Old World monkeys and apes, scaled anteaters, one hyena, two rhinoceroses, wild pigs, wild cattle, and Old World porcupines and bats are also shared, at least at the family and genus level, with the Ethiopian region.

The *Ethiopian region* is notable for its big game mammals such as giraffes, African elephants, zebras, a great abundance of antelopes, two rhinoceroses, hippopotamuses, Old World porcupines, bats, elephant shrews, Old World monkeys, two great apes (chimpanzee and gorilla), hyenas, and many cats. In essence, the Ethiopian region includes all of Africa except for its northern corner above the Tropic of Cancer. Some authorities exclude part of South Africa because of its temperate nature, and others exclude Madagascar because of the unique presence there of various primitive prosimian mammals and strange bird groups. Thus the Ethiopian region is mainly tropical. There is a large block of rainforest in equatorial West Africa and many small patches of rainforest elsewhere, particularly on the great isolated mountains of East Africa. Much of the rest of the continent (and hence the Ethiopian region) is covered by dry or seasonal thorn scrub and grassland, grading into desert northward and southwestward. Some of the species here, as noted previously, also occur in the Oriental region or have close relatives there.

The two remaining major biogeographic regions are considerably more distinctive than any of those already mentioned. The *Neotropical region* is composed of South America, Central America, and the tropical lowlands of Mexico. Some authorities also include the West Indies, Trinidad, and other Caribbean islands. Across South and Central America, huge tracts of primeval rainforest were originally present, as well as regions of drier forests, and in the southern part of South America is a south-temperate zone reaching a southern latitude equivalent to that of England or Labrador. Extensive areas of grassland and savannah also occur in the Neotropics, as do some of the highest mountains in the world, with a succession of temperate forests and grasslands as elevations rise. A great many rodents, such as guinea pigs, squirrels, pocket gophers, and pocket mice, and many other families are typical inhabitants of Neotropical areas. Opossums, sloths, armadillos, a bear, canids (members of the dog family), many cats, tapirs (found elsewhere only in the Oriental region), deer, true anteaters, and New World monkeys form part of the distinctive Neotropical mammalian fauna. The greatest relict concentrations of vertebrate species lie in the central tropical portion of South America. In fact, the whole existing Neotropical vertebrate fauna seems to be a mixture of surviving parts of an old endemic fauna dating back to Tertiary times and species resulting from more recent invasions of northern groups. The biogeographical relationships of the present Neotropical vertebrates were presumably determined more by extinctions and survivals in this continental area than by directions and sources of dispersal. The northern boundary of the Neotropical region abuts on the southern boundary of the Nearctic, from which portions of its recent fauna have been received.

The *Australian region* is even more distinctive than the Neotropics in fauna and flora. Geographically, it comprises Australia and New Guinea, including also Tasmania and certain smaller islands that have essentially depauperate continental faunas. While New Guinea is wholly within the tropics, the mainland Australian area and islands to the south, such as Tasmania, are mostly temperate, with areas of grassland, deserts, and some wetter forests. The only amphibians in the Australian region are frogs (elsewhere salamanders, toads, and odd caecilians occur in addition to frogs). The birds of the Australian region include representatives of about 58 families that live in terrestrial and fresh-water habitats. Twelve of these families are exclusively found in the Australian region, or occur primarily there, including cassowaries, emus, the mound-builders, or megapodes, lyrebirds, honeyeaters, flowerpeckers, bell magpies, bowerbirds, and birds of paradise. These families comprise some 318 species, including a few extra-Australian ones. Many pigeons and parrots occur here, and the proportion of Australian birds in endemic or nearly endemic families is larger than in any other continental faunal region except the Neotropical.

In contrast to the extra-Australian geographical relationships of most of the bird families, the native land mammals of the Australian region consist principally of marsupials and all of the species belong to families that occur nowhere else in the world. (The two families of living marsupials in North and South America, including the familar opossum, are not present in Australia.) A few monotremes, including the duck-billed platypus and the echidna, or spiny anteater, along with some bats and rodents, complete the mammalian fauna. The very limited mammal fauna is strikingly primitive. Thus it seems that the Australian vertebrate fauna is made up of very few old and endemic groups that have survived or radiated in isolation

FIGURE 17–7

FIGURE 17–7

The inferred shape and world distribution of continental land masses at different times in earth's history: (A) the supercontinent Pangaea in the late Precambrian; (B) the land masses of the Jurassic period in the Mesozoic era; (C) the present configuration of continents. Extensive geological and paleontological documentation of the process of continental drift has provided new insights into patterns of evolution and diversification of plants and animals.

A

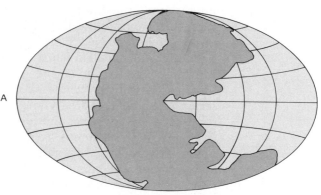

640 million years ago

B

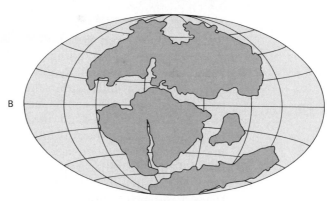

140 million years ago

C

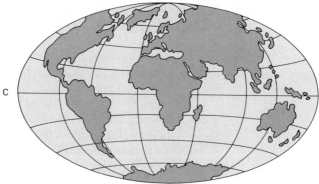

The present

in Australia for a long time. There has also been an accumulation of additional groups of salt-tolerant fishes and frogs, reptiles, birds, rodents, and bats. The native placental mammals — bats, rats, and a dog — are mostly of distinct species or genera confined to this region.

How are these patterns of vertebrate geographical distribution explained in a historical sense? The fossil record does not help us much because sometimes earlier faunal relationships were quite different from those of today. Millions of years ago, a northern part of the Neotropics such as Guatemala and Yucatán may have been considerably cooler and therefore the resident mammals of that period were of northern affinities, whereas in South America the isolated continental fauna of tropical species was quite distinctively diversified even at that time. Today, the living mammal faunas in Central America and South America are quite close because they share common vegetational formations, and considerable migratory movement of faunal elements has occurred through the nearly continuous rainforest of lowland Central and South America.

Thus organisms have been developing on a constantly changing earth, with climates varying, mountains arising, and seas advancing and retreating where today there is land. Movements in the earth's crust have caused seas now separated by land and lands now separated by oceans to have been united in the past. The interconnections among the presently defined continents and major seas have had profound effects on the distribution of organisms among these biogeographic regions. Historical biogeography is concerned primarily with this spread of new species into a region or community from elsewhere; and the history of these changes in the development of new biotas is intimately bound up with the changes in the geography of the earth we have mentioned, for geological changes opened and closed routes of dispersal. Let us look at some of the changes in land forms and routes of dispersal available for changing biotas.

Continental drift

Until the last few decades, successive generations of geologists studied the rocks of the earth and the fossils contained within these rocks and largely envisaged the changes they hypothesized *as taking place on and around continents of comparatively fixed positions.* Thus the geological and biological changes observed in their reconstructions of the events of earth history had been assumed to occur on continents solidly anchored in the positions they now occupy. However, the "geologic revolution" of recent years, based on modern evidence that seems to favor the idea of **continental drift** (Figure 17–7), has resulted in a profound change of attitude.

The concept of continental drift is not a new idea, and its beginnings may go back as far as the days of Francis Bacon (1561–1626), when the first reasonably accurate maps of the world revealed astonishing parallels between the outlines of the western border of the African continent and the eastern border of South America. However, it has only been in recent years that hard geologic and fossil evidence has accumulated to make this a favored theory today. The basic concept is that there was once a single, immense southern continent – which has been called Gondwanaland – composed of Africa, South America, Antarctica, Australia, New Zealand, and peninsular India, all intimately joined. A single, immense northern continent, christened Laurasia and composed of North America and Eurasia, joined Gondwanaland to form a huge supercontinent that has been named Pangaea. These parental continents then broke asunder, their fragments drifting across the earth through time to become the modern continents shaped and positioned as we know them today. The German scientist Alfred Wegener was a pioneer in the development of the concept of drifting continents – the theory of continental drift. His basic idea was that the position of our continents had not been fixed through geologic time but rather that they had drifted across the face of the globe to their present positions. Just as Darwin developed in detail the theory of evolution, Wegener, who published his principal work in 1912, truly developed the idea of continental drift. His ideas were largely ignored for the next several decades because the proponents of continental drift and Gondwanaland in particular often stretched the evidence beyond the limits of what then seemed to be credibility. The idea of continental drift did not seem to be respectable, worthy of serious scientific investigation.

In the last two decades, however, a modern revival of the theory of continental drift has been prompted by much evidence from geophysical and oceanographic work. Especially important has been our modern knowledge of the character of the bottom of the sea, including profoundly deep oceanic trenches bordering the various continental blocks, the configurations of the edges of these blocks, and paleomagnetic studies that indicate a lateral spread of the sea floor as well as shifting movements of great blocks of the earth's crusts. Principal biological evidence for continental drift was provided by the first discovery of fossils on the continent of Antarctica. In January 1968, Peter Barrett, a New Zealand geologist, brought a portion of the lower jaw of a labyrinthodont amphibian to the famous paleontologist Edwin H. Colbert, who pronounced it the first vertebrate fossil to be found in Antarctica. This land-living amphibian, found on a continent now completely isolated by oceans, had been found in rocks of early Triassic age, which would make it more than 200 million years old. Colbert, realizing the great significance of this study, led collecting expeditions back to the Antarctic during the 1969–1970 and 1970–1971 field seasons and discovered numerous fossils of the key amphibian *Lystrosaurus* as well associated land-living vertebrates (Figure 17–8). These fossil bones gave us a glimpse of animals that many millions of years ago had inhabited a presently frozen continent almost as large as North America.

Lystrosaurus had been previously found as fossil remains in southern Africa, India, and western China, always in association with tropical plant fossils, and the new remains found in Antarctica were so nearly identical to the petrified bones of this ancient and interesting reptile that it seems evident that in early Triassic times the several populations of *Lystrosaurus* were inhabiting a region in which they could move freely, thereby maintaining an identity of form and function among themselves. Apparently there were no mountain ranges or oceanic straits separating these populations that could allow differences to develop. From fossil evidence this reptile seems to have been a lover of lakes and rivers — an animal that might be regarded ecologically as a sort of miniature reptilian hippopotamus. Like the hippopotamus, it could not cross great stretches of ocean but had to move from one region to another by a land route, in particular a path of migration having considerable fresh water.

Such a pattern of fossil distribution fits in well with the belief that the similarities between present-day faunas of certain continental areas can be traced to historically continuous land masses. The present continents and hence their basic faunal constituents resulted from the breakup of the great continents of Gondwanaland and Laurasia. Rifts occurred where great outpourings of basaltic lavas gushed forth along the edges of the blocks that were to become separate continents. These first events took place perhaps about 200 million years ago.

The movement of the African block away from the northern part of Pangaea continued through the Jurassic and Cretaceous periods; and the separation between Africa and South America seems to have been well under way in the Cretaceous period, so that by the close of

FIGURE 17–8

Fossil remains and an artist's reconstruction of Lystrosaurus, *a key reptile fossil from the lower Triassic of Africa, Asia, and Antarctica in giving evidence for continental drift.*

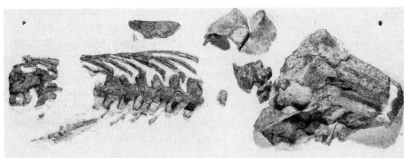

the Cretaceous (63 million years ago) the extensive South Atlantic Ocean was created between Africa and South America (though still closed at its northern end). The subsequent opening of the Atlantic Ocean by continental drift affected the faunas now separated by the North Atlantic as well as the ocean to the south. Thus following the close of the Cretaceous, a considerable gap is believed to have existed between the northern part of the African continent and the eastern seaboard of North America, although Laurasia probably still remained connected to North America by a land connection across what was later to become Greenland. This was the broad land bridge that allowed an abundant flow of giant Upper Jurassic dinosaurs between North America and the Eastern Hemisphere.

While South America was drifting away from Africa, and North America was becoming separated from its African connection as Eurasia rotated or moved somewhat eastward, the Antarctic and Australian block of Gondwanaland also was probably beginning to drift away from the African center. These now-separate continents were at that time still connected with each other and with South America through the Antarctic peninsula. The continental blocks continued to move slowly through some 60 million years of Cretaceous history, and by the end of the Cretaceous period all of the continents as we know them would seem to have been substantially blocked out.

This, then, is a summary of the possible chain of continental drift events taking place during the middle and late Mesozoic times after the breakup of the original Pangaea continent. Narrow land bridges still allowed land-living vertebrates to move from one region to another. During the Cenozoic, the Age of Mammals, the faunal connections between the continents were alternately interrupted and reestablished by climatic cycles causing the raising and lowering of the oceans. What could be the ultimate cause of these great divisions and movements of massive continental areas?

The modern concept of **plate tectonics** divides the surface of the earth into a number of huge plates, each including not only a continental mass but portions of ocean basins as well. In the past, these plates have been constantly moving in relation to each other through sea-floor spreading—that is, the movement of oceanic crust away from the mid-oceanic ridges. It is thought that these plates are still actively in motion today. There is considerable convincing evidence being amassed yearly that supports the plate tectonic theory of continental drift.

The idea of moving continents, though, affects organisms only many millions of years ago. What about more recent changes in distribution that were not affected by continental drift but are still the subject of historical biogeography?

Dispersal and barriers to distribution

FIGURE 17–9

Major dispersal routes for the world's fauna: barriers, filters, and corridors. The characteristic features of the geographic history of land faunas, especially mammals, are best interpreted by considering the main ocean barriers as blocks between continental masses. These barriers have been crossed with varying degrees of ease by different animal groups, depending on whether variable filter bridges and corridors exist between continents (open circles) or a variable sweepstakes route (dark circle), as over the passage from Eurasia to Australia.

Many organisms, especially plants and single-celled animals, are able to disperse by passive means through floating in the air and water currents. Insects, spiders, and other animals are often blown for long distances by wind, and vegetation is carried downriver, especially during times of floods. Swimming and wading birds carry eggs and larvae of many small aquatic animals on feet or under feathers and deposit them elsewhere during their aerial wanderings. Thus there are many agencies of dispersal aside from continental drift or the animal's or plant's own locomotory ability.

For salt-water-intolerant plants and animals such as mammals, birds, and reptiles, a land path is the most readily transversed dispersal route. It is relatively easy for land plants and animals to move across the continent if the physical and climatic conditions permit, and species found in western Europe today may move all the way east to northern China, causing the natural communities in both areas to look remarkably alike. If the chances are good for the spread of many or most species of the flora and fauna, the route of dispersal is called a *corridor*. The Eurasian route mentioned above has been a corridor for Palearctic floras and faunas.

Other dispersal routes tend to be increasingly selective for groups of organisms; some species are able to migrate readily along them and others are not (Figure 17–9). This kind of a route is called a *filter*

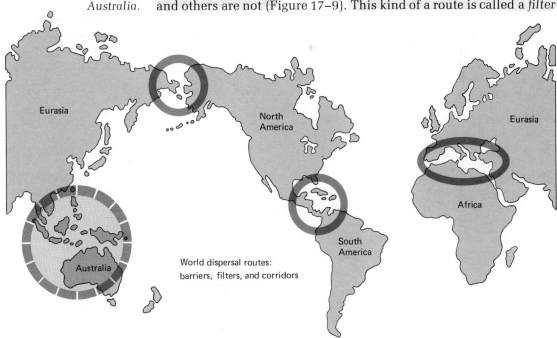

World dispersal routes:
barriers, filters, and corridors

because parts of biotas are able to pass and others are not. The distinction between corridor and filter is based on the percentage of the whole biota that follows the route. The connection between North and South America across the Isthmus of Panama is a good example of a filter route. For the past 15 million years, Central America has provided a land bridge between the two continents and thus a potential pathway of migration for land plants and animals. During the geologic periods that the isthmus has been above sea level, North and South American floras and faunas became extensively mixed by dispersal on this route. In fact, at the height of the interchange near the end of the Cenozoic, both continents became richer in land mammals than they had been. Subsequently, competition between animals with the same ecological roles (having evolved convergently on the originally separate continents) led to widespread extinction. Even today, however, many of the old endemic plants and animals of South America as well as a considerable number of species (e.g., grassland and mountain species) found in North America are not able to migrate across the land bridge, probably largely because of the lack of a continuously suitable habitat on this narrow land connection for their particular ecological requirements.

The dispersal route may be a barrier to many species if physically or ecologically it is unsuited for the organisms impeded by it. Thus grasslands provide barriers against forest animal dispersal, and mountain ranges are dispersal barriers to species adapted to lowland conditions on either side of them. In a situation where there is a strong barrier, the concept of *sweepstakes*, or *waif*, *dispersal* is applicable (Figure 17–10). Barriers in this case represent impossible

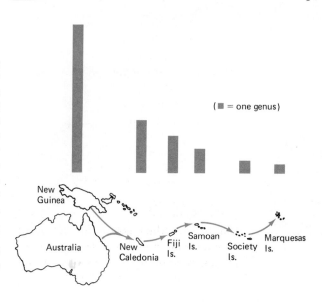

FIGURE 17–10

A sweepstakes route, showing the progressively smaller number of genera (equals decreasing diversity of fauna) of the weevil family of beetles from New Guinea toward the mid-Pacific Islands. (The bar graph above indicates the number of genera at each location.) The overall picture is of a great funnel fitted with graded filters of ocean barriers, excluding more and more genera the further one travels from the largest source areas of land.

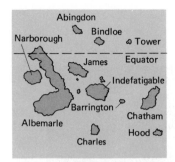

FIGURE 17–11

The Galápagos Islands in the Pacific, off South America's west coast: home of Darwin's finches and other unique animals whose ancestors originally reached the archipelago from the mainland by sweepstakes dispersal.

habitats for particular organisms to survive in, so that species may not spread across them by any normal processes of population expansion or migration. If the barrier is crossed at all, it is by single individuals doing it in one jump, not by gradual expansion of a population. During the long history of life, such waifs have crossed even the strongest barriers of this kind and founded new populations. Thus the Hawaiian Islands, one of the most remote island chains in the world and surrounded by an oceanic barrier, have never been connected to continental areas, but they have a luxuriant native flora and fauna on land. Sweepstakes dispersal was the means of ultimate origin for all the thousands of species of Hawaiian plants and animals. Likewise, the biota of the Galápagos Islands originated by waifs reaching them on drifting rafts of vegetation or other means from the continental areas of North and South America (Figure 17–11).

Summary

In biogeography we are concerned with distributional patterns of life and the factors that bring about this particular faunal and floral geography. Ecological determinants of the geography of biotas include the dominant role of plants, especially terrestrial species, because of the influence of producer organisms on later links in all food chains. The distribution of plants, and hence of animals, is especially sensitive to the varying physical environmental factors of temperature, precipitation, and solar radiation at different latitudes and altitudes. Historical determinants of biogeography are reflected in the species composition of the great terrestrial faunal regions of the world: Holarctic (divided into the Palearctic and Nearctic), Oriental, Ethiopian, Neotropical, and Australian. These faunal patterns were generated by changing climates and changing connections between the great land masses, particularly those caused by continental drift in the Mesozoic era when mammals and birds first began to diversify.

The behavioral and physiological means available for active dispersal of an organism may be insufficient to overcome barriers to distribution, in which case isolation results. Passive dispersal of seeds and tiny animals by air and sea currents, or by birds carrying seed material on their bodies, increases the distribution of many species. Easily traversed land routes for dispersal are called corridors; routes that allow only certain species to successfully pass are called filters. If the dispersal route crosses major physical or ecological barriers that prevent the passage of most organisms except by chance, the few species that make it successfully are said to have shown sweepstakes, or waif, dispersal. The native biota of remote oceanic islands result from this type of dispersal.

18 Basic principles of behavior

In looking at biology at the population level, we have examined the ecology, genetics, evolution, and distributional patterns of life on earth. One of the most interesting areas of biology observed at both the population and individual levels is animal behavior. The behavior of an animal, like any other characteristic of the organism, can be studied at various levels. Most of us tend to think of behavior as *what an animal does,* placing primary emphasis on the level of organization that we call the organism. But behavior can also be viewed as the *effect* of individuals on others of the same species — that is, the population. or even the entire biological community, which includes all of the various populations in a given area. Here the complicated flow of materials and energy between many species of organisms effects behavioral relationships across an extraordinarily broad spectrum, from predator-prey contacts to cooperative symbioses. In this chapter we shall learn about behavior at all these levels, with primary emphasis on the behavioral attributes that will enable us to understand social behavior, the most highly developed form of behavior (Chapter 19).

FIGURE 18–1

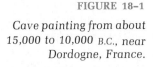

Cave painting from about 15,000 to 10,000 B.C., near Dordogne, France.

The development of the study of behavior

The beginnings of the study of animal behavior are lost in the prehistory of our earth. The beautiful cave paintings in the Dordogne Valley of France — the earliest recorded history — portray small antelopes, a rhinoceros, huge spotted bulls, rows of horses and antlered deer heads, and even a wounded bison with its entrails dragging (Figure 18–1). The unexpected sophistication both in the detailed observation and drawing of the animals and in the use of color in these paintings at first led many modern experts to believe that this was not a work of Paleolithic peoples. However, the association of Cro-Magnon tools and other cultural artifacts with these cave paintings indicate their great age and prehistoric source. The Cro-Magnon artists were most certainly aware of the various feeding, sleeping, and migratory behaviors of the animals that served as their hunting prey and were essential for their survival. While these paintings were probably done originally for magical religious purposes to promote hunting success, they do show a keen observation of animal behavior.

Through the subsequent ages, people have always studied and experimented with animal behavior whether it be with studying the wild habits of cattle and sheep with an eye toward domestication or developing the Chinese art of sericulture, which involves the domesticated rearing of silk worms in order to harvest and weave silk.

Today we would call these people **ethologists,** that is, those who study the behavior of animals in their natural context.

Although a practical understanding of animal behavior is perhaps as old as the human species itself, the development of the theoretical foundations now underlying its study has been more recent and has been entwined with some of the greatest philosophical and scientific controversies in history. For the most part, these modern foundations were laid during two specific periods in the history and philosophy of science: (1) the beginning of the scientific revolution in the sixteenth and seventeenth centuries, and (2) the rapid development of evolutionary theory in the mid nineteenth century.

The scientific revolution brought a "new birth" of interest in the natural world via a revival of interest in classical Greek and Latin literature and the rise of a spirit of questioning and reformulation of world views. The invention of the printing press and discovery of the New World further stimulated this renewed intellectual interest. The medieval view of nature had been teleological, that is, in natural occurrences people looked beyond nature to the "final causes" of events. These final causes, which always had some direct connection with the supernatural, explained the true nature of the event. Thus the prevalent system of reasoning used at this time was **deduction,** that is, reasoning from accepted general principles to specific examples. On the other hand, the scientist-philosophers of the scientific revolution began to place greater emphasis on the event itself, which could be perceived with the senses, rather than the idea or purpose behind the event. The universe came to be viewed as a great machine whose intricate workings could be uncovered by investigation and the application of mathematics and the theoretical foundations of physics rather than explained as a manifestation of divine will. This empirical approach gave rise to what we generally call the *scientific method,* which relies on observation and experimentation to discover facts that are then arranged into general principles. Reasoning from specific facts to the general principle is known as **induction;** although the idea was prevalent among learned people at this time, Francis Bacon, using it as a common theme in all his writings, is credited with showing the importance of the inductive method in scientific investigation.

These general historical developments also greatly influenced the specific areas of psychology and animal behavior. Human beings were viewed as automatons. The greater part of human behavior could be explained by mechanical processes, that is, automatic responses to external stimuli that were controlled by the nervous system. Between 1780 and 1850, more and more behavioral phenomena in animals and human beings were explained physiologically.

Advances in understanding of body chemistry and the structure and function of the nervous system substantially supported ideas of physiological mechanisms.

These developments set the stage for the advent of the theory of evolution, which a number of scientists preceding Darwin had anticipated. A mechanism for evolutionary change — natural selection — was Darwin's contribution. As we would expect, the theory was welcomed by the mechanistically oriented scientists of the time. This theory supplied for the first time a completely natural explanation for the origin of living things that did not require the introduction of the supernatural. Darwin's evolutionary theory had its most direct impact on the subject of animal behavior by supporting the notion that human beings were not different from the animals qualitatively, but only in *degree*. Darwin's second important contribution was that the behavior of an organism is as *adaptive* as any morphological feature. In other words, Darwin included behavior as part of the phenotype.

Comparative behavior

One of the basic doctrines in evolutionary theory is that during the course of evolution, there has been an increase in the complexity of organisms from simple to complex, lower to higher, primitive to advanced. We should be able to understand advanced organisms, including behavioral aspects, then, in the light of the lower organisms from which they are thought to have evolved. Thus the field of *comparative behavior* was born. By studying behavior in a great variety of organisms and comparing the results, investigators thought they would be able to find the underlying bases for all behavior, which in turn would eventually be reduced to mechanics.

Obviously this kind of study could start at either end of the biological continuum, which is what happened. Some scientists began to study the simplest organisms, searching for those basic principles that would unlock the secrets of human consciousness. Others continued to study human beings and their complex behavior and applied their results to the simplest organisms. Today we understand more clearly that the modes of behavior used by an organism are a direct function of the complexity of that organism and the environmental problems that it faces. Modes that provide maximum adaptability at one end of the spectrum of complexity may not provide the same adaptability at the other end. Thus if researchers restrict themselves entirely to the study of complex organisms, for example, their results may not correlate with those produced by the study of simple organisms. The great *nature versus nurture* controversy between ethologists working in nature and comparative psychologists working in the laboratory was initiated in this way.

Comparative psychology and ethology were both founded in the physiological mechanism and evolutionary theory that had preceded them; however, each developed a different approach to the study of animal behavior. *Ethology* sprang from the work of European zoologists who believed that observation of animal behavior was more accurate in the context of the organism's natural environment where it was exhibiting normal behavior. They chose birds, fish, and insects as common subjects for study. *Comparative psychology* arose primarily among North American psychologists who were especially concerned with the study of learning (Figure 18–2). They worked in a laboratory situation in order to control as many variables in the environment as possible. As might be expected, more complex organisms were chosen for this kind of study. The favorite experimental organism of the early comparative psychologists was the white rat, although they also used other mammals. When the results of these two lines of investigation were compared, considerable confusion ensued. The ethologists greatly emphasized the importance of the **instinct** (genetically programed behavior), whereas the comparative psychologists greatly emphasized **learning** (behavior modified by experience). Fortunately this debate has for the most part been resolved and both approaches to the study of behavior have been seen to complement each other. In conjunction, they have greatly increased our understanding of animal behavior at all levels of organization.

Before we begin to survey the basic principles of behavior, we should know the kinds of questions students of animal behavior are currently attempting to answer. The well-known questions that follow were originally compiled by Niko Tinbergen, a 1973 Nobel laureate, for his pioneering work in animal behavior:

1 What are the mechanisms underlying behavior?
2 How does behavior develop in the individual?
3 What is the adaptive significance of a behavior?
4 What is the phylogeny (evolutionary history) of the behavior?

The first two questions are largely organismic, and we shall consider them in more detail in the last section of this book (Chapters 29–31, 33, and 34). The adaptive significance of a behavior and its evolutionary history are best understood at the population level, and after our introductory survey of the basic levels of expression of behavior, we shall consider behavioral interactions at that level.

A   B   C

FIGURE 18–2

Two famous ethologists and a comparative psychologist: (A) Niko Tinbergen in his laboratory at Oxford University, (B) Konrad Lorenz, and (C) B. F. Skinner in his office at Harvard.

Types of behavior

From a phylogenetic point of view, behavior seems to have evolved from very rigid, stereotyped patterns to more complex behavior that is variable between individuals of a species. The lack of variability in a species and the performance of simpler behavior patterns are thought to be due largely to the inherited properties of the nervous system. Because of the high degree of nervous system control, these behaviors are said to be *innate*. In them, the stimulus seems to simply trigger a response or set of responses in the nervous system of these animals and the behavior is said to be *stimulus bound*. In more complex species of animals, behavior is less stereotyped, and the individual history of the organism plays a more significant role through learning. At this level, behavior is said to be *acquired* to a larger extent. Behavior is more variable because it is less stimulus bound and more dependent on the differential experiences of the organisms. Figure 18–3 depicts the relative dependence of each class of animals on a particular mode of behavior. The behaviors listed along the vertical axis are arranged in order of increasing importance of information acquired from the environment, that is, learning. It is easily seen that the mode of behavior known as **taxis,** which is highly stimulus bound, plays an important role in simpler organisms. In human beings and the other primates, behavior is essentially entirely acquired through experience, and thus *learning* is the dominant behavioral mode. Although any such scheme is necessarily speculative, Figure 18–3 is valuable in that it presents an approximation of the dependence on each of the behavioral modes in each phylum. Let us now look at these kinds of behavior in more detail.

FIGURE 18–3

FIGURE 18–3

The evolution of adaptive behavior can be interpreted by comparing the differences among primitive and advanced animals alive in the world today. The most simply organized animals rely mainly on inherited behavior: reflexes and instinct. More advanced animals show increased reliance on learning and reasoning.

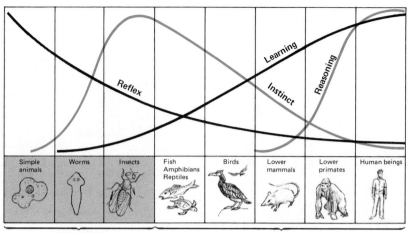

Flight of butterfly blinded in one eye

Normal flight (butterfly with normal vision)

Taxes

Taxes are probably the most basic type of behavior. A taxis is the continuous orientation by an organism to a specific stimulus in its environment. Thus an organism may orient toward the stimulus of light (*phototaxis*), gravity (*geotaxis*), or other environmental factors. Often this orientation occurs because an animal is trying to maintain equal stimulation on each member of a pair of sense organs, such as eyes. A European satyrid butterfly, *Eumenis senele,* will try to escape predators by flying into the sun (Figure 18–4). If one eye is blinded, the escape reaction of the butterfly consists of flying in circles. Apparently the butterfly tries to equalize the light in its orientation for the sunward escape flight, but cannot do this with one eye blinded. The type of behavior known today as taxis was called tropism by early workers. Today the term *tropism* is generally reserved for the orientation of plants by differential growth. Thus most plants are said to be positively phototrophic, that is, they grow toward the source of light.

FIGURE 18–4

An example of taxis control of flight behavior. A butterfly that is newly blinded in one eye will fly in circles, being unable to keep constant the amount of light hitting both eyes as it could do with two normal eyes. It can no longer orient or maintain a constant flight direction with regard to the position of the sun.

Reflex behavior

Reflexes are a somewhat more complex type of behavior but are also regarded as innate because of the strong reliance on inherited nervous mechanisms. These behaviors are highly stereotyped with little variability between members of a species. Reflexes differ from taxes in that they involve a part of the body rather than the entire organism.

A common example of the reflex is the human knee jerk. When the tendon is struck below the knee cap, a nerve impulse is transmitted by a sensory neuron to the spinal cord. Rather than sending this impulse to the brain for processing, a decision is made in the spinal chord and a signal is transmitted to a motor neuron to raise the lower leg. The adaptive value of reflexes is clear. A great deal of time is saved by processing the stimulus at the level of the spinal cord. In dangerous situations every second saved in responding to the danger increases the probability of escape. Thus when a child puts her hand on a hot stove a great deal of damage is avoided because the reflex causes her to jerk the hand away.

The general arrangement of nerves that brings about a reflex reaction is called a **reflex arc.** Reflex arcs were thought at one time to be the basis of all behavior. Experiments in which stimuli were associated with reflexes, such as Pavlov's experiments in teaching dogs to associate the sound of a ringing bell with their feeding time, showed that reflexes could be acquired (through conditioning in the Pavlov experiments) as well as inherited. Therefore it was thought that these types of associations could account for the higher behavioral phenomena of thought and learning. Further study, however, has shown the behavior of higher organisms to be more complex than a simple chainlike grouping of reflexes. Generally, reflexes are not an important means of human behavior since our highly developed brain cortex affords us a large amount of voluntary control over our behavior. Nevertheless we still find, even in human beings, differences between what has been called instinctive behavior and learned behavior.

Instinctive behavior and learned behavior

The concept of instinct has long been a bone of contention between psychologists and ethologists. Unfortunately, much of the misunderstanding that has arisen is probably due to a misapplication of the term rather than genuine scientific disagreement. The scientific concept of instinct originally indicated the type of innate behavior that could not be modified by learning. The word instinct is now commonly used to describe any behavior that seems to be characteristic of a particular species, usage to which psychologists in particular have objected. For example, the behavior of a cat killing a mouse is often referred to as "instinctive." However, if kittens are reared apart from adult cats, they do not kill mice. Apparently, this so-called instinct must be learned by observing adult cats in the process of killing mice (Figure 18–5).

Experimental investigations of instinctive behavior have revealed some interesting facets of these complex patterns. The European ethologists have preferred to conduct experiments with behaviors in the natural environment. As naturalists, much of their early work in this field originated from their interest in the display patterns of birds. These behavior patterns appeared to serve a function of communication between members of a species. Frequently, these initial displays were followed by a courtship ritual between the two individuals that was highly stereotyped and could be elicited time and again by the member of the opposite sex. Thus this particular courtship behavior was thought to be "released" by the previous display. Because the response was so stereotyped, it was considered "instinctive." Thus in ethology the part of the environment that causes an instinctive action to be produced came to be called a **releaser.** In its original use, the term releaser was applied to the ac-

FIGURE 18–5

In the presence of its mother, the kitten learns to catch and eat rats. If the kitten never observes an adult cat catching rats, it may play with rats as if they were other kittens. Much of the perfection of a cat's hunting behavior is achieved through observational learning of this type.

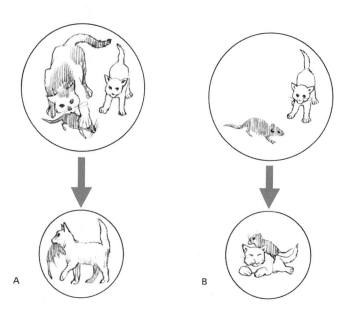

A

B

FIGURE 18–6

An oystercatcher reacting to a giant egg. (A) The female reacts to the sign stimulus of correct size, shape, and pattern to incubate her normal egg instead of a nearby rock or other object. (B) The female chooses a giant egg model (supernormal stimulus) in preference to her own egg.

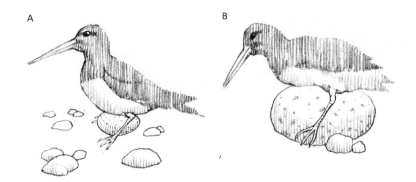

A   B

tions or characteristic of a companion animal that caused an instinctive act to be performed.

Ethologists also discovered the role of so-called sign stimuli in releasing or inducing particular behavior acts. The term *sign stimulus* was used to describe the aspect of the environment that elicited an "instinctive" behavioral pattern in a manner similar to releasers, but was not associated with a companion organism. The relative components of a sign stimulus can be experimentally enlarged to form a *supernormal stimulus*, which has a greater releasing effect on the behavior of an animal than the normal sign stimulus (Figure 18–6).

When we move from what is clearly innate behavior into behavior that is acquired by learning through experience, we find a high variability in behavioral attributes within a species. This is only logical, for *learning* is primarily a biological adaptation. It enables an organism to benefit from its past experiences and thus enhance its own survival and that of its offspring. Learning provides a more rapid assimilation of information for the individual than does the long selective process needed to evolve innately determined instinctive behavior. It makes possible the recognition of specific landmarks, locations, and individuals, and hence allows long-lasting social behavioral relationships to develop between members of a single species. Of course, differences in the ability to learn can also be expected between species as well as within species because of differences in genetic constitution and adaptive strategies. Learning is perhaps best defined as an enduring modification of behavior that is probably physically evidenced as a change in the nervous system and that arises through individual experience. It usually demonstrates clearly adaptive value in the organism's environment. Many types of learning have been distinguished.

Habituation is often cited as the simplest kind of learning that

FIGURE 18–7

Habituation: city pigeons in front of the New York Public Library lose their fear of human beings by repeated feeding free of any harm.

occurs in animals. It is the progressive waning of response to an insignificant stimulus that is repeatedly presented (Figure 18–7). For example, a loud noise may cause a number of reactions in an individual, including a muscular jumping, an increase of heart rate and an orientation toward the stimulus source. If the loud noise is repeated a number of time, these reactions will gradually wane in their intensity, and the organism is said to habituate to the stimulus. A female bird that builds a nest in a tree next to a house will at first be startled by the frequent slamming of a screen door. After a while she will cease being startled by this regular, but harmless, noise. This type of learning is of great value to an animal in an environment with young animals. Relevant environmental stimuli can be culled from the large volume of sensory information available to the animal, and the animal learns to respond to only significant stimuli. Thus it filters out irrelevant stimuli through its sensory systems. Habituation becomes a learning *not* to respond to insignificant stimuli, in contrast to the other types of learning, which result in a strengthening of responses to stimuli that are of significance.

Classical conditioning was discovered by the Russian physiologist Ivan Pavlov and is a modification of a normal stimulus-response pattern of behavior. In his classic experiments, Pavlov blew meat powder into the mouth of a harnessed dog and measured the amount of saliva that was produced (Figure 18–8). He then sounded a bell shortly before giving the meat powder, and after a number of trials the sound of the bell alone came to elicit salivation. The unconditioned stimulus of meat powder elicited an unconditioned response of salivation; however, when the conditioned stimulus of the bell was repeatedly paired with the meat powder, the unconditioned presentation of the meat powder and eventually presentation of the conditioned stimulus alone was able to elicit a conditioned re-

FIGURE 18–8

Classical conditioning in the salivary response of a dog to the sight of food. Designed by the Russian physiologist and Nobel prize winner Ivan Pavlov, this apparatus permits a light (as the stimulus to which the dog is conditioned to expect the delivery of food) to appear in the window and meat powder (as the stimulus normally eliciting salivary flow in the mouth) to be delivered to the food bowl. The harness simply holds the dog in a standard position, and the tube into the cheek measures salivary flow (ultimately recorded on the rotating drum graph at far left).

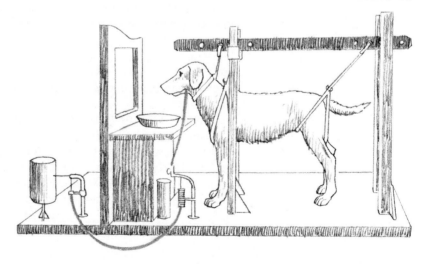

FIGURE 18–9

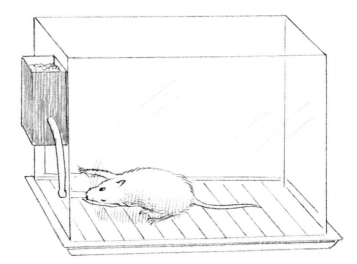

Operant conditioning: a rat in a Skinner box. When the rat pushes the bar, a food pellet is delivered from the hamper. The delivery of the pellet constitutes the reinforcement of the bar-pressing response. Unlike classical conditioning (e.g., salivation), however, the reinforced behavior bears no resemblance to the behavior normally elicited by the reinforcing stimulus (e.g., pressing a wire bar is not a rat's normal response to food). Operant behavior of this sort often appears to be spontaneous rather than a simple direct response to a specific stimulus. The word operant *comes from the fact that the behavior "operates" on the environment to produce some effect. Thus going to the bar and pressing it are operant acts that lead to the producing of food within the cage.*

sponse: salivation. Of course, the conditioned stimulus will lose its power to elicit the conditioned response if it is not intermittently paired with the unconditioned stimulus: there will be no reinforcement.

This type of learning is highly significant to animals in nature. If a bird suffers nausea from eating a distasteful monarch butterfly, it will learn to avoid preying on butterflies of that species. Furthermore, the bird may avoid eating similar-looking species as the result of **stimulus generalization** (a previously unencountered stimulus similar in nature to the conditioned stimulus can come to elicit a conditioned response). Classical conditioning has regained much interest recently because of the rise in popularity of biofeedback devices that allow people to condition themselves into relaxed states through listening to their own brain waves.

Operant conditioning is involved in learning situations in which desired responses are rewarded and undesired responses are not rewarded. As a result of the reward, the desired behavior occurs more and more often until the organism is conditioned to emit these responses. In psychology laboratories operant conditioning is used in training white rats to press bars, a response that can be employed in a number of interesting experiments (Figure 18–9). (Actually, all animal trainers use a similar procedure, whether with dogs or tigers.) A hungry rat is introduced into a cage and allowed to adapt to the surroundings. A few pellets of food are supplied to acquaint the animal with the position of the food cup, which is usually near the bar. Food pellets may then be provided if the animal looks at the bar or goes near it. Such rewards increase the probability of this behavior occurring again. The rat is rewarded only as it gets close to the bar and eventually presses it. Soon the animal may be pressing the bar

Biological clocks and circadian rhythms

One of the fundamental properties of living matter is an underlying rhythmicity. It has often been suggested that this property be added to the list of the four basic characteristics of life: growth, reproduction, movement, and irritability (or responsiveness). Natural cycles vary in length from the seventeen-year growth and flight cycle of a periodical cicada to the staccato of individual neurons firing at the rate of a thousand times per second. Activity and inactivity in organisms also shows a rhythmic periodicity that is often associated with geophysical alternation of light and dark. Animals that are active during the day and rest at night are called *diurnal*; organisms that restrict their activity to the dark are referred to as *nocturnal*. Some animals, including many rodents and birds, are crepuscular, i.e., active primarily at dawn and dusk. It has been suggested that these cycles of activity and rest are not merely passive responses of the organism to changing environmental conditions; rather, they are controlled by an internal timing mechanism denoted as the *biological clock*.

One of the most convincing arguments for the existence of the biological clock is the maintenance of rhythmicity when the organism is removed from diurnal influences and placed under constant conditions. During the daylight hours, the bean seedling *Phaseolus multiflorus* lifts its leaves to face the sun. This active movement is the result of osmotic changes in pressure within specialized cells located at the base of each leaf and is accompanies by a lowering of the leaf at night. When the plant is cultivated in the laboratory without changes in light and temperature, the movement persists approximately at 24-hour intervals. Similar results are obtained when the activity of rodents is measured under constant laboratory conditions. Activity periods recur approximately in 24-hour intervals despite the lack of rhythmicity in environmental influences on behavior.

Further support for an internal timing mechanism is suggested by the fact that animals can be trained to anticipate a meal when it is offered at the same time each day. After food is made available at a feeding dish for several days at the same time, bees will continue to look for food on successive days at that time even if food is no longer being supplied. The family dog, if fed regularly at a particular time, will show considerable anticipation as that hour approaches. Our own stomachs will show considerable gastric secretion and movement as our regular dinner time nears.

Although biological rhythms occur in annual, seasonal, and monthly cycles, perhaps the most thoroughly studied have been the *circadian rhythms*, which approximate the length of the 24-hour solar day. The word "circadian" is derived from two Latin words: *circa*, about; and *diem*, day.

The fundamental unit of the rhythm is the cycle, which is measured in a length of the time known as the *period*. By definition, circadian rhythms are composed of cycles of activity and inactivity with periods of approximately 24 hours. The period of the circadian rhythm is usually measured from the onset of one active period to the onset of the next. Circadian rhythms appear to be innate rather than learned or imprinted on the organism by the 24-hour day-night light and temperature cycle caused by the earth's rotation. The biological clock that controls these rhythms is suspected to be associated with a biochemical system. However, it displays some odd characteristics not normally associated with biochemical reactions. Circadian rhythms are surprisingly unaffected by sublethal doses of metabolic poisons, substances that usually interfere with metabolic changes involving chemical reactions. Furthermore, although the rate of most biochemical reactions is approximately doubled for every 10°C change in temperature, circadian rhythms maintain their frequency with little or no change during a rise in temperature.

The source of the rhythmicity of this internal biological clock is more controversial. Two general hypotheses have been advanced to explain the basis of timing: an *endogenous* timer mechanism that is an intrinsic part of the physiology of the organism, or an *external* timer mechanism based on some aspect of the external environment, such as geomagnetism or the solar flux of electromagnetic radiation.

The *endogenous timer hypothesis* states that the biological clock is autonomous from the environment. Rhythmicity is thought to be an endogenous characteristic of organisms that has arisen during the course of evolution in a rhythmic environment. This capacity is an adaptive feature enabling organisms to anticipate favorable and unfavorable periods for activity. According to this hypothesis, the sole function of environmental cues is to entrain the rhythm to precisely 24 hours.

Advocates of *external (exogenous) timing mechanisms* suggest that the timer is present in the form of subtle geophysical forces that are part of an animal's sensory world. Rhythmicity is proposed to have a

dual nature, with one rhythmic system underlying another. Overt physiological rhythms can be modified in the laboratory by changes in the length of the light of dark periods, temperature changes, or feeding times. On the other hand, geophysically dependent rhythms are a result of forces that the experimenter cannot control, and these rhythms cannot be modified. Thus the rhythmicity of activity that is noted under constant conditions is viewed as a response to changes in magnetic field, barometric pressure, or ionization of the air. The organism's biological clock merely comprises a capacity to respond to these subtle external cues.

The external timer mechanism was originally proposed as a result of observations on marine organisms. Oysters have a characteristic behavior of opening their shells during high tide and closing them as the tide goes out. When some of the crustaceans were shipped from the East Coast to a laboratory in Illinois for study, their rhythm of shell opening gradually drifted out of phase with the tides of their home areas. In fact, this behavior assumed a distribution that corresponded with the lunar cycle in Illinois, as if it were a seacoast town! Thus it appeared that the gravitational field of the moon, which controlled the tides on the East Coast, was also the major influence on the behavior of these oysters in Illinois. The oysters were responding to this environmental variable, perhaps

through its effect on barometric pressure, a factor that had not been controlled in the lab.

It is misleading to talk about biological clocks as though a single mechanism exists. Multiple clocks may exist in an organism for different functions. Crabs display a color change that is coupled to the 24-hour light-dark cycle, yet have a rhythm of locomotor activity that follows the lunar day of 24 hours, 50 minutes. The activity cycle of human beings is inextricably linked to the solar day, yet other rhythms such as the menstrual cycle may be linked to the phases of the moon. The metabolism of any organism is characterized by circadian fluctuations of biochemical substances and organ functioning.

Finally, we should stress the importance of biological clocks as an adaptive response. Not surprisingly, Darwin was the first to point out that rhythmicity had to have adaptive advantages to arise as a result of natural selection. The cyclic alternation of activity and inactivity possibly enables the organism to "anticipate" optimal periods for activity. Similarly, variations in the actual periods of activity in different sympatric species serves to reduce interspecific competition for the same food and other resources.

FIGURE 18–10

Insight learning in chimpanzees. The experimenter hung bananas from the ceiling, out of reach of the chimpanzees, and then put boxes of various sizes inside the room. The animals soon discovered that boxes could be piled on top of each other to reach the fruit.

hundreds of times for a single food reward. A key point in operant conditioning is that the animal must be motivated to seek the reward. In this case, it is motivated by hunger to work for a food reward. The response that the animal makes is always voluntary in operant conditioning, whereas classical conditioning may result in both voluntary and involuntary responses. The operant mode also involves the presentation of a reward, and there is no pairing of stimuli as in the classical conditioning outlined above.

In **trial and error learning,** a number of different solutions are applied unsystematically to a problem until a suitable one is found. Conditioning is thought to play a role in trial and error learning because the successful solution results in reinforcement, which increases the probability of the animal repeating the successful behavior. Trial and error behavior is common in the wild; young animals undergo such a process when learning the kinds of food that are edible. Choosing edible foods is reinforced by alleviation of hunger whereas choosing inedible foods may be negatively reinforced by the production of nausea.

The highest forms of learning are known as insight and reasoning. **Insight learning** is the sudden perception of relationships that were heretofore unnoticed. This is often referred to as the "aha!" reaction because this interjection embodies the spirit of insight. The classic example of insight in animals is that of a chimpanzee placed in a room where bananas are suspended from the ceiling (Figure 18–10). After surveying the room, the chimpanzee suddenly perceives the solution to the problem and stacks boxes on top of one another to reach the food. Subsequent trials show that learning has occurred because these animals immediately solve the problem in any new similar situation and do not proceed in a trial and error manner. If a connection is made between heretofore unlinked experiences that are put together in order to solve a problem, we say that the animal has a *reasoning* ability. This has been found in many mammals, but there is no evidence for reasoning in less complex animals.

In the natural environment it is often difficult to separate these higher processes of reasoning and insight learning from trial and error learning. The construction and/or use of tools by some animals seems to indicate the action of a higher learning process; however, such tool use has been found in only a few mammals and birds. Gorillas and chimpanzees throw sections of branches at intruders. Chimps may prod edible termites out of a nest with a long straw fashioned from a grass stalk. California sea otters use stones to knock abalones loose from rocks. They also regularly balance a small, flat stone on their upturned stomach, which serves as a pounding "anvil" for smashing shellfish. Young otters apparently learn to use the stone as a tool by watching their mothers feed. In the Galápagos

Islands, woodpecker finches and mangrove finches use cactus spines to pry beetle grubs from tree branch burrows. Egyptian vultures in Tanzania use their beaks to fling stone missiles against ostrich eggs to fracture the tough shell and feed on the contents (Figure 18–11).

Whether the use of twigs and straws by chimpanzees to probe for insects in crevices is the result of trial and error learning or the higher process of reasoning is not clear. However, an interesting case of higher learning occurring in the natural environment has recently been observed in the Japanese macaque monkeys. One female of the species acquired the habit of washing sweet potatoes before feeding. This habit gradually spread through the population of the monkeys by observational learning.

Imprinting is a special type of learning that has been found in many different kinds of animals. It was first described in ducklings and goslings by the Austrian ethologist and Nobel laureate Konrad Lorenz. These animals appear to be sensitive to any moving stimulus during a short period of time after birth (known as the *critical period*), and they will follow any object that moves during this time (Figure 18–12). Normally, these baby birds would imprint to their

FIGURE 18–11

Trial-and-error learning by the woodpecker finch in the Galápagos Islands. This member of the Darwin's finch subfamily perfects its use of a cactus spine or twig as a probing tool for boring for insect larvae through experimentation early in its adult life.

FIGURE 18–12

Imprinting: baby goslings have imprinted on Konrad Lorenz after hatching and will follow him even if their natural mother is presented to them a day or two later.

mother, as the first living object they see after hatching. However, Lorenz was able to imprint newly hatched ducklings with his voice and movements, and they followed him throughout life. In fact, when these animals reached reproductive age, they tried to mate with his hand, indicating that early imprinting associations may play a role in species recognition. This kind of early learning behavior is obviously of high survival value to most organisms in natural populations.

Summary

Animal behavior can be studied at different levels of organization, from the genes ultimately involved in coding for a particular behavioral response to the externally visible activities of the individual whole organism in its population and the biological community. The dominance of the various modes of behavior, from inborn traits to learning and reasoning, differs with increasing biological complexity in animal groups. Diverse human approaches to the current study of behavior and its causes may be traced primarily to historical foundations laid during the beginning of the scientific revolution in the Renaissance and the development of evolutionary theory in the past century. The known types of innate behavior range from highly stereotyped taxes and reflexes through instinctive acts and motivated behavior that can be extremely variable in nature. Though a tendency toward a specific behavior can be inherited in higher animals, the expression of this behavior depends more on the external environment, hormone levels, nervous influences, and past learning experience. The presence of a highly developed brain cortex in human beings and other primates has enabled them to reason and thus exert voluntary control over their behavior. Learning is a biological adaptation that allows animals to modify their behavior through the benefit of individual experience. The different types of learning range from habituation, conditioning, and trial and error learning, which are common in many vertebrates and higher invertebrates such as insects, to insight and reasoning, which are restricted primarily to mammals. Learning plays a greater role in the behavior of higher organisms, and this observation led to the essential resolution of the nature-nurture controversy, an argument on the origins of behavior that divided ethologists and comparative psychologists for most of the past century.

19   Social behavior

To the majority of us, social behavior among members of animal populations is probably the most intriguing aspect of animal behavior. Virtually all animals show social behavior at some point in their lives, especially at the time of courtship and mating. A much smaller number of species show sociality as the dominant feature of their life; however, the highly coordinated cooperative efforts shown by such species, especially among the primates, represent the highest levels of complexity in the evolution of adaptive behavior. In this chapter we shall emphasize the behavioral adaptations that affect the cohesiveness and coordination of animal groups. Before we begin to look in detail at the nature of social behavior, let us consider the meaning of several useful terms.

Concepts of social groupings

Social behavior refers to any behavioral interaction between two or more individuals of a species. Such interactions may result in or lead to the promotion of a social grouping that may differ quite a bit from a true society. Many animals, for instance, form coincidental

A

B

aggregations because they are attracted to a common resource such as food or water. Moving or migrating animals are affected by physical factors. Thus over the mountains of the narrow Isthmus of Panama, huge numbers of broad-winged hawks, genus *Buteo*, are concentrated by geography and their behavioral preference for following ridges, where they can glide on updrafts as they migrate south from the United States and Mexico to South America for the winter. Rainforest butterflies will aggregate around a pile of bird droppings to suck the nitrogenous juices; although these butterflies may be mostly of the same species, the aggregation is temporary and has no permanent social basis (Figure 19-1). A true **society** involves a stable group whose members communicate with each other extensively and maintain some permanent social relationship to one another. A society may be highly organized and last for many years, such as those of the ants and other social insects, or they may be relatively loosely organized and be made up of individuals from diverse genetic backgrounds. All degrees of intermediate relationships occur between these extremes.

Jerram Brown has recently provided a useful summary of grouping patterns in animal societies that we shall largely follow. Although certain biological relationships are shared by all of these social

A   C

FIGURE 19–2

Kin groups: (A) a bull southern sea lion guarding a harem and newly born pups in a coastal colony on the South Atlantic in Patagonia; (B) a family of the Blue-footed Booby in the Galápagos Islands; (C) African elephants (cows and babies) in Kenya.

groups regardless of origin, the ecological and evolutionary factors that cause or predispose animals to form groups and stay in them are exceedingly varied.

For example, **kin groups** are social groups whose members are closely related genetically because many have the same parents (Figure 19–2). They may be **families,** where groups are formed by one or two parents and their most recent offspring, or **extended families,** where groups are formed because many of the offspring fail to leave their parents on reaching maturity. Prairie dogs, some primate groups, Mexican bluejays, and goose and swan families are examples of animal societies brought together by kinship.

Mating groups are social groups that depend ultimately on the mutual attraction of the sexes to each other for mating. Thus the smallest such groups, of course, are **pairs** formed of monogamous groups of two individuals, as in most birds. **Harems,** groups in which a male attempts to keep females together and away from other males, may be commonly found among mammals, especially grazing herbivores and seals and sea lions and their relatives. Other mating groups include **leks,** which are formed by the attraction of males (and subsequently females) to a communal mating ground; however, the eggs from females that are mated in such communal settings are

FIGURE 19–3

Many Hawaiian fruitflies, such as this Drosophila grimshawi *male, exhibit a peculiar "lek" display behavior, when a number of males gather on a tree fern or other plant to display a particular posture that attracts passing females to stop and mate.*

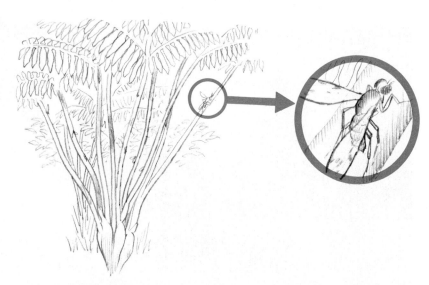

laid elsewhere (Figure 19–3). Many butterflies congregate around the tops of hills to seek mates and go through courtship displays, but then the females disperse to lay their eggs elsewhere. Other types of mating groups include spawning groups, where groups of both sexes form at localized spawning grounds and do not provide subsequent care for their young. Many fishes and amphibians form such spawning groups.

Colonial groups are social groups formed by the colonial nesting of pairs or one-male harems; the young are raised at the nest site by the adults radiating out from the colony location to obtain food and other necessities (Figure 19–4). Many sea birds nest in colonies on remote islands where protection is achieved from major land predators. Bats nest in cave colonies where few predators can reach them on high rocky ceilings.

Still other social groupings are those that Brown has called **survival groups** — groups formed by the aggregation of randomly related, usually nonbreeding individuals who are mutually attracted by each other. Thus foraging flocks of birds, migrating and feeding flocks of ducks and geese, herds of mammals, and schools of fish probably result from the increased survival rates of groups as opposed to solitary individuals, and therefore these species tend to form large groups.

The names of these groups, of course, emphasize the ultimate, or **evolutionary,** causes of grouping behavior. The immediate, or **proximate,** reasons that bring about spacing patterns within a group include forces of both attraction and repulsion. Aggression, in the form of threat and attack, is the primary repulsive force. The forces

FIGURE 19-4

Colonial nesting in sea birds: a Cape Gannet colony along the coast of South Africa.

of attraction are much more varied and may include communication through acoustic, visual, chemical, and tactile means. Sexual attraction clearly operates in some groups but not necessarily in all. Submissive behavior allows closer relationships between dominant and subordinate mammals or birds, soliciting or begging behavior also brings animals together for social purposes. The attraction between parent and the offspring (the kinship bond) is obviously important in groups of higher animals. The advantages of feeding together, as we shall see, may also offer principal attractive reasons for social grouping. Let us look now at some of the communication systems and cooperative factors that promote the development of social ties, and hence of social behavior.

Communication and social behavior

To survive and reproduce, an animal must contain within its repertoire various behaviors involving interaction with its environment. Survival depends in large part on the appropriateness of these behaviors, as the animal interacts with other individuals of its own species as well as other species in the area. The communication of information from the individual to other individuals and the reception of new data are important aspects of adaptive behavior that affect the outcome of such interactions. The methods used to transmit information between individuals can be grouped according to the basic kinds of stimuli striking receptor organs; namely, visual, acoustic, chemical (taste or smell), and tactile. The specific types of communication used in social behavior are as diverse as the varieties and combinations of the basic kinds of receptors that receive these signals.

Visual communication

In *visual communication*, signals involving movement, color pattern, posture, and body form transmit information. The nature of these cues tends to restrict this kind of communication to diurnal animals, yet some nocturnal insects have luminous signals, as do certain deep-sea organisms. The importance of body postures and movements in the behavior of the Black-headed Gull is illustrated in Figure 19–5. The so-called long-call posture is part of the gull's recognition behavior, and the vertical posture is part of the ritualistic courtship behavior. Immediately prior to mating, individuals will exhibit the upright posture. All of these visual signals are used in conjunction with auditory signals in the gull.

Many fish, birds, and reptiles change color to intensify visual signals during the courtship season and reproductive period. Nocturnal visual signals are illustrated by the flashing of lights of fireflies. Different species of fireflies will exhibit slightly different flash patterns that act as releasers only for their species (Figure 19–6). Thus the male and female can find one another even in the dark. Higher organisms that possess the ability to learn may also frequently use visual signals to communicate, but in these cases the signals are not usually innately fixed. In any organism that exhibits a high degree of sociality and structure within the population—for example, human beings and chimpanzees—many visual cues will be learned. However, some human expressions are so universal that they may also be part of innate behavior. Expressions of joy, anger, hate, fear, pain, and grief are among many visual signals that now are thought by

FIGURE 19–5

A series of four agonistic displays given by the male Black-headed Gull when another male approaches his territory.

Long call

Upright

Choking

Facing away

FIGURE 19-6

Visual communication in various firefly species of the eastern United States. Males of different species show different flash durations and repetitive patterns as they fly on a summer evening. Only females that recognize their species' flashing pattern will normally respond to the appropriate males.

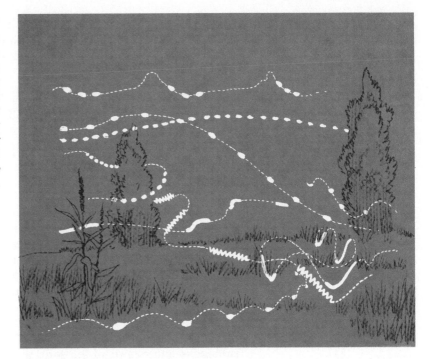

some behaviorists to suggest innate components. Research based on films of men, women, and children around the world has revealed the similarity of many visual expressions among cultures and population groups. Some of these same expressions can be found in chimpanzees. For example, the smile, nod, kiss, and eyebrow flash can be found within their repertoire of greeting behavior. Chimpanzees also use many hand gestures in communication (Figure 19-7).

FIGURE 19-7

Social communication by gestures and contact in wild chimpanzees: (A) a mature male scratching when frustrated; (B) socially high-ranked males threaten each other; (C) a mature male hides his face with one arm in response to sudden movement by an observer; (D) a young juvenile (with "pout" face) tries to persuade a 9-month-old infant to cling to the front of her body.

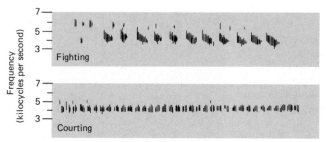

FIGURE 19–8

Acoustical communication in crickets. The male cricket produces his song by moving his wings from an open to a closed position, dragging a scraper (reinforced segment of cuticle on the edge of one wing) across a file (series of ridges on the underside of the other wing). Both wings are similarly equipped so that either wing can be on top. The scraping movement causes the wing to vibrate at about 5,000 cycles per second, producing a remarkably pure tone.

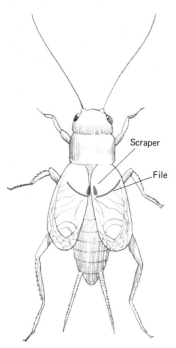

Acoustic communication

In *acoustic communication*, sounds are sent through air, water, and even ground to communicate information. Sounds are important in reproductive behavior, the communication of signals for species recognition, and as warning signals for imminent danger. The repeatability of their transmission gives them an advantage over visual signals, but a greater advantage is that sounds are independent of available light.

Although sounds may be produced by vocal cords, as in human beings, this is not always the case. Cricket songs are produced by rubbing together parts of the body (Figure 19–8), and during mating seasons a male woodpecker will drum with his bill on hollow objects in order to attract a female. Even in organisms that can vocalize, at times sounds may be produced in other ways, such as a gorilla beating his chest or a beaver slapping the water with his tail.

The complex sound-communication systems of echo location, shown particularly well by bats and aquatic mammals, assist in locating objects during feeding, navigation, and avoidance of barriers. Porpoises, for instance, have a much greater range of hearing than human beings and are able to respond to many vibrations in the area of ultrahigh frequencies. Their complex system of echo location does not solely sense the presence of an echo, but involves considerable reasoning. This auditory communication requires intelligence and previous learning to interpret, evaluate, and identify the echo—an avenue of perception quite beyond the capacity of ordinary human beings. In the aerial environment, bats use a similar type of ultrasonic auditory signal system to navigate in total darkness. To a lesser extent, they use audible lower-frequency signals (heard as squeaks and chirps to human ears) to communicate, especially when danger threatens. Many tropical leaf-nosed bats of the fruit- and nectar-feeding family, Phyllostomatidae, have large expanded noses and external ear appendages to aid in reception and discrimination of sonar signals that they employ to locate fruit and other food sources (Figure 19–9).

Chemical communication

Chemical communication is also widespread among all groups of animals. The major chemical communication in insects is by means of externally released hormones called **pheromones.** Some pheromones act as releasers of immediate behaviors; others produce physiological changes in the recipient without stimulating an immediate response. The female silkworm moth releases a chemical capable of attracting a male from a distance of two miles or more, though the amount of chemical released is less than 0.00000001 gram. On initially sensing the chemical, the male will begin flying upwind until the female has been found. A great many insects use sex attractants of this type.

Another classic example of pheromone use is trail markers employed by ants. The first ant to go from the nest to a food site releases tiny amounts of this chemical trail substance by touching its abdomen to the ground. Following the chemical path made by this foraging ant, other workers find their way to the food (Figure 19–10). As they return they also secrete some of the chemical. Workers continue to mark the trail as long as they find food. Without food to carry back to the nest, they secrete no markers and within a few minutes the pheromone trail largely disappears. These releasing pheromones have been shown to be highly species-specific; that is, a pheromone trail left by one species of ant will not be followed by another species. It is interesting that chemical signals do not require that the sender and receiver be coordinated in time. Apparently, even the most volatile chemical substances used for communication will persist in the environment relatively much longer than visual or auditory signals, which are usually measured in at most a few seconds. This property certainly extends the overall effectiveness of the communication for social purposes.

Tactile communication

Tactile communication is generally restricted to social animals and hence tactile signals play a major role in primate life. Mutual social grooming in monkeys and apes is perhaps the most familiar example of communication by touch. These primates part their companions' fur with their hands, removing fine particles and parasites such as ticks with fingers or lips. A male chimpanzee or baboon will begin social grooming of a female as a symbolic gesture after she presents

FIGURE 19–9

The head of a leaf-nosed bat from Costa Rica, showing the huge external ears and the elaborate folding of the nose for sound reception.

FIGURE 19–10

Ant trails are maintained by external secretions of chemicals (pheromones) onto the ground, such that other ants can follow the original "scent" ant by an odor trail.

FIGURE 19–11

Reassurance, friendliness, and other emotional states are communicated between chimpanzees by tactile (touching) communication.

her anal region to him in appeasement-greeting behavior. Chimpanzees also embrace each other and kiss with the touching of lips when they meet an individual they know. At other times, chimpanzees greet by shaking hands the way people do. The lower ranking animal presents its hand to the higher ranking animal while holding the palm up in a begging fashion. In response to this gesture, the higher ranking animal gives its hand, which in turn calms the other (Figure 19–11). Tactile communication is commonly experienced as adults sit or even sleep together in furry clumps. The young of all primates are carried for long periods on the bodies of their mothers, requiring even the infant to grasp the adult female securely as she bounds through the savanna or forest. Considerable nuzzling and patting is included in daily primate behavior, as well as the agonistic tactile communcation of pinches, bites, and even kicks.

Combined signals

In social insects as well as other groups, *combined signals* of several types are frequently essential for communication. The "language" used by bees in locating a source of nectar has been, since its discovery by Karl von Frisch, a classic example of the use of combined signals in a social insect. Let us look at this language of the bees to see how such signals are utilized for communication.

If a feeding dish filled with sugar solution is placed in an open field, the first bees to arrive seem to find it by chance. Soon thereafter, bees are seen flying along a straight line from the hive to the feeding dish. Von Frisch, a pioneer ethologist who has worked with bee behavior for more than half a century, postulated that some type of communication must be occurring within the hive to convey this information regarding food location from the first returning "scouts" to the new recruits. On examining the behavior of scouts when they return to the hive, von Frisch noted that a peculiar dance was performed on the combs of the hive. Two variants of the dance were subsequently found to communicate information about the quality, distance, and direction of the food source (Figure 19–12).

The nature of the food source (e.g., species of flowering plant) is indicated by the scent clinging to the body of the scout, and the actual distance of the food is indicated by visual and auditory signals. If a food source is less than 50 to 100 meters away from the hive, bees will perform a *round dance*. If the distance is greater than 100 meters, a *"waggle" dance* in the form of a figure eight is performed. The precise information on distance is conveyed by the time spent waggling on the vertical climb up the center of the sideways figure eight, the tempo of the dance, and possibly the sounds produced

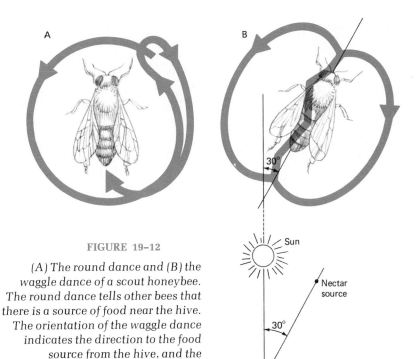

FIGURE 19–12

(A) The round dance and (B) the waggle dance of a scout honeybee. The round dance tells other bees that there is a source of food near the hive. The orientation of the waggle dance indicates the direction to the food source from the hive, and the intensity of the dance indicates the distance (slower for long distances).

during the straight part of the waggle dance. If the scout bee had to go upwind or uphill in transit, his dance indicates that the other bees will need to fly over a greater distance than the actual linear length from one point to the other.

In the waggle dance, directional information is conveyed by the angle that the "straight run" (Figure 19–12) makes with a perfectly vertical line on the comb. This angle corresponds to the angle between the sun, the hive, and the food source. Some scientists have pointed out that in the darkness of many hives such information may have to be transferred by means of olfactory or tactile stimulation—perhaps by bees placing their antennae on the dancers. How this information could be transferred into communication of physical distance and direction is perhaps more mind-boggling than the original von Frisch model of visual communication of these data! There is some evidence that olfactory cues may be sufficient communication in the absence of dancing to recruit a few bees for foraging out to the scout-reported food source, but the directional information contained in the usual waggle dance by the scout seems to be by far the primary source of communication.

The round dances, on the other hand, do not appear to indicate direction to any significant degree. The bees seek food on all sides of the hive (but within 100 meters) when the scout bee performs the round dance.

This communication of complex information about spatial orientation within the bee's environment obviously plays an essential role in the maintenance of the social organization of the hive.

Cooperation: the development of social ties

When a group of animals spends time together, as in an aggregation, the advantages of grouping must benefit the individuals composing the group or sociality would not persist. This grouping response to the forces of natural selection answers the question: Why do animals come together at all? Sometimes the adaptive advantage comes strictly from the physical size of the group. Thus in the Antarctic, aggregations of adult emperor penguins aid in incubation of the eggs laid during the height of the winter (Figure 19–13). These penguins huddle together as they stand over their eggs, conserving heat in the group. Single birds or pairs attempting to incubate their eggs on their own could not maintain the proper temperature during the fierce winter gales. Protection from predation is another advantage of grouping. Colonial nesting birds, including blackbirds and gulls, join together in defense against an invading predator, such as a fox, by their coordinated diving attacks. Within troups of baboons living on the East African savannas, several adult males will move out as a

FIGURE 19–13

King penguins, Aptenodytes patagonicus, *incubating their eggs on South Georgia Island.*

unit to defend the rest of the group against a prowling leopard or other predator.

The activities of the group may favorably modify the local environment, so much so that all individuals find better forage or more suitable conditions for survival. Thus in parts of North Dakota and Colorado, the grazing activities of blacktailed prairie dogs keep the weed height low, enhancing visibility around their burrows and at the same time allowing the smaller herbaceous plants on which they feed to grow on the loose soil of their mounds. Many birds in temperate and tropical woodlands forage together in flocks, the action of the moving flocks scaring up various insects as the flock moves through the undergrowth. Hunting wolves depend on cooperation between one another in their highly organized social packs to hunt caribou, moose, or deer, running them to the ground by spelling one another during a great circular chase of the chosen prey. When the prey animal reaches the point of exhaustion, the full pack (composed mostly of rested animals) moves in for the kill.

Another selective advantage of grouping is that the visual and auditory displays offered in a large group of birds or mammals may help to stimulate and synchronize breeding efforts. Considerable field work has also shown that prey species that have offspring synchronously and flood the breeding area with a sudden flush of eggs and young will tend to swamp predators, who can only take so much prey each day. Thus breeding and synchronization reduce the chances of an individual nest and its contents being attacked. Social stimulation in the center of the breeding area also seems to help the females lay more eggs and raise more young successfully (Figure 19–14). Birds occupying more peripheral nesting sites in a colony are often less successful in breeding.

The development of cooperation in an aggregation of individuals of the same species, then, may lead to the evolution of social ties. As these social relationships become more developed, they make sociality an adaptive feature of the population that transcends in value the formerly important attributes of individual space and dispersion throughout the available environment.

FIGURE 19–14

A female masked booby in the center of a nesting colony on Hood Island in the Galápagos.

Social rankings and dominance hierarchies

In observing a small flock of chickens, ducks, or geese in a barnyard, one usually notices that a definite peck order develops among the confined birds. As the flock moves around, especially to the source of food or water, the birds periodically dispute among themselves for precedence in line. Gradually one will emerge as the dominant bird in the flock, which can displace all the other birds at food or in nighttime roosting place. Below the dominant bird is the second-ranking hen or cock, which can dominate all the flock except the top bird; and so on down through the group to the bottom-ranking bird, which can be displaced by all the other birds in the flock. A peck order of this type is called a **dominance hierarchy** of social ranks. While considerable fighting may take place to establish the dominance hierarchy, this linear array of dominance relationships becomes fairly permanent once it is established, and subsequent fighting is rare. Subordinate animals will normally give way without question at the approach of animals that are more dominant in the hierarchy.

A simple dominance-subordinance relationship in this linear fashion is widespread among vertebrate social groups that involve large flocks or bands (Figure 19–15). However, sometimes several animals share the highest ranks of dominance, and these animals cooperate to assert their dominance over others. Sometimes a fairly large and generalized group of young males and females may be uniformly subordinate to the older animals in a society. In herds of grazing animals and in troops of primates such as baboons, a dominance hierarchy of males is established at the time of the breeding season, with the most powerful males gathering small, temporary harems of females around them. Juvenile and mature subordinate males are forced to forego breeding by the aggressive activities of the dominant males; however, dominant males also assume much of the responsibility for the defense of the group in such species (e.g., baboons), and from an evolutionary point of view it is not difficult to understand why the strongest and most aggressive animals provide this defense. More often then not, they are protecting their offspring.

FIGURE 19–15

Dominance in the highly gregarious African impala is established during the rutting season, when older males must fight other old males and younger males to establish and maintain harems. Each buck has a harem of 15 to 20 "ewes," sometimes up to 100. Males that do not win the rutting season fights rank at the bottom of the impala dominance hierarchy and form bachelor herds (at right).

Many dominance hierarchies are reflected in territorial arrangements. A dominant male during the breeding season will stake out larger and generally more central territories in the breeding group than subordinate individuals. This behavior usually ensures that he obtains first choice of females; or if the female chooses among the males that are holding territories, she reacts to this larger area and the more aggressive behavior and courtship of the dominant male. Let us now look at several of these territorial arrangements in more detail.

Territoriality

A **territory** is simply any defended area. Territories may be set up by invertebrates such as ithomiine butterflies, genus *Mechanitis*, which perch on leaves along jungle trails during the course of a day and fly out aggressively attacking any passing butterfly. Territoriality takes many forms besides the temporary spatial holdings of certain insects. In large carnivores such as lions, leopards, or cheetahs, the territory is often held as a feeding preserve. Social herbivores that stay together in groups, such as howler monkeys in the lowland forests of the New World tropics, will often maintain a core of territory against all other groups of the same species. The early-morning group howling of these monkeys warns other groups of howlers of the presence of a defended territory, and hence the howler groups space themselves through the forest and avoid contact through reasonably exclusive though apparently transitory feeding territories (Figure 19–16).

Many territories are associated with the breeding season and are set up only at this time. In many bird species, each territorial male will defend a substantial area in the spring that is adequate to supply food for himself and eventually his mate and young. Territories of adjacent birds closely follow the boundaries of the first bird's holding, and frequently mutual adjustments are necessary. Thus nearly all available habitat space is occupied by the most aggressive breeding males and later by their consorts.

It is important to note here that although superficially these male individuals appear to be solitary, each restricted to his own territory, the entire collection of territories represents a social organization. In nesting sea bird colonies, these territories are very small compared with those of warblers and other land birds, and they become re-

FIGURE 19–16

Aggressive posture in a male African baboon in Kenya: most dominance hierarchies are established by aggressive behavior but often little actual fighting is involved.

FIGURE 19–17

A tightly packed colony of nesting kittiwakes utilizes every available rocky shelf on a coastal cliff.

stricted to just the small space around the nest (Figure 18–17). Because the territories are tightly packed, the birds form a socially coherent colony. Within the territory the owner has complete dominance. The boundary of a territory marks the point at which the dominance of one bird begins to give way to the dominance of its neighbor.

The advantages of the dispersion of the social group that results in territorial isolation are threefold. First, interindividual competition in the social group occupying these territories is reduced. After the territories are established, actual fighting is minimal or nonexistent, and the males can concentrate on the courtship of passing females. Second, energy that would be spent in antagonistic fighting if all the individuals were close together is considerably decreased during critical periods, such as the soon-to-follow mating season. If there were no territorial organization and males were not restricted to their own areas, every courtship would probably be interrupted repeatedly by other males crowding in around the first displaying male to try to abduct the interested female. Finally, territoriality prevents overcrowding and exhaustion of the food supply for a social group. Thus the evolutionary advantages of territoriality for social animals are considerable.

Social behavior in the primates

A remarkable number of field studies with social vertebrates have focused on groups of birds and primates. The early ethologists concentrated their field work on birds because these were large, relatively easily observed animals in the northern temperate-zone countries where most of these scientists lived. In recent years, studies on the primates have accelerated at a great rate because of their intrinsic interest and in the hope that they will provide information on the origins of human societies and social behavior. The majority of primates are arboreal and not as easy to observe in a tropical forest as the ground-dwelling primates in savanna areas. We shall look at examples of each type: the howler monkey, which is an arboreal primate, and the African baboon, which is a ground-dwelling type.

C. R. Carpenter studied bands of howler monkeys on Barro Colorado Island, which lies at the center of the Isthmus of Panama in the Canal Zone. This island is about three miles in diameter and separated from the mainland by 300 yards of deep water in Gatun Lake. Covered by tropical rainforest and a few areas of scrub and brush, the island has provided a refuge for rainforest animals since the time of the building of the canal and the creation of the lake around it. When Carpenter initiated these studies over 40 years ago, they took almost entirely the character of a strictly observational field study. In more recent years, other behaviorists have banded these monkeys or used other identification to carry out studies in Panama and Costa Rica.

Howlers live in small, cohesive bands (Figure 19–18). While the size of a band varies, its optimum number is 17 members, including on the average three adult males, seven adult females, three infants, and four juveniles. The bands are territorial and aggressive toward outsiders, but relations within the band are characterized by peaceful cooperation and a lack of competition. In fact, there are remarkable communal ties within the band; while moving, it will wait for slower or injured members, and adults cooperate in the rescue of young that have fallen from a tree. About 20-odd bands inhabit Barro Colorado Island, comprising several hundred monkeys. Bands maintain territories that are sharply defined but may or may not be exclusive. That is, the territory may belong solely to one band or the territories of several bands may overlap or even be almost identical. Identical territories may result if the bands split to form two, and they amicably stay in the same general area. Such splitting and growth results in a dynamic situation on the island, with shifts of territories from year to year.

The size of the territory is related to the size of the band, but it varies in three dimensions with the type of terrain. When two bands meet, they engage in verbal battles with great cries and much

FIGURE 19–18

A small band of howler monkeys in the tropical rainforest of Central America.

howling, but no actual physical contact or fighting was observed by Carpenter or later observers. These meetings seldom occur, in fact, and Carpenter believes that the primary function of the morning and evening howling for which the monkeys are so well known is to broadcast the location of each band so that contact between them may be kept to a minimum. Hence these vocalization displays play a significant adaptive role.

It is clear from the above discussion that the band is an important social unit to the howlers. Its movements and reactions are closely coordinated. The band is directed by the males that may lead or follow at the end of the progression. Occasionally a female may be seen leading, but the male will still determine the actual direction and rate of progress. Lack of dominance among the males is striking. Juveniles are helped over difficult spots in the band's path. The total impression is of a highly integrated and coordinated arboreal group that spends most of its time eating fruit and leaves in the forest canopy and avoiding contact and aggression with other groups through its exclusive territories maintained by howling.

Primate species living in open grassland or plains in the tropics tend to have large social groups that wander quite widely. In dense forests, a large group would be unable to maintain its cohesion and thus arboreal monkeys such as the howlers tend to have small bands. The baboons are an excellent example of ground-living primates with highly developed social behavior. Troops of baboons on the plains of East Africa reach 100 or more individuals. The group moves in a rather specific arrangement (Figure 19–19). There is a central group of females and young occupying the protected center portion and surrounded by larger, powerful males. On the periphery of the troop are adults of both sexes and play groups of juveniles. The band feeds on the plains during the day, gathering roots and seeds and occasionally meat in the form of a young gazelle or other small mammals surprised in the grass. Large males act as sentinels, sitting on high ground to scan the countryside as the troop passes.

When a predator is spotted by a baboon, it gives a loud bark, and the troop members cluster more closely together. The males continue to give the alarm with quick barks as the troop prepares to leave or to attack the predator. If they decide to attack, the large males will split off from the troop and move in a concerted effort toward the lion, leopard, hyena, or other threat. Normally, the show of aggressiveness by these formidable males is sufficient to drive the predator off without difficulty or physical conflict. At night the troop moves to concentrations of suitable trees on hills or by water sources and sleeps in the trees.

The large adult males, in addition to directing the movements of the baboon troop and protecting the females and young against predators, act to prevent fighting within the troop (Figure 19–20). If two females, for instance, begin to fight over some food, a male has only to draw near and grunt or bark to break up the conflict. This behavior enhances the maintenance of the stability of the social system, and the animals are protected from injuring each other. In baboon troops as well as sheep flocks and lion prides, two or three males will frequently be codominant at the top of the dominance hierarchy. With troop acceptance of the authority of several top dominant animals that are dispersed through a large social unit of 100 or more baboons, the group's stability and general security is increased.

Baboons are abundant and widely distributed over the African continent. Much of this evolutionary success is the outcome of becoming successful in a widespread African habitat, the savanna, as Stuart and Jeane Altmann have pointed out. The behavioral traits

FIGURE 19–20

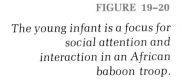

The young infant is a focus for social attention and interaction in an African baboon troop.

that we have examined enable baboon troops to defend themselves from attack by large terrestrial predators, and the aggressive yet cohesive structure of the large baboon group makes it possible for baboons to successfully displace any other monkeys they might encounter. The Altmanns aptly summarized the ecological success of the baboons:

On the savannas, the success of baboons depends upon their ability to exploit a wide variety of plant and animal food sources, and to feed selectively on some of the most concentrated sources of nutrients in their environment. The efficiency of this exploitation and indeed the survival of the animals depends, in turn, on the fact that through intimate familiarity with one particular area, the baboons of a group are able so to distribute their activities that they have adequate access to the essential natural resources of their home range without exposing themselves to excessive risks.

The adaptive benefits of social behavior in the baboon can be directly extended to the small groups of hunting-gathering Australopithecines and *Homo erectus* that inhabited these same savannas, starting several million years ago.

Summary

Social behavior refers to any behavioral interaction between two or more individuals of a species. True social groupings are distinguished from temporary aggregations by their relative stability of membership, extensive communication between individuals, and often permanent social relationships among the members, as in kin groups. Individual distance refers to the buffer zone of space that almost every animal normally maintains around it. Sociality modifies the dispersion and individual distances of a group's members. The development of social ties is dependent on adaptive benefits resulting from cooperation, such as protection from poor weather and predation, aid in obtaining food, and stimulation and synchronization of breeding. The development of social ties is promoted by communication through visual, acoustic, chemical, and tactile signals; these may only occur at special times, as in courtship displays, or they may be a daily feature of life among social insects and higher social animals. Social rankings in the form of dominance hierarchies may develop and partition or otherwise influence the use of resources. Territoriality, the defense of a feeding or breeding territory, is a widespread component of social behavior. Comparisons of social systems in arboreal and ground-dwelling primates offer insights into the evolutionary role of adaptive behavior in assuring the ecological success of a species.

PART FOUR

The biology of individual organisms

As we have shifted our focus from the biosphere to the ecosystem, community, and population, our attention has become increasingly centered on the activities and role of the individual organism. In this final part we pursue the structure and functioning of this organism in finer and finer detail, culminating with the molecular and atomic interactions that are beyond our ability to observe directly but that we can infer experimentally. A consideration of the levels of organization in the individual (Chapter 20) takes us to the microscopic structures of plant and animal cells and their specific functions (Chapter 21). Reproduction at the cellular level (Chapter 22) and the accompanying reorganization of the hereditary material leads us to the principles of genetics and the study of the inheritance of character traits, including sex (Chapter 23). The discovery of the actual mechanisms of gene action (Chapter 24) is one of the landmark achievements in modern biology. The exciting story of the acquisition of energy for the living cell through photosynthesis and cellular respiration (Chapter 25) enables us to understand more fully how energy intake and waste release is regulated in the whole organism (Chapter 26). The internal transportation routes of food and other materials are treated in Chapter 27, including the circulation of blood in human beings and the medical complications that sometimes arise when this circulation is altered. Then we take up the problem of respiration and gas exchange in both plants and animals (Chapter 28) and examine how body temperature is regulated in diverse groups under the great variety of environmental conditions on earth (Chapter 29). The integration and coordination of bodily processes through chemical hormones and the nervous system are treated in Chapters 30 and 31, where we explore the most recent discoveries on the nature of learning, memory, and human sleep. In Chapters 32 and 33 we study the stages and factors involved in the growth and development from egg to adult of plants and animals, especially human beings. Finally, we reach the end of our exploration of the biological world as we observe "the passing of the torch" through the process of reproduction (Chapter 34). The perpetuation of life was our starting point, which has involved evolutionary change and ecological relationships from the biosphere down to the ontogeny of the individual.

20 The levels of organization in the individual

The structure and functioning of the individual organism depends on its constituent parts as well as the world around it. We have been looking at the influence of the biological and physical environments at the population level, and the organization of these populations into communities and into the great biomes of the earth. Yet when we look at the organization within a single organism, we find complexity of systems and parts equal to that at higher levels.

The properties of life

The properties we associate with life are reflected as the total of the activities of organisms. In the ultimate sense, these activities depend on the properties of chemicals in living systems. These chemicals and the molecules that they form are not simply distributed at random within the organism but are contained within discrete structural units called cells, which in turn are organized into tissues, organs, and organ systems. Thus if we took all the chemical constituents of a mouse in precisely the quantities present in the living animal and dumped them all together in a dish, they would not make a mouse. The *arrangement* of the various component molecules is all-important for the existence and activity of the organism. Only the organism possesses the ability to maintain this arrangement of molecules as a discrete functioning entity through time.

The various chemical activities going on in an organism are collectively called its **metabolism;** here again is a principal difference between a living creature and an inorganic object, namely, that the living creature is able to carry on metabolic activities and the nonliving object is not. As a result of these metabolic activities, the organism exhibits additional traits that are not shared by nonliving things, namely, *growth, responsiveness,* and *reproduction.*

A crystal can *grow* from the outside by attracting similar molecules to itself from the environment; but only an organism can grow from the inside by engulfing and reconstituting molecular substances from the environment, breaking them down to release the energy stored in their chemical bonds, and reorganizing some of these molecules to build them into additional parts of its body.

Responsiveness is the ability to respond to changes in the environment. By and large, as we have seen in preceding chapters, these responses are the adaptive result of natural selection, and they enable the organism to protect itself and feed. In other words, responsiveness involves adaptive organism-determined reactions that nonliving objects are not capable of making.

The excess material that an organism takes in from its environment—that is, material normally necessary for its maintenance and growth—may be energetically used in *reproduction.* As we have traced in preceding chapters, variability among the offspring of parents makes it possible for evolutionary forces to mold this raw material, leading to new ways of meeting the problems of existence and developing the diversity of life that we see today.

Organizational levels in the individual

When we look at the composition of living matter, then, we see that it is composed of chemical atoms organized into molecules, and these molecules are grouped into cells (Figure 20–1). There are only 92 naturally occurring elements in the matter composing the universe, and these basic substances are composed of units—namely,

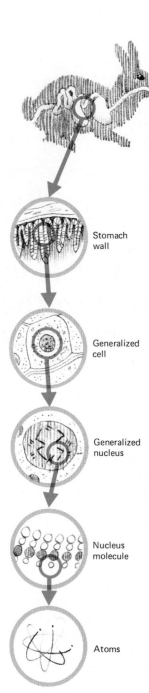

Stomach wall

Generalized cell

Generalized nucleus

Nucleus molecule

Atoms

FIGURE 20–1

The organizational levels in an animal individual: from atoms to organ system.

atoms — indivisible by ordinary chemical means. The atoms of a particular element differ in substantial ways from the atoms of other elements; each has a different number of electrons, for example. Atoms of one element are able to make bonds in particular ways with those of other atomic elements to form molecules. **Molecules** containing at least two different atoms may be classified as either organic or inorganic compounds. The organic compounds contain the element carbon; however, many inorganic substances — such as water, molecular oxygen, and nitrates — are as basic to the chemistry of life as carbon compounds.

A little later in this chapter we shall look at the basic molecular composition of living matter in more detail.

The basic unit of organization in a living organism is the **cell.** The collection of chemical molecules means little without this organic organization, for only intact cells are capable of carrying on the range of metabolic activities necessary for life. Cellular structure and function is surprisingly complicated, as might be expected from the fact that some organisms are composed entirely of a single cell and must maintain all the properties of life through the subunits contained within the boundary of the cellular membrane. We shall look at the basic structures of plant and animal cells in more detail in Chapter 21. For now, let us note that the living cell is an extraordinarily complex unit that contains within it numerous subunits that are themselves relatively complex.

When we look at the bodies of multicellular organisms — those composed of many cells — we find that these plants and animals are organized on the basis of tissues, organs, and organ systems. A **tissue** is a group of many similar cells that are bound together by intercellular material. Because these cells are not randomly clustered but specialized to perform particular functions, they share similar structure and metabolic activities. The basic plant and animal tissues are relatively few in number, yet diverse in character.

An **organ** is a structural unit composed of several different types of tissues, joined together to perform a particular function that requires the cooperation of different kinds of cells. Thus the lungs form an organ that has the function of conveying oxygen into the circulatory system, for distribution to all the cells of the animal body. Their structure includes a lining of epithelial tissue, and supportive cartilage as well as nervous tissues are found within their thin walls. Organs in turn are arranged into systems. An **organ system** is simply a group of interacting organs that function together during the life processes of the organism. Thus the esophagus, stomach, and intestine are separate organs that make up part of the digestive system. The major human systems include the respiratory, digestive, excre-

FIGURE 20–2

Four basic types of tissue in human beings.

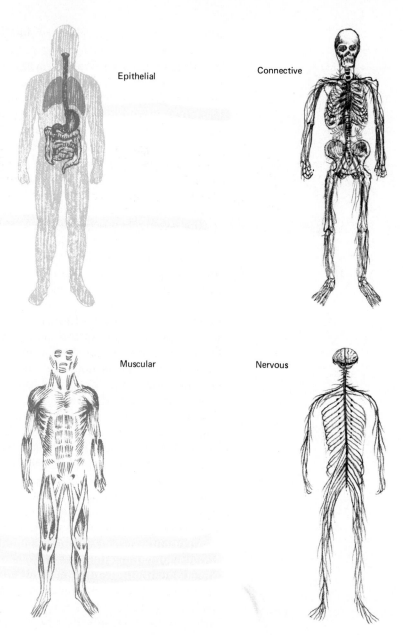

Epithelial

Connective

Muscular

Nervous

tory, reproductive, skeletal, muscular, circulatory, hormonal, and nervous systems, which function together to bring about the well-being of the whole body (Figure 20–2).

Let us look at some of the basic systems and structural units that are involved in all the levels of organization in plants and animals.

Basic tissues and structures in plants

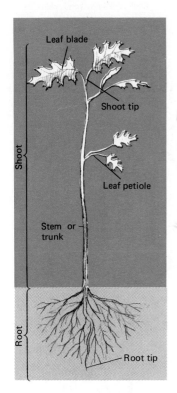

FIGURE 20–3

The plant body and its principal parts, as represented by a young oak. The basic tissues — vascular, supportive, protective, and meristematic — are distributed throughout the plant body. Although meristematic tissues are found mainly at shoot and root tips, the lateral meristems found in dicotyledonous seed plants occur all along the main stem and side shoots.

Plant tissues tend to be somewhat more difficult to classify than animal tissues because they may change in structure and function during the course of their lives (Figure 20–3). Consequently, the tissues formed from such cells may also change. However, plant tissues can be divided into two major categories: meristematic and permanent tissues. The **meristematic** tissues are composed of cells that are actively dividing throughout their lifetime. The **permanent** tissues, composed of the more mature and differentiated cells, are frequently divided into three subcategories — namely, *surface, fundamental,* and *vascular* tissues.

Each of these types of meristematic and permanent tissues has different functions. Meristematic tissues are restricted largely to areas where production of new cells is important, such as the growing tips of shoot stems and roots. These *apical meristems* are responsible for the increase in length of the plant body at both ends. Also, new leaves are produced at the shoot apex from this rapidly dividing group of embryonic cells. Additional meristems in side branches remain active during the life of the plant, producing twigs and leaves. An apical meristem at the root tip gives rise to new root cells and produces a protective cap over the growing root. The *lateral meristems* are responsible in the higher plants for an increase in girth. These various meristems continue their activity indefinitely (except in annual plants), leading to an unlimited system of growth where the only restrictions are those imposed by environmental conditions and the genetic background of the plant. The typical meristematic cell is considered to be the least differentiated of any plant cell. All its dimensions are usually of the same size; vacuoles (membrane-bounded inclusions) in the **cytoplasm** (protoplasm outside the nucleus) are quite small and easily overlooked under the light microscope; and in general, the cell's minimal structural complexity well matches its role as primarily a reproductive factory for new cells by cell division.

Of the permanent tissues, the *surface* tissues provide the protective outer covering of the plant body. Permanent tissues of this type include the thin **epidermis,** which is the surface tissue of all leaves, stems, and roots, and the **periderm,** a corky type of bark found on the stems and roots of plants with active lateral meristems. The periderm functions as an additional protective outer covering of the plant beyond the epidermal tissues in most seed plants.

The *fundamental tissues* are the simple tissues found in the middle layers of leaves, the **cortex** and **pith** of stems, and the cortex of roots. In other words, these are the common interior cells that are

the outer layers or bark   soft tissue located in center of stem.

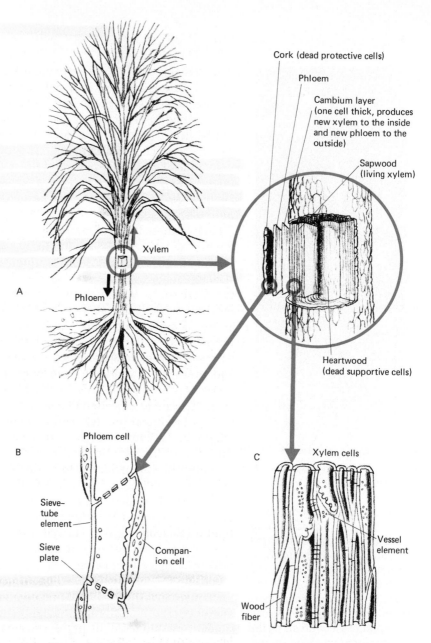

FIGURE 20-4

(A) The types of vascular tissue (xylem and phloem) in a tree and their relationships to other basic plant tissues; the heartwood is composed largely of dead xylem cells; (B) the detailed structure of a phloem tube; (C) details of living xylem tissue.

Cork (dead protective cells)

Phloem

Cambium layer (one cell thick, produces new xylem to the inside and new phloem to the outside)

Sapwood (living xylem)

Heartwood (dead supportive cells)

Xylem

Phloem

A

Heartwood (dead supportive cells)

B

Phloem cell

Sieve-tube element

Sieve plate

Compan-ion cell

C

Xylem cells

Vessel element

Wood fiber

often least different from the parental meristematic cells. Yet they play a key role in the plant body, for they handle photosynthesis, energy storage, and vital other physiological functions, in addition to giving structural strength to the plant body. Commonly, the cell wall is thin and pits dot the wall, allowing interconnections of protoplasm between adjacent fundamental cells. If these cells are stimulated by growth hormones to undergo cell division again, as in response to wound healing and in the production of new buds and roots, the fundamental cells are capable of returning to the meristematic condition. In laboratory tissue cultures, plant physiologists have produced entire plants from groups of fundamental cells or even from single isolated cells, as, for example, from a carrot stick piece.

Finally, the *vascular tissues* form the "plumbing system" of the higher plants (Figure 20–4). Some of these cells, the **xylem,** transport water and dissolved minerals upward from the roots to the rest of the plant body. Part of the xylem forms woody fibers and other thick-walled cells. **Phloem,** the second type of vascular tissue, conducts materials in all directions through the plant and is particularly important in the transport of soluble organic materials such as sugars and other carbohydrate foods produced in the leaves, amino acids, and hormones. Long cells with sievelike ends move these substances from leaf to root and in the opposite direction. Maple trees are known for the sweet sap contained in their phloem cells in the winter; aphids regularly drive their hollow mouth parts into phloem tubes in stem and leaf to draw out sugar-containing sap.

The four basic tissues in plants are grouped into only two major organ systems, the *root* and the *shoot*. These two systems can be readily distinguished on the basis of many different morphological characteristics, and, of course, they are functionally different as well. The roots of a plant are used to procure inorganic nutrients such as minerals and water and to transport them to the rest of the plant. Certain roots also function in the storage of nutrients, and almost all roots are important in anchoring the plant to the surface on which it is growing. The shoot contains a greater number of organs, in particular the basic stem itself, the leaves, and the reproductive organs such as flowers, cones, and spore capsules.

Basic tissues and organ systems in animals

There are also four basic categories of animal tissues: epithelial (surface), connective, muscular, and nervous tissues.

As in plants, the *epithelial tissues* form the covering of all external body surfaces, but also line many internal surfaces such as those of the lungs and the digestive tract. Mitosis frequently occurs in epithelial tissues.

Connective tissues have cells embedded in an extensive, nonliving intercellular *matrix* lying between the actual living cells, such as in cartilage and bone or the fluid matrices of blood and lymph. Most connective tissues, then, such as cartilage, bone, and loose connective tissue, may be collectively described as supporting tissues. They provide a framework for many other tissues and indeed often bind them together. They include **ligaments,** which are tough, fibrous bands that connect bones or support internal organs, and **tendons,** which join muscles to bones or other structures. Blood and lymph are made up of cells derived from other connective tissues such as bone marrow, and they are suspended in a liquid, nonliving matrix. Connective tissue cells have frequent cell divisions. The cells in the *muscular tissues* possess a greater capacity for contraction than most other cells. Muscular tissues are responsible for most movement in higher animals, and they form the major structural parts of many internal organs such as the heart, intestinal walls, and the diaphragm portion of the respiratory system. The elongate muscle cells are usually bound together in sheets or bundles by connective tissue. Once formed and differentiated, the muscle cells do not divide further on a regular basis.

The *nervous tissues* have a remarkably highly developed property of irritability, and the cells are highly specialized for the reception of stimuli and the transmission of nervous impulses. Simple cell bodies have one or more long, thin extensions called **fibers,** which are able to serve as conductors of messages over long distances. A group of nerve fibers bound together by connective tissue constitutes a **nerve.** In all multicellular animals above the sponge level, the organ systems enervated by the nervous system, and functionally moved or controlled by muscular tissue, confer the essential adaptive ability of rapidly responding to stimuli. Nerve cells do not normally divide again, once they have matured.

The molecular composition of living matter

The basic materials in the body of an organism are grouped into a complex mixture of substances called **protoplasm.** Despite the great diversity of life across the surface of the earth, there is a remarkable similarity in the chemical constituents and organization of protoplasm. About 40 elements are essential to life and hence are regularly found in this basic life substance.

The most abundant protoplasmic element is oxygen, making up about 62 percent by weight of the typical cell's contents. Carbon, the characteristic constituent of all organic molecules, makes up about

TABLE 20–1

The elemental composition of the human body

Elements	Percentage of total body weight	Percentage of total atoms in body
Four most abundant elements		
Hydrogen, H	10	63
Carbon, C	18	9
Nitrogen, N	3	1
Oxygen, O	65	26
TOTAL	96	99
Some additional elements		
Calcium, Ca		
Phosphorus, P		
Potassium, K		
Sulfur, S		
Sodium, Na	4	1
Chlorine, Cl		
Magnesium, Mg		
Iron, Fe		

18 percent, hydrogen 10 percent, and nitrogen about 3 percent. Two other elements, phosphorus and sulfur, play a particularly important role in proteins and hence occur in all living creatures. Other elements such as calcium, sodium, potassium, magnesium, and iron are essential to most organisms (Table 20–1).

To a great degree, these elements are deployed in protoplasm in the form of chemical compounds. Each compound is made up of two or more atoms of one or more elements bound together to form a new and more complex aggregate. The attraction that holds two or more atoms together in this way is called a chemical bond. Because of the way that a chemical bond is brought about through the sharing of parts of each atom, each bond represents a certain amount of potential chemical energy. This energy was required to form the bond originally and can be released on the later breaking of that bond. Atoms bond together in specific ways, and only a certain maximum number of such bonds can be formed by the atoms of a particular element. Both inorganic and organic compounds are held together by similar kinds of bonds. The important distinction is that inorganic compounds do not contain carbon, and organic compounds include any compounds containing carbon, whether or not they occur in organisms. The study of organic chemistry, then, can include compounds not necessarily present in nature, such as nylon, a synthetic carbon compound.

Chemical bonds

Molecules are formed when atoms are linked together by chemical bonds. While there are several kinds of chemical bonds, each requires energy to form. It is during the process of forming and breaking such bonds between atoms that the chemical potential energy of a molecule is released. In one sense, a chemical bond is like a spring joining two atoms: the stretch of the spring represents the stored potential energy, which is released if the bond is broken.

The formation of a chemical bond involves an exchange or sharing of electrons between two atoms. Since atoms have an equal number of positive and negative charges, they are electrically neutral; thus any transfer of an electron from one atom to another results in one atom becoming more positive and the other more negative. The attraction between the positively and negatively charged entities forms the basis of the chemical bond.

Electrically neutral molecules are formed whenever an electron associated with one atom becomes partially incorporated into the electron structure of the opposite atom. With the forming of the *covalent* chemical bond, the electrons are shared by both atoms, and electric interaction holds the atoms together. The number of bonds that can be found with a given atom depends on the structure of its outermost electron orbital shell. Thus hydrogen with one elec-

tron is capable of forming only a single bond, whereas oxygen can form two bonds, nitrogen three, and carbon four. In methane (CH_4) the four chemical bonds possible for carbon are formed with four separate hydrogen atoms. Sometimes a double bond is formed between two atoms, as in carbon dioxide (CO_2), which can be written $O=C=O$, where carbon still forms four chemical bonds and oxygen only two.

A chemical bond is generally represented as a line joining two atoms, two lines indicating a double bond. A molecule that is composed of a number of atoms, especially carbon atoms, can have many branches. The bonds are distributed around a given atom in a definite pattern that may or may not be symmetrical, but the result is a definite geometrical configuration that often plays an important role in the molecule's function in living organisms.

If the electrical charge within the molecule is not uniformly distributed, with the result that one portion of the molecule is relatively more positive and another relatively more negative, the molecule is said to be polarized. *Polar* bonds result when the electrons are not equally distributed on the joining of two different atoms. In the water molecule, for instance, the negative electron of the hydrogen atom is more strongly attracted to the positive oxygen nucleus, containing six protons, than to the single proton of the hydrogen nucleus, and therefore the oxygen atom is slightly nega-

tive and the two hydrogen atoms slightly positive. The hydrogen-oxygen bonds of the water molecule are polarized, and thus water is said to be a polar molecule.

Some bonding interactions between atoms result in the complete transfer of an electron from one atom to another, producing atoms known as *ions* that have a net charge. Thus when the ionic bond allows the formation of common table salt, sodium chloride, an atom of sodium transfers an electron to an atom of chlorine, forming a positive sodium ion containing 11 protons but only 10 electrons. On gaining an electron the chlorine atom becomes a negative chloride ion containing 18 electrons but only 17 protons. Ion formation is an extreme case of a polarized chemical bond in that there has been a complete transfer of charge between two atoms. When such an ionic bond is dissolved in water, charged atoms (ions) result. Thus hydrochloric acid, HCl, dissolved in water produces a hydrogen ion, H^+, and a chloride ion, Cl^-, the electron of the hydrogen atom being transferred to the chlorine atom. When an ionic bond is broken like this, the molecule is said to undergo *ionization*. Many small to large molecules in living organisms are able to undergo ionization and produce charged atoms, or even charged molecules if the original large molecule was composed of one or more sizable subunits.

Inorganic compounds

FIGURE 20–5

Two water molecules, with a weak hydrogen bond between them shown in color. The peculiar physical properties of water include the following: It reaches its maximum density at a temperature (4°C) above freezing. At very low temperatures (below 4°C), it expands. Considerable heat energy (called the heat of fusion) is necessary to rupture enough hydrogen bonds between adjacent water molecules to turn ice into a liquid. Considerable heat energy (called the heat of vaporization) is needed to break still more H-bonds to turn the liquid into a vapor. Hence water is a very good solvent and will dissolve appreciable quantities of almost any molecule that carries a net charge.

The most important inorganic compound as far as life is concerned is the water molecule, which makes up about two-thirds of the weight of the human body and probably a major fraction of the weight of almost all other species of organisms (Figure 20–5). A molecule of water consists of two atoms of hydrogen bonded to one atom of oxygen, written in symbolic form as H_2O. As a general rule, roughly 80 to 90 percent of protoplasm (excluding cell walls, bone, and other hard substances) is water. Inside the cells water provides a transportation medium for cellular substances to float to the tiny **organelles,** or structures in the cell that carry on metabolic activities. Free water outside the organism is necessary for life—not only because it provides a source of water that can be taken up for internal use but because chemical reactions in the soil and the air largely take place within a water medium. Water is among the best solvents known, allowing a great many substances to dissolve in it and in greater quantity than almost any other liquid. Thus it facilitates chemical reactions both inside and outside organisms. Water is almost certainly the medium within which life arose on earth, and vast numbers of organisms in marine and fresh-water habitats still live most or all of their lives in water. Water absorbs heat slowly and consequently changes temperature slowly, thus playing a crucial role in stabilizing temperatures across the surface of the earth. Through the greenhouse effect, which we studied earlier, water vapor in the atmosphere exerts a warming effect on the earth by absorbing infrared radiation re-emitted from the earth's surface. Because water is less dense when it is frozen than when it is a few degrees (4°C) above the freezing point, it has its greatest density at 4°C. Consequently, ice will float. Bodies of water will freeze from the top down instead of from the bottom up, protecting the organic life beneath the layer of ice during the winter or in perpetually cold Arctic ocean areas.

Another vitally important inorganic compound is carbon dioxide (Figure 20–6), formed by the union of two atoms of oxygen to one atom of carbon (CO_2). While this substance contains carbon, it is generally classified as an inorganic substance because of its simplicity of structure and lack of associated hydrogen atoms in the molecule. Atmospheric carbon dioxide is the principal inorganic source of carbon for living matter; it gets into living systems mainly through the process of photosynthesis, in which green plants manufacture complex organic compounds from the simple raw materials of carbon dioxide and water.

FIGURE 20–6

The carbon atom and the carbon dioxide molecule. Important facts of interest about carbon include the following: The carbon atom has 4 electrons in its outer orbital shell and therefore needs 4 more to complete that shell. Hence as shown in the carbon dioxide molecule, carbon can share 2 pairs of electrons with such atoms as oxygen and complete the outer shells of each atom. This bond, formed by the sharing of 2 pairs of electrons, is known as a double covalent bond. Carbon's ability to form so many covalent bonds (up to 4 on a single atom) makes it able to serve as the "backbone" atom for a great variety of complex organic compounds.

Carbon

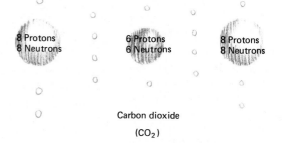

Carbon dioxide

(CO_2)

Organic compounds

The organic compounds of particular importance in organisms fall into four major classes: carbohydrates, fats and other lipids, proteins, and nucleic acids.

Carbohydrates

The **carbohydrates** are usually principally composed of carbon, hydrogen, and oxygen. In simple carbohydrates such as sugar molecules, the hydrogen and oxygen are in the same proportions as in water—that is, two hydrogen atoms for each oxygen atom—and usually matched with a single carbon atom. Consequently, the basic structure of a carbohydrate is written (CH_2O). More complicated carbohydrates are built as multiples of this basic unit (Figure 20–7).

Simple sugar molecules are those that cannot be broken down to form smaller sugar molecules, and they can be oxidized to supply energy for the organism or can be combined by condensation to form very long branched chains called *polysaccharides*. Plant starch, and glycogen, or animal starch, are examples of such long-chain

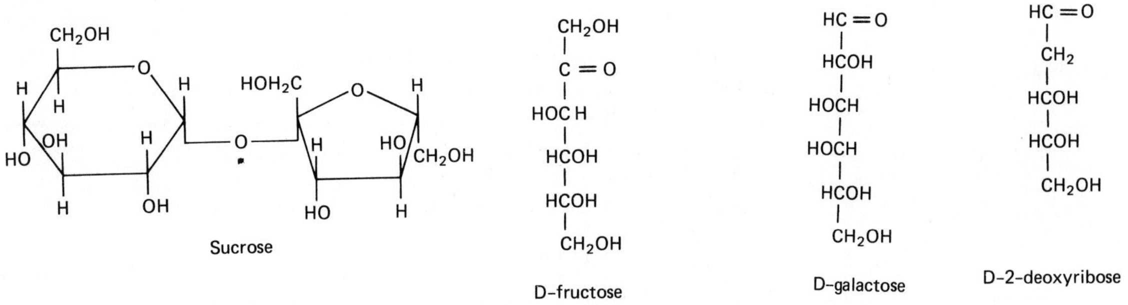

CH$_2$OH

C = O

HOC H

HCOH

HCOH

CH$_2$OH

D–fructose

HC = O

HCOH

HOCH

HOCH

HCOH

CH$_2$OH

D–galactose

HC = O

CH$_2$

HCOH

HCOH

CH$_2$OH

D–2–deoxyribose

Sucrose

FIGURE 20–7

A sampling of the diversity of carbohydrate molecules.

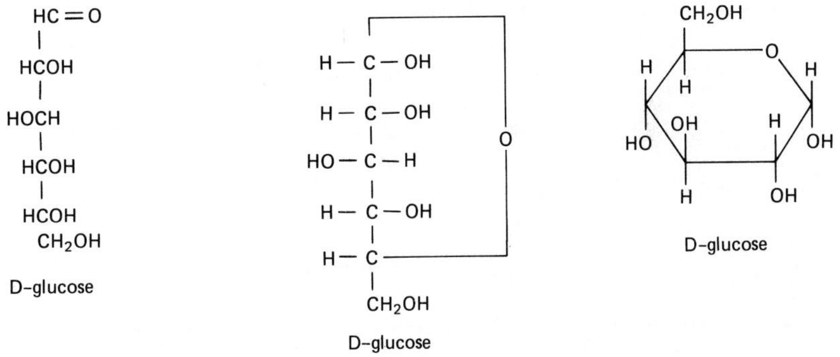

HC = O

HCOH

HOCH

HCOH

HCOH

CH$_2$OH

D–glucose

H — C — OH

H — C — OH

HO — C — H

H — C — OH

H — C

CH$_2$OH

D–glucose

CH$_2$OH

HO

OH

OH

OH

D–glucose

polymers of simple sugar molecules. Most of the biologically important sugars are five- or six-carbon sugars such as glucose and fructose. These two molecules can be condensed together to form a disaccharide called sucrose, or common table sugar.

In general, carbohydrates are storage molecules that are useful for holding sugars in reserve until they are needed as food by the organism. Certain polysaccharides, such as cellulose, are highly insoluble materials that serve as a major support constituent of cell walls in plants, analogous to the role of supportive tissues such as bone found in animals.

Lipids

Lipids include fats, waxes, oils, and steroids, all of which are characterized by having carbon, hydrogen, and oxygen. They are essentially insoluble in water and differ from carbohydrates in that they contain a much smaller proportion of oxygen for the same amount of hydrogen. Sometimes they contain other elements such as phosphorus and nitrogen (Figure 20–8).

Fatty Acids

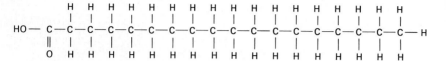

Stearic acid (from beef fat)

Important source
of food energy;
constituent of
other lipids

Glycerides

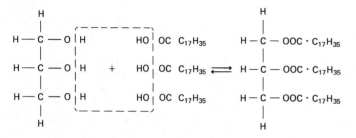

| 1 molecule glycerol | + | 3 molecules stearic acid | | 1 molecule fat | + | 3 molecules water |

$$\text{1 molecule glycerol} + \text{3 molecules stearic acid} \longrightarrow \text{1 molecule fat} + \text{3 molecules water}$$

Source of food energy;
fats (triglycerides) serve
also as thermal and
mechanical insulators

Phospholipids

Glycerol with one hydroxyl esterified to
phosphate (which may make a second
ester bond to another constituent) and
fatty acids at its other hydroxyls

Major constituent of all membranes

Waxes

Fatty acid ester of any alcohol except
glyercol

Source of energy; the harder waxes
(beeswax) may also be structural
components

Terpenes

Polymers of isoprene

$$CH_2 = \overset{\overset{\displaystyle CH_3}{|}}{C} - CH = CH_2$$

Isoprene

Several vitamins (A, E, K) and some
oils are terpenes

Steroids

Constituent of membranes; some of
the hormones (e.g., sex hormones)
are steroids, as is vitamin D

Diffusion and osmosis

In the physical world, including solutions in the form of gases, liquids, and even solids, the molecules of a dissolved substance will tend to move about randomly because of molecular collisions. Over time, the molecules of particular substances (e.g., red mercurochrome) should disperse themselves more or less uniformly throughout the carrier medium (e.g., air, water, or gelatin). The movement of dissolved materials from an area of high concentration to one of lower concentration is called *diffusion*. This passive process of molecular movement (i.e., it requires no special energy expenditure) may speed up at higher temperatures and will change with other environmental parameters, and it represents a highly important way of getting substances from one area to another inside all organisms.

Osmosis is the diffusion of water (or, more generally, a solvent of any kind) through a membrane in response to a gradient of concentration of a dissolved substance. Thus, for example, when the concentrations of a dissolved substance (called the solute) are different on the two sides of a cell membrane, water moves in or out of the cell until the concentration of solute is equalized between the outside and the inside of the cell. If the membrane is only permeable to solvent molecules (such as water), it is said to be *semipermeable*. A membrane that lets molecules of one material, such as salt, pass through more readily than another, such as sugar, is said to be *selectively permeable*; this selectivity may be due to the actual physical size of pores through the membrane (screening out large molecules) or to the chemical composition of both the membrane and the solute (as regards ability to diffuse through the cell membrane layers). The selective permeability of cell membranes to certain substances more than others is currently a subject of intensive scientific investigation, not only for its intrinsic interest in broadening our basic knowledge of cell functioning but for its biomedical applications in understanding the effects of various chemicals, such as aspirin or birth control pills, on the human body.

FIGURE 20–9

(A) Amino acids. (B) Poly-peptide composed of amino acids shown in (A). (C) Tertiary structure of protein: the bending or folding of the polypeptide chain over itself.

NH$_2$
|
H — C — H
|
C = O
|
OH

Glycine

NH$_2$
|
H$_3$C — C — H
|
C = O
|
OH

Alanine

CH$_3$ NH$_2$
| |
H C — C — H
| |
CH$_3$ C = O
       |
       OH

Valine

NH$_2$
H |
C — C — H
H |
   C = O
   |
   OH

Tryptophan

A

|
C = O
|
NH
|
H — C — H
|
C = O
|
N — H
|
H$_3$C — C — H
|
C = O
|
CH$_3$ NH
| |
H — C — C — H
| |
CH$_3$ C = O
     |
     N — H
H |
C — C — H
H |
     C = O
     |
     NH
     |
     |

B

C

Proteins

The **proteins** are groups of similarly structured organic acids and are fundamental to both the structure and the function of living material. Proteins consist of a relatively few simple building-block compounds called **amino acids.** Each amino acid is composed of four essential elements — carbon, hydrogen, oxygen, and nitrogen — arranged in the form of an acid group, COOH, connected by one or more carbon atoms and side chains to an amino group, NH$_2$. About 20 different amino acids are commonly found in proteins. One amino acid binds itself to another by condensation reactions between the acid group on the first molecule and the amino group on the second. These bonds are called **peptide bonds,** and chains that result from such bonding are called **polypeptide chains.** As many as

50 to 50,000 or more amino acids may combine to form a single protein molecule. The potential variation involved in protein structure is obviously great, and the possible number of different proteins could be said to be endless (Figure 20–9).

The three-dimensional structure and form of these protein molecules is also extremely important in their activity. As we shall see later, many of the most important enzymes are globular proteins whose activity depends on retaining a particular three-dimensional structure. If exposed to high temperatures, radiation, excessive acidity, or various chemical reagents, this structure may be disorganized, and the protein is said to be denatured. When this occurs, the activity of the enzyme is lost, and it is incapable of performing the reaction that it normally catalyzes (i.e., speeds up without being consumed itself).

Nucleic acids

The fourth major class of organic compounds — the **nucleic acids** — is also composed of building-block units, called **nucleotides.** The more familiar names for these acids are *deoxyribonucleic acid* (DNA) and *ribonucleic acid* (RNA). The nucleic acids control the proteins that an organism is able to make. DNA forms the basis of the genes on the chromosome that specify the structure of each protein, and RNA controls the actual manufacture of the proteins on the ribosomes. Nucleotides contain carbon, hydrogen, oxygen, phosphorus, and nitrogen, and the particular sequence in which the several different known nucleotides are arranged gives each nucleic acid its own distinctive properties (Chapter 24). DNA, the basic constituent of chromosomes in all organisms except certain viruses that use RNA, is the hereditary substance that carries the coding information for synthesizing proteins from parent organism to offspring.

Thus many kinds of molecules are essential to organisms in order to maintain their functioning. They provide the basic structural units for the ultimate nature and composition of the various levels of organization in the individual. Other important molecules, such as the energy-carrying substances, including *adenosine diphosphate* (ADP) and its more highly energized relative *adenosine triphosphate* (ATP), serve vital functions in the organism. For instance, a molecular form (ATP) can serve as temporary storage of energy, releasing this energy whenever and wherever the extra phosphate bond is broken by the proper enzyme. There are also many pigments in plants and animals such as the **cytochromes** (proteins joined to pigment molecules) that play an important role in the transfer of energy and the construction of new molecules to build the living organism. We shall consider these in more detail in the chapters that follow.

Summary

The structure and functioning of the individual organism may be studied at a number of levels of organization — from atom, molecule, cell, tissue, organ, and organ system to the intact animal or plant. The properties of life — a specific arrangement of molecules, metabolism, growth, responsiveness, and reproduction — may be found only in the cell and higher levels. The basic tissues in plants are the meristematic, surface, fundamental, and vascular tissues. Only the meristematic tissues regularly participate in cell division. In animals, the four basic tissues are the epithelial (surface), connective, muscular, and nervous tissues. The cells of epithelial and connective tissues regularly divide. The molecular composition of living matter includes vital inorganic compounds, such as water and carbon dioxide, and organic compounds, especially the carbohydrates, lipids, proteins, and nucleic acids.

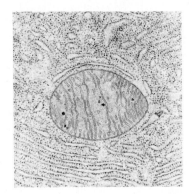

21 Cellular structure and function

The idea that cells are the common units of structure and function in all organisms is one of the three great conceptual foundations of modern biology, the other two being the theories of evolutionary change and ecological relationships. Biologists did not realize the cellular nature of life until the invention of the microscope. A Dutch worker, Anton van Leeuwenhoek (1632–1723), and his contemporaries developed microscopes from magnifying lenses in the mid-seventeenth century, for the first time producing optical instruments capable of a simple scientific investigation of structures smaller than the eye could see (Figure 21–1). The first observations of dead and living cells were soon made by Leeuwenhoek and by Robert Hooke (1635–1703), who first described cellular organization in plants in 1665, in a lecture before the Royal Society of London. Hooke observed repeated vesicles, which he called "cells," in cork and other plant tissues (Figure 21–2). Leeuwenhoek was the first to see minute single-celled animals in a drop of pond water.

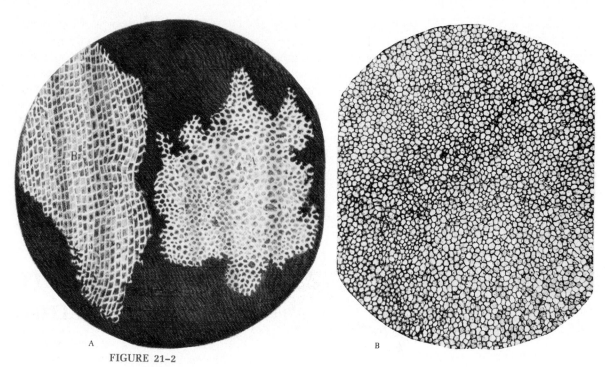

FIGURE 21–2

(A) Hooke's drawing of empty cork cells in cross section; (B) a modern photomicrograph of empty cork cells from the oak (Quercus saber).

The cell theory

It was not until the nineteenth century that microscopic observations began in earnest and the first formulations of the cell theory were made. The German zoologist Theodor Schwann put forth the generalization that "all organisms are composed of essentially like parts, namely, of cells." Whether we are dealing with a lowly flowering plant among the meadow grasses, or a large and complex organism such as a human being, the whole body may be seen to be an aggregation of cells. The cell theory is credited to both Schwann and another German investigator, the botanist Matthias Jakob Schleiden, who published their ideas in 1839 and 1838, respectively. A few years later, in 1858, Rudolf von Virchow of Germany said that all living cells arise from preexisting living cells — that is, life comes from life — and there is no spontaneous creation of cells from nonliving matter. This became known as the theory of biogenesis.

Only a year later, Darwin's theory of evolution was published. Simultaneously, the great French scientist Louis Pasteur (1822–1895) began experiments between 1859 and 1861 to prove that no living thing arises except from other living things. Up to this time, surprisingly enough, scientists as well as laymen believed in spontaneous generation—that lower organisms can originate from nonliving matter, especially from the flesh of dead animals. This idea can be traced back to Aristotle (384–322 B.C.), who taught that fleas and mosquitoes originate from putrefying matter. Pasteur's classic experiments disproved every known example of supposed spontaneous generation of living things (Figure 21–3). For instance, it was generally thought that even if wine, milk, meat broth, and other organic substances are protected from flies and other apparent sources of infestation, microorganisms will eventually appear and multiply. Pasteur showed that high heat kills any microorganisms present in a subtance, and that if the sterile broth or other nutrient media is then protected from freely circulating air, no new microorganisms will appear. Today we call this heating process *pasteurization*.

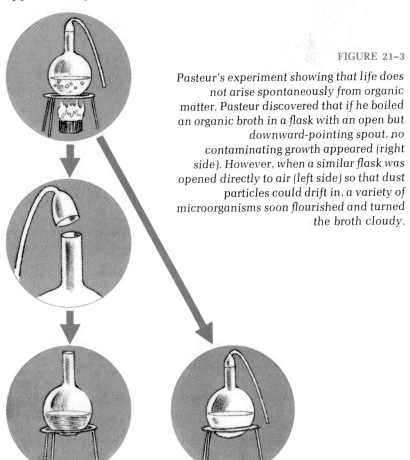

FIGURE 21–3

Pasteur's experiment showing that life does not arise spontaneously from organic matter. Pasteur discovered that if he boiled an organic broth in a flask with an open but downward-pointing spout, no contaminating growth appeared (right side). However, when a similar flask was opened directly to air (left side) so that dust particles could drift in, a variety of microorganisms soon flourished and turned the broth cloudy.

FIGURE 21–4

Redi's experiment showing that fly maggots did not appear spontaneously in rotting fresh meat but that their presence was due to earlier visits by egg-laying adult flies. When a jar was screened, no fly maggots appeared. This study helped to disprove contemporary beliefs in the spontaneous generation of life from nonliving matter.

Actually, another scientist in Italy, Francesco Redi (1627–1697), had already disproved the idea that maggots arise spontaneously in rotten meat (Figure 21–4). With careful screening, he showed that maggots never appear unless flies have laid eggs in the meat; contrary to popular belief at the time, he concluded that even flies have parents. It remained for Pasteur to demonstrate two centuries later that microorganisms also do not originate spontaneously from nutrients exposed to the environment. Thus biologists in general believe that the principle of biogenesis as a derivative of the cell theory holds true for presently existing conditions; that is, it is apparent that spontaneous generation of life from nonliving matter does not occur today, but it may have at some time in the distant past (as we have considered in Chapter 14), when life first arose on the primitive earth.

Living things, then, are characterized by a cellular pattern of organization capable of reproducing itself. The generalized "cell," however, is perhaps as difficult to describe as a generalized "organism." It is the minimal level of organization that is able to show the basic characteristics of life. There are smaller organisms known, such as the viruses, but the viruses are all dependent on living cells for their own reproduction. When they infect a living cell, they take over the cellular machinery of the host to produce new copies of the old virus, and in the process the whole cell is destroyed. Thus the virus is incapable of self-reproduction since it can reproduce only through the mechanisms provided by host cells. Likewise, the virus particle does not possess the capacity for self-regulation that cells do. Today, therefore, the **cell** is said to be the minimum unit possessing the characteristics that we associate with living things, such as metabolism, reproduction, irritability, and movement.

The structure of cells

Our description of the structure of a cell is based on relatively modern observations resulting from better and more powerful microscopes than were available to the early investigators. Microscopes using ordinary light for illumination can magnify the apparent size of the object being viewed about 2,000 times. However, the resolution ability of such microscopes — that is, the capacity to separate adjacent objects distinctly — is only about 500 times better than that of the unaided human eye. The use of stains in modern microscopy provides contrast, enabling us to distinguish one part of a cell from another by the colored substances that particular cell parts will take

FIGURE 21-5

The diversity of animal cells includes incredibly complex and quite large cells that are entire animals, such as the Paramecium (a protozoan) shown here, and rather simple cells lacking even a nucleus, such as the human red blood cell (which has a depressed central portion and can slip through tiny capillaries).

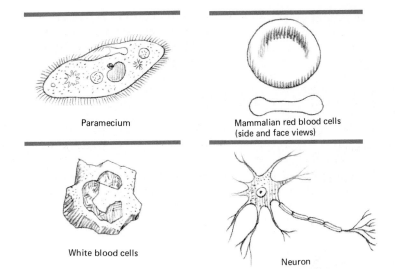

Paramecium

Mammalian red blood cells
(side and face views)

White blood cells

Neuron

up; and it is possible to stain different parts with various dyes, depending on the affinities for these dyes. In the past several decades, however, the greatest boon to our understanding of cell structure has been the use of the electron microscope. This instrument utilizes a beam of electrons instead of light as a source of energy (see Microscopes, p. 370), and its resolution ability is some 10,000 times better than that of the unaided human eye. The smallest components of the cell were not evident before the invention of the electron microscope and hence were unknown to the earlier investigators.

Most of the cells of the animal and plant groups are microscopic in size (Figure 21–5). A few cells such as bird-egg cells and large single-celled amebas can be seen with the naked eye. Some human nerve cells may be as long as three or four feet, although extremely thin in diameter. The average diameter of a cell in a human being or similar multicellular animal is about 10 microns, that is, 0.01 millimeter (a micron is one-thousandth of a millimeter, or one-millionth of a meter). Certain bacterial cells are as small as 0.4 micron in diameter, which is approaching the limit of vision with an ordinary microscope. The largest known cell is the egg of an extinct bird (*Aepyornis* of Madagascar), whose yolk (the only part of a fertilized bird's egg cell that can be seen until the embryo begins developing; the outer shell is not part of the egg cell) had a capacity of more than two gallons, and thus in volume was the largest known cell.

The shape of cells varies greatly, from elongate spindle-shaped or fiberlike cells to large globular masses of constantly shifting dimen-

Microscopes

The ordinary *light microscope*, invented over three centuries ago, is still the workhorse of the biologist investigating the microscopic world. The earliest light microscopes were composed of very crude magnifying lenses that could magnify as much as 50 times natural size. Today the same basic arrangement of ocular lenses and objective lenses is used, but a magnification of up to $2,000\times$ natural size can be achieved. The smallest object resolved (seen clearly and distinctly from similar-sized objects separated by like small distances) is determined by the wavelength of light. Under optimal conditions of maximum contrast in the structure under observation, the smallest object that can be seen is theoretically about one-third the wavelength of visible light, or about 0.13 to 0.2 micron in blue light. The principle of the light microscope is simple. The objective lens, or lenses, is closest to the object and this lens, or group of lenses, produces an image at a certain point inside the microscope tube. The eyepiece or ocular lens is used as a magnifying glass to view this projected image of the object. Thus the total magnification power of a light microscope (e.g., $10,000\times$) is measured by multiplying the power of the objective lens (e.g., $100\times$) by the power of the ocular lens (e.g., $10\times$).

The *phase contrast microscope*, which has the same resolving power as the ordinary light microscope, enables one to achieve greater contrast of structural details in the living cell by taking advantage of the slight differences in refractive index between any two structures to improve their visibility. The phase contrast microscope does this through the use of special optical devices to cause part of the light going through an object to be a quarter wavelength out of phase, producing a diffracted image as well as a direct image. The phase difference that is not seen by the eye is converted to a darker or lighter intensity difference that we do see. It is not necessary to kill and stain the cell to achieve these effects, which is a considerable advantage over work with the ordinary light microscope.

The *electron microscope*, which was developed on the eve of World War II, has proved a powerful tool for analysis of proteins and nucleic acids, viruses, and structural details of the cell. In the electron microscope, lenses are replaced by magnets that focus beams of short-wavelength electrons instead of light. The object must be viewed on a fluorescent screen or photographed. The electron microscope has a resolving power 50 to 200 times that of the light microscope. The main disadvantages involved in use of the electron microscope are the required extreme thinness of sections (which must be coated with a metal) and the necessity of studying them dry (in a vacuum), which means water and salts must be removed and therefore structures are likely to be altered from their precise natural arrangement in the intact cell.

FIGURE 21–6

Some common organelles and microscopic structures in a generalized cell.

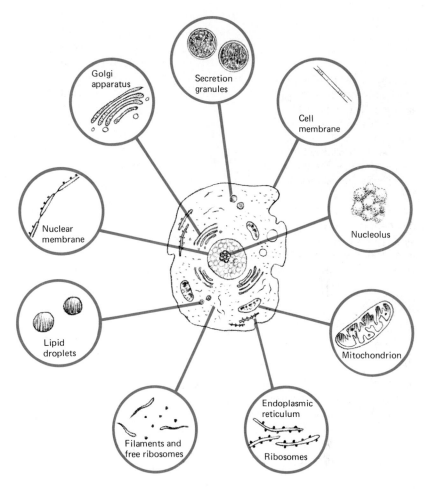

sions, such as the single-celled ameba. Human blood cells are biconcave disks, with depressed centers on either side; many skin cells are cubelike in dimensions. It is impossible to present a truly generalized cell because of the great variety of size and shape in cells throughout all organisms—especially since electron microscopy has disclosed the fine structure of subcellular particles. We can look at a "generalized" cell diagram in Figure 21–6, as long as we understand that this illustration shows the features that almost all cells possess in common and does not depict any particular cell. Let us consider now in more detail the structural components of this hypothetical generalized cell.

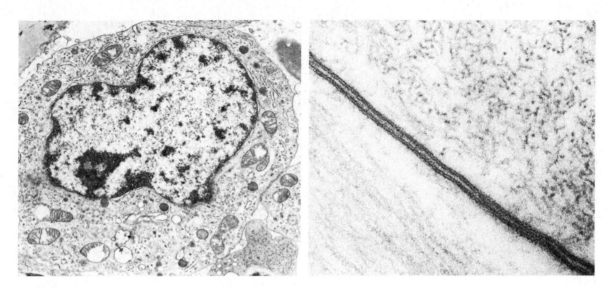

FIGURE 21-7

An electron microscope view of a cell (above, left).

FIGURE 21-8

An enlargement of the cell membrane area under the electron microscope (above, right).

The cell membrane

The typical animal or plant cell has a large amount of cytoplasm bounded by a **cell membrane,** also called the **plasma membrane.** This cell membrane is so thin—approximately 80 angstroms thick (an angstrom is a ten-thousandth of a millimeter, or a ten-millionth of a meter)—that it cannot be seen with the normal light microscope, although the boundary between the cell cytoplasm and its environment can be detected readily by the observer. The invention of the electron microscope was necessary before the extremely fine structure of the cell membrane could be resolved (Figures 21–7 and 21–8).

Today, cell biologists believe that the boundary membrane of a cell consists at least in part of a double layer of phospholipid molecules, the outer side of each layer consisting of lipids mixed with globular proteins. The protein molecules project beyond the outer surface of the membrane. In other words, we have a sandwich of two layers of lipid molecules, each having their fatty acid chains projecting *toward* each other in the center and little molecules of globular proteins scattered among the lipids on the outer surfaces (Figure 21–8). However, this is only one of the popular current models for describing membranes in the living cell. In the earliest model, for instance, constructed by J. F. Danielli in 1940, the membrane consists of inner and outer layers of protein with two layers of lipid between them. It is at least known that most membranes appear to be composed of lipids and proteins, that they vary considerably in

FIGURE 21–9

Cell walls in the woody tissue of black locust (Robinia pseudoacacia).

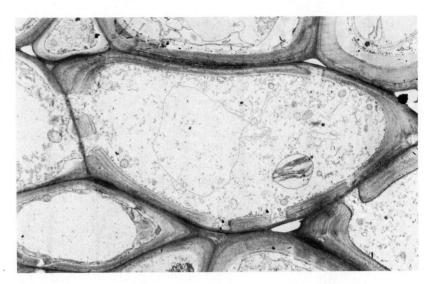

thickness from 50 to about 100 angstroms, and that the ratios of the various types of lipids in the membrane vary widely from one membrane to another. Tiny pores through the membrane appear at intervals along the cell surface.

The membrane is a highly important part of the cell and not simply a passive boundary layer. It functions in selectively controlling the entrance and exit of certain substances between the cell and its environment. For instance, the cell membrane can pass potassium ions into the cell and force sodium ions out. It can actively transport certain large molecules such as sugars and particles from the cell surface into the cell interior; this ability varies from one type of cell to another and from time to time in the same cell. The chemical structure of the cell membrane layers is believed to control this selective movement of substances in and out of the cell.

The membranes are very flexible structures and do not provide a rigid framework for the fluid, largely aqueous protoplasm inside. In single-celled animals or plants, the membrane may provide sufficient support for independent life with a definite form, but a large multicellular organism would collapse from its own weight if there was not a supporting skeleton either externally or internally. Plants lack such mechanical support from wholly separate skeletal systems and instead depend on a thick **cell wall** around each cell membrane (Figure 21–9). Thus one of the most striking differences between the cells of plants and animals is the conspicuous cell wall of the former. The wall is much thicker than the membrane that lies within it and plays a smaller part than the membrane in controlling the passage of materials in and out of the cell. Its primary function is mechanical. The cell wall is composed principally of the complex polysaccharide *cellulose* and other related compounds. The long, threadlike structures of cellulose are mixed with pectin, a complex polysaccharide that helps to bind the primary walls of two cells together. The cell walls that Hooke observed in cork were good examples of the relatively simple structure of the wall and the arrangement of these walls, which are often squarish or rectangular when viewed in cross section.

The study of cell membranes has recently become one of the most revolutionary fields in biology. The workings of this fragile biological wrapping are being explored in the hope of discovering how sperm recognize the egg and steer themselves to it; how cells becomes specialized during embryonic development; how white blood cells (one of the body's defense systems) are summoned to the site of an infection or wound — there to attack foreign bacteria or other substances; and how cancer cells get started, how they manage to elude the body's immune systems, how they defy the limits to growth imposed on normal cells, and how they clump together in tumors. In one sense, the membrane that encloses every living cell is the biological equivalent of China's Great Wall — a structure designed to keep unwanted biochemical invaders out while at the same time holding useful elements in. Only in the last few years have scientists begun to regard the membrane to be as important a factor in biological processes as the other two major parts of the cell: the **nucleus,** which holds the essential genetic material of the organism; and the **cytoplasm,** the warehouse of biological materials and special bodies that enables the cell to operate. The present view of the mechanism of exchange with the environment between the interior of the cell and the outside of the membrane is that the proteins (with short-length sugar molecules attached) that stud the membrane provide this critical service, admitting fresh materials and chemical signals to the cell's interior and passing out waste products.

The cytoplasm

The cytoplasm forms the fluidlike interior of the cell and contains a large number of membranes and granulelike organelles. Collectively, these form a rich and diverse system of membranes and other elements that could not be seen by the light microscope but have been revealed by the electron microscope. All protoplasm inside the cell membrane except for the nucleus, then, is generally called the **cytoplasm.**

Tiny microtubules apparently formed of aggregates of 10 to 14

subunits are also scattered through the cytoplasm. These microtubules may be responsible for changes in solidity of the cytoplasm,
and they play a role in the movements of which many cells are capable. They also are involved in the processes of cell division.

The nucleus

Inside the membrane of every plant and animal cell, except those of
bacteria and blue-green algae, the nucleus is usually the largest and
one of the most conspicuous structures. It forms the control center of
the cell and plays the central role in cellular reproduction, the
process by which the genetic material is duplicated and a single cell
forms two new daughter cells. Its genetic material controls the differentiation process that a cell undergoes from the time of its origin
to the form that it exhibits at maturity, and, of course, this material
forms the ultimate control center for all the metabolic activities of
the living cell. In other words, the nucleus serves as the guidance
center for the life processes of the cell, the basic unit of organization
(Figure 21–10).

The nuclear structure is surrounded by a double membrane similar to the cell membrane, and the protoplasm inside is called
nucleoplasm. Long, tangled threads composed of nucleic acids and
proteins—the *chromatin*—are scattered throughout. At the time of
cell division, the DNA threads forming this network of chromatin
will condense into discrete rod-shaped bodies, the **chromosomes.**

FIGURE 21–10

(A) The nucleus of a bat liver
cell has a dark nucleolus area,
but is relatively uniform at
this magnification, in contrast
to the surrounding cytoplasm
filled with mitochondria and
endoplasmic reticulum folds.
The cell membrane may be
seen across the bottom and in
the upper left corner. (B) A
highly magnified detail of the
nucleus with a very large,
dark nucleolus, from a corn
root tip cell.

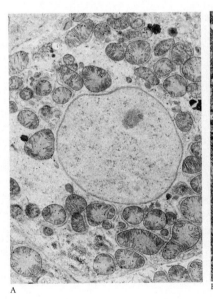
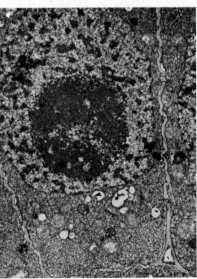

A

B

Also within the nucleus, one or more **nucleoli** are embedded in the mass of somewhat granular-appearing nucleoplasm. These dark-staining, generally oval bodies apparently associate themselves with specific regions of specific chromosomes and may play an important role in sending instructions for protein synthesis out to the cytoplasm. We shall discuss the functioning of the chromosomes and the nucleoli further in Chapter 24, when we consider the structure and functioning of the genetic material.

The nuclear membrane is interesting in itself because it is interrupted at intervals by fairly large pores, where the inner and outer membranes of the nuclear envelope are continuous. Despite the presence of these pores, the nuclear membrane is highly selective in allowing materials to cross the cell membrane into the cytoplasm and in the reverse direction. Another interesting and highly significant feature is that the nuclear membrane is continuous at some points with an extensive cytoplasmic membrane system called the **endoplasmic reticulum.** Materials may flow out of the nucleus through the pores into certain channels in this endoplasmic reticulum, and thus the connection is likely of great significance for the transport of materials and the functioning of the cells.

The endoplasmic reticulum

The **endoplasmic reticulum (ER),** discovered in 1945 by the use of the phase contrast microscope (capable of greater acuity for observation of living material than the ordinary light microscope; see Micro-

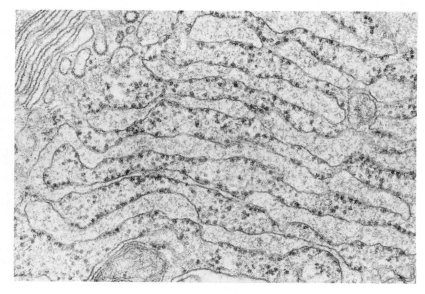

FIGURE 21–11

An electron microscope view of the smooth endoplasmic reticulum (top left) and rough ER with ribosomes (dark granular objects) from the tracheal cell of a rat.

scopes, p. 370), is a network of membrane-enclosed spaces extending throughout the cytoplasm. In structural appearance the ER looks like a set of branching membrane-lined tubules that connect in continuous fashion the plasma membrane and the nuclear membrane. An ER has been found in all living cells with nuclei, although it is often not as well developed in protozoans as in higher organisms. The endoplasmic reticulum transports cell fluids and chemicals throughout the cell. It also forms a structural framework for the ribosomes, which we shall consider next (Figure 21–11).

There are two types of endoplasmic reticulum: smooth ER with no granules and rough ER with ribosomal granules on the outer surface. It is believed that the ER functions as a cytoplasmic framework, providing manufacturing surfaces for the cell across its enormous surface area, and, of course, its channels serve as routes for the transport of materials between the various regions of the cytoplasm or between the nucleus and specific parts of the cytoplasm.

Ribosomes

Ribosomes are tiny granules tht occur on the outer surface of rough endoplasmic reticulum or float seemingly free in the cytoplasm. The typical ribosome is composed of ribonucleic acid and protein. The function of the ribosome is in the synthesis of proteins such as enzymes and blood proteins. Ribosomes may form chains called polyribosomes for the synthesis of large protein molecules. We shall look at their cellular functioning in more detail in Chapter 24.

The Golgi apparatus

Another interesting cytoplasmic organelle is the **Golgi apparatus,** or body, discovered in 1898 by Camillo Golgi in Italy by the use of certain chemical stains and the light microscope. The Golgi apparatus is a system of membranous vesicles, usually occurring between the nucleus and the cell membrane in one particular area of the cell. Composed of membrane-lined channels and expanded spaces, the Golgi apparatus makes lipids, assembles collagen molecules (fibrous proteins) for connective tissues, and packages secretions such as hormones from endocrine gland cells. Thus it is not surprising that the Golgi apparatus is particularly well developed in cells that secrete enzymes and produce collagen; and, in fact, as the level of the secretory activity of these cells changes, corresponding changes occur in the structure and size of the Golgi apparatus. The Golgi apparatus also apparently makes polysaccharides from sugars and may join them to proteins (Figure 21–12).

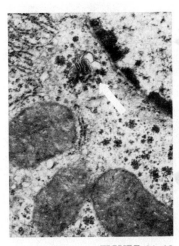

FIGURE 21–12

The Golgi apparatus (see arrow) is located just outside the nuclear membrane and surrounded by large, dark mitochondria in this view of a mouse liver cell.

Lysosomes

Lysosomes are membrane sacs randomly scattered in the cytoplasm. In essence, a lysosome is a package of powerful digestive enzymes bounded by a membrane layer; it breaks up large molecules into smaller ones by releasing its enzymes. The lysosome membrane is impermeable to the outward movement of these enzymes and is also capable of resisting their digestive action. However, if the sacs are broken inside the cell by injury to the cell membrane and contents, the lysosome functions as a "suicide packet" and the contents of the lysosome spill out, destroying the molecular integrity of the cell. Its normal role is apparently to serve as the cell's digestive system, processing some of the bulk materials taken in through the cell membrane.

Mitochondria

Mitochondria are small bodies of variable shape that are found in all types of cells and are scattered randomly in the cytoplasm. They are generally round or oval bodies, with a two-layered membrane bounding the outside. The inner layer of this membrane has folds called shelves, or *cristae*. On the cristae are located enzyme systems that make adenosine triphosphate (ATP) molecules, an energy source for cellular activity. Some enzymes active in energy production also occur in solution in the fluidlike matrix between the cristae. Thus the mitochondria have been called the "powerhouses of the cell." Active cells tend to have more mitochondria than less active cells (Figure 21–13).

FIGURE 21–13

A highly magnified mitochondrion, surrounded by endoplasmic reticulum, in a bat pancreas cell. Note the infolding cristae inside the mitochondrion.

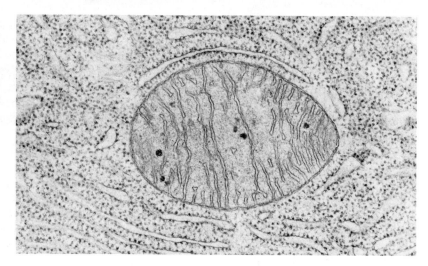

While they are much larger than many other kinds of cytoplasmic organelles, the mitochondria are too small for the light microscope to resolve much detail about their morphology. It was not until the invention of the electron microscope that their structure and function could be examined in detail. Each cell has, on the average, about a thousand mitochondria. Here is where most of the food taken into the cell is converted into usable energy. The magnitude of complexity of a mitochondrion is only beginning to be realized. Recent discoveries have shown objects composed of three parts, called tripartite units, standing upright on the cristae that resonate 1,300 times per minute. An average of 40,000 to 50,000 tripartite units are found in each mitochondrion. The structure and functioning of the mitochondrion will undoubtedly provide fruitful areas for research for many years to come.

The central body

The **central body,** or centrosome, is a region of specialized cytoplasm located just outside the nucleus of most animal cells, which contains two dark bodies called **centrioles.** These structures, oriented at right angles to each other, aid in cell division, and if they are lacking for some reason cell division cannot occur. We shall discuss the structure and function of the centrioles in the next chapter, when we consider their role in the process of cell division.

Peroxisomes

In some cells, particularly those of the kidney and liver, organelles similar to little lysosomes occur. These membrane-bound vesicles, called **peroxisomes,** contain an assortment of powerful enzymes that tend to be oxidative (taking off hydrogen atoms) rather than reductive (adding hydrogen atoms to break bonds), like digestive enzymes. They catalyze the removal of amino groups from amino acids. Some peroxisomes contain three enzymes that catalyze reactions producing hydrogen peroxide (H_2O_2), a very toxic substance. A fourth enzyme in these organelles is capable of decomposing hydrogen peroxide. Other peroxisome enzymes have the capacity for changing fat to carbohydrate.

Plastids

Most plant cells contain large cytoplasmic organelles called **plastids.** They are bounded by a membrane and the internal structure consists of stacks of membranes. There are two principal categories of plas-

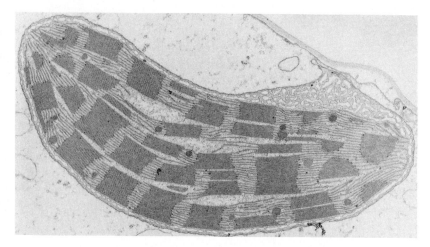

tids: The colored plastids, the *chromoplastids*, contain the pigments that give color to plant parts; those in leaves that contain the green pigment, chlorophyll, are called chloroplasts and represent extremely important units for the maintenance of all life. Here photosynthesis, the fundamental process of converting the energy of sunlight into the energy of chemical bonds, takes place. It should be mentioned that not all pigments are restricted to plastids; some types are found free in the cytoplasm. Chloroplasts and mitochondria alike contain DNA and ribosomes and are able to manufacture some of their own proteins as well as control their own reproduction, a key property for the two classes of organelles that serve as cellular powerhouses. Other plastids in the cell, called leucoplasts, are colorless and function in the formation and storage of complex molecules such as starch, oils, and protein granules. Carrots and potatoes are roots that have a high content of starch-filled plastids (Figure 21–14).

Vacuoles

Vacuoles are fluid-filled spaces in the cytoplasm bounded by membranes. They are found in both plant and animal cells, but are more common and highly developed in the cells of higher plants. The latter contain a large central vacuole that pushes the other contents of the cell within a short distance of the plasma membrane. In protozoa, specialized vacuoles called contractile vacuoles play an important role in excreting excess water and wastes from the cell cytoplasm. The central plant vacuole contains a liquid called cell sap and serves as a storage area for waste products from metabolic processes in the plant cell (Figure 21–15).

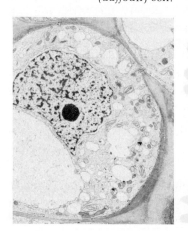

Cilia and flagella

From the tiny one-celled protozoan to the complex multicellular human organism, we find cells with one or more movable hairlike structures projecting from the outer surface. If these structures are short and numerous, they are called **cilia.** If there are only a few of them and they are relatively long, beating with undulating motion rather than a simple bending movement, they are called **flagella** (Figure 21–16). Structurally, the basic components of a cilium and a flagellum are very similar, and in many cases the terms may be used interchangeably. Their function is either to propell the cell along, as in single-celled organisms, or to propel substances across the surface of the cell, as in cells lining the intestinal gut where their beating aids in moving materials through the passageway. The sperm cell travels to the egg in most animals and many plants by means of a waving flagellum.

The stalk of a cilium or flagellum consists of groups of microtubules; two intertwined microtubules are located in the center with nine pairs arranged in a circle around the periphery. The whole cylindrical stalk is actually an extension of the cell membrane. At the base of the stalk, within the main portion of the cell, is a basal body, a structure that closely resembles a centriole.

FIGURE 21–16

(A) A flagellum on a bacterial cell; (B) cilia on two conjugating ciliated protozoans.

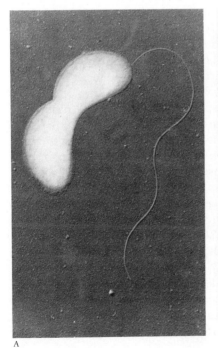

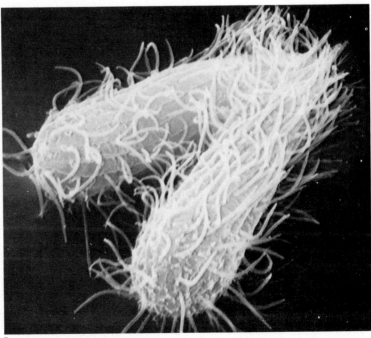

A

B

Cellular theories of aging

A particularly active area in biology currently is the study of senescence, that is, the biology of aging. The primary goal of research on this subject is identification of the cause or causes of the physical and mental deterioration that afflicts the aged. While a unifying hypothesis that explains the aging process has not yet emerged, it is clear that cellular and molecular mechanisms are of primary importance in understanding these changes. If investigators are able to identify the cause or causes of aging, these discoveries may permit not just an extension of the maximum human life span but also — and more importantly — may prevent the declining vigor that accompanies increasing years.

For a long time it was thought that aging is a property of the whole complex organism and that individual cells, if properly cultured in the laboratory, are potentially immortal — that is, there is no limit to the number of times they can divide. However, investigators working with cells of human beings and other species have found that this notion is apparently not correct. Leonard Hayflick found that cultured human *fibroblasts* from connective tissue double in culture only a limited number of times before they deteriorate, become senescent, lose their capacity to divide, and finally die. In fact, the number of cell doublings that a fibroblast is capable of going through in the tissue culture is roughly related to the age of the original donors of those cells and to the typical longevity of the species involved. Thus in the case of fibroblasts taken from early human embryos, the cells will go through about 50 divisions. However, those taken from persons after birth divide only 20 to 30 times. The work of several other investigators has also shown that, in general, the longer the life span of the species, the greater the number of times its cells will divide in culture. The dependence of this rate of division is not linked to loss of the "right conditions" (which would act as a limiting factor), for when Hayflick cultured young and old cells (distinguishable by chromosome markers) in the same flasks, the old cells died out and the young cells divided the expected number of times. The few cells that apparently *do* have the capacity to divide indefinitely in nature are the single-celled protozoans and cancer cells. Thus a better understanding of why normal cells from multicellular animals have a finite lifetime may contribute to the solution of the causes of cancer.

A number of investigators think that the problem of human aging is explained in part by the losses of cell function that occur before cells reach their maximum division limit. As cells malfunction, organs or even whole systems suffer adverse effects, and eventually the individual dies. Hayflick and other investigators do not think that people age because some of their cell types lose the capacity to divide. Most investigators think that an organism's genes determine, at least partially, how long it will live, that aging is *intrinsic*, or *programed*. Different species, after all, have characteristic and inheritable life spans; thus senescence is part of the cell's genetic program, as are other changes during development.

Other investigators think that *extrinsic*, or *environmental*, forces cause aging, including those mutational forces that damage the cell's DNA. An accumulation of environmental alterations, however small, may result in death. Some theories have both intrinsic and extrinsic components. The relative contribution of heredity and environment to senescence of both cells and whole organisms remains a central issue of aging research. It is clear that mutational damage to DNA could be particularly serious in cells that do not divide after they have differentiated to their mature forms. These include brain and most muscle cells. If the damage to DNA is too subtle for the DNA repair system to detect, or if it accumulates faster than the repair system can handle it, the cell will gradually become defective, and the essential control systems or enzymes will lose the battle for successful maintenance. If cells such as those in the brain and muscle tissues function poorly or die, they are not replaced. Mutational damage might be less serious for regularly dividing cells, such as those of the surface skin, the liver, or the lining of the gut.

The possibility is strong that several mechanisms are at work in the process of aging and that different cells age for different reasons. The abundance of current theories on cellular aging provides a rich source of ideas to test in future research; there yet remains much to be learned about the internal structure as well as functioning of the cells and the influence of the well-being of the cells on the total body of an organism.

Summary

All organisms are composed of cells and all living cells arise from preexisting cells; there is at present no evidence for the spontaneous generation of cells from nonliving matter. These statements, known as the cell theory, form one of the three great integrating generalizations of biology, the others being the theories of evolutionary change and ecological relationships. The structure of cells has been progressively well defined by the light, phase, and electron microscopes. A typical cell is bounded by a double-layered cell, or plasma, membrane and contains a large nucleus. Plants have a relatively rigid cell wall outside the cell membrane. The nucleoplasm contains the chromatin threads of DNA, which condense into rod-shaped chromosomes at the time of cell division. The cytoplasm contains microtubules, a network of membranes known as the endoplasmic reticulum, ribosomes (protein-manufacturing centers), the Golgi apparatus, lysosomes, mitochondria, the central body (centrosome and two centrioles), mitochondria, peroxisomes, plastids (including chloroplasts and starch-storage plastids in plants), vacuoles, cilia, and flagella. Cell biologists have recently shown that changes in cellular and molecular mechanisms are of primary importance in understanding cancer and the physical and mental deterioration that afflicts the aged.

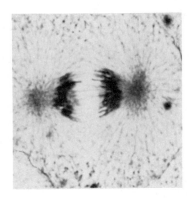

22 Cellular reproduction

The diversity of cells is almost endless, but all cells share one characteristic: They reach a time when they must divide themselves into two new cells. The larger a cell becomes, the longer it takes for materials to cross through the cytoplasm by diffusion, and an upper limit to the cell size is soon reached because of this physical limitation in diffusion rate. It may either stop growing at this point and lose its ability to divide, or divide into two daughter cells. If the two daughter cells remain closely associated, the organism *grows* through an increase in cell number, and the daughter cells gradually increase in volume again to the size of the single cell that gave rise to them. If, however, the daughter cells separate to give rise to new organisms, **reproduction** has occurred rather than merely growth; this ability of an organism to reproduce copies of itself is one of the basic attributes of life, as we noted in Chapter 20. Uncontrolled cellular growth results in tumors or cancers.

If cellular reproduction involves only the production of new daughter cells from the equal division of a mother cell, reproduction is said to be **asexual.** If cellular reproduction results in the formation of a specialized cell that can combine with another specialized cell to form the new fertilized egg and hence a new organism, reproduction is said to be **sexual.** In both types of reproduction, the impor-

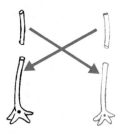

FIGURE 22–1

The influence of the nucleus on morphological development in the large single-celled alga, Acetabularia. *On the left is A. mediterranea, a species with a smooth cap. On the right is A. crenulata (in color), a species with a rayed cap. When the nucleus-containing "root" portion of one type is grafted onto the stalk segment of the other, the "hybrid" plant produces a new cap characteristic of the species contributing the nucleus.*

tance of nuclear control over the cytoplasm is evident. The nucleus contains the cell's instruction, the hereditary material. If the nucleus is removed from a simple animal such as an ameba, the cell will not continue to grow but will become spherical, and its visible activity in the form of movement greatly reduced. After some days or weeks, the cell appears to run down, and, finally, death results without further growth or reproductive division. However, if a new nucleus is inserted into the cytoplasm of the cell within the first several days after the initial removal of the old nucleus, the ameba will resume normal activity. If nuclei are transferred between two species of other simple single-celled organisms, the new nucleus will change the basic character of the cytoplasm of the recipient cell and make it form new structures characteristic of cells present in the nucleus-donor species (Figure 22–1).

Clearly, then, the time of cell division must involve critically important activities in the nucleus, for it would be extremely important that the hereditary material contained in the nucleus be duplicated and evenly distributed to each new cell. Let us look now at the processes involved in asexual and sexual cell division.

Asexual cell division

When an ameba or other simple single-celled organism divides, it seems to split itself into two halves by fission (Figure 22–2). Investigators have found that the nuclear material making up the chromosomes is duplicated prior to this physically evident splitting, and a precise qualitative and quantitative division of this material follows. Each daughter cell resulting from the original mother cell contains exactly the same amount of chromatin that the mother cell originally had. The same general process of asexual cell division goes on constantly in multicellular organisms. In fact, from the time of formation of the fertilized egg to the death of the organism, cell division of this type goes on in essentially every part of the body. This kind of cell division where the same kind and amount of chromosome material is passed to each daughter cell is called **mitosis.** It is the normal process of cell division throughout the plant and animal kingdoms and is responsible for growth and asexual types of reproduction.

The ordinary result of mitosis is two new diploid daughter cells, each identical to the diploid parental cell. Cell biologists today often describe the process in terms of the *cell cycle.* A complete cell cycle usually consists of five stages, labeled G_1, S, G_2, M, and D. The first

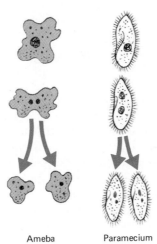

FIGURE 22–2

Asexual reproduction by binary fission in two single-celled protozoans.

Ameba   Paramecium

three stages were originally grouped together as a resting phase, or *interphase*, during which the cell supposedly had no activities relating to cell divisions. In the G_1 or "first gap," stage, the cell is in a pre DNA synthesis phase; however, synthesis of most proteins and RNA continues then and throughout interphase. DNA synthesis occurs during the S (*synthesis*) stage. The cell then enters a G_2, or "second gap," stage of post DNA synthesis. Up to this point, the cell (if it is a mammalian cell in laboratory culture) has spent about 10 hours in G_1, 9 hours in S, and 4 hours in G_2. Now it enters the relatively rapid periods of M (*mitosis*), during which daughter nuclei form, and of *cytokinesis* (or division of the cytoplasm and the whole cell), labeled D for division. Together, the M and D stages may take only an hour, if the animal cell is repeating this sequence every 24 hours. Cells not destined for an immediate repeat of the division cycle are commonly arrested at the G_1 phase. The original nomenclature of *interphase* and four phases of *mitosis* is generally retained for most of the structures and events involved, and we shall use it to describe mitosis.

Mitosis: ordinary cell division

The process of mitosis involves a series of five arbitrarily defined stages in which nuclear as well as cytoplasmic division occurs (Figure 22–3). The genetic material is precisely duplicated in the first stage and then a complete set of the material is distributed to each daughter cell in the next four stages.

FIGURE 22-3

The stages of mitosis in an organism with a diploid (2n) number of six chromosomes.

Middle prophase

Early prophase

Interphase

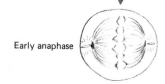

Late prophase

Metaphase

Early anaphase

It used to be said that a cell not actively dividing was in a resting stage, but it has been found in recent years that this is actually the busiest time for the cell. All the normal metabolic activities take place between actual divisions. In fact, following the process of cell division, some cells, such as those in nerves, may never again divide, though they may live on for many years.

The most obvious parts of the nucleus involved in cell division are the *chromosomes*, on which the units of hereditary material, the genes, are carried. The process of mitosis ensures that the number of chromosomes, and hence gene content in the body cells, remains constant throughout growth, for the number of chromosomes is usually the same in all normal body cells of all individuals of the same species.

The overall process of mitosis may take as little as five minutes in some species to as long as several hours in cells of other kinds of plants and animals. Environmental conditions, such as temperature, will also influence the rate of the mitotic process. Let us now look at the five principal stages of the mitotic cycle.

Interphase

Interphase is the nondividing stage of the cell, when all other functions except nuclear division are going on. Thus it falls between the completion of one cell division and the first appearance of the obvious changes in the nucleus that indicate a subsequent cell division. The chromosomes are present as scattered threads of chromatin material and are normally not visible to the eye even under the light microscope. At least one developed nucleolus is present, and the nuclear membrane is distinct and well organized. Along the edge of the nucleus in the cytoplasm of animal cells, one can see the special

Late anaphase

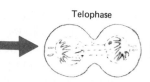

Telophase

Interphase

centrosome area that contains two small cylindrical bodies, the *cen-trioles*, oriented at right angles to each other. Centrioles are found in some of the lower plants such as algae, fungi, and mosses and ferns, but have not been found yet in the cells of most seed plants. The centrioles play an active role in animal nuclear division in organizing the mitotic apparatus of the dividing cell (Figure 22–4).

The cell grows during interphase to the maximum size of the parental type of cell, and at this point there is a precise duplication of the nuclear material forming the threads of chromatin. Also near the end of interphase, in cells of some animal species, each centriole gives rise to a small daughter centriole that lies at right angles to it. During cell division, each mother-daughter centriole complex will move as a unit, however, and this unit will end up in a new daughter cell, forming the two centrioles necessary for subsequent division of that daughter cell. Other animal species wait until the last stage of mitosis to duplicate their centrioles.

Thus the interphase, though a period between mitotic divisions, is actually the most active phase of the cell cycle, carrying out the innumerable activities of a functioning cell such as protein synthesis, growth, differentiation, and respiration, and also duplication of the genetic material in preparation for the next division sequence.

Prophase

In *prophase*, the longest stage of mitosis, the nucleus is prepared for the crucial event of mitosis: the separation of two complete sets of chromosomes into two daughter nuclei. In the typical animal cell, the two centriole complexes begin to move at the beginning of prophase toward opposite ends of the nucleus. While they move, they appear to be connected by long fibers called spindle fibers (really microtubules of protein). Around each centriole, similar fibers radiate out into the cytoplasm, forming an asterlike structure. Plant cells do not contain these asters.

While the two centriole complexes are moving apart, the originally indistinct threads of chromatin begin to condense into visible *chromosome threads*, which become apparent as progressively shorter and thicker rods. Each chromatin thread apparently is tightly coiling, and at this point it can be seen that each chromosome is made up of two strands, one representing the duplicated copy formed during interphase. In this shorter form, the chromosomes are able to move about freely in mitosis without becoming entangled with each other, for as many as 220 kinds of chromosomes ($n = 220$ in the haploid cells producing gametes in the testes and ovaries, and $2n = 440$ in regular body cells) are known from some animal cells

FIGURE 22–4

Mitosis in the whitefish (this page) and the African blood lily (facing page). In the whitefish cells, the stages shown are prophase (A), metaphase (B), anaphase (C, D), and telophase (E, F). In the African blood lily, the chromosomes proceed from interphase through prophase to metaphase (end of second row), anaphase (third row), and telophase (bottom row). Note the cell plate forming in the telophase stage of mitosis in the plant cell.

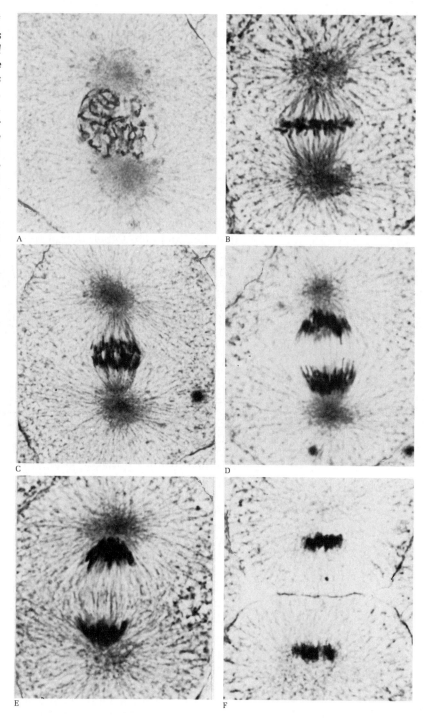

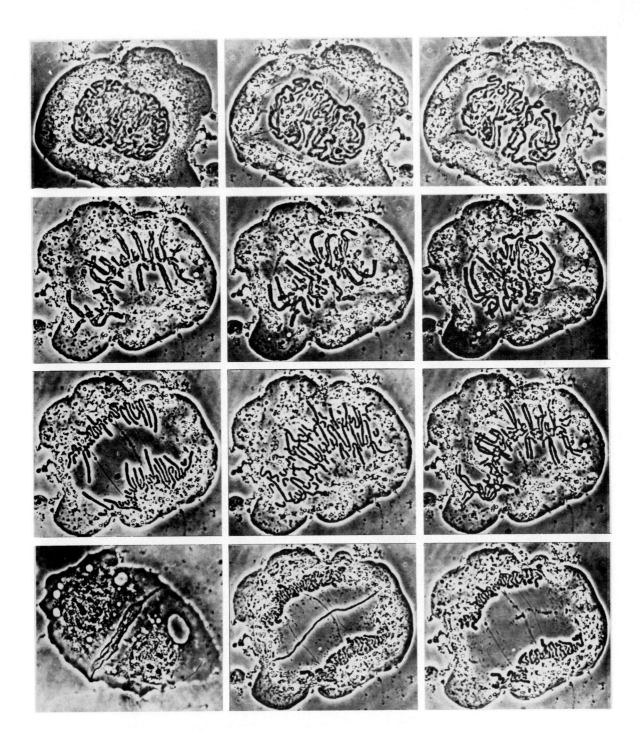

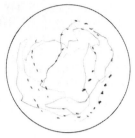

In interphase

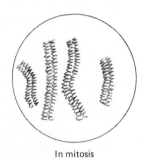

In mitosis

FIGURE 22–5

At the beginning of mitosis, the chromosomes coil tightly and become shortened, thick rods. At this time, the duplicated chromosome is also apparent next to each original chromosome. These are not homologous pairs connected only at the centromere, as found in meiosis. They are duplicate strands, resulting from DNA replication in the preceding interphase, and the duplicates will be separated during mitosis.

(certain lycaenid butterflies in Spain), and chromosome numbers of 600 to 1,200 are reported from certain ferns (Figure 22–5).

The two duplicated strands representing each chromosome are called **chromatids** and are joined at a region called the kinetochore, or **centromere.** This small body may be located at any point along the double strand but is commonly found at the center; the location is characteristic for that particular chromosome in that species of organism. In Chapter 24, we will consider how the replication process during interphase produced one of the chromatids by using the other as a model. For the moment let us simply note that the two chromatids of a prophase chromosome are identical. Spindle fibers radiating out from each centriole attach themselves to the centromere and the chromosome strands move onto the equatorial plate, the region of the spindle midway between the centrioles.

At the end of prophase, the nuclear membrane has disappeared, and each chromosome is located at the midpoint of one of the spindle fibrils. The nucleolus has also disappeared by this point.

Metaphase

During the brief *metaphase* stage, the chromosomes line up on the equatorial plane of the spindle in parallel fashion and form a line across the middle of the spindle. At this point, a fibril from either centriole is attached to each side of the centromere holding two duplicated chromosome strands together. In fact, it is the centromeres that are lined up across the equator, while the arms of the chromosomes may flop in any direction. The end of metaphase comes when each chromosomal centromere divides and each of the former chromatids becomes a separate, single-stranded chromosome with its own centromere as a result.

Anaphase

Just as the division of the centromere between each double strand of chromosomes marks the end of metaphase, the separation of the chromatids themselves in each chromosome marks the process described in the stage called *anaphase.* Here, the chromatids of each duplicated chromosome separate and move toward opposite poles of the spindle, each centromere appearing to be pulled by a spindle fibril. Both the centromere and the fibril appear to be necessary for the movement; otherwise the chromosome will be left behind on the equatorial plate of the spindle.

By late anaphase, the cell will have two widely separated groups of chromosomes, each cluster having almost reached a pole, or centriole, of the spindle. At this point, division of the cytoplasm, **cytokinesis,** begins. In animal cells this stage is indicated by a furrowing or infolding of the plasma membrane around the center of this cell, approximately in the location of the equatorial plate of the spindle. This division of the cytoplasm appears to be mediated or caused by the contraction of a band of microtubules that apparently forms under the control of the aster.

Telophase

Telophase, the last phase of the mitotic cycle, is essentially a reverse of prophase, reconstructing each of the daughter cells in the form of the original parental cell.

The cluster of chromosomes at each end of the old spindle is surrounded by fused membrane sacks to form the double-layered nuclear membrane while the spindle itself disappears. The chromosomes start to uncoil, become indistinct, and gradually assume the character of the old chromatin network. The mother-daughter centriole complex left with each new cell separates, and each of these two centrioles now gives rise to a new daughter centriole if the mother centriole did not already divide in the preceding interphase. A new nucleolus forms in each daughter nucleus and the process of cytoplasmic division is completed, physically dividing the original cell into two daughter cells. The end product, of course, is two daughter cells that exactly resemble the mother cell from which they arose. They may be somewhat smaller at first, but growth will normally ensue.

An interesting variation of cytokinesis occurs in plant cells because of the relatively rigid cell wall that cannot develop cleavage furrows. Instead of a pinching-in of the plasma membrane, a special membrane called the **cell plate** forms halfway between the two nuclei, lying across the equator of the spindle (Figure 22–6). The cell-plate mechanism is characteristic of plant species in which cytokinesis occurs at the same time as the anaphase and telophase nuclear stages.

FIGURE 22–6

The development of the cell plate in telophase of plant cell mitosis.

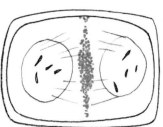

Sexual cell division

In the reproductive processes we have been considering, the daughter cells almost invariably exactly resemble the mother cells, particularly in unicellular organisms where half of the cellular constituents and structural characteristics of the mother cell are passed precisely

Cancer

Cancers are of many types and expressions, from superficial skin cancers induced by excessive sunlight exposure to breast cancers requiring major surgery. But cancerous growths all share at least one characteristic: uncontrolled reproduction of abnormal cells. In one widely held current theory, the human body, in which all cells except muscle fibers and neurons are continually replicating themselves, produces up to hundreds of genetically abnormal and potentially cancerous cells each day. This may occur by radiation or other mutagenic agents damaging the DNA or by accidents in transmission of the correct number and kinds of chromosomes to each daughter cell in mitosis. Ordinarily, the body's immune system of natural defenses against disease recognizes these biological fifth columnists as "foreign" because they are genetically different; it destroys them with scavenger cells such as lymphocytes and macrophages, which chemically digest or even devour the intact foreign cells before they begin dividing and reproducing. But when the immune defense mechanism is weakened for any reason, it fails to dispose of the errant cells, either because it cannot recognize them or because it is incapable of attacking them. Under such conditions the mutant cells (which are apparently not under the same genetic restraints on mitotic timing as normal cells) have the opportunity to run wild. They reproduce themselves at an extremely rapid rate, invade normal tissues, and, if not destroyed by radiation, cut out by surgery, or arrested by chemotherapy (the three principal cancer treatments in use today), eventually kill the host body. Considerable evidence indicates that cancer thrives when the immune system is defective, as in the aged or very young or those with immunodeficiency diseases, who are unable to resist infection. One promising approach, therefore, in the treatment of cancer patients is the use of immunotherapy: stimulating the body to turn on an inactive immune system against foreign bodies and destroy cancer cells as well.

on to each new cell. Unless a genetic mutation occurs by chance in one of the chromosomes, reproduction by mitotic cell division continues to produce replicate cells and asexual organisms, generation after generation. Since sexual reproduction usually involves the fusion of two germ cells from *different* parents, an opportunity is provided for variability and new combinations of characters not possible with asexual reproduction. The paragraphs that follow will help us to understand the importance of the process of cell division that gives rise to germ cells.

The body cells of most higher plants and organisms contain two of each type of chromosome in the nucleus, that is, chromosomes occur in pairs. The members of such pairs of chromosomes are called *homologous chromosomes* because the members of each pair resemble each other in genetic constitution and morphology. The homologous chromosomes actually have the same order of genes on each member of a particular pair. Cell nuclei that have these paired chromosomes are said to be *diploid* (2n).

The explanation for this diploid condition is simple. One member of each chromosome pair was received at fertilization of the egg from the maternal parent and one from the paternal parent. The problem, then, is how to control the number of chromosomes from one generation to the next. Since diploid parents give rise to diploid offspring, the egg and sperm gametes (the two types of germ cells that join to produce the fertilized egg) must have only half the normal number of chromosomes, that is, a *haploid* (n) number; only one representative of each particular type of chromosome is present in that cell.

Meiosis: sex cell formation

The process of producing haploid gametes that can combine to form a diploid zygote is called **meiosis;** in this type of cell division, the daughter cells receive only half the number of chromosomes present in the parent cell. In animals, the process of meiosis occurs simultaneously with the time period in their lives when gametes are produced. In plants, meiosis may occur at other times, but it always involves the formation of the tissue that will eventually produce the gametes. Thus in some plants, meiosis produces spores that grow up into haploid tissue and this tissue in turn produces the haploid gametes. The interesting variations and strategies of reproduction will be considered at greater length in Chapter 34.

The purpose of meiosis is to separate the members of homologous pairs of chromosomes from one another. Thus meiosis consists of two successive divisions of the cytoplasm and nucleus but only one division of the chromosomes. The separation of members of a pair occurs during the first division of meiosis. The specialized cell produced by meiosis is known as a **gamete,** or germ cell; in the male this is called the **sperm** and in the female the **ovum,** or egg.

Human spermatogenesis

To see how the process of meiosis works in these two successive division sequences that result in four new haploid cells, let us examine the process of human sperm formation. The first division reduces the number of chromosomes; the second division is essentially equivalent to a mitotic one (Figure 22–7). In each of the two divisions the nuclear events follow the same general stages as in mitosis—namely, interphase, prophase, metaphase, anaphase, and telophase. In sperm formation, or **spermatogenesis,** large cells located in tubules in the testes form the active cellular reserve from which

FIGURE 22–7

(A) Human spermatogenesis and (B) a mature sperm cell.

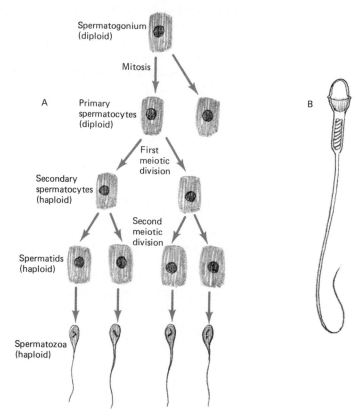

Spermatogonium (diploid)

Mitosis

A   Primary spermatocytes (diploid)

First meiotic division

Secondary spermatocytes (haploid)

Second meiotic division

Spermatids (haploid)

Spermatozoa (haploid)

B

sperm cells may be made. These cells are actually the direct descendants through mitotic divisions of primordial germ cells that were present in the developing embryo. These large cells, called spermatogonia, continue to divide mitotically throughout most of the life of the male and utlimately give rise to cells called *primary spermatocytes,* which will go through the two meiotic divisions.

The first meiotic division

In the resting stage, or interphase, of meiosis, the primary spermatocyte undergoes changes that resemble those in the interphase of mitosis. In prophase the chromatin material appears as long, slender threads that arrange themselves side by side in double units, corresponding to homologous chromosomes. This process is called **synapsis.** As in mitosis, each chromosome has already doubled in interphase, so that this synapsing pair forms a four-stranded structure known as a tetrad. The threads begin to shorten and thicken as the chromosomes coil in mid prophase. By late prophase, the centromeres of the two homologous chromosomes have moved apart somewhat; the chromatid strands are still wrapped around each other and sometimes exchange parts. The cross-shaped figures formed by chromatids crossing over each other are called **chiasmata** and may represent an actual physical exchange of genetic material (called a **cross-over**) that affects variation in the mature organism, as we shall see in the next chapter. Thus the chief difference between the prophase of meiosis and mitosis is that in meiosis the members of each pair of homologous chromosomes move together in this process of synapsis and form an intimate association in which genetic exchange may occur (Figure 22–8).

FIGURE 22–8

Crossing-over (exchange of genetic material) between the two homologous members of a chromosome pair.

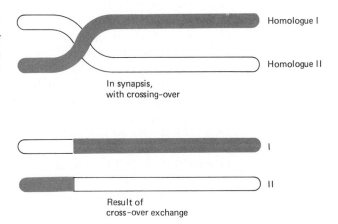

Homologue I

Homologue II

In synapsis, with crossing-over

I

II

Result of cross-over exchange

Toward the end of prophase, the synaptic pair of chromosomes moves to take up the metaphase position on the equator of the spindle. During anaphase, the two chromatids that are connected by a centromere separate from the other two members of this tetrad, and thus the homologous chromosomes of a pair each end up at a different pole by the end of anaphase. This *reductional* division results in one member or one representative of each kind of chromosome (either the original maternal *or* paternal homologue) being present at a pole, rather than copies of both members of each chromosome pair, as in mitosis. The two new nuclei formed in telophase will resemble those formed in ordinary mitosis except that each of the two new meiotic nuclei has only half the number of chromosomes present in the parental nucleus. Each of these chromosomes, however, is double-stranded. The new daughter cell, known as the *secondary spermatocyte*, normally does not undergo nuclear reconstruction, and it proceeds directly into the second meiotic division sequence. Rather than an interphase, then, we have *interkinesis*, referring to the period between the two meiotic divisions. No doubling of the chromatin material takes place during interkinesis because this division took place prior to the first meiotic division. Hence, no new chromatids are formed since replication is unnecessary at this time.

The second meiotic division

The second division sequence of meiosis that follows interkinesis occurs in the secondary spermatocytes and is essentially similar to normal mitosis. It is called an *equational* division. There is no synapsis of the chromosomes since the nucleus is haploid and there are no longer homologous chromosomes present. During prophase, the two duplicate strands of each chromosome are connected at the centromere, although they may flop out from each other elsewhere along their length. When the spindle forms, each double-stranded chromosome moves onto the spindle independently, and its centromere attaches to a fibril.

At the end of metaphase in this second meiotic division, the centromeres divide. The new single-stranded chromosomes, each attached to its own centromere, move away from each other toward opposite poles of the spindle during this following anaphase division. The nuclei are reconstructed in telophase, and the cell itself divides. The nucleus of the original primary spermatocyte has now given rise to four daughter nuclei located in four separate sperm cells, each of which contains one representative from each pair of homologous chromosomes.

The result of these two meiotic divisions is that the first division produced two haploid cells containing double-stranded chromosomes, and the second division separated these double strands to make a total of four new haploid cells containing single-stranded chromosomes.

The development of sperm cells

Each of these four cells that result from the two meiotic divisions of the original primary spermatocyte is a *spermatid*. No further division occurs in the spermatid, and it develops a long flagellum that makes the male gamete, or sperm cell, a motile gamete. The enlarged head area contains the nucleus and a small amount of cytoplasm, along with many mitochondria. The sperm cell is then able to generate sufficient energy from sugars broken down in the mitochondria to provide energy for the swimming activity carried out by means of the long, whiplike flagellum. As many as several hundred million sperm cells a day may be produced in the mature human male, and the quantity of sperm produced in most other animals is similar. Thus biologists interested in studying nuclear divisions and chromosome structure often use cellular material gathered from tissues undergoing spermatogenesis for their research.

The development of egg cells

In the female, the process of meiosis (**oogenesis**) is the same but gamete formation differs (Figure 22–9). In a sperm cell, there is a reduced amount of cytoplasm, only one-quarter the amount present in the original sperm mother cell or primary spermatocyte. But the egg must provide stored *food* for the developing embryo as well as a complete copy of nuclear material. Whereas in the male, four functional sperm are formed from each primary spermatocyte, in the female only one egg develops from each primary oocyte. This results from a deliberately unequal distribution of the cytoplasm in the two meiotic divisions that give rise to the egg cell. Three of the daughter cells are extremely small, containing only a nonfunctional nucleus and a minimum of cytoplasm. The fourth cell receives the great bulk of the stored cytoplasmic food material. The three odd structures resulting from meiosis in the female are called polar bodies and are nonfunctional.

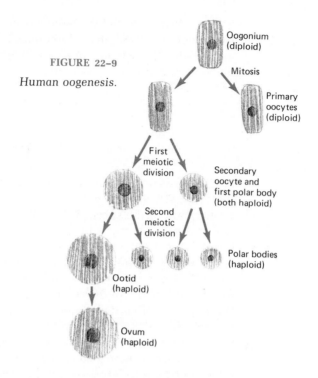

FIGURE 22–9

Human oogenesis.

Summary

Asexual cellular reproduction, or mitosis, involves the production of two daughter cells from the equational division of a mother cell. It is the basic cause of growth in multicellular organisms and the usual means of asexual reproduction in single-celled organisms as well as some higher plants and animals. Mitosis involves five arbitrarily divided stages. Interphase is the nondividing but metabolically active phase. The chief nuclear event is the duplication of the chromatin material. In prophase, the spindle forms between the centrioles and the chromatin threads condense into distinct chromosomes, which are composed at this stage of two duplicated strands called chromatids, joined at one point by the kinetochore, or centromere. Each centromere moves toward the equatorial plate of the spindle. The nucleolus and nuclear membrane disappear. In metaphase the chromosomes line up in parallel fashion on the center of the spindle. During anaphase the spindle fibrils pull the chromatids of each duplicated chromosome apart, toward opposite poles (the centrioles) of the spindle. Division of the cytoplasm (cytokinesis) begins by either furrowing of the plasma membrane (in animal cells) or the laying down of a central cell plate (in plant cells). The last stage, telophase, involves reconstruction of each of the daughter cells in the form of the original parent cell. A new nuclear membrane forms, the spindle disappears, the chromosomes uncoil, the nucleolus forms again, and the process of cytoplasmic division is concluded. In sexual cell reproduction, or meiosis, there are two consecutive divisions that reduce the normal diploid number of chromosomes present in cells to the haploid number (one homologue representative of each parental pair) needed for the gametes. In the first meiotic division, the members of each pair of homologous chromosomes move together and form an intimate association (synapsis) in which genetic exchange through cross-overs in chiasmata may take place. In anaphase and telophase of the first division, the homologous chromosomes of a synapsed pair are separated and end up at opposite poles of the original spindle. This reductional division produces two daughter cells, each with half the number of chromosomes present in the parental diploid nucleus. Each daughter haploid cell then promptly divides again in an equational division, separating the double-stranded chromosomes and resulting in a total of four haploid cells, each containing one single-stranded representative of each original homologous chromosome pair. Meiosis in the male (spermatogenesis) produces four small but functional sperm for each original primary spermatocyte. Meiosis in the female (oogenesis) produces one functional gamete with the bulk of the original cytoplasm (the ovum, or egg), and three nonfunctional cells called polar bodies, which contain virtually nothing but nuclear material.

23 Heredity and reproduction: the principles of genetics

"If two plants which differ constantly in one or several characters be crossed, numerous experiments have demonstrated that the common characters are transmitted unchanged to the hybrids and their progeny; but each pair of differentiating characters, on the other hand, unite in the hybrid to form a new character, which in the progeny of the hybrid is usually variable. The object of the experiment was to observe these variations in the case of each pair of differentiating characters, and to deduce the law according to which they appear in the successive generations."

—Gregor Mendel, 1865

Often in human history, the greatest philosophical concepts have arisen from the most unexpectedly humble environments. The moral truths and spiritual relationships brought forth in the three short years of the ministry of Jesus in a remote Middle Eastern province of the Roman Empire have lasted 20 centuries without diminishing their impact on the human spirit and modern culture. Charles Darwin, a 21-year-old amateur naturalist, by sheer good fortune set sail on the exploring ship *Beagle* in 1831 and from his experiences derived the concept of evolution by natural selection that transformed the scientific world some 28 years later. Albert Einstein was a 26-year-old Swiss Patent Office minor official, in the obscure town of Speichergrasse, when in 1905 he rocked the established world of physical science with the publication of his theory of relativity. The theoretical foundation stones of the science of genetics were firmly laid by the garden experiments of an otherwise ordinary Austrian monk, Gregor Mendel (Figure 23–1), whose fundamental contribution, published in 1865, went unrecognized until many years after his death. A later generation rediscovered Mendel's brilliant work and honored his contribution by applying the term "Mendelian genetics" to the fundamental ways that a majority of genes are inherited.

Monohybrid inheritance

Between 1856 and 1863 Mendel crossed thousands of ordinary garden pea plants in his effort to understand the inheritance of specific traits. He selected seven clear-cut character traits to follow in detail. Each of these seven characters was expressed in two alternative ways:

1   *Form of seed* (round smooth or irregular wrinkled).
2   *Color of interior of seed* (yellow or green).
3   *Color of seed coat* (white or gray).
4   *Form of ripe seed pod* (fully inflated or wrinkled and constricted).
5   *Color of unripe pod* (green or yellow).
6   *Position of flowers* (distributed along stem or bunched at top of stem).
7   *Length of stem* (long stem of 6 to 7 feet or short stem of ¾ to 1½ feet).

FIGURE 23–1

Gregor Mendel.

By carefully selecting developed horticultural strains and by growing each strain of plants alone for a long time (allowing reproduction to occur only within that particular strain), Mendel obtained pure-line strains, each showing a particular character state for the seven traits. Taking plants with alternate trait states (e.g., long versus short stems), he made crosses between these purebred stocks. Only one trait (long stems, in our example) of the alternate traits for a character (stem length) showed up in the resultant offspring of any particular monohybrid cross. When he crossed these hybrid offspring, however, the next generation (second generation removed from the parents) showed a peculiar distribution of character traits. Three-fourths of the plants exhibited one state of the character (e.g., long stems) and one-fourth exhibited the alternative condition (e.g., short stems). In order to explain these results, Mendel made the following hypotheses:

1   There must be *two factors* controlling the expression of each character in a plant.

2   The two factors were derived from the parents, one coming from each parent.

3   When gametes are formed in a parental plant, the pair of factors segregates from one another and each gamete receives only *one* of the factors.

4   If the two factors that end up in a fertilized egg (through the union of two gametes) are different, one (*dominant* factor) will express its character trait to the exclusion of the other (*recessive* factor).

Figure 23–2 shows one of Mendel's crosses that dealt with stem length in peas. This diagram uses the same symbols that Mendel used to label each generation: P_1 for parental generation, F_1 for the first filial generation, and F_2 for the second filial generation. Mendel also introduced the use of capital letters to stand for a dominant factor (e.g., T for tall or long stem length) and lowercase letters to stand for a recessive factor (e.g., *t* for dwarf stem length).

FIGURE 23–2

A diagrammatic representation of Mendel's cross of tall and dwarf parental plants and the resultant offspring through the next two generations.

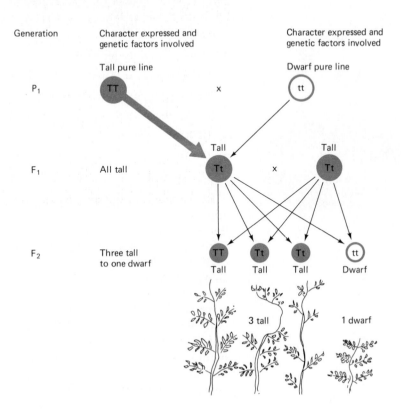

Generation	Character expressed and genetic factors involved	Character expressed and genetic factors involved

Tall pure line · Dwarf pure line

P_1 TT   x   tt

F_1 All tall   Tall Tt   x   Tall Tt

F_2 Three tall to one dwarf   TT Tall · Tt Tall · Tt Tall · tt Dwarf

3 tall · 1 dwarf

FIGURE 23-3

A Punnett square for the four possible combinations of two different sperm cells and two different egg cells. Three of the plants resulting from these fertilized eggs will exhibit the dominant tall trait (colored circles).

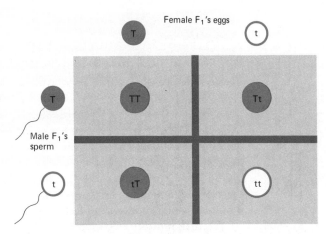

A purebred tall plant (hence, "*TT*") and a purebred dwarf plant (hence, "*tt*") are used as parents and pollen grains (carrying male sperm) are transferred from a *TT* flower to a *tt* flower where the sperm (each carrying a *T* factor) travel to that plant's eggs (each containing a *t* factor). In Chapter 22 it was shown that at meiosis the number of hereditary units in each sperm or egg cell is reduced to half the normal number. The fertilized eggs become *Tt* in factor composition and these F_1 progeny will grow tall, reflecting the *dominant* (*T*) trait. The recessive trait does not appear in this F_1 generation. When flowers of this uniformly tall F_1 generation are crossed together, however, the recessive trait (dwarf stem length) does appear and it is found in reduced numbers compared with the tall plants. In fact, tall plants outnumber dwarf plants in about a 3 to 1 ratio no matter how many second-generation offspring are raised. From our knowledge of meiosis, we can see why the factors present in the parental plants for each generation segregate and why they recombine in the F_2 generation in a 3 to 1 ratio. The gametes produced by the F_1 plants are of two types: *T*-carrying eggs or sperm, and *t*-carrying eggs or sperm. Figure 23-3, called a Punnett square (after an early English geneticist), shows the ways in which they can combine by chance in fertilization (pollination). Each of the double symbols in the four boxes represents a possible combination of egg and sperm. All are equally possible, since half of the sperm and half of the eggs are carrying either *T* or *t* factors. In the characters expressed by the offspring, however, we see that three out of four of the possible combinations (namely, *TT, Tt,* and *tT*) will be tall, and the fourth (*tt*) will be dwarf, all because of the presence or absence of dominant factors.

This is an appropriate point to introduce more modern terminology into the discussion. Today we call a factor that controls a partic-

If

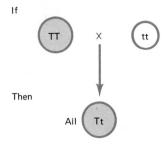

Then

But if

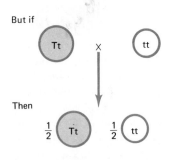

Then

FIGURE 23-4

Mendel's test cross of tall plants whose genotype is unknown. The results of a test cross of a tall plant to a (recessive homozygous) dwarf plant indicate the genotype of that tall plant.

ular character (e.g., stem length) a *gene*, and the alternate states of the gene that produce different expressions of the same character are called *alleles* (e.g., *T* and *t* are alleles, or forms, of the same gene). We say that an organism is *homozygous* or that a character is in a homozygous state when both alleles present in the plant or animal are identical (e.g., *TT* or *tt*). If two different alleles are present together, we say that the organism is *heterozygous* for that gene or character. The *phenotype* refers to the visible attributes of the particular individual involved (e.g., tall or dwarf); the *genotype* refers to the hereditary constitution of the individual (e.g., *TT* or *Tt* or *tt*). Two pea plants that have the same "tall" phenotype may differ in genotype, one being homozygous and the other heterozygous. This is seen in the F_2 generation just described.

To differentiate between homozygous and heterozygous tall individuals, Mendel carried out *test crosses*. In a test cross, the dominant-phenotype plant (tall) whose exact genotype one wishes to know (*T*-) is crossed to a homozygous recessive plant (*tt*). One always uses the homozygous recessive test stock because it is the only phenotype with a known genotype that can differentiate between homozygous and heterozygous stocks with a dominant phenotype. If the tall parent in this example is homozygous (*TT*), all the offspring will be tall in phenotype (and *Tt* in genotype). If the tall parent was heterozygous (*Tt*) one would expect that half of the offspring would be tall (*Tt*) and half would be dwarf (*tt*). The two possible outcomes are indicated in Figure 23-4.

Geneticists after Mendel were able to extend the laws he discovered to other organisms, and in many cases they found these plants and animals more amenable for study. The common fruitfly, *Drosophila*, has been exhaustively studied in laboratory crosses, for instance. It has a short generation time because of its rapid growth (going from the egg stage to reproductive adult in two weeks); it has a small body size that allows hundreds to be raised in a single pint-sized bottle; it may be cultured easily on mashed bananas or an artificial diet; and *Drosophila* species have a great many characters that can be reasonably easily followed in crossing experiments.

Human beings are not so easily studied because they have a long generation time and relatively small families of offspring (making it often difficult to determine character ratios among progeny), and do not arrange crosses (in marriage or other relationships) to suit the convenience of an interested geneticist! Nonetheless, the inheritance of many human traits has been studied through tracing family pedigrees across several generations. Albinism, for instance, is inherited in human beings as a simple recessive character (Figure 23-5). If an

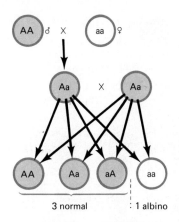

FIGURE 23-5

The inheritance of albinism in human beings, a lack of pigmentation controlled by a recessive (a) allele.

albino woman (*aa*) marries a homozygous normally pigmented man (*AA*), their F$_1$ offspring will be normally pigmented but heterozygous (*Aa*) in genotype. If this child should marry another person heterozygous for albinism, *their* children (F$_2$ generation) will be normal in three out of four cases, but albino (*aa*) one-fourth of the time. Should the F$_1$ heterozygote marry a homozygous albino (similar to his mother), the F$_2$ generation will be composed of both normal and albino children in a 1 to 1 ratio.

Dihybrid and trihybrid inheritance

Mendel did not stop with the study of the simple inheritance of just one character in any particular pairing of parents. He followed many cases of *dihybrid* crosses, where two distinct characters were involved. Figure 23-6 shows Mendel's observations of the inheritance of seed *shape* (round versus wrinkled) and *color* (yellow versus green) genes simultaneously. He crossed homozygous round, yellow-seeded parents with wrinkled, green-seeded parents. From the fertilized flowers, he obtained exclusively round, yellow seeds (F$_1$). When these F$_1$ plants grew to maturity, he crossed them among themselves and found four different combinations of phenotypic characters among the F$_2$ progeny. Furthermore, these four phenotypes occurred in a particular ratio.

FIGURE 23-6

Dihybrid cross by Mendel, involving seed coat texture (round or wrinkled) and seed color (yellow or green).

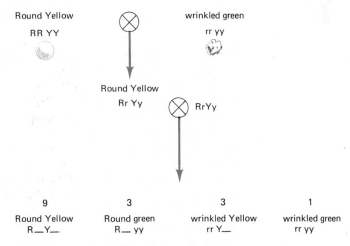

Round Yellow
RR YY

wrinkled green
rr yy

Round Yellow
Rr Yy

RrYy

9	3	3	1
Round Yellow	Round green	wrinkled Yellow	wrinkled green
R—Y—	R— yy	rr Y—	rr yy

FIGURE 23–7

The results of a cross between two pea plant parents heterozygous for two different genes affecting seeds. Dominant gene R causes round, fully packed seeds; it is dominant to its recessive allelic gene, r, which causes a wrinkled, unfilled seed when the genotype is homozygous rr. Dominant gene Y leads to yellow seed color; its recessive allele, y, results in the normal green color when it is homozygous.

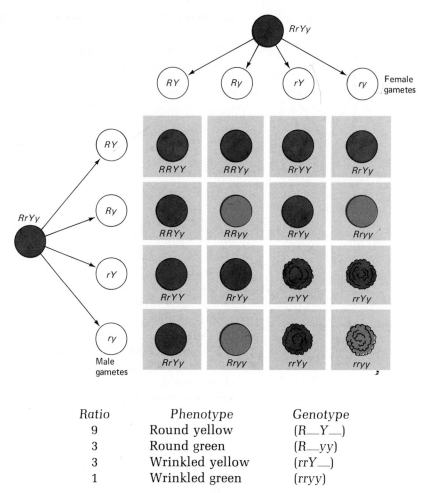

Ratio	Phenotype	Genotype
9	Round yellow	(R__Y__)
3	Round green	(R__yy)
3	Wrinkled yellow	(rrY__)
1	Wrinkled green	(rryy)

The phenotypic ratio of a dihybrid cross between two heterozygous F_1 parents was 9-3-3-1. When we use a Punnett square to see how the gametes have recombined to yield this ratio (Figure 23–7), it is clear that random segregation of alleles of each gene, and independent assortment of the alleles for the two genes, yields quite a few (8) genotypes. In totaling up the phenotypes yielded by the different combinations, though, only four phenotypes emerge, and they are in the 9-3-3-1 ratio.

If we attempt to follow the inheritance of alleles for each of three different genes in a *trihybrid* cross, the problem begins to become formidable. For instance, we could try to follow the results of a cross of a tall, round, yellow plant (*TTRRYY*) with a dwarf, wrinkled, green (*ttrryy*) plant. To use a Punnett square again would involve a checkerboard of 64 combinations, 8 possible combinations of the alleles for 3 genes being on *each* side of the square. A genuine short cut is to use the mathematics of probability theory for finding the frequencies of phenotypes and genotypes.

Probability theory in genetics

The key factor that permits the use of probability theory in genetics is the independence of assortment of the alleles for one gene from the alleles of the next gene. The product rule of probability theory states that *the frequency with which two or more independent events occur together is equal to the product of the frequencies with which each event occurs individually.* Thus, in a $F_1 \times F_1$ dihybrid cross of $RrYy \times RrYy$, we can ask: How many offspring will be round and yellow in phenotype? The dihybrid cross is simply treated as involving two independent events: that is, two single-factor crosses.

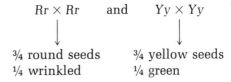

$$Rr \times Rr \quad \text{and} \quad Yy \times Yy$$

¾ round seeds	¾ yellow seeds
¼ wrinkled	¼ green

The chance of getting a round seed is ¾, and the probability of getting a yellow seed is ¾. Thus by the product rule the probability of getting both round and yellow states in the same seed is $¾ \times ¾ = ⁹/₁₆$. The answer is that 9 out of every 16 seeds will be round and yellow, or the probability of any one seed being both round and yellow is ⁹/₁₆.

In a complicated trihybrid cross, $TtRrYy \times TtRrYy$, we can ask what will be the frequency in the offspring of tall, round, yellow plants. The answer is the product of the independent probabilities for each gene's character state: $¾ \times ¾ \times ¾ = ²⁷/₆₄$. That is, 27 out of every 64 offspring will show the dominant trait for each of the three characteristics involved in this $F_1 \times F_1$ trihybrid cross.

Intermediate inheritance

In the seven basic characters that Mendel worked on in peas, clear-cut dominance of one alternate state of each character was always present. In *incomplete dominance*, the heterozygote shows signs of expression of the repressed recessive allele and takes on an intermediate appearance between the homozygous dominant and the homo-

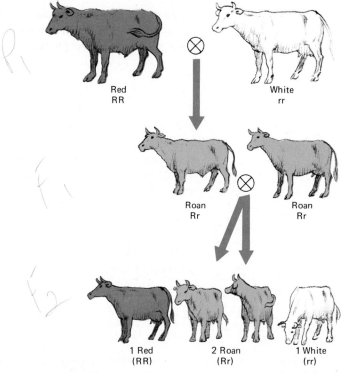

zygous recessive individuals (Figure 23–8). This type of blending inheritance is seen in the breeding of certain types of cattle. When a red-coated animal (*RR*) is crossed with a white-coated animal (*rr*), the F₁ offspring are roan in coat color. This roan color is produced by an intermingling of colored and white hair in the heterozygote (*Rr*). When two roan heterozygotes are crossed, the F₂ offspring show a ratio of 1 red (*RR*) to 2 roan (*Rr*) and 1 white (*rr*) animals. This latter result shows again that inheritance is particulate and that even though blending of parental phenotypes occurs in the F₁ (and subsequent heterozygotes), the genes remain intact and fully capable of producing a pure effect when placed again in a homozygous relationship (the red F₂ animal and the white F₂ animal, respectively).

Gene interaction

Often in organisms we find that more than one gene controls the appearance of a particular character. In the dihybrid and trihybrid crosses we have just considered, each gene still controls a separate character. However, in *polygenic*, or multiple-gene, inheritance, the alleles of two or more genes interact to produce a particular phenotypic character. In continuous variation, running from one extreme to another, a number of genes are commonly involved and interact in an additive fashion. Thus in human height, for example, we might hypothesize that 10 or more genes are involved, each adding an increment of height if it is in an allelic state that contributes to growth. When none of the genes is in the growth-fostering allelic form, a person reaches only minimum height. If 1 of the 10 genes is in the proper allelic state to increase height, the person is slightly taller. The greatest height a person could achieve would be produced if all of the height-controlling genes were in the growth-stimulating allelic state. Thus no distinct phenotypic classes appear in this type of inheritance — only a continuum.

Sometimes several genes control the expression of a trait and an allelic variation in one of these genes can prevent the others from being expressed. This type of gene interaction is called **epistasis** (Figure 23–9). In human beings, the alleles for albinism do not

FIGURE 23–9

Polygenic inheritance of the darkness of skin pigmentation in human beings. A number of genes are involved, all expressing intermediate inheritance rather than a dominant-recessive relationship. Each gene for heavy deposition of melanin makes the skin a little darker. If a child receives all the genes for a heavy deposit of melanin from both parents, the child's skin will be much darker than either parent's.

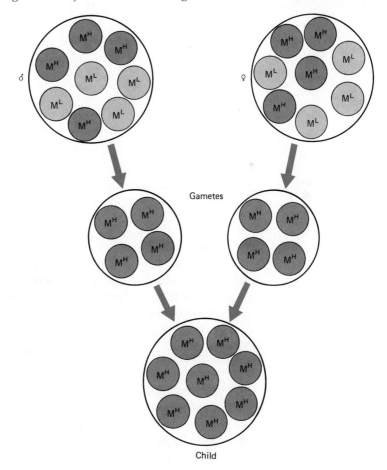

Gametes

Child

FIGURE 23–10

Gene epistasis resulting in a 13 to 3 ratio. The Punnett square checkerboard shows the expected composition of the F₂ generation from a cross between two varieties of chickens, one with dominant white plumage and one with recessive white plumage. Gene locus I is epistatic to locus C since it allows the production of pigment by C only when locus I is in the homozygous state.

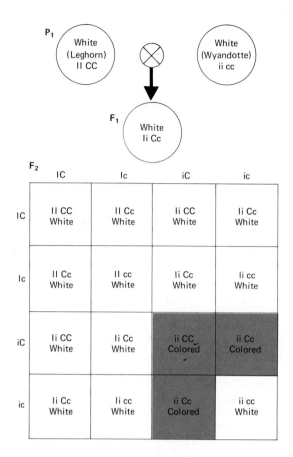

produce an enzyme needed to produce the pigment melanin in the albinic homozygote. There are other genes affecting skin color that produce a heavy deposit of melanin in the skin, yet they will have no influence on skin color unless the gene for albinism is in the heterozygous or homozygous normal state instead of homozygous recessive. Thus the gene for albinism is said to be epistatic to those that control the intensity of melanin deposition, because it can have the effect of masking the expression of these other genes (Figure 23–10).

In *gene collaboration*, two different genes interact to influence the same character and in certain allelic combinations they can produce a novel character state that neither can produce alone. Thus the genes do not behave independently like those studied by Mendel in dihybrid pea crosses. One interesting example of this type of gene interaction is provided by Bateson and Punnett, who studied the inheritance of comb type in poultry strains. One gene, in its dominant state R, produces rose comb whereas the recessive allele r produces single comb as a homozygote. A second gene produces pea comb in

FIGURE 23–11

Gene collaboration (interaction) in two independent pairs of alleles affecting comb type in poultry. The 16 possible combinations of the F₁ gametes and the genotypes and phenotypes of the resulting zygotes are drawn in the F₂ checkerboard.

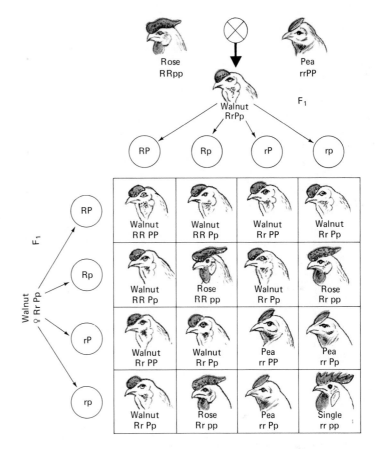

its dominant state, *P*, and single comb in the recessive state, *p*. Single comb is the normal state in these chickens and is merely the double recessive homozygote, *rrpp*. If one gene has at least one dominant allele present, the comb assumes the phenotype produced by that allele. However, if *both* genes have at least one dominant allele (*R* and *P*) present, the comb becomes a unique fourth phenotype, called walnut comb (Figure 23–11). A cross between pure rose comb and pure pea comb chickens yields this series of progeny:

$$
\begin{array}{lcccc}
P & RRpp & \times & rrPP & \\
 & \text{rose} & & \text{pea} & \\
 & & \downarrow & & \\
F_1 & RrPp & \times & RrPp & \\
 & \text{walnut} & & \text{walnut} & \\
 & & \downarrow & & \\
F_2 & 9R\_P\_ & 3R\_pp & 3rrP\_ & 1rrpp \\
 & \text{walnut} & \text{rose} & \text{pea} & \text{single}
\end{array}
$$

Multiple alleles

Up to this point in our study of genetics, we have assumed a maximum of two allelic forms of any particular gene. However, it is possible for a gene to have a whole series of allelic forms. In other words, for the gene controlling flower color in a particular wild rose, the normal allele might produce pink-colored flowers. But different plants may have different alleles of that gene that could make lavender, red, yellow, orange, or other pigments. An individual plant can

FIGURE 23–12

Multiple alleles: ABO blood types in human beings. (A) The possible blood group phenotypes and genotypes resulting from crosses involving the three alleles that determine ABO blood types. (B) The reactions of red blood cells of O, A, B, and AB individuals to blood serum transferred during transfusions. Individuals of blood type O are universal recipients (can receive blood serum of any group), and those with blood type AB are universal donors (can give blood successfully to a recipient of any blood group). If the wrong antibodies are present in other combinations, the serum antibodies will clump the red blood cells of the recipient and cause death.

Allele contributed by parent 1

Allele contributed by parent 2

	i	I^A	I^B
i	Type O ii	Type A $I^A i$	Type B $I^B i$
I^A	Type A $I^A i$	Type A $I^A I^A$	Type AB $I^A I^B$
I^B	Type B $I^B i$	Type AB $I^A I^B$	Type B $I^B I^B$

A

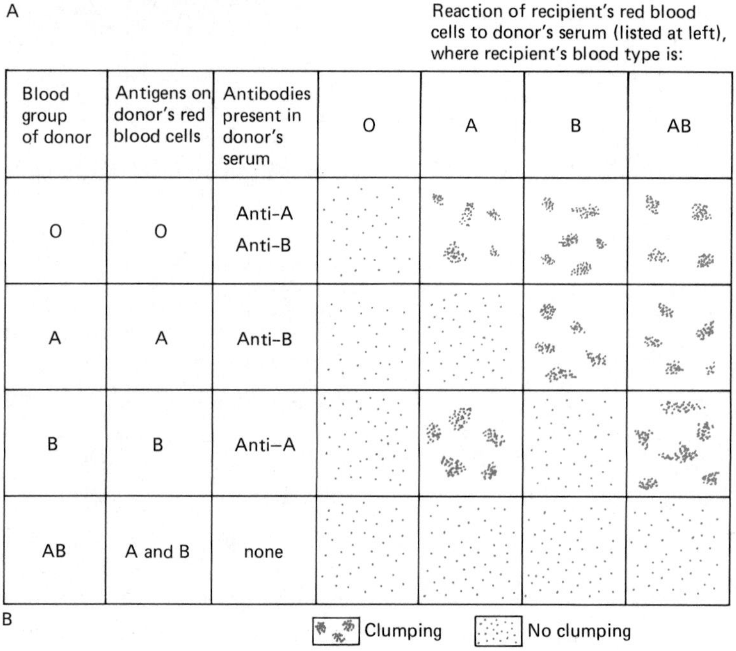

Reaction of recipient's red blood cells to donor's serum (listed at left), where recipient's blood type is:

Blood group of donor	Antigens on donor's red blood cells	Antibodies present in donor's serum	O	A	B	AB
O	O	Anti-A Anti-B		clumping	clumping	clumping
A	A	Anti-B			clumping	clumping
B	B	Anti-A		clumping		clumping
AB	A and B	none				

B

Clumping　　No clumping

be homozygous for any one allelic form or heterozygous for any two of them, but no more than two forms can be carried in the same individual.

Multiple alleles are found in the four main human blood groups: O, A, B, and AB (Figure 23–12). These groups are genetically determined by a series of three alleles — *A*, *B*, and *O* — at one gene site. The alleles *A* and *B* are both dominant to *O*; the heterozygote *AB* produces a blood type that shows the effects of both alleles. The blood groups, then, are produced by these specific combinations of alleles:

Phenotype	Genotype
Group O	OO
Group A	AA or AO
Group B	BB or BO
Group AB	AB

In paternity cases, for example, a knowledge of the inheritance of these blood types is often used in determining the defendant's innocence or guilt. For example, a man whose blood type is AB cannot be the father of a type O child; he also cannot be the father of a type AB child if the mother is type O.

Mutations

New allelic forms of a gene arise by **mutation,** that is, a change induced in the chemical structure of the gene. The mutation-producing agent, or *mutagen*, may be ionizing radiation, a strongly reactive chemical such as chlorine gas, or even ultraviolet light. Since the organism in question has had a long evolutionary history of adjustment to its environment and any change in a finely tuned and balanced physiological system will probably upset its functioning, usually such mutations are deleterious. A *lethal* allele will kill the organism when its effect is expressed.

Sometimes deleterious or lethal alleles resulting from mutation have physiological benefits in the heterozygous state. One such case was found with a human gene that is needed in the production of hemoglobin, the vital molecule found in every red blood cell that carries oxygen. A deleterious allele alters the structure of the hemoglobin molecule in such a way that the molecules clump together in long chains, especially when there is little oxygen in the blood. The

FIGURE 23–13

Inheritance of the sickle-cell anemia disease in human beings. When both parents are heterozygous and are carriers of the sickle-cell trait, their children will tend to fall into a phenotypic ratio of 1 to 2 to 1, especially under conditions of low oxygen pressure (such as at high altitudes).

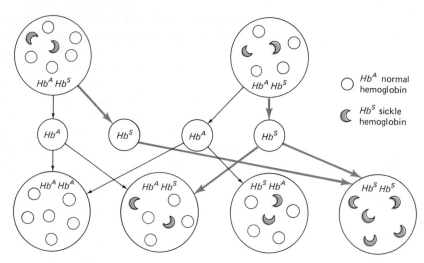

blood cell itself becomes distorted into a sickle shape (Figure 23–13) and cannot carry oxygen effectively; thus the cells jam in the vessels. A person homozygous for the sickle-cell allele develops so much sickling that he has *sickle-cell anemia.* Death results at an early age, usually before sexual maturity. A heterozygous person has among his normal cells anywhere from 20 to 50 percent sickle cells; he may live a normal life unless he is exposed to low levels of oxygen, as at high altitudes, when sickling is stimulated. In African tribes, an extraordinarily high level of heterozygosity was found for this sickle-cell hemoglobin allele, despite rapid elimination of the allele in every generation through the death of homozygotes. The explanation for the maintenance of this allele in the form of heterozygotes is that children who possess the sickle-cell trait are resistant to malaria and tend to survive better than normal individuals. Thus the allele that is disadvantageous and even lethal in a homozygous state becomes beneficial when it is in a heterozygous state. In this state the sickling trait is not expressed to the lethal level and a definite immunity against the fatal effects of the malarial blood parasite is conferred.

In the United States, where malaria is not present today, sickle-cell carriers (heterozygotes) still make up about 9 percent of the black population. The West Africans brought in as slaves some 12 generations ago (300 to 350 years) had about 22 percent carriers in their population. Thus a significant fall in the frequency of the trait seems to have occurred in the absence of the malarial advantage to the heterozygote, and eventually it is hoped that the sickle-cell allele will be eliminated to the point where it is only reintroduced by re-current mutation. In the meantime, it has been recently discovered that carriers and people homozygous for sickle-cell can be prevented from developing fatal sickle-cell anemia through taking small daily doses of the compound *urea*.

Chromosomal abnormalities

Sometimes mutations occur on the chromosomal level rather than merely in one gene. A small piece of the chromosome or even a whole chromosome may be added or lost and greatly influence the development of the offspring having the abnormality. These chromo-somal mutations are of five general types (Figure 23–14).

1. In *duplication*, a portion of a particular chromosome is ac-cidentally repeated in DNA replication during mitosis, such that the genes occurring in it are duplicated on one of the chromosomes.

2. In *deletion*, a section of a chromosome is lost, usually from one of the ends. Cells in the bone marrow of people with chronic granulocytic leukemia have one chromosome (chromosome number 21), which is about 29 percent shorter than its homologue. Since the discovery of this deletion was made in Philadelphia, it has become known as the Philadelphia chromosome, and it has been found as-sociated with this type of human leukemia all over the world. An-other uncommon deletion (on human chromosome number 5) causes the vocal cords to fail to close at the bottom, and an afflicted baby will cry much like a cat mewing (called in French the *cri du chat*, or "cry of the cat," syndrome).

3. In *translocation*, the part broken off one chromosome becomes attached to the end of another, nonhomologous chromosome. The genes contained in it are now part of another chromosome and are inherited independently of the genes contained on the remainder of the original chromosome. The effects of this change of physical posi-tion vary, depending on the genes involved, but it may cause physio-logical malfunctioning of some of the genes if they require a specific association of neighboring genes (as in the intact original chromo-some) for successful operation.

4. In *inversion*, a segment of a chromosome becomes reversed. This can happen at the end of the chromosome or anywhere along its length and results in the same *position effect* on neighboring genes that occurs in translocation: activities of the genes may be modified as a result of their new positions in the chromosomes. Additionally,

FIGURE 23–14

Five types of chromosomal mutations. The order of particular gene loci along each chromosome is indicated by the lowercase letters.

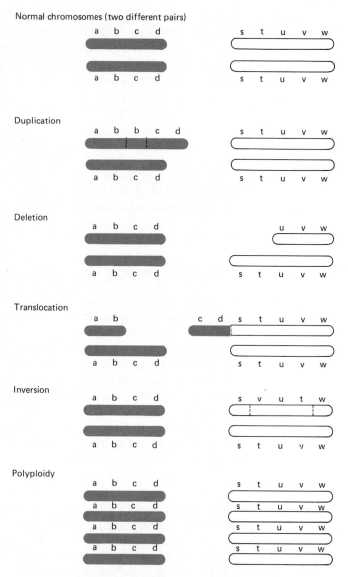

inversions inhibit the occasional (and normal) course of exchanges of genetic material between homologous chromosomes.

5. In *polyploidy,* instead of a duplication of material within one section of a chromosome, an extra copy of the whole chromosome is accidentally made. The organism then has a *karyotype* (collection of chromosomes) composed of the normal *2n* number plus one extra chromosome. This condition is called **aneuploidy,** the addition of

extra chromosomes to the normal set; it also includes situations where one or more chromosomes may lack a partner, instead of having an extra homologue for the already existing pair. In about one out of every 700 human beings in the United States, an extra chromosome number 21 occurs and causes *Down's syndrome,* or *Mongolism.* That person has 47 chromosomes instead of the normal human diploid number of 46, and the presence of the complete set of extra chromosome number 21 genes causes subnormal mentality (Mongolian idiocy), a slanting fold of the skin in the upper eyelids, and physical retardation. A woman between 20 and 29 has one chance in 1,500 of having such a child; if she is past 40, the probability increases to about one in 80. If she waits to have a child until after 45 years of age, there is one chance in 44 that the baby will have an extra chromosome number 21 and hence be mongoloid.

Sex linkage and the determination of sex

Among the set of chromosomes present in every diploid cell in higher animals are usually two *sex chromosomes* (Chapter 22). Each of the other pairs of chromosomes (**autosomes**) has quite similar homologues, but one or the other sex will have homologues in one chromosomal pair that markedly differ from each other in shape and size. The other sex will have two similar looking sex chromosomes.

In human beings, the female has two morphologically similar sex chromosomes (called X chromosomes), whereas the male has one X chromosome and one short chromosome bearing few genes (called the Y chromosome). This XY chromosomal constitution for males and XX for females is usual in higher animals. The genes determining maleness in human beings are located on the Y chromosome. The genes located on the X chromosome are also said to be sex-linked, but (as with some of the genes on the Y chromosome) they are not necessarily involved with sexual characteristics. They are said to be sex-linked merely because their inheritance follows the inheritance of the chromosomes.

The sex of a child is determined at the moment of fertilization, and it depends entirely on the chromosome constitution of the male parent's sperm cell that penetrates the egg. At meiosis in the female's ovary, the two X chromosomes segregate and go into different haploid eggs. At meiosis in the male's testes, the two sex chromosomes segregate; the X chromosome enters about half the haploid sperm cells, and the Y chromosome enters the other half of the

Hemophilia: a sex-linked disease

Hemophilia is characterized by easy bruising and prolonged, excessive bleeding that may cause death from a simple cut or an operation such as circumcision or tonsillectomy. Clotting of the blood may be delayed as long as two hours. Bleeding also may affect the internal organs; intracranial hemorrhage from blows to the head is especially dangerous. Joint deformities are also characteristic.

Persons with hemophilia are deficient in a part of the blood plasma needed for normal clotting. The gene for the disease is recessive and carried on the X chromosome. Females, who have two X chromosomes, rarely have the disease (because of the low probability that *both* parents will have the rare mutation and pass it to their daughter), although they may carry the abnormal allele as a heterozygote and transmit it on an X chromosome to their children. Males, however, have only one X chromosome and invariably develop hemophilia if that chromosome carries the abnormal gene. Children whose mother is a carrier have a 50 percent chance of inheriting the gene; each son who receives it will be a hemophiliac and each affected daughter will be a carrier. However, up to 25 percent of all cases of hemophilia appear in families with no history of the condition; many of these may result from new mutations.

Hemophilia is one of the oldest recognized hereditary conditions (see illustration). The Jews had noticed the fatal effect of hemophilia on circumcised boys in the second century A.D. Rules laid down in the Talmud at this time demonstrate that their originators clearly understood the facts, though not the causes, of sex-linked inheritance. Thus a baby boy was not to be circumcised if his mother had previously lost two boys at circumcision, even if she had married another man in the intervening time. But if a man had previously lost two sons at circumcision, his sons produced by a different wife still were required to be circumcised. Thus the experts of that early period realized the fact that a boy inherits the hemophilic disease from his mother and not his father.

Because of its high incidence in many of the inbreeding royal families of Europe since 1791, hemophilia came to be known as the "royal disease." The most famous case was that of Alexis, the Tsarevich of Imperial Russia. The desperation of his mother, Queen Alexandra, about her son's condition brought her under the influence of the notorious Rasputin and contributed materially to the onset of the Russian Revolution in 1917. Alexandra was the granddaughter of Queen Victoria of England, who was herself a carrier (and possibly the first — via a mutation — in her family). Hemophilia, however, is not restricted to royalty. An estimated 20,000 individuals in the United States alone currently have moderate to severe forms of the disease.

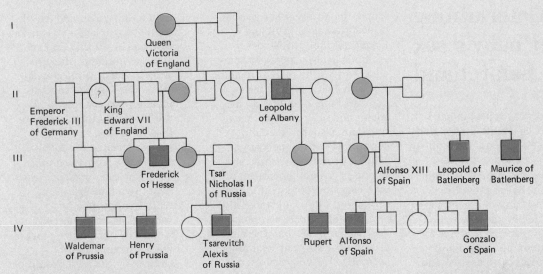

Since the 1930s it has been known that the blood-clotting defect that causes classic hemophilia can be temporarily corrected by infusing preparations of a material—called antihemophilic globulin factor (AHG or AHF), or factor VIII—found in normal blood plasma. Persons with hemophilia lack the normal allele on the X chromosome that produces this factor of the blood plasma needed for normal clotting. Males, with only one X chromosome and the normal allele, produce as much AHG as females who have two. With the hemophiliac allele, a male produces no AHG. The few homozygous recessive females are always hemophilic, since AHG is not produced; in heterozygous carrier females, however, there is considerable variation in AHG levels. In a few female carriers, the AHG concentration rises as high as in normal women; in other female carriers, who may be bleeders, the AHG concentration drops as low as in hemophilic men. Recent research has shown that the diagnosis of hemophilia carriers can be reliably made by testing for a deficiency of functional AHG factor in the blood. A woman from a family with a history of hemophilia who wants to know her chances of having a hemophilic son can now receive adequate genetic counseling.

The inheritance of hemophilia (X chromosome-linked mutation) in the royal families of Europe since 1791. In a pedigree, circles represent females and squares represent males. The colored circles indicate heterozygous female carriers of the recessive allele for hemophilia, and the solid squares indicate bleeder sons. Later generations included many more individuals than are indicated here.

Determining your baby's sex beforehand

Through the ages people have wanted to be able to predict and even select the sex of their offspring. Folklore prescriptions range from the Hebrew Talmud suggestion that placing the marriage bed in a north-south direction favors the conception of males to Aristole's advice to have intercourse in the north wind if males were desired and in the south wind if females were desired. In Germany, husbands in the Spessart Mountains take an ax to bed to produce males and leave it in the woodshed to produce females; and in some rural American communities men still hang their pants on the right side of the bed if they want a male and on the left side if they want a female.

As we have seen in this chapter, the sex of the fetus is determined by which sperm does the fertilizing: a sperm carrying the X chromosome will yield a female and a sperm carrying the Y chromosome will yield a male. Under the micro

scope, the Y-sperm has a round- or wedge-shaped head and longer tail than the X-sperm, which is oval-headed and somewhat wider. The male-producing Y-sperm swim slightly faster than the female-producing X-sperm and are much more sensitive than X-sperm to acidic conditions within the vagina. (The environment within the vagina is generally acidic, whereas conditions within the cervix and uterus are generally alkaline. The closer a woman approaches ovulation, the more alkaline her cervical secretions become.) The Y-sperm often far outnumber the X-sperm in the semen of the average male, perhaps to compensate for this physiological inferiority. These attributes, in combination with proper timing and certain techniques, offer a chance to choose a baby's sex before conception settles the issue. (It has been possible for some time to detect the sex of an *already conceived* baby through amniocentesis, a procedure in which a needle is inserted in the mother's abdomen and amniotic fluid surrounding the fetus is sampled to determine the sex chromosome constitution of loose cells.)

Dr. Landrum B. Shettles, at Columbia College of Physicians and Surgeons, recommends two procedures for prospective parents—depending on the sex desired:

For A Female Child

1. Cease intercourse two or three days before the time of

ovulation (which may be pinpointed with a gynecologist's assistance). At this time, an acid environment still prevails in the cervical area; and while the X-sperm can survive for two or three days, the Y-sperm rarely last more than 24 hours.
2. Immediately *before* intercourse, the woman should use an acidic douche consisting of two tablespoons of white vinegar to a quart of water. (The acidic environment will immobilize most of the Y-carrying sperm.)
3. Orgasm by the wife should be avoided. (It increases the flow of alkaline secretions that would neutralize the acid environment that helps the chances of the X-sperm to reach the egg first.)
4. Shallow penetration by the male at the time of male orgasm will increase the acid exposure of the sperm as they swim to the cervix.
5. Frequent intercourse *prior to* the final intercourse two or three days before ovulation will aid in keeping a low sperm count, which increases the possibility for female offspring.

For A Male Child

1. Time intercourse to come as closely as possible to the moment of ovulation. (This reduces the time of womb exposure for the Y-sperm, and the natural cervical secretions are also more alkaline.)
2. Immediately *before* intercourse, the woman should use an alkaline douche consisting of two tablespoons of baking soda to a quart of water, which has been allowed to stand for 15 minutes before use. (The alkaline environment will be kind to both types of sperm and generally enhance the chances for fertilization, but the Y-sperm can swim faster now than the X-sperm and should reach the egg in a Fallopian tube first.)
3. Normal orgasm by the wife should be encouraged and her husband should allow it to occur first or to coincide with his.
4. Deep penetration by the male at the time of male orgasm will decrease the distance that the faster Y-sperm have to swim to the cervix.

5. Abstinence from intercourse is necessary from the beginning of the monthly cycle until the day of ovulation. This period maximizes the sperm count, which favors Y-sperm.

Success rates in sex selection by these procedures range from about 85 to 90 percent, Dr. Shettles reports.

Techniques are also being developed for culling Y-sperm from X-sperm as they swim through a dense albumin solution (as much as 85 percent of the sample that can be used in artificial insemination consists of Y-sperm). Several researchers are investigating the use of immune reactions on the Y- and X-sperm to select in favor of either male or female offspring. Others are interested in the *Guinness Book of World Records* report that in the highest inhabited village in the world, located in the South American Andes, the villagers can produce only daughters, unless they move down from the mountain temporarily to have their children. Perhaps the low oxygen density or low air pressure adversely affects the physiological functioning of the less robust Y-sperm.

In any event, continued research on sex selection offers the admirable possibility of helping to achieve zero population growth, for many peoples around the world place particular emphasis on having a male heir or a daughter and continue having children until their personal goal is achieved. Sex selection also offers the possibility of a male-oriented imbalance in the sex ratio of the human population with attendant possibilities for increased social problems (e.g., increased crime, which is predominately carried out by males) and opportunities (e.g., greater numbers of males available for spinsters and widows). Perhaps the projected scarcity of females, though, would shift the selected ratio back toward neutrality after a short period of imbalance. The outcome for society of this parental choice for the sex of their offspring will be fascinating to watch in the decades ahead.

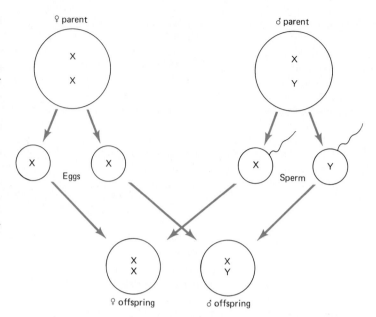

FIGURE 23–16

The determination of sex. Gender, in human beings and many other animals, is determined by the presence of a pair of morphologically different sex chromosomes. A normal female mammal has two X chromosomes (XX in all cells) and produces only X gametes (eggs) as diagramed. The male is XY and produces both X and Y sperm. The gender of the offspring, then, is determined by which type of sperm gains entry to the egg.

sperm cells (Figure 23–15). Thus it is largely a matter of chance as to whether a Y-carrying sperm or an X-carrying sperm reaches the X-bearing egg first. If the Y-carrying sperm penetrates and fertilizes the egg, the offspring will be male (XY in sex-chromosome constitution); if the X-carrying sperm penetrates first, the child will be female (XX in sex-chromosome constitution).

Conclusions

In our review of the basic principles of genetics, we have covered only the aspects of heredity that perhaps most directly affect us in everyday life—that is, genetic phenomena that we can observe in our own experience and whose explanations may be of direct benefit to us in planning marriage, timing childbirth, and understanding inheritance in general. Aside from the pioneering work of Mendel, we have not gone into the fascinating and rapid historical development of the field of genetics in recent years; for example, the actual chemical constitution and physical structure of the gene has only been worked out in the past 20 years! Since much future work in genetic manipulation of phenotype will be tied to the chemical nature of the gene and the mechanism of gene action, we shall take up this topic in detail in the next chapter.

Summary

Gregor Mendel laid the foundation of modern genetics with his 1865 publication of the results of extensive crossing experiments with garden peas. In any particular individual, there are two factors, or allelic genes, controlling the expression of each character. The two genes were derived from the parents, one from each parent. When gametes are formed in a parental plant, the members of a pair of genes segregate from one another and each gamete receives only one of the genes. If the two allelic genes that end up in a fertilized egg (through the union of two gametes) are different (heterozygous) and there is a dominance relationship between them, one (the dominant allele) will express its character trait to the exclusion of the other (the recessive allele). In intermediate inheritance, the heterozygous individual shows an intermediate appearance through blending of the effects of the two alleles. Alleles of genes for different characters will assort independently in dihybrid, trihybrid, or more complicated crosses, if they are located on different chromosomes. In polygenic (multiple-gene) inheritance, the alleles of two or more genes interact to produce a particular phenotypic character, often in additive fashion. In epistasis, one gene acts as a switch to control the expression of other genes. In gene collaboration, two different genes interact in certain combinations of alleles to produce a novel character state that neither can produce alone. A gene can have many alleles, in which case multiple-allelic inheritance will produce a number of phenotypes. New allelic forms of a gene arise by genic mutation. Chromosomal mutations include duplication, deletion, translocation, inversion, and polyploidy. Sex in human beings is determined by the two sex chromosomes (XX in women and XY in men). Individuals who receive a Y chromosome from their fathers become males; those who receive an X chromosome from their fathers become females. The genes located on the sex chromosomes are said to be sex-linked because their inheritance follows the inheritance of these chromosomes rather than any of the ordinary autosomes.

24 The mechanisms of gene action

In this chapter we shall look at the physical and chemical nature of the genetic material. What is a gene exactly, and how does it function? Let us start with the chromosomes, since we know from evidence discussed in the previous chapter that they carry the genes.

The chemical nature of the genetic material

Chromosomes were found by chemical analysis to contain proteins, deoxyribonucleic acid (DNA), and ribonucleic acid (RNA) — all being substances that could possibly function as the genetic material because of their complicated nature and hence capacity to store a large amount of information. The principal nucleic acid in the nucleus, DNA, was discovered in 1869 by Friedrich Miescher, a Swiss biochemist who was working with the sperm cells of fish. In the mid 1920s Robert Feulgen, a German chemist, discovered through a method of selectively staining DNA with dyes that nuclear DNA is restricted to the chromosomes. However, this did not prove that DNA was the nuclear genetic material; indeed, many biologists thought at the time that only the protein material associated with chromosomes could be the genetic material because the DNA did not have the chemical complexity of a great many kinds of subunits that might be necessary to code so much information.

The first strong clue to the importance of DNA in heredity came by accident through studies done by Fred Griffith, an English bacteriologist working with pneumonia-producing bacteria (Figure 24–1). He discovered several strains of *Pneumococcus* bacteria; one was virulent (producing pneumonia symptoms), and another was avirulent (a strain not fatal to mammals). The living virulent strain killed mice that were injected with it with pneumonia, but the avirulent strain was not fatal to them. When Griffith injected living bacteria of the avirulent strain along with some heat-killed virulent strain cells, he discovered that the mice died and that the bacteria removed from the dead mice were all virulent. The heat-killed virulent strain cells **had** *transformed* the avirulent strains. Griffith discovered that the **same** bacterial transformation could be done in a test tube, using an **extract**

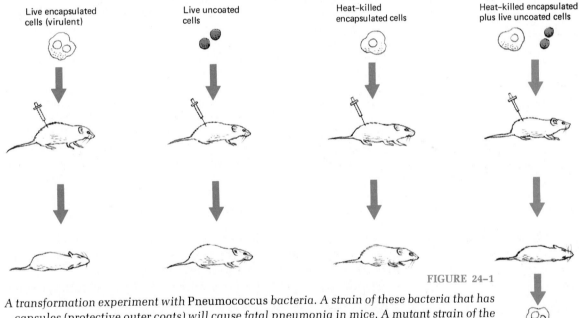

Live encapsulated cells (virulent)

Live uncoated cells

Heat-killed encapsulated cells

Heat-killed encapsulated plus live uncoated cells

FIGURE 24–1

A transformation experiment with Pneumococcus *bacteria. A strain of these bacteria that has capsules (protective outer coats) will cause fatal pneumonia in mice. A mutant strain of the same bacteria lacks the capacity to make capsules and does not cause disease. Frederick Griffith found that either strain alone could be injected into a mouse and cause the expected effect (pneumonia or no disease). If the virulent, capsule-bearing strain was heat-killed before injection, it, too, was harmless. But if Griffith injected dead virulent bacteria and live harmless bacteria into the same mouse, the animal died. He subsequently found that a chemical agent from the dead virulent strain had transformed the live harmless strain into live virulent (capsule-bearing) bacteria, which caused fatal pneumonia. On isolation, this chemical proved to be DNA.*

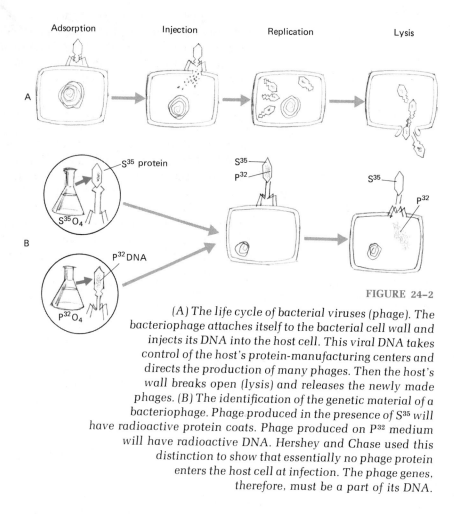

Adsorption   Injection   Replication   Lysis

A

B

S³⁵ protein

S³⁵O₄

P³² DNA

P³²O₄

S³⁵
P³²

S³⁵
P³²

FIGURE 24–2

(A) The life cycle of bacterial viruses (phage). The
bacteriophage attaches itself to the bacterial cell wall and
injects its DNA into the host cell. This viral DNA takes
control of the host's protein-manufacturing centers and
directs the production of many phages. Then the host's
wall breaks open (lysis) and releases the newly made
phages. (B) The identification of the genetic material of a
bacteriophage. Phage produced in the presence of S³⁵ will
have radioactive protein coats. Phage produced on P³² medium
will have radioactive DNA. Hershey and Chase used this
distinction to show that essentially no phage protein
enters the host cell at infection. The phage genes,
therefore, must be a part of its DNA.

from the virulent cells. A chemical analysis of the active component
in the extract showed that over 97 percent of this active material was
DNA. The remainder was a trace of protein.

This strong correlation of the presence of DNA with the action of
genetically controlled characteristics did not constitute a proof of
the presumed relationship. The confirmation that DNA is the genetic
material came from a group of studies dealing with *bacteriophages*,
viruses that infect bacterial cells. Certain of these viruses, studied by
Alfred D. Hershey and Martha Chase, of the Carnegie Laboratory of
Genetics, have a head containing a large amount of DNA inside an
outer protein coat. The virus particle attaches itself by its tail to the
bacterial cell wall (Figure 24–2) and then injects its own genetic material into the cell to take over the bacterial cell's metabolic machinery and manufacture more virus particles.

The question Hershey and Chase were able to answer was whether the DNA in the center of the virus particle or the protein in the outer coat of the virus particle actually contained the genetic instructions for the characteristics of the virus. DNA, being rich in phosphorus and lacking sulfur, was labeled distinctively with phosphorus[32], a radioactive tracer substance. Protein, being rich in sulfur but lacking phosphorus in this particular case, was labeled with radioactive sulfur[35]. This labeling of the respective molecules was done by growing the virus in bacterial cultures (their hosts) that were rich in S^{35} and P^{32}; the P^{32} became incorporated into the DNA core and the S^{35} became incorporated into the protein coat.

These labeled virus particles were then allowed to attack nonradioactive bacterial cells. When Hershey and Chase washed off the material left on the outside of the bacterial cells, they found that the material was almost entirely labeled with sulfur[35] (97 percent S^{35} and 3 percent P^{32}). On the other hand, about 97 percent of the bacterial cells now containing the genetic material of the virus were labeled with P^{32} (but essentially none with S^{35}), indicating that only DNA had been injected into them by the virus. That is, the P^{32}-carrying DNA was the hereditary material that was being injected into the bacterial cells, not the S^{35}-labeled protein.

The results of this experiment, reported in 1952, confirmed the identity of the genetic material as DNA, and the rapid progress in other laboratories brought about a description of the molecular structure of this genetic material in the following year.

The properties of the genetic material

Before we look at its structure, let us consider the *basic properties* that the genetic material (DNA) must have. First, a great deal of genetic information must be carried between the generations in the form of egg and sperm cells. It has been estimated that all the sperm cells that went into creating all the human beings living today could fit into a teacup. As the physiologist John McCrone has aptly said, "That's something to think about for awhile!" Actually, this fact simply points out the extremely *condensed character* of the transmission of genetic information between generations. Second, the genetic material must have the ability to precisely *duplicate* itself. If DNA was not able to replicate itself, the genetic material would be halved at every cell division. Third, the genetic material must be a *stable* substance. From generation to generation, people have people and cows have cows, every kind reproducing its own kind within certain limits of variability. Fourth, and seemingly contradictory to this last point, the genetic material must be capable of *change*. If there were no ways by which it could be altered, we would not have mutation or the property of genetic variability in populations and hence there would be a lack of different genotypes and phenotypes for evolutionary forces to work on. These four properties we can now look at in some detail, as we consider the structure and functioning of DNA.

The molecular structure of DNA

Deoxyribonucleic acid (DNA) is a *polymer*, a macromolecule made up of many chemical subunits; the four basic subunits are known as nucleotides. Each *nucleotide* is composed of a phosphate group, a five-carbon sugar molecule, and a nitrogenous base (Figure 24–3). The difference between the four nucleotides found in DNA is centered in the structure of the nitrogenous bases. Adenine and guanine are both *purines* and have a double-ring structure; thymine and cytosine are both *pyrimidines* and have a single-ring structure (Figure 24–4).

The discovery of how these nucleotides are arranged in the DNA molecule is one of the milestones in the history of biology. Maurice H. F. Wilkins, of King's College, London, and James D. Watson and Francis H. C. Crick, working at Cambridge University, shared the Nobel prize for their discovery in 1953 of the spatial configuration of the DNA molecule. Wilkins supplied the exceptionally sharp X-ray diffraction pictures of patterns of the crystalline DNA molecule; Watson and Crick developed a model of the structure of the DNA molecule by combining what was known about its chemical nature with the physical interpretations that could be made from Wilkins' excellent photographic patterns.

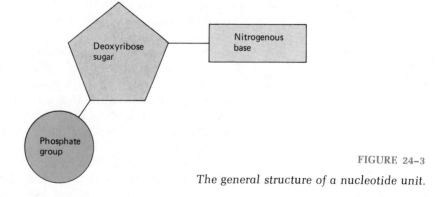

FIGURE 24–3

The general structure of a nucleotide unit.

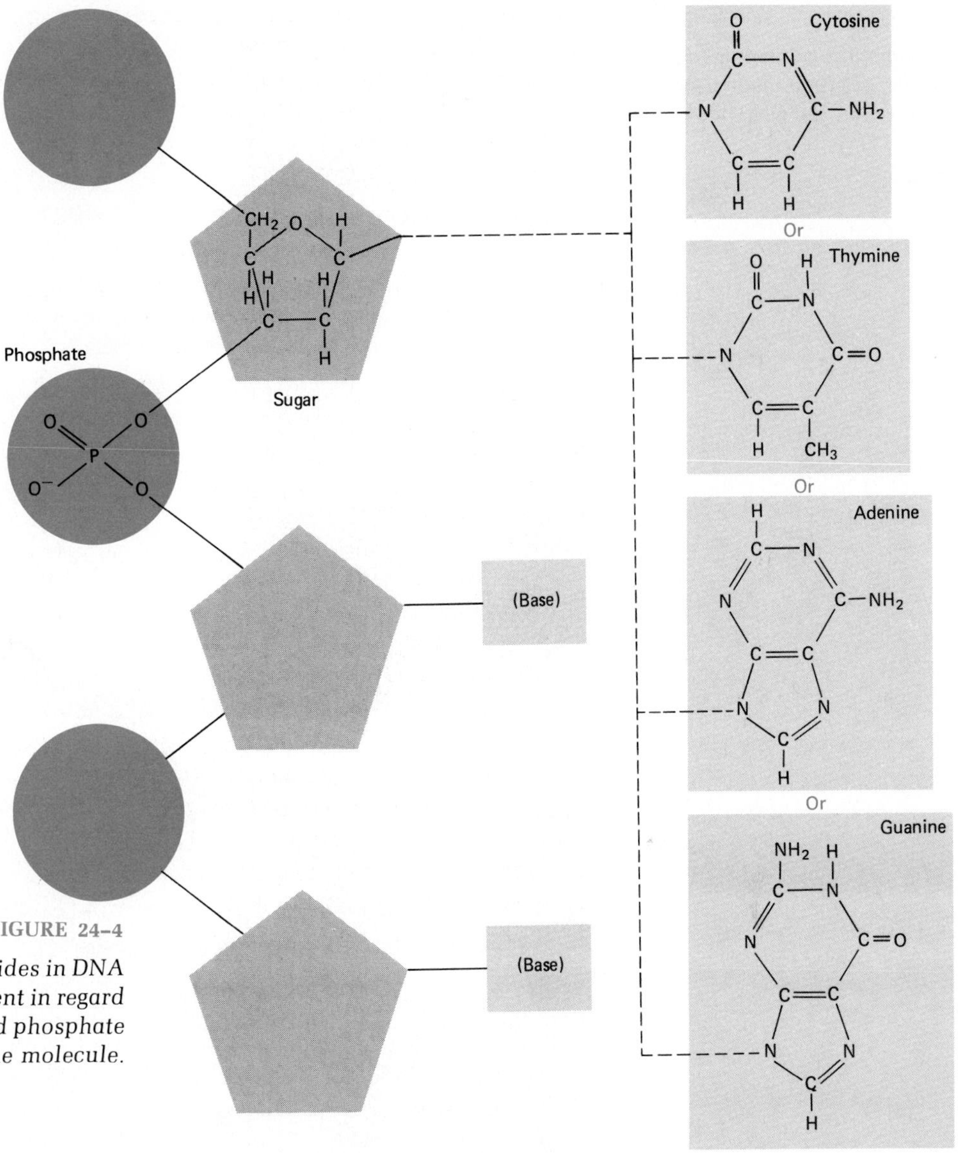

FIGURE 24-4

The four nucleotides in DNA and their placement in regard to the sugar and phosphate backbone of the molecule.

The Watson-Crick model, as it soon became known, proposed that the DNA molecule is composed of two connected strands of nucleotides. Each strand is formed by bonding the phosphate group of one nucleotide to the sugar of another nucleotide. This results in a chain of alternating sugar and phosphate groups, with the various nitrogenous bases oriented to one side. The complete DNA molecule

FIGURE 24-5

The hydrogen bonds (dashed lines) between the two strands in the DNA molecule.

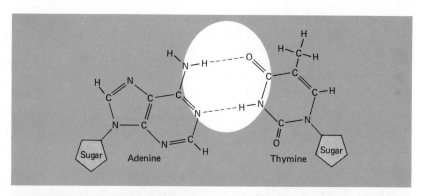

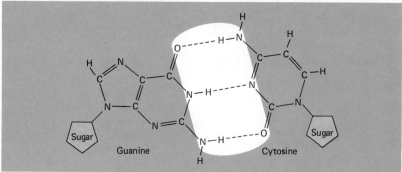

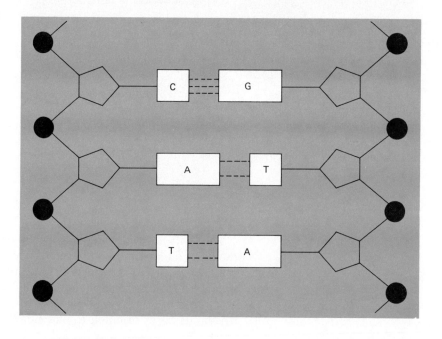

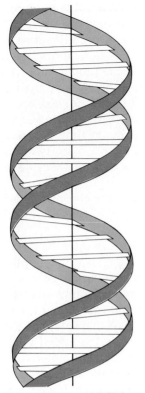

FIGURE 24–6

The Watson-Crick model of the helix structure for the DNA molecule. The two colored bands indicate the outside phosphate-sugar backbones of the molecule, with the paired nucleotides shown as colorless connecting bands.

consists of two of these strands held together by weak hydrogen bonds between their respective lateral branches of nitrogenous bases (Figure 24–5). Thus the DNA molecule resembles a ladderlike structure made up of two chains of nucleotides joined together by these hydrogen bonds between the bases of one chain and the bases of the other.

As a consequence of the difference in structural configuration of purines and pyrimidines, and the number of hydrogen bonds that can be formed between them, an adenine base can pair only with a thymine base, and a guanine base with a cytosine base. Thus every "rung" of the molecule will be either adenine-thymine or guanine-cytosine. Another feature of the Watson-Crick model is that this whole molecule is arranged as a helix and thus resembles a twisted ladder that has been wrapped around the outside of a cylinder. The sides represent the two sugar-phosphate backbones and the rungs represent paired bases (Figure 24–6). The helix was approximately 20 angstroms wide and had turns 34 angstroms long, matching the periodicity discovered by Wilkins in his X-ray diffraction photographs.

The chemical evidence found by other investigators matched this model well. The amounts of adenine and thymine in any DNA molecule are always equal because of this pairing requirement; the same is true of the quantities of guanine and cytosine. Yet there is no restriction on the order of the bases within a *single* chain of the DNA ladder. In other words, a series of adenines can be arranged along a chain, then a thymine, several guanines, cytosines, and back to some other molecule, as long as this sequencing takes place on one side of the ladder. The other side will contain the proper matching bases.

The replication of the genetic material

One of the features of the 1953 Watson-Crick model for the structure of the DNA is that this ladderlike structure can readily replicate itself. The requirement that particular bases pair with each other means that exact duplicates of the ladder can be made in some fashion, as long as there are sufficient nucleotides in the cell nucleus to line up next to this old DNA ladder at the time of chromosomal replication. However, the precise mechanisms in which this replication occurs did not become clear until a number of years later.

When the chromosome prepares to duplicate itself, apparently the DNA molecule begins to split in half at one end. The weak hydrogen bonds separate between the two chains of bases forming the rungs of the ladder. As the chain is unwound in the nucleus, each of the chains builds for itself a new partner from the surrounding base molecules floating free in the nucleus (Figure 24–7). The free nucleotides in the nucleus come into position with the proper nucleotides still present on the respective chains, and as they join by hydrogen bonds to the old chain, a new chain is built in parallel fashion with the old half. Each of the two original chains, as they are unwinding, builds a new replicate chain next to it. Thus by the time the original molecule has completely finished unwinding, there are two duplicate DNA molecules sitting in the nucleus, each of them double-stranded and containing one strand of the original molecule and one new strand. During cell division, then, each of the chromatids being separated on the mitotic spindle contains one of these duplicated double-stranded molecules of DNA.

Apparently, this replication takes place during interphase, when the DNA molecule is fully extended. In prophase, the duplicate chromatids, each containing its own double-stranded DNA molecule, are already formed and appear as tightly coiled structures. Some appreciation for the task of replication may be gained from the fact that a typical mammalian cell contains DNA that would be about one meter (39 inches) long if extended. This DNA is folded into the cell nucleus whose diameter is only about 0.001 centimeter. Recent investigators have found from X-ray diffraction studies that the DNA double helix may be twisted on itself to form a larger single helix or springlike structure that these researchers call a super coil. Small beads, formed of globular clusters of a special group of proteins, the *histones*, are located along this DNA molecule, and the string evident between these beads consists of short stretches of bare DNA between the histone clusters. There is some evidence that these positively charged proteins could play a role in developing the super-coil structure of DNA because DNA contains negatively charged phosphate groups. Thus histones could bind electrostatically to DNA and pull the molecule into the larger coiled form of a "condensed" chromosome.

The genetic code and protein synthesis

It has been known for more than 70 years that some simply inherited Mendelian-type genes control the production of enzymes and other substances in animals and plants. People who lack the proper nor-

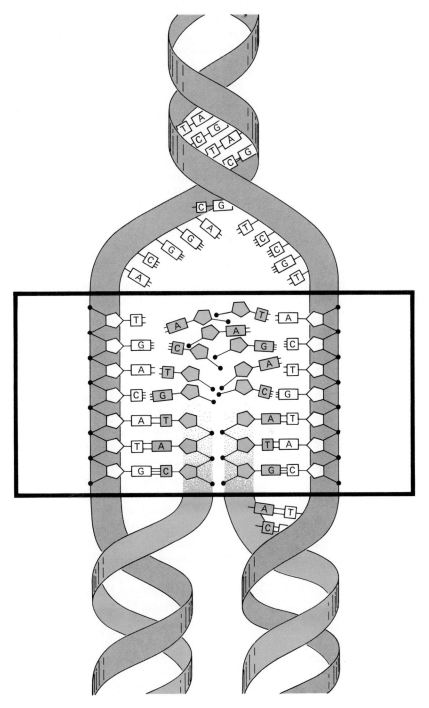

FIGURE 24–7

The replication of DNA occurs by specific A-T or C-G base pairings between free nucleotides and the bases of each original DNA strand of an unraveling helix molecule. Thus two new DNA molecules result, each containing one old and one new nucleotide strand.

mal genes for such a characteristic will develop deficiency diseases or other symptoms because they lack the particular chemical that the gene normally produces. In the 1940s George W. Beadle and Edward L. Tatum, of Stanford University, put forth a hypothesis that one gene in an organism controls a particular enzyme. Genes, then, would control cellular function by controlling the most important intermediaries of cellular function, the *enzymes* — protein molecules that catalyze biological reactions throughout the known organic world. These reactions in turn would determine the phenotypic characteristics of the organism.

We know from Chapter 21 that enzymes and other proteins are synthesized on the ribosomes found on the endoplasmic reticulum or free in the cytoplasm. Yet the DNA material is essentially restricted to the chromosomes of the nucleus. The problem that molecular biologists faced following the announcement of the Watson-Crick model and subsequent research on the genetic characteristics of DNA was to find out how the DNA in the chromosomes of a nucleus controls the synthesis of proteins in the ribosomes located in the cytoplasm. In particular, how does the sequence of four different types of nucleotides on the DNA molecule become transcribed or translated into the sequence of some 20 different amino acids located in the chains of hundreds of such units composing protein molecules?

The answer to this question lies in the activities of several types of the second nucleic acid, ribonucleic acid (RNA), which are found in both the nucleus and the cytoplasm. First, we should note that the chemical differences between RNA and DNA are threefold. The sugar in RNA has an additional oxygen atom and is called ribose, not deoxyribose. Also, the base thymine is replaced by the very similar base, uracil, which has the same bonding relationships as thymine. Thus an RNA uracil nucleotide will bond with adenine if it approaches a DNA strand. Finally, RNA is usually a single-stranded molecule rather than a two-stranded molecule (Figure 24–8).

The story begins with DNA in the nucleus acting as a template for the synthesis of RNA there. The particular type of RNA that is formed next to the DNA strand is called *messenger RNA* (m-RNA). The two strands of a DNA molecule uncoil and then the ribonucleotides are attracted to their respective complementary bases along the DNA strand. Once they are arranged, an enzyme bonds them together, and the m-RNA molecule can separate from the nuclear DNA strand, passing probably through a nuclear pore and out into the cytoplasm. It is still not known which of the two strands of the DNA molecule is copied in the formation of the m-RNA strand, nor

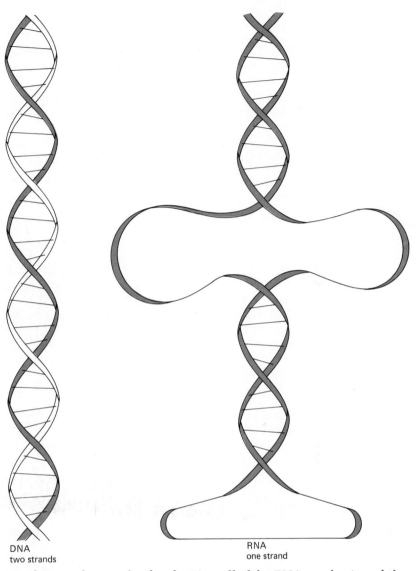

FIGURE 24–8

The single-stranded structure of RNA, as compared with the double-stranded structure of DNA.

DNA
two strands

RNA
one strand

are the cues known for the shutting off of the RNA synthesis and the beginning of DNA synthesis in preparation for a new cell division process in the nucleus. Nevertheless, the m-RNA model has been well established by many investigators, and it is clear that it has a base content complementary to that of the DNA master molecule.

When the m-RNA strand reaches the cytoplasm, it combines with one or more ribosomes, linking them into a so-called polyribosome, and it is on this m-RNA–polyribosome complex that protein synthesis will occur (Figure 24–9). The ribosome itself appears to function as an unspecialized device that is capable of synthesizing any protein called for by the m-RNA strand. *Ribosomal RNA (r-RNA)* is originally produced by DNA genes located in the nucleolus, within the nucleus.

As we have noted, there are thousands of these manufacturing machines of ribosomes located throughout the cytoplasm on the rough endoplasmic reticulum and occasionally scattered in the cytoplasm. Each of the several ribosomes moves along the m-RNA strand and reads the message encoded in it to build a polypeptide chain of amino acids. As many as four or six ribosomes may be seen moving along the m-RNA strand at any particular time. This second

FIGURE 24–9

The overall scheme for protein synthesis. (A) The genetic code is transcribed from the master template of DNA in the nucleus to the form of a complementarily coded messenger-RNA strand, which then leaves through the nuclear membrane for a site in the cytoplasm. (B) There the m-RNA strand becomes associated with one or more ribosomes and small transfer-RNA molecules bring specific amino acids to the ribosome-messenger RNA complex for translation of the m-RNA nucleotide code sequence into a particular amino acid sequence, forming a polypeptide (protein) chain.

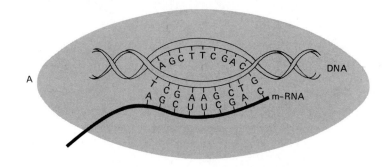

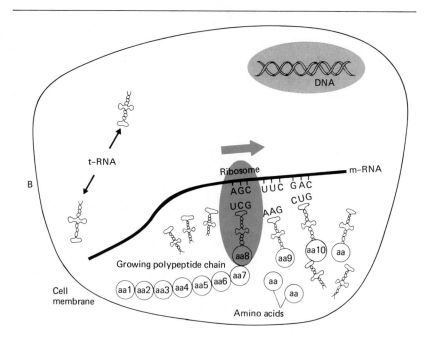

type of RNA — r-RNA — is by far the most abundant kind in the cell, yet surprisingly little is known about it as far as its functioning is concerned. It is known that ribosomes are quite stable and remain functional probably for the life of the cell, whereas m-RNA is less stable and will be broken up by enzymes after synthesizing a few of the kinds of protein enzyme molecules it is encoded to make.

A third type of RNA, *transfer RNA* (t-RNA), brings the amino acids to the proper place on the template of m-RNA, where the ribosomes are actively synthesizing the polypeptides. In 1957, it was discovered that this small type of cytoplasmic RNA attaches itself to a single molecule of amino acid and transports it to the ribosome — hence the name t-RNA. There is at least one type of t-RNA for each of the 20 known amino acids, and often several types are encoded to combine with a particular amino acid and carry it to the ribosome. When the t-RNA molecule reaches the m-RNA and the ribosomes, it attaches to the proper base on the m-RNA molecule by the base-pairing mechanism described in the Watson-Crick model. The amino acid on the other, exposed end of the t-RNA molecule is then synthesized into the polypeptide chain by the ribosome when it comes along the m-RNA strand. Each of these t-RNA molecules is comprised of about 75 to 80 nucleotides.

Obviously, this complex process of protein synthesis demands a coding correspondence between a certain combination of nucleotides in nucleic acids and the amino acids to be incorporated into the polypeptide chain. Through a great deal of research, it has been found that the genetic code is a linear triplet code of three nucleotides along the DNA molecule: a group of three bases in a row (ACT, GAG, and so on) codes for a particular amino acid. With four different nucleotides available to be combined into sequences of three units, each coding for a separate amino acid, it is possible to have as many as 64 three-unit combinations from these four basic units. Thus a triplet sequence of nucleotides on the DNA molecule codes for a particular amino acid, and the total sequence of triplet codes along a particular segment in the DNA molecule ultimately determines the sequence of amino acids in a particular polypeptide — the protein product of that particular gene.

The actual translation of this genetic code to produce polypeptides is accomplished with the aid of these three types of ribonucleic acids that we have been observing, including the ribosomes of the rough endoplasmic reticulum. First, the single strand of m-RNA forms in the nucleus alongside the DNA molecule and transfers the sequence of nucleotides composing a gene into its own nucleotide sequence (uracil, of course, substituting for thymine wherever its complementary base — adenine — appears in the DNA molecule).

This m-RNA molecule then leaves the nucleus through a pore in the nuclear membrane and becomes associated with one to five ribosomes on the endoplasmic reticulum (Figure 24–9). These ribosomes contain r-RNA, and with the aid of the t-RNA molecules bringing in amino acids to specific sites (also encoded by the nucleotide sequences) on the m-RNA molecule they assemble a coded sequence of amino acids, forming the polypeptide. This polypeptide may be utilized as one of the many different structural proteins or physiologically active enzymes.

We have already noted that there is a specific transfer of RNA for each of the 20 or so amino acids that are used as building blocks for all animal and plant proteins, including enzymes. The t-RNA molecule resembles a cloverleaf because it is folded back on itself. On one of its exposed loops is the functional triplet called the *anticodon* (the bases being complements of those in the m-RNA *codon* because of the base-pairing phenomenon), which is the specific code word for that amino acid. At the other end of the molecule is the point of attachment for the amino acid itself. As the ribosome moves along the m-RNA strand, then, the anticodon of the t-RNA pairs with the appropriate three complementary bases of the m-RNA molecule. There is enough room on the ribosome for two or three triplets to be translated almost simultaneously. Therefore the amino acids are brought close together and the peptide bonds form between them to produce the polypeptide chain. Thus *each m-RNA* carries the code for *one polypeptide* and continues to produce this one product as long as ribosomes move along its length. The polypeptide products then move along the smooth endoplasmic-reticulum channels and finally into membrane-coated vesicles, which incorporate the product into vacuoles capable of transporting them to other areas of the cell or to other tissues and organs outside that particular cell. These enzymes and structural proteins, ultimately under the control of the DNA material, control the development and maintenance of the body, its functions, and its characteristics, or phenotype. The sum of the characters produced through this mechanical translation of the genetic code in any given organism sets its lifelong morphological and behavioral limits.

DNA has been shown to be a remarkably stable substance that can be boiled, put under high pressures, and submitted to other extreme conditions that may inactivate it but will not change the DNA chemically. Yet mutagenic agents such as radiation and certain powerful chemicals can change the physical arrangement of base pairs or even convert one base to another, allowing a heritable mutation to a new allelic form of the gene to occur. Thus while DNA normally is stable, it is capable of change—two properties needed for genetic variability and hence evolutionary change in populations.

The benefits and hazards of DNA research

One of the chief benefits of DNA research may be its usefulness in health-related problems. For instance, a gene may be replaced by *genetic engineering*, using a virus particle or some other means to carry a beneficial form of a gene into a cell to replace a deficient one. Genetic engineering might someday be done on a large scale in an organ or a whole human being, if it can be brought to such a point successfully. This might reduce the impact of hereditary diseases, such as hemophilia and diabetes, on the human population.

On the other hand, there is considerable potential danger in using recombinant DNA molecules for genetic manipulation of this type. In fact, the possible hazard to society appears so great that in July 1974 a group of leading biochemists and other researchers appealed to the world's scientists to temporarily halt all research on certain kinds of experiments that involve the genetic manipulation of living cells and viruses. Their concern is that genetically altered bacteria being created with the new techniques could present a health hazard. Using a newly discovered class of enzymes, scientists could introduce particular genes of other species into living cells such as bacteria. The fear is that experiments might be conducted involving insertion into bacteria of bacterial genes that confer either resistance to antibiotics or ability to form bacterial toxins, and also that genes or viruses might be inserted into bacteria that would form a potentially monstrous combination. The bacteria endowed with genes that do not normally occur in nature might escape and infect the human population, particularly since the standard bacterium used by molecular biologists is *Escherichia coli*, a common inhabitant of the human gut. Another type of experiment involving the insertion of animal genes into bacteria could produce public health dangers because of the fact that any type of animal-cell DNA contains sequences common to RNA tumor viruses. Thus the prospects of genetic manipulation of DNA sequences are both promising and hazardous, and experiments are continuing very cautiously.

Summary

The genetic material of the cell was shown to be DNA through biochemical staining of DNA in chromosomes, transformation of genetic characteristics of live bacterial cells with DNA from dead cells, and radioactive labeling of DNA and proteins in DNA-containing viruses that attack bacteria. The basic properties of importance for DNA's activities as the genetic material are its greatly condensed complexity, ability to precisely replicate itself at every cell division, stability, and capability for mutational change. DNA is a polymer with four nucleotides, each composed of a ribose sugar, a phosphate group, and a nitrogen base (adenine, guanine, cytosine, or thymine). The Watson-Crick model proposes that the DNA molecule is composed of two interconnected strands of nucleotides. Each strand is formed by bonding the phosphate group of one nucleotide to the sugar of another nucleotide. This results in a chain of phosphate and sugar groups, with the various nitrogenous bases oriented to one side. The complete DNA molecule consists of two of these strands held together by weak hydrogen bonds between their respective lateral branches of bases. The ladderlike structure is twisted into the form of a helix. It can readily replicate itself by splitting in half and building up a complementary strand next to each of the two old strands. A group of three bases on one strand is the coding unit for a particular amino acid that will go into the final protein product of the gene. It is matched on a complementary m-RNA strand, which travels out into the cytoplasm to a ribosome area on the endoplasmic reticulum. There, t-RNA molecules bring the required amino acids to the m-RNA strand, and the associated ribosomes help to bond the amino acids together into a polypeptide. Genetic engineering involves attempts to introduce new or replacement genes into the already-existing genome of a cell or organism. This field of research may become a major medical tool in the future, although there is also considerable risk of creating unique bacteria or other organisms that would not respond to antibiotics or other known control agents.

25 Photosynthesis and cellular respiration: energy for the living cell

All life here on earth depends on the thermonuclear reactions taking place deep in the sun. There, hydrogen atoms are being used to form helium, releasing great amounts of radiant energy. This radiant energy reaches our earth through the atmosphere, where it enters the world of life. Green plant cells are able to capture the radiant energy of the sun through the use of the pigment chlorophyll and convert this radiant energy into chemical bonds that form glucose sugar molecules.

This reaction of converting the simple materials of carbon dioxide and water into glucose and oxygen is only one part of a complementary cycle. In both plants and animals, the process of cellular respiration converts glucose in the presence of oxygen to carbon dioxide and water, thereby releasing the chemical energy of the original bonds in a form that can be used by the cell for work. The waste products of cellular respiration—namely, carbon dioxide and water—can in turn be used again by green plants in the environment to continue the process of photosynthesis (Figure 25–1).

This orderly series of complicated chemical processes, which maintains the organized living cell, is possible only because of the outside supply of solar energy. Otherwise, as we have seen early in this book, by the laws of thermodynamics a highly organized system

FIGURE 25–1

The overall cycling of materials in photosynthesis and cellular respiration.

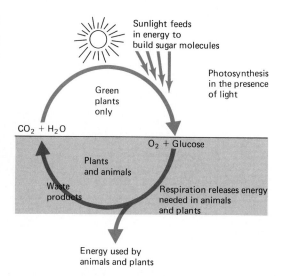

such as the cell has a tendency to move toward greater disorder. Only by the constant use of energy, and, of course, the acquisition of this energy in usable form in the first place, is a living cell able to continue to exist as an orderly arrangement.

The exchanges of living cells, whether in photosynthesis or cellular respiration, usually involve the movement of electrons from one so-called energy level to another. As we know from studies in physical science, electrons are found at certain fixed distances from the nucleus of an atom and are arranged in electron shells that have different energy levels (Figure 25–2). These negatively charged electron particles are moving at a particular distance, then, from the nucleus, and the outermost electrons participate in chemical bonding reactions.

Energy conversions in chemical systems involve the movement of electrons from one energy level to another. If there is an input of energy, the electron can be raised to a higher level. At that level it possesses potential energy, and when it is returned to a lower level its potential energy is released. Thus the fall of electrons from a high level to a lower level provides this source of energy for biological work. The photosynthetic reactions push electrons to a higher level by means of light energy. Conversely, in the processes of cellular respiration, energy is released by allowing electrons to drop to a lower level in a series of chemical reactions. As the electrons move down to lower energy levels, their energy is converted to forms of chemical energy that can be widely used by all kinds of cells. Within the cell, the release of energy supplies the necessary power for cell division, metabolic processes and general maintenence of the cell, the bio-

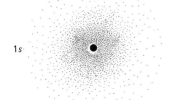

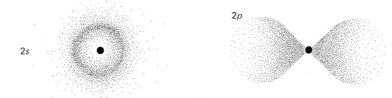

FIGURE 25–2

Representations of the first three electron orbital shells out from the central nucleus (black). The orbitals are actually three-dimensional. The density of the dots is proportional to the probability of finding the particular electron in the region around the nucleus. The orbitals of s electrons are approximately spherical; those of p electrons are roughly dumbbell-shaped. The numerals associated with s and p indicate the energy level or shell; the further out from the nucleus, the higher is the energy level of the electron.

FIGURE 25–3

The chlorophyll a molecule.

synthesis of new compounds, the transport of materials across cell boundaries, and the other life processes of the particular cell type. Most of this occurs through the methodical breakdown of the sugar glucose.

Photosynthesis

The process of photosynthesis is the indirect source of most of the organic matter on earth; it provides the primary source of the energy in all living things and supplies the free oxygen that makes respiration in animals and higher plants possible. In taking in some 128 billion tons of carbon dioxide and 52 billion tons of water a year, plants release some 52 billion tons of oxygen into the atmosphere and synthesize some 87 billion tons of organic molecules beyond what is used up in their own respiratory processes. The energy required is provided ultimately from the light energy absorbed by the various plant pigments.

The key molecule involved in the process of photosynthesis is the green pigment *chlorophyll* (Figure 25–3). The structure of this pigment is similar to the inorganic iron groups of hemoglobin in red blood cells and the cytochrome pigments used in cellular respiration; but instead of having a central atom of iron as in hemoglobin and cytochrome molecules, chlorophyll has one of magnesium. The electrons in the molecular orbitals near the central magnesium atom are pushed into an excited state, enabling an electron of the chlorophyll molecule to be raised from its normal stable energy level to a higher energy level through absorption of a packet of light energy striking the molecule. The electron at this higher level is relatively unstable and will soon fall back to its normal level, giving up the absorbed energy. This released energy then is used to split a water molecule through a rapid series of steps as it moves along a series of carrier molecules adjacent to the chlorophyll and is eased from the excited state down to the normal state.

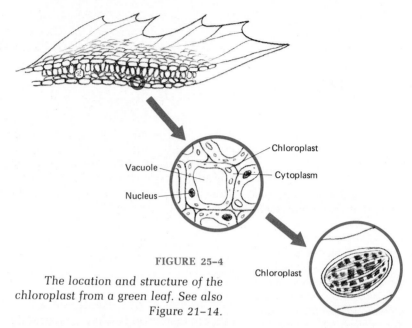

FIGURE 25–4

The location and structure of the chloroplast from a green leaf. See also Figure 21–14.

The chlorophyll molecule is synthesized by plants in the presence of light and is arranged in stacks of membranes, the *grana*, which are situated in special organelles called chloroplasts (Figure 25–4). These chloroplasts resemble mitochondria in general sausage-like shape, but differ considerably in internal membrane structure. The chlorophyll molecules as well as enzymes and electron transport systems needed for photosynthesis appear to be organized along the internal membranes in the chloroplasts.

By the early nineteenth century, the overall reaction for photosynthesis had been determined and could be summarized by this simple equation:

$$6CO_2 + 6H_2O \rightarrow C_6H_{12}O_{6+} 6O_2\uparrow$$

which simply means that carbon dioxide plus water yields, in the presence of chlorophyll and light, a glucose molecule plus six molecules of oxygen gas. However, it required intensive work by biologists, particularly since the 1930s, to work out the actual source of the oxygen released. Rather than carbon dioxide simply being split, with carbon being added to the water and oxygen released, it was discovered that all of the oxygen released came from the water and none from the carbon dioxide. In order to secure the release of six molecules of oxygen for each molecule of glucose produced, 12 molecules of water must be split. Then we have to assume that the chemical reaction also produces water as an end product. This is new water and not the same as that used as a raw material. The overall reaction that best describes photosynthesis today can be summarized as:

$$6CO_2 + 12H_2O \rightarrow C_6H_{12}O_6 + 6O_2\uparrow + 6H_2O$$

The processes involved in photosynthesis, then, include (1) a light reaction, whose products drive (2) a dark thermochemical process involving the uptake of CO_2 from the air, and (3) the reactions concerned with the evolution of oxygen from the split water molecules.

Light reactions

The unique part of photosynthesis occurs in the *light reactions,* which require the presence of chlorophyll. The energy of light absorbed by chlorophyll is used for activating the electrons in chlorophyll molecules:

chlorophyll + light energy $\rightarrow$ activated chlorophyll

The light that is most effective in this reaction is primarily light in the blue-violet and red regions; green, yellow, and orange light is absorbed only very slightly. Other accessory pigments such as carotenoids assist in absorbing light in these regions of the spectrum and then pass the energy on to chlorophyll in the form of excited electrons.

When chlorophyll is activated, the two outer electrons are raised to a high energy level and then released, passing through a stepwise series of tranfers between molecules that gradually release their potential energy. Eventually, the electrons return to the chlorophyll molecule from which they were derived. Most of this released energy is utilized to separate water molecules into hydrogen (H) and hydroxyl (OH) radicals (with unpaired electrons, but not net charges). About 4 percent of the total light falling on leaves is absorbed by the chlorophyll and used in this reaction of splitting water:

activated chlorophyll + $4H_2O \rightarrow$ normal chlorophyll + $4H + 4OH$

The free hydrogen released by this process in the photosynthetic cells is temporarily attached to the electron acceptor, nicotinamide adenine dinucleotide phosphate (*NADP*). This NADP molecule receives excited electrons via the chain of chlorophyll electron transfers; and instead of passing this pair of energized electrons along to another acceptor molecule, it apparently pulls two hydrogen protons away from water and forms a temporary molecule:

$$NADP + 2H \rightarrow NAPDH_2$$

This $NADPH_2$ is now able to donate hydrogen in the process of reducing carbon dioxide to carbohydrate. In the meantime, the hydroxyl groups are recombined into new water, simultaneously with the release of molecular oxygen gas:

$$4OH \rightarrow 2H_2O + O_2 \uparrow$$

The other process of donating hydrogen to change carbon dioxide into carbohydrate takes place under either dark or light conditions and does not require the presence of chlorophyll.

Dark reactions

The process of storing the energy of sunlight in the form of carbohydrate molecules can take place in the dark in a complicated series of reactions that vary from one species of plant to another. This series of *dark reactions* has been well clarified by the Nobel prize-winning research of Melvin Calvin at the University of California, working with a unicellular alga, *Chlorella* (Figure 25–5). The basic research procedure involved suddenly introducing a radioactive isotope of carbon dioxide, namely, $C^{14}O_2$, into the culture media where the *Chlorella* algae were growing. Calvin found that after only five seconds of photosynthesis in the presence of air enriched with 5 percent $C^{14}O_2$, five different organic compounds in the algae became radioactive. By suddenly adding the radioactive gas and then killing (by boiling) the algal cells at specified intervals after exposure, it was possible to identify the series of compounds involved in the reactions that changed free carbon dioxide eventually to glucose sugar.

The basic process usually involved in the dark reactions is that free carbon dioxide diffuses into the cell and combines with ribulose, a five-carbon sugar, to form an unstable six-carbon compound. This large molecule promply splits into two smaller molecules of a three-carbon compound, phosphoglyceric acid (PGA):

$$CO_2 + C_5 \rightarrow 2C_3$$

The PGA has phosphorus added to it by the highly energetic molecule ATP. Then hydrogen atoms are added by $NADPH_2$, which, as

Calvin's experimental apparatus for studying the reactions in photosynthesis. Chlorella algae were placed in the flat center disk and removed at the desired time after the initiation of the reaction by the stopcock valve at the bottom.

we have seen, comes out of the light reactions, and the result is a three-carbon sugar, phosphoglyceraldehyde (PGAL), and water:

$$PGA + NADPH_2 + ATP \rightarrow PGAL + H_2O + NADP + ADP$$

The NADP is then free to go back and accept more hydrogens from other excited chlorophyll molecules. The ATP that provides the phosphorus atom for PGA was previously produced in the light reactions, and the ADP end product can now be returned to the light-reaction process for making more ATP from ADP and free phosphate ions.

At this point, the three-carbon sugar PGAL can be used in three different ways. Most of it is used in the synthesis of more ribulose, some of it is respired for immediate energy, and some of it can be combined to form a six-carbon sugar, *glucose*. The six-carbon glucose sugar ($C_6H_{12}O_6$) may be stored in the cell where photosynthesis is occurring or transferred to other cells throughout the body and

FIGURE 25-6

The dark reactions in photosynthesis: a summary of the Calvin cycle. Three molecules of ribulose diphosphate, a five-carbon compound, are combined with three carbon dioxide molecules, yielding six molecules of phosphoglycerate, a three-carbon compound. The series of conversions and rearrangements that follows will produce, in six revolutions of the cycle, a six-carbon sugar, such as glucose. The energy that "drives" the Calvin cycle comes from ATP and NADP produced by the light reactions. Thus the reduced form of $NADP_{red}$, for instance, is used to reduce the carbon (adds hydrogen) and is changed to the oxidized form, $NADP_{ox}$, in the process. ATP does the same reducing in a preceding step and becomes ADP.

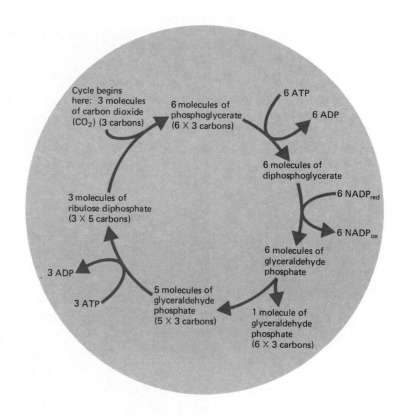

used in a variety of ways by the organism. This series of dark reactions is summarized in Figure 25–6.

This is the best available summary to date of the light and dark reactions of photosynthesis, but by no means is it a final summary; revisions may be expected in the future, with more research on the subject. It should also be remembered that the process of photosynthesis is not restricted to the higher plants but occurs in even the microscopic, single-celled algae and, in fact, in these plants the process is by far the most important as far as life across the earth is concerned. It has been estimated that *90 percent* of all photosynthesis is carried on by algae in the oceans, whereas only about 10 percent is conducted by land plants. For this reason, ecologists such as Paul R. Ehrlich have stressed the extreme danger that the world faces if oceans are polluted to the point where algal populations are seriously affected. The loss of these vast algal communities would mean a devastating loss of oxygen regeneration in the atmosphere for the use of animals, and hence probably lead to our own rapid demise.

Chemosynthesis: an alternate evolutionary pathway

A unique group of microorganisms is able to synthesize all its protoplasmic requirements from carbon dioxide, an inorganic nitrogen source, and other inorganic salts. The energy for these syntheses is derived from the oxidation of inorganic compounds rather than the light energy from the sun. The similarity of these organisms to photosynthetic organisms does extend to the fact that they use carbon dioxide as their sole carbon source, but they are not photosynthetic since all their metabolism can take place in the dark.

The first of these organisms were discovered in the latter part of the nineteenth century, when a biologist noted that some bacteria were able to live in waters that contained a high concentration of hydrogen sulfide. These so-called sulfur bacteria were found to oxidize *sulfur* to sulfate, that is, add oxygen molecules to inorganic sulfur in the presence of water and produce sulfate (SO_4) compounds. Soon two other groups of chemosynthetic bacteria were recognized (Figure 25-7). One group derives its energy from the oxidation of inorganic *nitrogen* compounds and plays an extremely important role in the movement of nitrogen within the biosphere. Some kinds are able to oxidize nitrite (NO_2) to nitrate (NO_3); others live by oxidizing ammonia to nitrite. The third group of chemosynthetic bacteria lives by oxidizing *hydrogen* to water ($2H_2 + O_2 \rightarrow 2H_2O$). All of these organisms are able to trap the small quantities of energy released by their oxidation of these inorganic raw materials, and this energy may be used in the synthesis of carbohydrates. While they seem on the surface to be bizarre forms of life, all of these bacteria contain the total range of amino acids found in all other cells, whether plant or animal; many other organic compounds such as vitamins are also present. In the total ecology of the natural world, the nitrogen bacteria play a role as fully as important as the algae of the sea do in producing oxygen (Chapter 4).

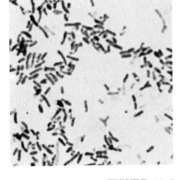

FIGURE 25-7

Chemosynthetic nitrogen-fixing bacteria, Rhizobium leguminsarum, isolated from soil.

Cellular respiration

The most important energy-releasing reactions in all plant and animal cells are those in cellular respiration. Photosynthesis binds light energy to make sugar molecules, and these sugars serve as a storage form of energy. However, the large energy-rich sugar molecules must eventually be broken down chemically by either the plant that made them or the animal that ate the plant and the energy

FIGURE 25–8

The structure of the ATP molecule.

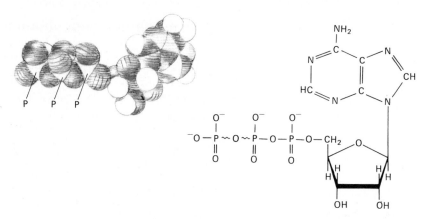

released to do useful work. The oxidative breakdown of these chemical molecules is called *respiration*. Since it occurs in every living cell, cellular respiration is a fundamental property of life.

The energy released from these cellular respiration processes is stored in a readily convertible chemical bond form in the ATP molecule. The function of ATP in temporarily storing this chemical energy may be understood better if we briefly look at its structure. ATP is composed of three types of subunits (Figure 25–8). One of these subunits we have seen already in DNA, namely, the compound *adenine*, which is a nitrogen base. Adenine may be connected to a second subunit, the five-carbon sugar, *ribose;* and attached to this ribose is *phosphate*, composed of an atom of phosphorus and surrounded by four oxygen atoms. The adenine, the ribose, and one phosphate group make up a nucleotide compound useful in DNA or RNA as well as other chemical combinations in the cell.

In ATP *three* phosphate groups are attached in a chain to the ribose sugar. The last two of these bonded phosphates are connected by so-called high energy bonds to the rest of the molecule. The high energy bond is one that will yield its energy readily when broken by an enzyme. Normally in the course of cellular energy reactions, the third or last-constructed outside phosphate bond of the ATP molecule is broken, making the molecule *adenosine diphosphate* (ADP). This ADP molecule may then be converted back into ATP again in areas where the oxidation of sugar or some other energy compound is releasing energy that needs to be captured. The primary processes involved in the breaking of the bonds of glucose require enzymes at first and will ultimately require cytochrome pigments and oxygen. At every step where energy is released, ADP is converted to ATP. The storage of the energy of these bonds in the form of ATP molecules, by this process of phosphorylation, is known as "respiration" because *oxygen* is required for the completion of the process.

The stages of respiration

The processes involved in cellular respiration are first centered in the cytoplasm and then the mitochondria, the powerhouses of the cell. Respiration occurs in three stages:

1. *Glycolysis.* In the glycolysis process, the glucose molecule produced in photosynthesis and taken in by the animal during feeding is split (Figure 25–9). This stage takes place in the cytoplasm in nine separate steps, during which two new molecules of ATP are formed from ADP. In all, little energy is produced by this process, but in some bacteria and other organisms that live under anaerobic conditions where no oxygen is present, this is their sole energy-yielding

FIGURE 25–9

Glycolysis: the breakdown of glucose units (which can be in the form of a glycogen or animal starch molecule) during cellular respiration.

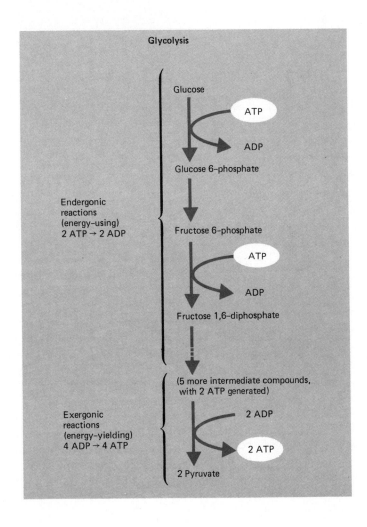

Glycolysis

Glucose

ATP

ADP

Glucose 6-phosphate

Endergonic
reactions
(energy-using)
2 ATP → 2 ADP

Fructose 6-phosphate

ATP

ADP

Fructose 1,6-diphosphate

(5 more intermediate compounds,
with 2 ATP generated)

2 ADP

2 ATP

Exergonic
reactions
(energy-yielding)
4 ADP → 4 ATP

2 Pyruvate

FIGURE 25–10

A summary diagram of the Krebs cycle reactions in cellular respiration, emphasizing the roles of the electron carriers NAD and FAD.

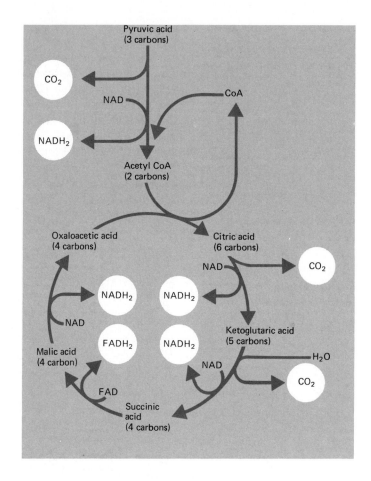

reaction. The other two steps require the presence of oxygen and must take place in the mitochondria.

2. *Krebs citric acid cycle.* In the Krebs cycle, the hydrogen atoms are removed from the carbon atoms of parts of the old glucose molecule and are combined with oxygen atoms in a complicated network of reactions (Figure 25–10). A series of carbon compounds with different numbers of carbon atoms (one being citric acid) are passed around in this overall cycle; the carbons donated by the first carbon compound are a portion of the old intact glucose molecule. These compounds are ultimately oxidized to carbon dioxide, and their hydrogen atoms are passed to electron carrier molecules such as NAD. The electrons accepted by NAD and its cohorts during this series of cyclic reactions are passed to a series of electron carriers located on the cristae of the mitochondria.

The electromagnetic radiation spectrum

The range of electromagnetic radiations in the universe, from the longest known radio waves to the shortest known cosmic rays, is called the spectrum. Approximately halfway between these two extremes lies the visible portion of the spectrum, or white light. While the sun emits ultraviolet rays of less than 2900 Å, they are absorbed by the earth's atmosphere and do not reach the surface of the planet. In the enlarged portion of the spectrum, shadings show the wavelengths of principal usefulness in photosynthetic activity.

This chart illustrates only the relative position of the various frequencies of electromagnetic radiation and does not show proper proportions of these classes.

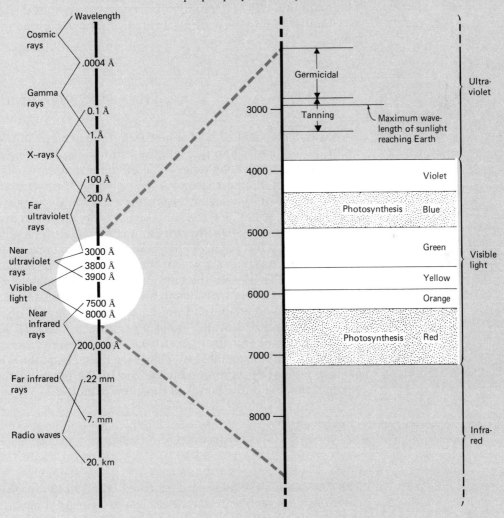

Wavelength

Cosmic rays

.0004 Å

Gamma rays

0.1 Å

1. Å

X-rays

100 Å

200 Å

Far ultraviolet rays

Near ultraviolet rays — 3000 Å
3800 Å
3900 Å

Visible light

7500 Å
8000 Å

Near infrared rays

200,000 Å

Far infrared rays

.22 mm

7. mm

Radio waves

20. km

Germicidal

3000

Tanning

Maximum wavelength of sunlight reaching Earth

Ultraviolet

4000

Violet

Photosynthesis — Blue

5000

Green

Yellow

Visible light

6000

Orange

Photosynthesis — Red

7000

8000

Infrared

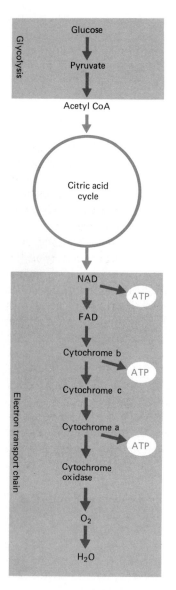

FIGURE 25–11

The electron transport chain in cellular respiration.

3. *The electron transport chain.* The electrons are passed downhill along the electron transport chain of various cytochrome molecules on the cristae, and the energy released is transferred to molecules of ATP created by binding a third phosphate group to ADP molecules (Figure 25–11). At the end of this chain of reactions, the electrons are accepted by protons, forming hydrogen atoms, and these energized hydrogen atoms are able to combine with oxygen and produce water. The terminal step — the transfer to oxygen by cytochrome oxidase — can be blocked by the cyanide ion that is more strongly attracted to cytochrome oxidase than oxygen, and once on it is not readily released. Hence a person who breathes cyanide fumes or drinks a cyanide solution will very rapidly die as cells throughout the body fail to complete the transfer of electrons in this cytochrome molecule chain.

Energy yields from respiration

A considerable yield of ATP comes from the reactions in the Krebs cycle and particularly in this latter electron transport chain, so that at the end of the whole series of cellular respiration reactions, an energy yield of 38 molecules of ATP is obtained from the complete breakdown of one molecule of glucose: 2 from glycolysis and 36 from the latter stages. Because this glucose is broken down in a series of small enzymatic steps, the energy in the molecule is able to be packaged in the form of high energy bonds in the molecules of ATP. Without the presence of oxygen, such as happens in fermentation with yeast cells, the product of glycolysis is ultimately alcohol. However, if oxygen is present, and the organism is able to use it as an electron acceptor at the end of the electron transport chain, the energy yield in terms of ATP molecules is much higher (Figure 25–12). This series of processes, when carried out to the oxygen acceptor stage, provides considerable advantages to the cell and the whole organism because of its more efficient use of the potential chemical energy available in the bonds of the glucose molecule.

Summary

The sun is the ultimate source for energy used in living systems to carry out their metabolic activities. The exchanges of energy in the living cells usually involve the movement of electrons from one

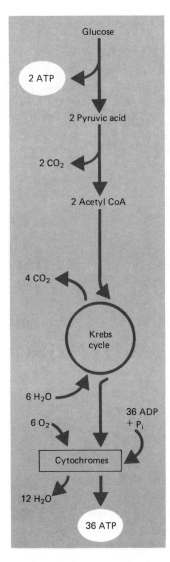

FIGURE 25–12

A summary diagram of the energy yields in formation of ATP "molecular currency" during the aerobic breakdown of glucose to carbon dioxide and water in cellular respiration.

energy level to another. On photosynthesis, the electrons of the chlorophyll molecule are excited to a higher energy state temporarily, and as they return downhill to a lower energy state, the release of energy is used to break up water into hydrogen and hydroxyl groups. This reaction can take place only in the light and in the presence of chlorophyll. The ATP molecules and NADP molecules produced during this process are used in the so-called dark reactions of photosynthesis, when carbon dioxide is reduced (hydrogen atoms added to the carbon backbone) to form a sugar, commonly glucose. This glucose sugar formed by a series of cyclical reactions provides the energy source for both plants and animals through the processes of cellular respiration. In cellular respiration, the glucose molecule is broken down in a series of small enzymatic steps, starting with glycolysis, in which the six-carbon glucose molecule is split into two three-carbon molecules. Two molecules of ATP result in this stage, the whole reaction taking place in the cytoplasm of the cell. If oxygen is present, these three-carbon molecules may enter the so-called Krebs citric acid cycle, which takes place in the mitochondria and breaks apart the old carbon groups in a series of reactions until finally carbon dioxide is produced. The high energy electrons taken off by NAD and its cohorts during the Krebs cycle are passed along the electron transport chain, a series of cytochrome pigments located on the cristae of the mitochondria. Each time a pair of electrons is passed from one acceptor to another in this chain, ATP is formed from ADP and phosphate. At the end of the entire process of cellular respiration, a total of 38 molecules of ATP are formed from the breakdown of one molecule of glucose, and most of these ATP molecules result in the two series of reactions located in the mitochondria.

26 Digestion and excretion: energy and waste control for the whole organism

The cells that compose the body of any plant or animal require adequate amounts of carbohydrates, proteins, fats, vitamins, minerals, and water. These materials are obtained in various ways from the external environment. Plants acquire nutrients such as magnesium for their chlorophyll molecules, nitrates for their proteins, and phosphates for ATP and DNA through their roots, and they acquire other needed substances such as carbon dioxide through their leaves, where photosynthesis manufactures simple sugars from carbon dioxide derived from the atmosphere and water pulled up by the roots from the soil. Some kinds of green plants, such as the carnivorous species, are able to obtain nitrogen-containing substances directly from prey items captured in highly modified leaves (Figure 26–1).

In general, the raw materials for photosynthetic reactions by the plant are few in number and are usually obtained from the atmosphere and the soil; the additionally needed carbohydrates, fats, proteins, and vitamins are manufactured by the plant cells themselves, through the energy obtained by cellular respiration of the glucose sugars produced by photosynthesis. On the other hand, animals need to obtain their raw materials from the external environment

Sundew

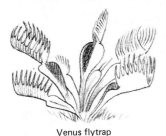

Venus flytrap

Pitcher plant

FIGURE 26–1

Some carnivorous plants found in North America.

through the food that they eat, for they lack the ability to use solar energy for nutritive or energy-producing activities. In a similar manner, heterotrophic plants such as bacteria and fungi must obtain their nutrient requirements by absorption from the external environment.

In this chapter we shall be primarily interested in seeing how digestion of raw materials occurs in human beings and how the body handles the problems of excretion of waste products, as well as maintaining the chemical and water balance essential for normal bodily activities.

The digestion and absorption of food

The function of the human digestive system and that of all other animals is to transfer food and water from the external environment to the internal parts of the body where they can be distributed to all the cells via the circulatory system. The typical digestive system includes the salivary glands, gall bladder, liver, and pancreas, and the entire gastrointestinal tract, including the mouth, esophagus, stomach, small and large intestines, and anus (Figure 26–2). Food is usually taken directly into the mouth as large pieces of material consisting of substances such as proteins and starch, which have large molecules incapable of crossing cell membrane. Thus before the substances can be utilized by body cells, these large molecules must be broken down into smaller portions such as amino acids and simple sugars.

The breaking-down process is called **digestion,** and it is accomplished by the action of acids and protein enzymes secreted into the gastrointestinal tract. The small molecules resulting from these digestive processes are able to be transported through the cell membranes of the intestinal cells; there they enter the blood and lymph circulatory systems, a process known as **absorption.** Digestion and absorption occur in a number of places along this tract while contractions of the smooth muscle composing its walls move the internal contents through the many feet of tubing.

In general, the greater the amount of food ingested, the more will be digested and absorbed. The amounts of materials absorbed from the gastrointestinal tract into the circulatory system are not regulated, and this relative lack of discrimination means that while the digestive tract supplies nutrients for the body, it does not itself control the concentrations of the various constituents going into the plasma of the blood and lymph.

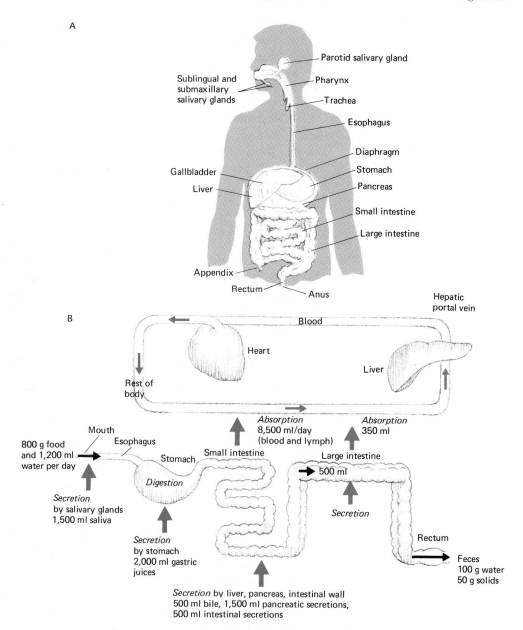

FIGURE 26-2

(A) The anatomy of the gastrointestinal system in human beings (B) A summary of daily gastrointestinal activity involving motility of food, secretion, digestion, and absorption into the circulatory system for distribution to the rest of the body.

The structure and activity of the human gastrointestinal tract

On a June day in 1822, at a trading post on the Michigan-Canadian border, a French-Canadian fur trader, Alexis St. Martin, was struck accidentally by a shotgun blast that tore a great gaping hole in his left side. A young army surgeon from a nearby fort, William Beaumont, treated him and dressed the wound as best he could. St. Martin, who was 18 at the time, survived and his wound largely healed, although a tunnel 2 1/2 inches in diameter remained open in his side, leading directly through skin and muscle into the stomach. After a period of convalescence, he was able to resume his vigorous outdoor activities, simply covering his wound with layers of gauze.

Similar openings into the stomach had been recorded in earlier times by doctors, but Beaumont was the first doctor to realize that this hole could be used to view the digestive system in action, an opportunity hitherto not grasped. While doctors knew that food was chewed and ground in the mouth and lubricated by saliva, passing from the throat to the stomach by way of a narrow 10-inch-long channel, the **esophagus,** they had no idea what happened with the food in the stomach. In fact, for centuries it was commonly believed that food simply putrefied there. Beaumont gained St. Martin's permission to study digestive activities in his stomach for the next eight years. St. Martin outlived Beaumont by 27 years, dying at age 76.

Beaumont carefully recorded his observations on stomach processes in a series of notebooks. "When he lies on the opposite side," he wrote, "I can look directly into the cavity of the Stomach, and almost see the process of digestion. . . ." One of Beaumont's techniques was to attach pieces of food to a length of thread and lower it through the open hole in the man's side into the stomach. By withdrawing the food periodically, he confirmed that the stomach secreted a powerful digestive juice that progressively digested the material. He also found that the stomach acted the same way for any diet, that food bulk as well as nutritive qualities were necessary in food intake, and that whisky and other alcoholic beverages produced deterioration or disease in the stomach. These conclusions provided the guidelines for the further extensive research on human digestive processes that followed Beaumont's pioneering observations.

Today we can summarize the gastrointestinal activity — involving the mobility of food, secretion of enzymes, digestion of large particles in the small intestine, and absorption of the small particles — as follows (Figure 26–2): The gastrointestinal tract (often called the GI tract) is a long, hollow tube running from the mouth to the anus. The tract has the same general structure throughout its length, being sur-

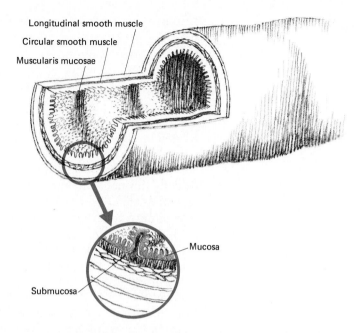

FIGURE 26–3

The structure of the mucosa lining of the small intestine.

rounded by two layers of smooth muscle. The muscle fibers of the outer layer run longitudinally along the tube, and the fibers of the inner layer run circularly. Thus the tube can contract and expand in all directions, causing the contained material to flow along the tract.

A very thin third layer of smooth muscle consisting of both longitudinal and circular fibers lies interior to these muscle sheets, and the cavity itself is lined by a thick layer of cells called the **mucosa** (Figure 26–3). The mucosa layer contains most of the gland cells that will secrete fluids and enzymes into the **lumen** (the central cavity of the gut). It also includes the surface epithelial cells involved in the absorption of material from the lumen into the blood vessels and lymphatic vessels passing through the mucosal and deeper muscle layers. The interior surface of this tube is generally not flat or smooth but is convoluted, with many ridges and valleys greatly increasing the total surface area available for absorption. This folding is most extensive in the small intestine part of the tract.

The processes in the gastrointestinal tract

The organs involved in the human digestive system include the *mouth* (including the teeth and salivary glands); the *esophagus*, which conveys the food to the *stomach*, where it is stored and mixed by muscular action; the long, *small intestine* into which the liver and pancreas secrete important digestive fluids; and the short, *large intestine* with its terminal opening to the outside, the *anus*. The GI tract is approximately 15 feet long in the living body; since it is contained within a head and trunk measuring only about four feet in length, the GI tract is folded over on itself many times. It is clear from Figure 26–2 that the food swallowed and passed through the GI tract is still *outside* the internal environment. To enter the internal environment of the body, it must be transported across the cellular walls of the GI tract and into the blood stream by the process of absorption.

The breakdown of food into simple molecules by digestion occurs because the inner walls of the GI tract and certain large glands associated with the tract secrete enzymes, which act as catalysts to increase the rate of chemical breakdown. Each enzyme is specific in its action, and the various digestive enzymes can be classed according to the basic foods on which they work.

With carbohydrates, including the sugars and starches, digestion starts in the mouth. Here the primary function of the teeth is to bite off and grind down chunks of food into pieces small enough to be swallowed. The incisors of an adult man can exert a force of 25 to 50 pounds and his molars can exert a force up to 200 pounds. Prolonged chewing of food does not appreciably alter the rate at which it is digested and absorbed, although, of course, it increases pleasure through the sense of taste. Thorough chewing, however, is important in order to prevent large particles of food from entering and possibly blocking the esophagus, causing choking if a particle lodges over the tracheal entrance into the lungs.

The mouth and esophagus

Saliva, secreted by three pairs of glands in the mouth area and containing the enzyme **amylase,** dissolves some of the molecules in the food particles. Amylase catalyzes the breakdown of polysaccharides into simpler disaccharides. While salivary amylase starts the digestive process in the mouth, food does not remain there long enough for much digestion to occur. The amylase will continue working on these complex starches even in the stomach, until inhibited by the hydrochloric acid there. During the course of a day, between 1 and 2 liters (2.2 to 4.4 quarts) of saliva are secreted. Most of this is swallowed and later reabsorbed into the circulation across the walls of the digestive tract. The most potent stimuli for salivary secretion are acid solutions, such as fruit juices and lemons, which may accelerate the secretion rate to 7 or 8 milliliters (ml) of saliva per minute.

The actual swallowing of the food down the throat starts when the tongue forces a mass of food into the rear of the mouth, where pres-

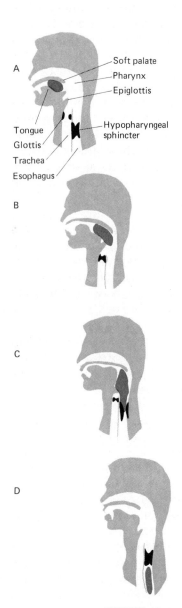

A
- Soft palate
- Pharynx
- Epiglottis

Tongue
Glottis
Trachea
Esophagus
- Hypopharyngeal sphincter

B

C

D

FIGURE 26-4

The movement of a bolus of food through the pharynx and upper esophagus during swallowing.

sure receptors in the wall of the pharynx are stimulated (Figure 26–4). These receptors send impulses to the swallowing center in the brain, which causes some 25 different skeletal muscles in the throat area to initiate swallowing. Once within the esophagus, the mass of food is moved along by waves of contraction that pass along the walls of the esophagus and move toward the stomach. These waves are known as **peristaltic waves** and are found in other parts of the GI tract.

At any particular point in the esophagus, the contraction and relaxation of the wall lasts about three to seven seconds; and the wave moves toward the stomach at a rate of about two to four centimeters a second, taking about nine seconds to travel the length of the esophagus. More forceful peristaltic waves may be generated if a larger sticky mass of food such as peanut butter becomes stuck in the initial wave.

The *gastroesophageal sphincter,* the last four centimeters of the esophagus before it enters the stomach, remains contracted when swallowing is not taking place and thus prevents the contents of the stomach from moving up into the esophagus. When a person gets motion sickness or has extensive stomach gas, the mass of food being digested in the stomach may be forced up through this sphincter. However, under normal conditions, as the peristaltic wave begins in the esophagus, this sphincter relaxes and allows the mass of food on its arrival to enter the stomach. When the mass is passed into the stomach, the sphincter contracts, resealing the junction.

Interestingly, during the last five months of pregnancy, because of the increased pressures in the abdominal cavity, there is a tendency for some of the contents of the stomach to be forced up into the esophagus. The hydrochloric acid from the stomach contents irritates the walls of the esophagus, causing contractile spasms of the smooth muscle layer, which are experienced as pain. This pain is known commonly as *heartburn* because the sensation appears to be located in the region over the heart. In the last weeks of pregnancy, heartburn often subsides as the uterus descends prior to delivery, decreasing the pressure on the abdominal organs. Overeating can cause the same type of pressure on stomach contents and result in heartburn.

The stomach

The stomach forms a large chamber between the end of the esophagus and the beginning of the small intestine. Probably the most important function of the stomach is its regulation of the rate at which ingested material enters the small intestine, where most digestion and absorption actually occur. Should the stomach be removed surgically and the esophagus attached directly to the small intestine, food moves so rapidly through the intestine that only a fraction of it has time to be digested and absorbed. The stomach also secretes hydrochloric acid and various digestive enzymes that, along with the salivary amylase, partially digest the food before it enters the small intestine.

The secretions of the stomach and the mechanical contractile activity of its smooth muscle walls reduce the size of large lumps of food and produce a thick mixture of food particles and secreted fluids known as **chyme.** The lower portion of the stomach has thicker and more powerful muscle layers, and their activity primarily controls the emptying of the stomach contents into the small intestine.

At the base of the stomach and the beginning of the **duodenum,** the first portion of the small intestine, is a ring of smooth muscle and connective tissue called the *pyloric sphincter.* This sphincter prevents material from the stomach from moving into the duodenum, although at rest the pressure in the cavity of the duodenum is normally great enough to prevent this material from moving. However, the sphincter also ensures that only small amounts of food at a time pass into the duodenum (Figure 26–5).

The stomach will empty at a rate proportional to the volume of material in it at any given time. Thus the more food in the stomach, the faster is the rate of gastric emptying. Fats, which are particularly hard to digest and absorb compared with most other components of food, inhibit gastric movement, giving more time for digestion and absorption. Thus substances with high fat content such as eggs and milk may inhibit gastric emptying to the point where some of the meal may be found in the stomach some six hours after swallowing. In contrast, a meal of meat and potatoes, which contains largely proteins and carbohydrates, may empty out of the stomach in four hours or less. For these reasons, coaches will usually have their football players eat a late breakfast or early lunch meal of steak and carbohydrates on the day of a game.

The stomach secretes a large amount of hydrochloric acid, which is passed on into the duodenum, where it is neutralized by sodium bicarbonate secreted by the *pancreas.* The normal human stomach secretes about two liters of hydrochloric acid solution a day. The very high acidity generated here denatures proteins and breaks up molecular bonds, particularly in connective tissue and cell membranes. Thus hydrochloric acid continues the process begun by chewing, namely, reducing large particles of food to smaller particles and eventually individual molecules. However, the acid is not

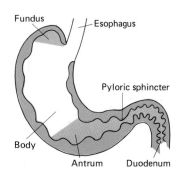

FIGURE 26–5

The general anatomy of the human stomach.

particularly efficient at breaking proteins and polysaccharides into amino acids and glucose. It does activate some of the enzymes secreted by the stomach and also serves to kill most of the bacteria that enter along with food. This last process is not entirely effective, and some bacteria do enter the intestinal tract; there they may release disease-producing toxins that reach the blood even after the bacteria have been killed.

The chief protein-breaking enzyme secreted in gastric juices in the stomach is **pepsin** (Figure 26-6), which catalyzes the splitting of bonds between particular types of amino acids in protein chains. Through its activity the products of protein digestion are primarily small peptide fragments composed of several linked amino acids. The activity of this enzyme is dependent on an environment having high hydrogen ion concentration, which is provided by the hydrochloric acid in the stomach. Later in the duodenum, where the stomach acid is neutralized by bicarbonate ions from the pancreas, pepsin becomes inactive.

Very little food is absorbed into the blood from the stomach because there are no special transport systems for salts, amino acids, and sugars in the stomach walls like those found in the intestine. In fact, most of the molecules reaching the stomach are still large,

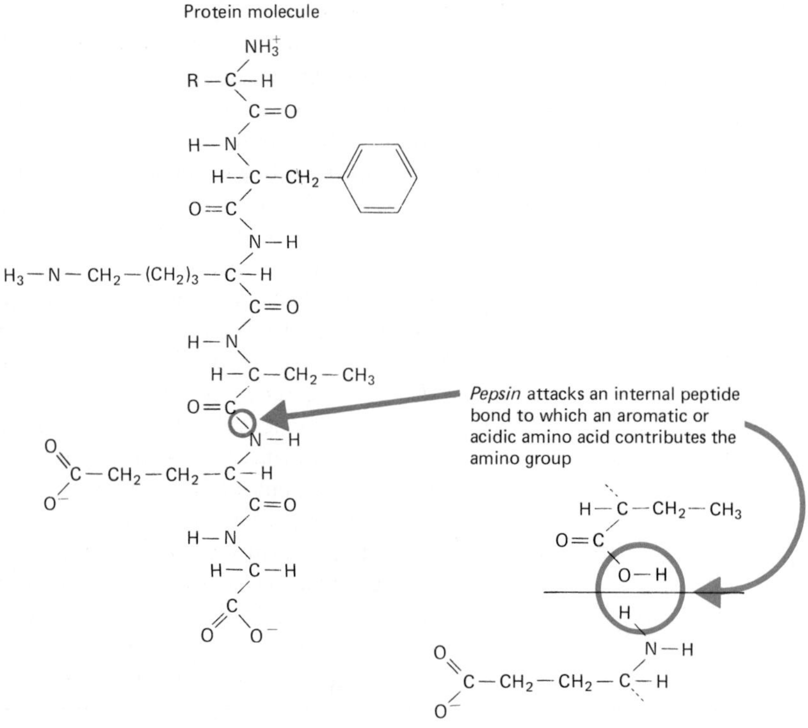

FIGURE 26-6

The action of the proteolytic enzyme pepsin in digesting proteins. Pepsin breaks the long, complex protein molecule into fragments by hydrolyzing (adding water to) peptide bonds between specific amino acids in the interior of the molecule. Other enzymes then complete the digestive process by chewing in from the ends of the fragments, removing one amino acid at a time.

highly charged, and ionized, so that they cannot diffuse across cell membranes. However, small lipid molecules have some degree of water solubility and are absorbed by the membranes lining the stomach.

Alcohol molecules can be readily absorbed and reach the blood stream through the walls of the stomach. However, alcohol is absorbed much more rapidly from the intestinal tract, which has a greater membrane surface available for absorption. If one drinks a glass of milk before a cocktail party or eats a lot of cheese dips and high-fat-content hors d'oeuvres while drinking, the rate of alcohol absorption is inhibited by the fats. However, this does not stop the ultimate absorption of the alcohol and its deleterious effects on the functioning of the brain.

Aspirin is also absorbed directly across the walls of the stomach. This weak acid (acetylsalicylic acid) becomes converted to its non-ionized form, which is lipid-soluble and can cross the cell membranes lining the stomach. Once it reaches the inside of these lining cells, the weak acid again ionizes, liberating a hydrogen ion that tends to make the cell interior more acid. If too much aspirin enters in a short time, the intercellular acidity may rise sufficiently to damage the cell. A few people are extremely sensitive to these damaging effects of aspirin on the gastric mucosa cells and may develop severe hemorrhages. The beneficial effects of aspirin occur in the blood vessels of the brain, where it is believed to primarily function in relieving headaches by relaxing the walls of constricted arteries and capillaries (constricted because of tension or excessive fatigue).

The contributions of the pancreas

Just below the stomach in physical proximity lies the **pancreas** (Figure 26–7). While not in the GI tract itself, this organ secretes two solutions involved in the digestive process, one containing a high concentration of sodium bicarbonate and the other containing a large number of digestive enzymes. These solutions are secreted into a series of ducts that converge into a single pancreatic duct, which joins the bile duct from the liver just before entering the duodenum. During the course of a day, the pancreas may secrete as much as 1,500 to 2,000 ml of sodium bicarbonate solution into the duodenum, netrualizing the acid entering from the stomach.

If a person suffers the loss of large quantities of bicarbonate ions from the intestinal tract, as during periods of prolonged diarrhea, the contents of the duodenum have a higher acid content, which leads to a greater acidification of the blood. Normally, however, there is no net change in the acid-base balance of the blood. A person who has excess acid because of an upset stomach may take a bicarbonate

FIGURE 26–7

The structure of the pancreas, with a detailed enlargement of a portion of the gland areas.

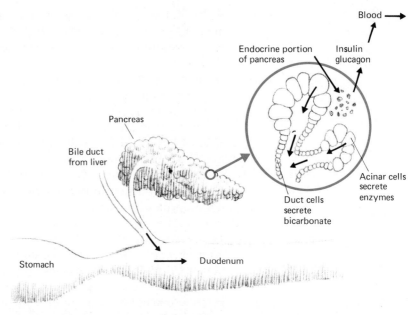

solution orally and the bicarbonate ions neutralize the acid in the stomach, just as bicarbonate ions secreted by the pancreas neutralize acid in the duodenum. Obviously, by taking bicarbonate orally, there will be a net increase of bicarbonate ions in the body and the blood will become alkaline. In this case, the acid-base balance of the body will be restored by the elimination of the excess bicarbonate ions through the kidneys.

The great number of enzymes secreted by the pancreas that are of crucial digestive importance include *trypsin* and *chymotrypsin*, which digest proteins, and other enzymes that break down proteins, lipids, polysaccharides, and the nucleic acids (Table 26–1). Another important secretion into the intestinal tract at this point is **bile,** coming from the liver cells at the rate of 250 to 1,000 ml per day. This bile empties into the duodenum from the bile duct, which comes from the gall bladder on the underside of the liver. The bile salts are the most important components of bile since they are involved in the digestion and absorption of fats in the duodenum. These salts emulsify fats (break them up into tiny globules, just as detergents do) so that the lipase enzymes can work on their digestion.

The small intestine

In the small intestine, which consists of about 9 feet of tubing (1 1/2 inches in diameter) coiled in the abdomen and leading from the stomach to the large intestine, most of the digestion and absorption in the entire GI tract occurs. (The longer lengths of the small intestine that have been reported — 20 feet or more — were taken from dead human beings. After death the smooth muscle in the intestine walls relaxes.) The food material is moved along the intestinal track by weak, contractile waves that travel only very short distances and at the rate of 1 to 2 centimeters per second toward the large intestine. The activities in the first part of the small intestine, the duodenum, are largely concerned with the pancreatic secretions.

Just before the large intestine another sphincter appears, which separates the last portion of the small intestine (the ileum) from the large intestine. Normally this is closed, but after a meal the contractions of the ileum increase, and the sphincter relaxes each time the last portion of the ileum contracts, thus allowing the food mass and waste to empty into the large intestine.

Anatomists and physiologists have divided the 9 feet of small intestine into three sections: the **duodenum**, about 8 inches long; the **jejunum**, about 3 feet long; and the **ileum**, about 5 feet long (Figure

TABLE 26–1

The pancreatic enzymes

Enzyme	Substrate	Action
Trypsin, chymotrypsin	**Proteins**	**Breaks bond between amino acids in the interior of long protein chains, forming peptide fragments**
Carboxypeptidase	**Proteins**	**Splits off the terminal amino acid from the end of a polypeptide chain having a free (unbonded) carboxyl group**
Amylase	**Polysaccharides, especially starch**	**Similar to salivary amylase: splits polysaccharides into simple sugars such as glucose and maltose**
Lipase	**Lipids (triglycerides)**	**Splits fat molecules into glycerol and free fatty acids**
Ribonuclease, deoxyribonuclease	**RNA, DNA**	**Splits nucleic acids into free single nucleotides**

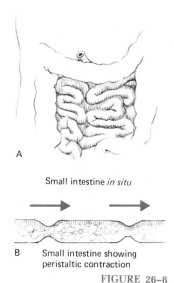

Small intestine *in situ*

B   Small intestine showing
peristaltic contraction

FIGURE 26-8

*The small intestine: (A) view
of the small intestine in situ;
(B) peristalsis movements
along the small intestine.*

FIGURE 26-9

*The internal structure of a
human intestinal villus.*

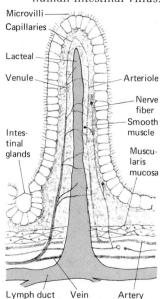

26-8). The mucosa of the small intestine is folded, and the surface of these folds is further increased by fingerlike projections known as **villi.** The surface of each villus (Latin for "shaggy hair") is covered by hundreds of epithelial cells, and the surface of each cell is even greater because of small projections known as microvilli. The total surface area of the human small intestine through this folded mucosa with its villi and microvilli has been estimated to be about 2,000 square feet, or approximately the area of a singles tennis court. The structure of a typical villus is illustrated in Figure 26-9. The rich capillary network of the mucosa enables the circulatory system to promptly pick up materials diffusing through the microvilli and surface cells into the interior of each villus.

Glands located in the mucosa of the small intestine apparently secrete about 2,000 ml of solution per day into the gut lumen, but the functions of these intestinal secretions are unknown. They were once believed to contain digestive enzymes, but it now appears that the enzymes found in the secretions are derived from the disintegration of the epithelial cells or other sources.

As we have said, most of the digestion and absorption of food and water occurs in the small intestine. In fact, by about the time that the mass of food has reached the middle of the jejunum, most of the contents of the small intestine have been absorbed. It is possible for a surgeon to remove about 50 percent of the small intestine without interfering with the digestive and absorptive processes. In people with chronic overweight problems, sometimes a portion of the small intestine is removed to reduce the amount of food actually digested and absorbed by the person. This usually results in a major weight loss. A great bulk of fluid is also absorbed across the walls of the small intestine and, in fact, only about 6 percent of the total volume of 9,000 ml (equivalent to about 7.5 quarts and weighing about 8.3 pounds) of ingested food and secreted fluids is passed on into the large intestine each day. These facts are rather astonishing when we consider that the average adult consumes about 2 pounds of solid food and 2.5 pounds of water per day (approximately 2,000 ml in total volume). This means that about 80 percent of the 22 pounds that the GI tract handles daily is merely secreted fluids from within the body.

The large intestine

When the processed food moves into the large intestine, or **colon,** it passes down a tube about 2.5 inches in diameter that forms the last four feet of the gastrointestinal tract (Figure 26-10). A blind-ended pouch, the **cecum,** occurs below the junction of the small and large

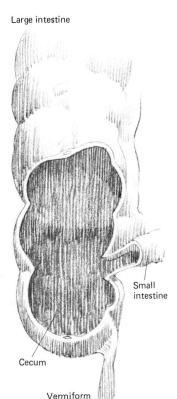

Large intestine

Small intestine

Cecum

Vermiform appendix

FIGURE 26–10

The large intestine, cecum, and vermiform (wormlike) appendix in human beings.

intestines. The **appendix,** a small fingerlike projection from the terminal point of the cecum, has no known function. In herbivorous animals, the cecum holds resident populations of bacteria that aid in digestion of cellulose and other complex carbohydrate compounds taken into the digestive tract.

The end of the descending colon empties into the rectum. No digestive enzymes are secreted by the large intestine, and its primary function is to store and control the release of fecal material. It is responsible for the absorption of only about 4 percent of the total intestinal contents per day. However, an important absorptive process in the large intestine is the active transport of sodium from the gut cavity to the blood and the simultaneous reabsorption of water. If fecal material remains in the large intestine for a long time, almost all the water is reabsorbed, leaving behind dry fecal pellets that cause *constipation,* the inability to move the bowels.

Intestinal gas is produced in the large intestine by bacteria, which have primarily anaerobic metabolism (partial breakdown of molecules without the terminal stops requiring oxygen; see Chapter 25) in this area of the gut because of the low oxygen content of the rear gut cavity. Bacterial fermentation produces gas in the colon at the rate of about 400 to 700 milliliters per day. This gas is largely a mixture of nitrogen and carbon dioxide, with small amounts of the inflammable gases hydrogen, methane, and hydrogen sulfide producing the odors of fecal material. With some organisms such as cows, the quantity of methane gas and hydrogen emitted by fermenting manure is sufficient to be at times a combustible source of energy.

Approximately 150 grams of feces are eliminated from the human body each day. This fecal matter consists of about 100 grams of water and 50 grams of solid matter, the latter being made up mostly of bacteria, undigested cell debris such as cellulose, bile pigments, and small amounts of salt. The exit of this material from the rectum is controlled by the internal and external anal sphincter muscles. The latter sphincter is under voluntary control from the brain, allowing a person to delay defecation as necessary.

Apparently there is no physiological necessity for having regular bowel movements, though constipation (the condition of having defecation only at prolonged intervals) can cause considerable pain and physiological distress. Nevertheless, toxic products are not produced by the retention of fecal material and bacteria in the large intestine. In fact, in several cases where defecation has been prevented for a year or more by blockage of the rectum, no ill effects from accumulated feces were noted except for the discomfort of carrying around the extra weight of 50 to 100 pounds of feces retained in the large intestine. Laxatives, which can be used occasionally to relieve constipation, act to stimulate elimination either by lubricating the hard, dry fecal material and easing defecation or by increasing the contractile activity of the intestinal muscles.

Excretion and regulation of liquid waste materials: the kidney

The gastrointestinal tract takes care of egestion of solid waste materials from the rectum; the primary regulation of salts and water in the body is carried out by the kidneys, which filter these materials from the blood (Figure 26–11). Water and salts enter the blood in the intestinal linings of the gut. The dissolved substances are transported in the blood to the kidneys, where they are passed through about a million tiny tubes called **nephrons.** These tubes are the actual sites of urine formation.

One end of each nephron is enlarged into a funnellike structure called *Bowman's capsule;* the other end empties into a larger tube called the *collecting tubule.* The water is carried to the area of the capsule by means of a capillary bed (called the *glomerulus*), which is contained within the capsule. This arrangement is shown in detail in Figure 26–11. After the blood leaves the glomerulus, it enters another capillary bed that surrounds the cells lining the tubules.

About one-fifth of the blood plasma filters through the capillary walls in the glomerular capillaries and is collected in the funnel-like structure of the capsule. The other four-fifths of the plasma continues on in the capillary vessels toward the tubular capillaries.

The pores in the glomerular walls are too small to permit the red blood cells or large protein molecules to pass through them, only the plasma itself. The filtered fluid then flows from the capsule into the tubule and on toward the collecting ducts. The fluid *entering* the tubule is very much like plasma, but the fluid *leaving* the tubule resembles urine and passes into the ureter; the composition of the fluid has been changed as it flows down the tubule. The primary function of the blood capillaries surrounding the tubular walls of each nephron is to reabsorb the many substances necessary for the body, such as water and salts, that were taken out en masse in the glomerulus and passed down the collecting tubule. Additional material such as small molecular wastes can also be passed from the blood to the tubules at this point. Thus the altered fluid that reaches the end of the collecting duct enters the ureter as urine. The blood cells, the large protein molecules, and other food substances are held

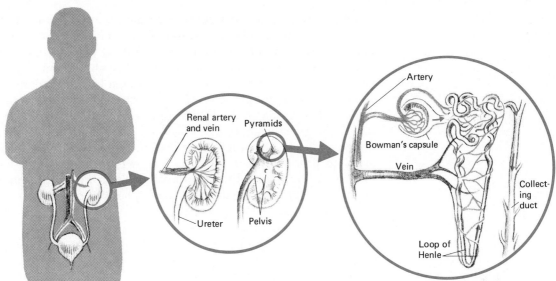

FIGURE 26–11

The structural details of the kidney portion of the human urinary system. The urine, formed in each kidney, collects in the renal pelvis and then flows through the ureter into the bladder, from which it is eliminated via the urethra. Note that the internal structure shows regional differentiation. The outer portion, which has a granular appearance, contains all the glomeruli (the cup portions of the nephrons). The collecting ducts form a large portion of the inner kidney, giving it a striped pyramid appearance, and drain urine into the renal pelvis. At right is the basic structure of a nephron (Bowman's capsule, loop of Henle, and collecting duct) and its associated circulatory vessels.

back in the blood and never appear in the urine because they are unable to pass through the filtration system in the glomerulus.

Besides regulating salts and water, the kidney also helps regulate the acidity of the blood. Cells in some of the tubules are able to transport hydrogen ions from the blood to the tubular fluid, and thus dispose of excess acid. The overall role of the kidney in excreting waste products is extremely important. These waste products become highly concentrated in the urine and can be safely passed out of the body without poisoning its cells, as would happen if they continued on through the circulatory system to the other parts of the body. Thus the kidney helps regulate the internal environment by forming and excreting urine, which contains waste products and variable amounts of salts, water, and acid. The formation of this urine in the kidney takes place in two stages as we have seen: filtration and selective reabsorption (or the opposite, secretion).

Summary

The function of the digestive system is to transfer food and water from the external environment to the internal parts of the body, where they can be distributed to all the cells via the circulatory system. Food is broken down in digestion into smaller portions, which can be absorbed into the cells lining the intestinal tract and enter the blood and lymph circulatory systems. The digestive system includes the salivary glands, gall bladder, liver, and pancreas, and the entire gastrointestinal tract, including the mouth, esophagus, stomach, small intestine (duodenum, jejunum, ileum), large intestine, and anus. Many enzymes and other secretions participate in the digestive process from mouth to duodenum. Most absorption takes place in the small intestine, where the surface area is greatly increased by the villi and microvilli. The large intestine stores and controls the release of fecal material, and its walls also actively absorb sodium and water from the gut cavity. The kidney functions in filtering salts, water, and excess hydrogen ions from the blood, and thus maintaining their proper level in the body. The nephron units of the kidney also filter out dissolved waste products, which are expelled as urine in the collecting ducts and, finally, out the ureter.

27 Internal transport in plants and animals

Whether we are looking at the production of photosynthetic food products in plants or the digestion of food particles in animals, the internal transport of nutrient materials as well as gases and fluids is of prime importance to the living cell and the whole organism. In the lower plants and animals, especially the one-celled species such as algae and protozoa, the movement of substances is relatively simple. Here the primary method by which substances are transported between the areas where they are synthesized or acquired and the areas where they are used in metabolic processes and eventually eliminated is simple physical **diffusion** (Figure 27–1).

The process of diffusion involves substances moving from an area of high concentration to an area of low concentration through random thermal agitation. In other words, particles of materials move about by random molecular motion (which is accelerated at higher temperatures) to distribute themselves equally throughout the fluid substrate in the typical plant or animal cell. When an organism is composed of only a few cells laid end to end, the process of diffusion can serve to transport these substances from cell to cell, providing there are ways for them to pass through the cell membranes. Thus even in higher plants diffusion plays an important part in the procurement of water by the root cells from the surrounding soil.

FIGURE 27–1

The diffusion of solute molecules from an area of initially high concentration to equal distribution throughout the container of water.

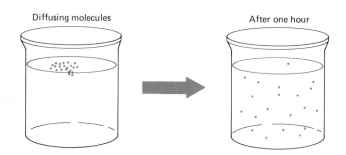

Diffusing molecules   After one hour

Diffusion also is of great importance in bringing carbon dioxide gas from the air into holes through the bottom of leaves to the internal layer of cells where photosynthesis actually occurs.

Unfortunately, diffusion is a slow process and its rate depends on the temperature of the medium. At cold temperatures it can be very slow, and over long distances, even at high temperatures, diffusion is a relatively inefficient transport mechanism. Thus it is not surprising that animals and plants that have higher metabolic rates for an active life have developed other means of transport. This need for better transport does not necessarily depend on size; some tapeworms 70 feet long or more depend on diffusion of food particles through their flattened bodies to bring in a food supply from the host's digestive tract. Nevertheless, most higher plants and animals have some sort of a circulatory system that moves water and dissolved materials through the body. We shall examine the physical structures involved in these internal transport systems, as well as the mechanisms that run them.

Transpiration: the movement of water and minerals in higher plants

In the so-called vascular plants, there are two types of circulatory tissue, the xylem and phloem (in Chapter 20 we discussed these tissues in terms of structure). These specialized internal-transport tissues supply plants as diverse as one-inch ferns and 300-foot California redwoods with the materials and water necessary for cells in all parts of the plant.

The *xylem* cells are primarily responsible for movement of water and dissolved minerals. These tissues are grouped in bundles of long cells laid end to end. The xylem bundles may be scattered throughout the stem structure, as in herbaceous plants with soft and succulent stems, or may be arranged in a continuous hollow cylinder, as in the woody plants.

In conifers and woody flowering plants, particularly in temperate-zone areas, the annual growth of new xylem cells produces a series of concentric annual rings of variously sized cells that are clearly visible when a cross section is cut through the stem. During the spring, the xylem cells produced are quite large; during the summer, growth slows and the new cells are somewhat smaller. Since each ring is made up of an inner area of this large-celled spring wood and an outer area of smaller-celled summer wood, a reasonably accurate estimate of the age of a tree can be made by counting the number of rings from the center out to the lateral meristem cells. In the older rings of xylem toward the center of the stem, the conducting cells become plugged with age, through the deposition of pigments, resins, gums, and tannins. While these materials prevent the older xylem cells from functioning in transport, they still remain important as a strong, supportive component of the tree.

The actual mechanism of movement of water and dissolved minerals in these xylem cells is a particularly fascinating problem. Considerable force is needed to translocate water into the topmost branches of a tree as high as 300 to 400 feet above ground level. If one applies a pressure of one atmosphere, it is possible to support a column of water approximately 34 feet high at sea level. If one wanted to support a column 400 feet high, a pressure of about 12 atmospheres would be required. Since the rate of upward-moving water is sometimes as fast as 40 or more inches a minute, occurring in a xylem cell system that offers some frictional resistance to the movement of the water, it has been calculated that a total force of at least 30 atmospheres is necessary to move water to the top of the highest tree.

Thus it is impossible for a simple physical force such as capillarity to move the sap this distance. (In a tiny capillary tube, water will creep up the sides of the glass and then surface tension will pull the center part upward as well.) The force required is many times that which capillarity could develop.

Various theories have been developed by plant physiologists to describe how this great pressure for the ascent of sap is generated, but no single theory has been accepted by all botanists. The best present theory is called the *transpiration*, or *cohesion pull*, *theory* (Figure 27–2), first developed by an Irish botanist, H. H. Dixon, and his physicist colleague, J. A. Joly, in 1894.

Starting at the base of the plant, a fair amount of pushing force would seem to develop by the process of simple diffusion of water across the tissue of the root, from the soil into the inner cells. How-

FIGURE 27–2

The transpiration-mediated transport of water through the vessel cells (at left) of the xylem. Water enters through the root hairs and other surfaces of the young root by diffusion, passes through the conducting vessels in the stem, and evaporates from the cell surfaces inside the leaf. The cohesion of the water molecules keeps each water column intact, and the transpiration loss of water from the leaf is believed to provide most of the "pull" on the upward rise of water in the plant's xylem tissues.

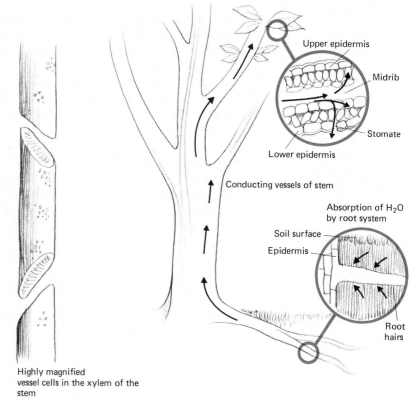

Highly magnified
vessel cells in the xylem of the
stem

ever, some active transport requiring the use of energy is necessary. To develop this pushing force (called *root pressure*), activity by living cells in the root is needed; for if the root cells are killed and diffusion continues to take place through the cell walls, root pressure disappears. Investigators have shown that the living root can develop pressures only as high as 6 to 10 atmospheres. Thus this pressure is probably not the principal motive force for the ascent of sap.

Dixon and Joly's primary contribution to the solution of the xylem transport problem is the hypothesis of water being pulled upward by transpiration from the leaves. They suggested that the water evaporating out of leaf cells raises the concentration of dissolved substances there, and consequently more water needs to be taken up from adjoining cells. These cells in turn withdraw water from cells adjacent to them. From the veins in the leaf, therefore, a pulling pressure extends all the way down the xylem to the roots, with the removal of water through evaporation from the leaves or through metabolic use *pulling* the column of water in the xylem tissue upward. The motive force of the water column is said to be primarily due to transpiration, then, which includes evaporation of water from leaf cells and the removal of water present in the leaf xylem for use in photosynthesis, growth, and other metabolic processes in the leaf as a whole organ.

The continuity—that is, high surface tension—between the water molecules throughout the system from leaf cell to root is essential for the xylem pathway to function properly. The great cohesive strength of water between the individual water molecules can reach as high as 15,000 atmospheres on a theoretical basis, and as high as 300 atmospheres in actual experimental checks of this theory. Thus it is apparent that the tensile strength of water is probably high enough to support the transpiration theory of the rise of sap. If the water in the xylem tissue in the roots is under this pulling force, water would tend to be drawn from the soil across the root tissues into the xylem with an active pull, rather than merely by diffusion. Thus the Dixon-Joly theory, which explains the movement of water and dissolved substances throughout the plant, is generally satisfying, although scientists continue the search for the best and most complete explanation for this still unanswered question: how water and minerals are moved actively through the vascular tissue.

Translocation in the phloem

The other form of vascular tissue in the higher plants, the *phloem*, moves sugar and other solute substances through the plant in all directions, not solely from root to leaf. The organic solutes consist of carbohydrates such as sucrose and organic nitrogen compounds, as well as some fats and related compounds. Inorganic ions such as those of calcium, phosphorus, and sulfur are also translocated from positions high in the plant to other locations through the phloem. All of these ions are translocated primarily through the xylem in their upward journey from the roots to the leaves. The problem of phloem function, therefore, is primarily concerned with how organic solutes are transported both up and down the plant, although the downward transport of minerals is also of importance.

Although several hypotheses on phloem functioning have been advanced, none completely satisfactorily answers the question of how the mechanism of flow actually occurs. One hypothesis is that *cytoplasmic streaming* between the open ends of sieve-celled plates of adjoining phloem cells allows materials to be carried the length of each cell, passed on into the next cell in the tube, and picked up by streaming cytoplasm in that cell (Figure 27–3). The passage between

FIGURE 27–3

The translocation of nutrient substances in the phloem: the cytoplasmic streaming hypothesis. This hypothesis assumes that solute particles diffusing into phloem sieve cells from other tissues (e.g., sugar molecules from photosynthetic leaf cells) are picked up by the streaming cytoplasm within each cell and are carried from one end of the cell to another. There the particles are believed to diffuse into the next cell through the tiny extensions of protoplasm in the holes of the sieve cell plates, and are picked up once again by the circulating cytoplasm.

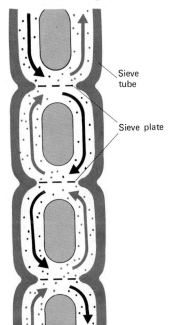

Sieve tube

Sieve plate

the sieve-plate ends of these adjacent phloem cells would be by simple diffusion, though it could involve active work by the cells (requiring energy). However, the velocities of streaming, if indeed it does occur in all phloem cells of this sieve-plate type, do not support the rapid known rates of solute movement through these phloem tubes.

A more widely accepted hypothesis is the *pressure flow*, or *mass flow*, *theory*, in which water and solutes flow through the phloem tubes along a pressure gradient. Cells high in the plants could achieve high concentrations of substances such as sugars through local photosynthesis, and thus water would be attracted to these cells through diffusion in an attempt to equalize osmotic pressure throughout the plant. This pressure tends to force substances from one phloem cell into the next cell, and hence the upper portions of the phloem tubes would receive material forced up by this mass flow.

Low in the plant or in actively growing tissues at the ends of branches, sugars are being used up for metabolic energy and hence the concentrations of these dissolved substances in the phloem tubes are lowered. A massive uptake of water by these cells is necessary because they already tend to lose water to the parts of the plant that have a higher concentration of sugar.

On the other hand, because the openings in the sieve plates between successive phloem cells are very tiny and because the protoplasm in the phloem cells seems to be rather viscous, there ought to be great resistance to mass flow. Thus while the mass flow hypothesis is the best concept proposed to date, botanists and plant physiologists are still searching for a completely satisfactory explanation of how sugar and other solute substances move around in the cellular sap.

Human circulation

Throughout the animal kingdom above the simple protozoan level, there are circulation systems for the internal transport of food, gases, and other substances. Unlike the plants, most of the higher animals have a true circulatory system in which blood is moved around the body in a fairly definite pathway. Most animals have some sort of pumping device, called a *heart*, but the blood does not necessarily flow through closed vessels during the entire circulatory course. In animals such as insects and other arthropods, as well as most

FIGURE 27–4

The open circulatory system
of an insect.

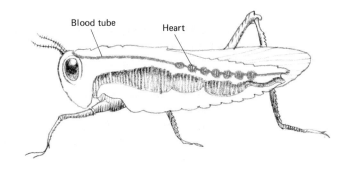

Blood tube   Heart

FIGURE 27–4

The open circulatory system
of an insect.

mollusks, there is an *open circulatory system*, where vessels open
into huge blood sinuses (or blood cavities) and allow the blood to
slosh unrestrictedly around the internal organs, being picked up
again by vessels that transport the blood to the heart (Figure 27–4).
Human beings and other vertebrates have a *closed circulatory system*, where the system of vessels extending out from the heart is continuous throughout the body and returns the circulating fluid, the
blood, to the pumping heart. Here the same small amount of blood is
repeatedly circulated. Let us look now at the human circulatory system in more detail.

The discovery of circulation

FIGURE 27–5

William Harvey.

The circulation of blood, which is the key to our basic understanding
of the entire circulatory system, was discovered only three centuries
ago. While early scholars in Greek and Egyptian cultures were aware
of the pumping action of the heart and the fact that vessels originate
in the heart, linking it to the rest of the body, they had no real understanding of the precise nature of the fluids being pumped through
these vessels and where the blood ends up in the body.

William Harvey, a seventeenth-century English scientist, was the
first to discover that arteries function as outgoing channels for the
blood and veins as incoming channels for the blood to the heart (Figure 27–5). His major achievement, after some two decades of vivisec-
tions of 15 different species of animals, was to prove that the blood
circulates as a result of the heart's mechanical action. He determined
the volume of blood pumped by the heart per hour and reckoned that
within an hour the heart pumps many times the total volume of
blood present in the entire body of a husky man. Thus, he concluded, the heart must pump the same blood again and again, and he
suggested that there might be a circular movement of the blood.

Today we know that there is a continuous network of vessels extending out from the heart in the form of the arteries, which eventually break down in the tissues into tiny **capillaries,** the smallest of all blood vessels—so slender that 10 of them together are no thicker than a hair. In the capillary network the exchange of materials occurs between the blood and the surrounding tissue cells. At this point, oxygen leaves the red blood cells for the tissues and waste products are put into the blood from the tissues; the latter chemicals then are circulated through the excretory organs, such as the kidneys, to dispose of these waste products.

The entire system moving the blood is called the *cardiovascular system.* The circulation of the blood in the body of a large mammal, such as a human being, is relatively simple. Within the heart there are four chambers: each of the two upper parts of the heart is called an **atrium;** each of the lower, more muscular parts is called a **ventricle** (Figure 27–6). The two vertically separated halves of the heart actually pump blood out through separate circuits. Blood is pumped by way of one circuit (the *pulmonary circulation*) from the right half of the heart through the lungs and back to the left half of the heart. The second circuit (the *systematic circulation*) pumps blood from the left half of the heart through all of the tissues of the body, except, of course, the lungs, and back by way of the veins to the right half of the heart.

In both circuits, the vessels carrying blood away from the heart are called **arteries,** and the vessels carrying blood away from the lungs

FIGURE 27–6

The circulation of blood through the human heart. The inferior vena cava and superior vena cava introduce deoxygenated blood from the body tissues into the right atrium. From there, it flows into the right ventricle, which pumps the blood out through the pulmonary artery to the lungs. After picking up oxygen in the lungs, the blood is returned to the heart's left atrium by the pulmonary vein and it passes into the left ventricle. This ventricle then pumps the oxygenated blood out to the body tissues via the aorta.

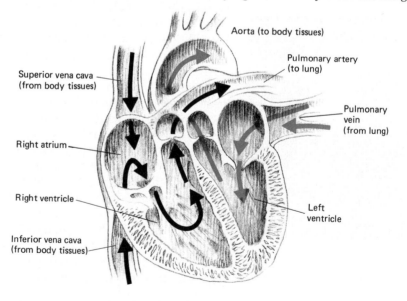

Aorta (to body tissues)

Pulmonary artery (to lung)

Superior vena cava (from body tissues)

Pulmonary vein (from lung)

Right atrium

Right ventricle

Left ventricle

Inferior vena cava (from body tissues)

and tissues back to the heart are called **veins.** The huge vessel extending from the left half of the heart to supply oxygenated blood to the rest of the body is a single large artery, the **aorta.** From the aorta, arteries branch out to conduct blood to all the various organs and tissues, breaking down progressively to the smaller arteries and, finally, the **arterioles,** which ultimately branch into a huge number of very small thin-walled vessels called capillaries. These capillaries unite to form larger vessels (*venules*), which in turn unite to form fewer and still larger vessels called veins. The veins in the lower portion of the body unite to form a large vein, the *inferior vena cava,* which empties into the right half of the heart in the same area that the *superior vena cava* brings blood from the veins in the upper half of the body.

Blood flow patterns

The pulmonary circulation takes blood from the right half of the heart via a single large artery, the pulmonary artery, which divides into two arteries, one supplying each lung. Within the lungs these arteries continue to branch into arterioles and, finally, capillaries, where oxygen is picked up from the tiny thin-walled sacs in the lungs. The oxygenated blood is then carried back from the lungs via the pulmonary veins, which empty into the left half of the heart. Now the oxygenated blood is pumped out again by the left ventricle and atrium to the capillaries of tissues and organs throughout the body, where much of the oxygen leaves the blood to supply oxygen to the tissue cells (Figure 27–7).

Normally, the right half of the heart pumps the same amount of blood as the left heart during a given time. The total amount of blood pumped by the heart is staggering. During rest, the heart beats about 70 times a minute and during each beat, each side of the heart pumps roughly 70 ml of blood. The amount of blood pumped each minute, then, equals 70 ml per beat times 70 beats per minute, or 4,900 ml per minute (making almost 5 liters, or 5 1/4 quarts, per minute). In strenuous exercise, the amount of blood pumped by each side of the heart may rise to as much as 25 liters per minute, and a highly trained athlete undergoing vigorous exercise may put out as much as 40 liters per minute through each side of the heart. Thus as many as 12 *gallons* per minute may be pumped during strenuous exercise or heavy physical labor.

The passage of blood through the heart, as well as the veins and ar-

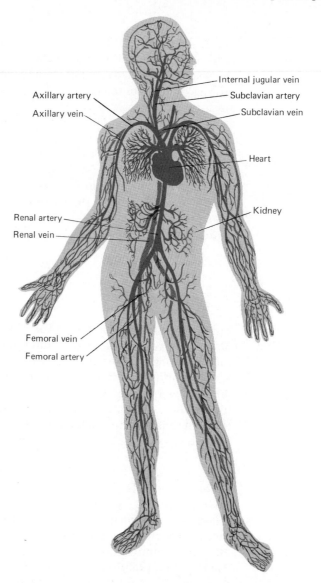

Axillary artery

Axillary vein

Internal jugular vein

Subclavian artery

Subclavian vein

Heart

Renal artery

Renal vein

Kidney

Femoral vein

Femoral artery

teries, is one-way. In Figure 27–8, showing the structure of the heart, there are small valve flaps between the atrium and the ventricle on each side of the heart. Blood pushes down on these from the atrium above, and pushes the flaps open to fill the ventricle. When the muscles in the ventricular walls of the heart start pumping by contracting and squeezing the blood out, these flaps close and blood cannot push its way back in the atrium; instead, it is forced into the arteries, where other sets of valves prevent the blood from flowing backward. Each time the valves open and close, they produce a

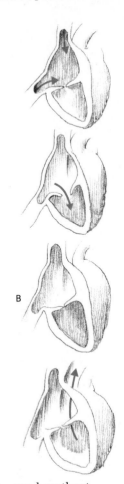

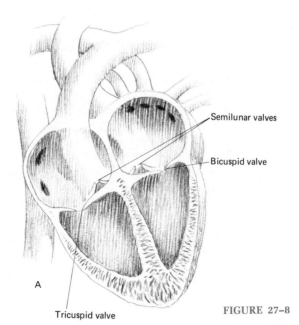

Semilunar valves

Bicuspid valve

A

Tricuspid valve

B

FIGURE 27–8

(A) The location of the main valves in the heart and (B) the mechanical functioning of the tricuspid valve as blood flows from the two vena cava into the right atrium, then into the right ventricle, and finally is pumped out through the pulmonary artery toward the lungs. The valve prevents backflow from the ventricle into the atrium. In the same way, the semilunar valves shown in (A) prevent backflow from the pulmonary artery and aorta.

sound, which is what we hear when we listen to our heartbeat.

There is also a system of valves in the veins themselves that prevents the blood from flowing backward, since it is under extremely low pressure after it leaves the capillary networks. The blood is moved back toward the heart by the contraction of muscles that squeeze the veins, and the closed valves prevent the blood from flowing backward, making the blood flow only toward the heart.

Blood pressure

The blood pressure throughout the main arteries seems to be about the same, although it varies during the course of each heartbeat. Each time the heart contracts, it thrusts blood into the arteries, which in turn stretches the elastic walls of the artery, causing an increase in pressure. When the heart relaxes, on the other hand, no blood enters the arteries, but some blood does leave the arteries through the arterioles. As a result, the amount of blood extending throughout the arteries is reduced, and the pressure tends to fall (but not to zero, because the next heart contraction starts too soon for that to happen). In a healthy, resting person the blood pressure rises to about 120 millimeters (mm) of mercury (the pressure of the distended artery walls will drive blood at a pressure equivalent to that needed beyond atmospheric pressure to push a column of mercury 120 mm high) during each contraction of the heart. Each time the heart relaxes, it falls to about 80 mm of mercury. This pulsation or oscillation of the actual pressure can be felt at several points on the surface of the body, such as the inner part of the wrist, the ankle, and the sides of the temple where arteries are close to the surface.

The composition of the blood

The function, of course, of the anatomical network is to circulate needed nutrients, fluids, metabolic products, and oxygen to the cells of the body. About 55 percent of the blood is a fluid called **plasma;** the remaining 45 percent is made up of three kinds of cells: red cells, or **erythrocytes;** white cells, or **leucocytes;** and platelets, or **thrombocytes.** Virtually all of these components (with the exception of one variety of white blood cells) are manufactured in the bone marrow.

The plasma matrix in which the cells float is a yellowish solution composed of about 92 percent water. The other 8 percent of the plasma represents nutrients essential to life such as glucose, fats, and amino acids; inorganic materials such as sodium, calcium, and potassium; special proteins, antibodies, and other defensive mechanisms to fight off viruses or other intruding cells in the body; and hormones such as insulin and adrenalin, the latter being able to speed up heart rate and metabolic rates whenever an emergency requires a greater blood flow to the muscles. The plasma plays a crucial role, then, in maintaining the body's chemical balance, water content, and temperature (Chapter 29), and, of course, it also transports food, oxygen, and wastes.

The transport of oxygen: red blood cells

The red blood cells have the exclusive task of picking up oxygen in the lungs, carrying it to the rest of the body, and conveying waste carbon dioxide from the body tissues back to the lungs. They out-

How blood pressure is measured

A fluid like blood flows through a tube in response to a graded difference in pressure between the two ends of the tube: the greater the *pressure difference*, the greater the total volume flowing per unit of time. One must also take into account the difficulty or *resistance* that the fluid experiences in flowing through the tube at any given pressure. Resistance depends on the nature of the fluid (viscous fluids such as maple syrup flow less readily because their molecules slide over each other only with great difficulty) and the geometry (length and diameter) of the tube. The lengths of blood vessels remain relatively constant in the body, but the resistance increases markedly as tube diameter decreases, especially in the tiny capillaries.

Artery walls are quite elastic and help to exert pressure to drive blood into the capillaries and through the tissues. Arterial blood pressure varies considerably depending on the phase of the heart beat cycle. During *systole*, the contraction of both ventricles ejects blood *into* the pulmonary and general systemic arteries. If, simultaneously, a precisely equal quantity of blood were to flow *out* of the arteries into the arterioles and capillaries, the total volume of blood in the arteries would remain constant and arterial pressure would not change. But this is not the case. A volume of blood equal to only about one-third the heart's stroke volume leaves the arteries during systole. Thus this excess volume in the arteries distends their walls and *raises* the arterial pressure. When ventricular contraction ends, the stretched arterial walls recoil passively (like stretched rubber bands on release) and the arterial pressure continues to drive blood through the arterioles. As blood leaves the arteries, the pressure slowly falls, but the next ventricular contraction takes place while there is still sufficient blood in the arteries to stretch them partially, so that the arterial pressure does not fall to zero. In this manner, the arterial pressure provides the immediate driving force for blood flow through the body tissues.

The maximum blood pressure is reached during peak ventricular ejection and is called *systolic pressure*. The minimum blood pressure obviously occurs just before the next ventricular contraction and is called *diastolic pressure*. Blood pressure is generally recorded as a combination of these two: systolic/diastolic, as, for example, in 130/70. These numbers refer to the millimeters of mercury (Hg) that could be supported in a column by the applied pressure (above the standard atmosphere pressure of 760 mm Hg at sea level). This blood pressure figure is essentially similar everywhere in the arterial tree of the body because the various larger arteries offer only negligible resistance to flow.

Both systolic and diastolic blood pressure are readily measured in human beings with the use of a sphygmomanometer. A hollow rubber cuff is wrapped around the arm and inflated with air to a pressure higher than systolic blood pressure. At this point, the arteries in the arm tissues completely collapse under the cuff and blood flows to the lower arm stops. The air in the cuff is now slowly released, causing the pressure in the cuff and arm to drop. When cuff pressure has fallen to a point just below the systolic pressure, the arterial blood pressure at the peak of systole is greater than the cuff pressure, causing the artery to expand and allow a brief flow of blood. During this interval, the blood flows very rapidly and with high turbulence and vibration through the small opening. These sounds can be heard through a stethoscope placed over the artery just below the cuff. The pressure (measured on the manometer gauge attached to the cuff) at which sounds are first heard as the cuff pressure is lowered is recorded as the *systolic blood pressure*. As the pressure in the cuff is lowered further, the sounds at first get louder and then become dull and muffled when the cuff pressure reaches the diastolic blood pressure, as the artery remains open throughout the cycle and allows continuous turbulent flow. Just below *diastolic blood pressure*, all sound stops because flow is now continuous and nonturbulent through the completely open artery. Thus systolic blood pressure is identified as the cuff pressure at which sounds first occur and diastolic pressure is identified as the cuff pressure at which sounds disappear.

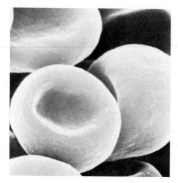

FIGURE 27–9

*Red blood cells at 20,000×
magnification under the
scanning electron microscope.*

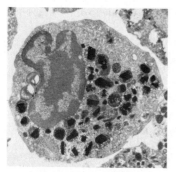

FIGURE 27–10

*A white blood cell leucocyte at
11,800× magnification under
transmission electron
microscopy.*

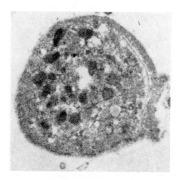

number the white blood cells 700 to 1. They live only about three or four months in the blood stream and are replaced by new cells sent into the blood stream from the bone marrow (Figure 27–9).

The red blood cells operate as an oxygen carrier because of their content of *hemoglobin,* a compound of protein containing iron that gives blood its red color. Hemoglobin is able to bind an oxygen molecule in areas of high oxygen content (such as the lungs) and then release it in areas of low oxygen content (the body tissues). Since oxygen does not dissolve well in plasma, this reversible binding power of hemoglobin is vital to a large, multicellular organism such as a mammal that requires considerable oxygen and cannot depend on diffusion alone to bring this oxygen to all its cells. If the hemoglobin content dips below minimum body needs, the result is **anemia,** that is, insufficient oxygen-transporting capacity.

Defense against disease: white blood cells

The white blood cells constitute the body's principal defense mechanism against invading bacteria, viruses, and decomposition of dead cells (Figure 27–10). Some white blood cells are able to pass out of the walls of the capillaries into the tissues and return. They congregate in great numbers wherever invading bacteria gain entrance into the body, engulfing and destroying them. Whenever white blood cells are called on to mobilize for such action, the body compensates by manufacturing more. Often, the number may double within hours in the blood. This rising white cell count serves as an early warning to a doctor of a dangerous infection.

Defense against bleeding: the platelets

The platelets, named for their resemblance to tiny plates, are perhaps the strangest of blood's three cellular components (Figure 27–11). They were discovered about a century ago, but their function was long a mystery. At that time it was observed that people with low platelet counts were very vulnerable to bleeding. The platelets were subsequently found to be vital to blood clotting. When they come into contact with the roughened surface of a torn blood vessel, they burst apart, releasing chemicals that set off a reaction in the blood and convert one of the plasma proteins, *fibrinogen,* into a network of crystallike fibers that enmesh the red blood cells— thereby forming a clot that seals the leak.

FIGURE 27–11

*Blood platelet (thrombocyte)
under transmission electron microscopy.*

Immunological phenomena

Immune phenomena represent an important set of physiological responses in higher animals. Included in immune phenomena is the beneficial protection gained against viral and bacterial disease agents that results either after a natural attack of the disease or after the deliberate injection (or other administration) of a modified form of the disease-producing agent, a process called *immunization*. Unfortunately, there are also deleterious effects of the immune phenomena, such as the rejection of tissue or organ grafts, the effect on babies conceived by parents of particular blood types, and certain diseases such as allergies or nephritis, characterized by an abnormal immune response. The most significant hallmark of the immune response is its high degree of *specificity*, which results from the three-dimensional form of weak interactions between two complex molecules, an antibody and an antigen. Thus a human being or other higher animal who recovers from one infectious disease is generally immune to that disease for at least some time, but is not immune to other infectious diseases.

The fundamental immune response consists of the appearance of specific proteins, *antibodies*, in the blood of higher organisms, in response to the entry of an injurious molecule, an *antigen*, into the animal. There are two types of antibodies. The first type appears several days after an antigen has been injected, is synthesized for another four to six days, and then disappears rapidly. The second type appears about one week after injection of antigen, increases in concentration for four to six days, and then remains approximately constant in concentration for a long period. Antibodies are produced principally in the spleen, lymph nodes, bone marrow, and other lymphoid tissue by lymphocytes (white blood cells) and blood plasma cells. The capacity to form antibodies belongs to *immunologically competent cells*, which appear to be mainly small lymphocytes. The small lymphocytes both circulate in the blood and reside in lymphoid tissue for variable periods. The lymphoid cell population constant moves from one lymphoid organ to another via the circulation. Antigen seems to stimulate increased division of immunologically competent cells and perhaps also to stimulate the maturation of immunologically competent lymphocytes into plasma cells. Plasma cells have a very elaborate endoplasmic reticulum and are thought to produce antibodies at high rates. Individual plasma cells appear capable of synthesizing antibodies to interact with more than one antigen.

Antigens capable of eliciting an immune response include a remarkably diverse group of substances: proteins, lipids, polysaccharides, nucleic acids, and various synthetic organic materials. However, not all naturally occurring substances (even proteins) are antigenic, and what is antigenic to one species may noy be antigenic in another organism. Effective antigens are, of course, foreign to the circulation of the animal to be immunized. If one injects an effective antigen at an *early* stage in the animal's development, the animal usually will not produce antibodies then or later in adult life if it is challenged with the same antigen; this phenomena is called *immune tolerance*.

The interaction between an antibody and an antigen involves a physical combining. They may precipitate out of solution as a solid mixture, or if the antigen is attached to the surface of cells, the antigen-antibody reaction results in a clumping or *agglutination* of the cells. The aggregate will then co-precipitate. The physical interaction of an antigen and an antibody requires a high degree of geometric complementarity at the combining sites of the two molecules, and this factor principally accounts for the great specificity of immune phenomena.

The lymphatic system

The lymphatic system is a parallel circulatory system to the blood (Figure 27–12).The watery fluid serves as a bridge by which oxygen and nutrients and wastes can pass between the capillaries of the blood stream and the body cells.

The lymph picks up the wayward proteins that escape from the blood stream under the considerable force of blood pressure and conveys them via a circulatory system of its own (composed of lymphatic capillaries and ducts) into blood veins in the shoulder area, which pass them back into the heart. A lymph duct also carries many chemical products, including droplets of fat and cholesterol that are absorbed during the process of digestion along the gastrointestinal tract. By the time the relatively sluggish lymphatic system brings these substances back to the heart, they have been greatly diluted and their potentially adverse effect on the body (e.g., in deposits building up on the inside walls of arteries) is reduced.

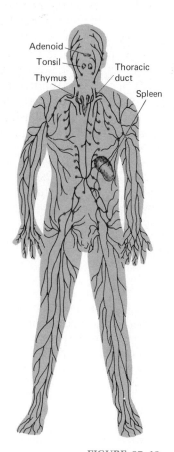

Adenoid
Tonsil
Thymus
Thoracic duct
Spleen

FIGURE 27–12

The human lymphatic system represents a second interconnecting circulatory system in the body. Lymphatic fluid, which transports some food such as fats, reenters the blood circulatory system through the thoracic duct on either side at the top of the chest.

Summary

Internal transport in the simplest plants and animals is by the random action of diffusion. Multicellular organisms generally require more rapid and specific conduction of materials throughout the body. Higher plants utilize xylem vascular tissue to move water and dissolved minerals from the roots to the stems and leaves; the motive force is believed to be a combination of diffusion pressure and transpiration pull that, with high cohesion properties of water, is sufficient to move water to the tops of 350-foot trees. Translocation of organic solutes and inorganic ions in all directions through the plant occurs via the phloem vascular tissue; the motive force is explained by a mass flow theory in which water and solutes flow through the phloem tubes along a pressure gradient generated by manufacturing (photosynthetic) and metabolic (respiratory) processes concentrated in different portions of the plant. Lower animals such as insects have an open circulatory system in which the blood does not circulate entirely in closed vessels. Vertebrates, including human beings, have a closed circulatory system where the system of vessels extending out from the heart is continuous throughout the body and returns the circulating blood to the pumping heart. William Harvey was the first to discover this circulatory pattern. The heart is composed of two functional halves, divided vertically, each half consisting of an upper atrium and a lower, muscular ventricle.

The right half handles the pulmonary circulation of the lungs, and the left half handles the systemic circulation to the rest of the body. Valves in the heart and the veins ensure that the blood flows in one direction. The blood carries nutrients, fluids, metabolic products, and oxygen to all the cells of the animal body. Its principal components are plasma, red blood cells, white blood cells, and platelets. The plasma carries antibodies, organic nutrients, inorganic materials, and hormones. The erythrocytes, or red cells, operate as oxygen carriers via their hemoglobin molecules. Leucocytes, or white cells, constitute the body's principal defense against invading foreign organisms and dispose of dead cells. Thrombocytes, or platelets, are vital to blood clotting. The lymphatic circulatory system carries fats and lipids as well as other chemical products and escaped blood proteins back into the cardiovascular system.

28 Respiration and gas exchange

In the organic world, most cells obtain the bulk of their energy from chemical reactions involving oxygen. In fact, over 90 percent of the ATP molecules produced by the metabolic breakdown of a glucose molecule result from the terminal oxidative phosphorylation steps in cellular respiration (Chapter 25). The presence of oxygen makes possible the complete oxidation of these and more complicated carbohydrates to carbon dioxide and water. At the same time, this complete breakdown introduces a problem: eliminating carbon dioxide, the major poisonous end product of these oxidations. A single-celled organism can exchange oxygen and carbon dioxide directly with the external environment through diffusion past the cell membrane, but in a complex organism with many cells, only a small fraction of the total cells is in direct contact with the external environment. Consequently, reliance on simple diffusion is impossible, and specialized systems for the supply of oxygen and the elimination of carbon dioxide have had to develop simultaneously with the evolution of these multicellular organisms.

FIGURE 28-1

Gas exchange by diffusion through the cell membrane in the aquatic green alga, Spirogyra.

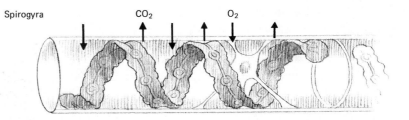

Gas exchange in plants

In single-celled plants and many small multicellular ones, particularly aquatic species, gas exchange with the external environment can occur directly across the cell membrane. Oxygen from the higher concentration in the water lying outside the cell penetrates in response to an oxygen deficit inside the cell; in the same way, the higher concentration of carbon dioxide generated by cellular respiration inside the cell moves across the cell membrane into the lower carbon dioxide concentration in the surrounding water. Of course, plant cells face the same general waste problem as animal cells, in that the breaking down of glucose in plant cells results in the waste products of carbon dioxide and water. These carbon dioxide molecules are not picked up immediately by the photosynthetic processes in the plant. Without a means of expelling CO_2, the plant would soon be poisoned by its own metabolic end products.

Sometimes fairly sizable plants, such as giant marine kelps that reach 200 to 300 feet in length, may still rely on simple diffusion if their bodies are thin and only several cell layers thick, thus allowing diffusion to operate efficiently in gas exchange (Figure 28–1). In terrestrial plants, however, more elaborate mechanisms have evolved to handle problems of gas exchange associated with both photosynthesis and cellular respiration. Leaves are particularly well adapted for this exchange process. Despite the usual waxy cuticle coating, openings (called **stomata**) in the epidermis allow gases to move easily between the surrounding atmosphere and the internal spaces in the loosely packed cells of the middle layers of the flattened leaf. Here the thin, moist membranes of the internal cells allow diffusion of gases in and out of the cells (Figure 28–2).

The total surface layer available for gas exchange is great because of the irregular shapes and amount of cell surface exposed through this loose packing of the middle leaf cells. Gases can move to each individual cell directly via the intercellular spaces; because the

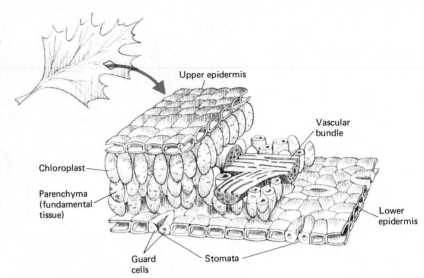

FIGURE 28-2

The structure of a typical leaf in connection with gas exchange. The guard cells allow oxygen and carbon dioxide to enter and leave through the lower epidermis when they open the stomata. The central parenchyma, whose cells contain chloroplasts, are loosely packed and are therefore surrounded by intercellular spaces, which, via the stomata, are continuous with the outside atmosphere.

exchange surface is internal, there is relatively little danger of mechanical injury, especially with the protective covering of spines, hairs, and waxes on the epidermis. Additionally, the exchange surfaces inside the cell can remain moist because of the retention of high humidity within these spaces; the protective outside tissues act as barriers between the dry outside air and the moist inside air.

The stomata are opened and closed by two highly specialized epidermal cells, called *guard cells,* one on either side. Since the leaf needs a large amount of carbon dioxide during the daytime when light is available, the stomata remain open during the day for free gas exchange. The internal cells constantly lose water by transpiration and since atmospheric gases are not needed for photosynthesis at night, the guard cells close the stomata after sunset, thus preventing excessive water loss. The actual opening and closing operation is mediated by photosynthetic activity in the chloroplasts of the guard cells. During the daytime, photosynthesis produces sugars in the guard cells, increasing the osmotic rate of diffusion of water into the cells and making the cells turgid. When filled by increased water pressure, the configuration of the guard cells is bowlike. At night, sugars are no longer produced by the chloroplasts and turgor pressure drops as they are metabolized or leave the guard cells with part of the water content. The guard cell walls relax and close the stomatal opening until the next morning.

In stems and roots there are numerous small areas of loosely arranged cells with many intercellular spaces between them. Scattered openings to the outside through the bark, called **lenticels,** allow free

FIGURE 28–3

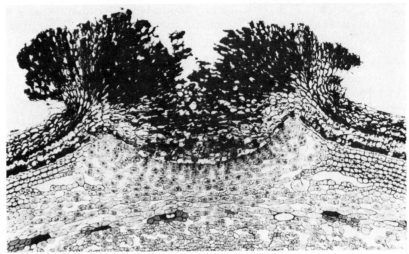

A lenticel opening in the bark, for gaseous exchange.

movement of gases into the interior of a stem (Figure 28–3). Roots, however, do not usually possess special structures for gas exchange and obtain gases by diffusion across the moist membranes of root hairs in well-aerated soils. If the air spaces in the soil become filled with water through flooding, the oxygen content of the soil may become so reduced that the plant will literally drown from lack of air circulation to its roots.

There is no special gas-transporting mechanism from the upper parts of the plant to the roots. Thus each area of the plant must have intercellular spaces in its tissues and use exchange with the outside through stomata and lenticels to supply oxygen to all the other tissues. Diffusion occurs extraordinarily rapidly through air compared with water (approximately 300,000 times faster), and thus even large redwood trees are able to supply sufficient oxygen to their interior cells through these loosely packed tissue areas filled with air.

Gas exchange in the lower animals

In the water, simple single-celled animals do not need any special gas-exchange devices. They use simple diffusion across their cell membrane. The coelenterates such as jellyfish and sea anemones are also able to use direct diffusion through their cell membranes to accomplish gas exchange, as do the thin-bodied flatworms. Sometimes in marine worms, the whole body surface may be supplied with

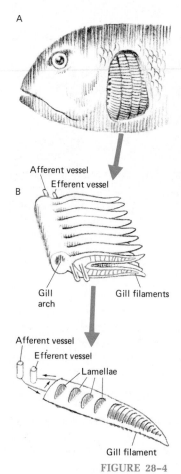

FIGURE 28–4

The gill respiratory system in a fish. The thin, flattened gill filaments lie in shingled layers, attached to arches. The fish takes in oxygen-bearing water through its mouth, passes it across the gills, and out through the opened gill covers. Each filament is made up of stacks of very thin lamellae, penetrated by capillaries coming from the afferent vessel at the filament's base and going back to the efferent vessel. Oxygen diffuses through the lamellar tissue to the capillaries.

many blood vessels and serve as a gas-exchange system. But most aquatic species of multicellular animals have developed some system of gill-like structures to take in oxygen and expel carbon dioxide.

The gill

A **gill** is a thin flap of tissue that is richly supplied with blood capillaries and hence exposes a large surface area of these tiny blood vessels to the diffusion of oxygen across the single-celled membrane (Figure 28–4). Likewise, CO_2 can be released through the membrane and be readily expelled from the thin gill surface. While gills have evolved a great number of times in different animal groups, ranging in diversity from aquatic insect larvae to clams, squids, and salamanders, all gills operate in the same general way. Oxygen moves by diffusion from the surrounding water through the gill surface cell to the wall of the vessel, where it is picked up by carrier molecules such as hemoglobin. The blood then carries the oxygen through the body where it is released in interior tissues that have a low oxygen concentration. At the same time, these molecules normally pick up carbon dioxide and continue through the circulatory system until they again reach the gill area, where carbon dioxide is released and diffuses through the gill surface to the outside, completing the cycle.

Many organisms have developed special means of passing water currents across a number of thin gill layers, and frequently the fragile gills are covered by hard protective extensions of the body. Other marine organisms such as sea cucumbers use a specialized system of internal tubes called *respiratory trees* (Figure 28–5). These are long-branched tubes coming off the rear part of the intestinal tract, which contracts and expands, expelling and drawing in water for the respiratory system. The thin membranes of these respiratory trees serve as gas-exchange surfaces, just as do the gills of other marine creatures.

The tracheae

In the terrestrial insects, an elaborate system of branching tubes called **tracheae** provides oxygen to all the cells of the body (Figure 28–6). Along each side of a typical insect are a number of openings called *spiracles*, which control the entrance and exit of gases into the interior. A large tracheal tube connects with this spiracle on the inside and separates into smaller and smaller branches until oxygen is brought into contact with every tissue area in the body. At these ultimate points, then, oxygen in the air diffuses through the tracheal membranes to the cells, and carbon dioxide moves from the tissue

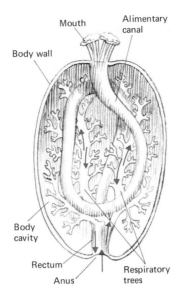

FIGURE 28–5

Respiratory trees in the sea cucumber. Oxygen-bearing water flows in the anus and into the respiratory trees where the oxygen diffuses through thin tissue walls.

cells back into the trachea where it moves by simple diffusion again to the outside. The closed tube system keeps the inside air quite moist and the cells lining the tracheae are covered by a film of water, through which gases dissolve and pass, as if the insect were living in an aquatic environment.

Tracheae are relatively efficient up to a certain body size, beyond which the rate of diffusion through the air is not fast enough to supply oxygen to an active animal. Hence the tracheal system, while useful for small animals and adaptively successful (as one can witness by the huge diversity of insects), has prevented the insects from developing into truly large terrestrial animals. Even the largest insect that ever lived, the giant dragonflies of the Mesozoic era, which reached close to 30 inches in wingspan, had relatively thin bodies with considerable air spaces in them.

The most advanced form of respiratory and gas-exchange systems is the lung, which ranges from a primitive sausagelike tube in land snails, lungfishes, and most amphibians, to a chamber with an inner

FIGURE 28–6

The tracheal system for respiration in a grasshopper. The spiracle openings, main tracheal trunks, and air sacs (which act as reservoirs) are shown for one side of the insect.

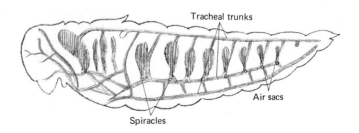

surface divided into many small pockets, or sacs, and increased numbers of circulatory vessels running throughout the exchange surface. As an example of the lung gas-exchange system, we shall examine in detail the human lung and its functioning.

The organization of the human respiratory system

The lungs allow us to take in oxygen and breathe out carbon dioxide, forming one of the three main gateways to the human body. The digestive tract and the kidneys, through which we take in solids and fluids and excrete wastes, were discussed in the preceding chapter. The trillions of cells in our bodies require so much oxygen that we need a very large surface area, which is supplied by the compartmentalization and vascularization of the membranes in the lungs.

It takes only several seconds for air to travel from the outside environment to the lungs, but it passes by an extensive filtering and testing system in the process. The *nose* forms the entrance (through the nostrils) for the air. Inside the nose the air enters the *nasal cavities,* which warm and moisten the air as well as trap dust particles and bacteria by mucus or tiny cilia that are beating in a direction opposite to the incoming air flow. The twisting passages of the nasal cavity cause eddies in the air stream and help to force the air against these cilia and mucus for the filtration process. Chemical substances

FIGURE 28-7

The human respiratory system, including the detailed structure of the throat region.

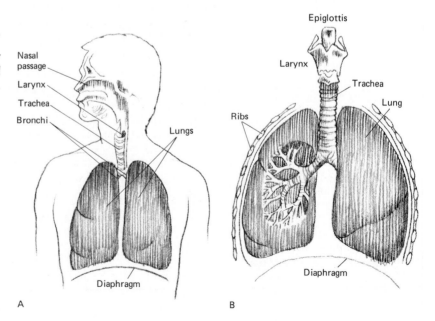

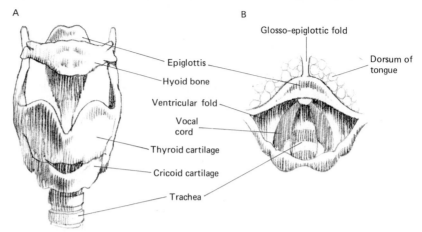

Glosso-epiglottic fold

Dorsum of tongue

Epiglottis

Hyoid bone

Ventricular fold

Vocal cord

Thyroid cartilage

Cricoid cartilage

Trachea

FIGURE 28-8

The human larynx and vocal chords: (A) a frontal view of the larynx and (B) a view of the vocal chords in a relaxed position.

in the air dissolve into the mucous layer, and receptors in the epithelium of the nasal passages give rise to the sense of smell at this point, as active smelling devices test the quality of the air.

The air soon passes from the nasal cavity into the **pharynx,** where it joins food that may be passing in through the mouth area (Figure 28–7). At this point, the air and food passages actually cross as well as join, for air enters the pharynx from the top and exits ventrally, whereas food enters from the bottom and exits dorsally into the esophagus. The ventral opening (the **glottis**) for the air channel passes into the **larynx,** which is connected to the trachea or windpipe. A flap of tissue called the **epiglottis** closes this glottis when the larynx is raised against it during swallowing. At the same time, the nasal cavities are closed off from the pharynx by the soft palate. This process normally prevents food from being inhaled into the trachea below the larynx, though not always. If food gets into the trachea, its walls are irritated and a coughing reaction begins.

When air enters the larynx, it passes through a cartilage-lined chamber with a pair of *vocal chords,* elastic ridges that vibrate when air currents pass over them (Figure 28–8). The pitch of the sounds emitted depends on the relative tension of the chords, which in combination with the work of the lips and tongue gives rise to speech and other articulations.

From the larynx the air passes into the *trachea,* which carries the air into the *thoracic,* or *chest, cavity.* Here again, the cilia of the epithelial lining beat in opposite directions from air flow, and carry bacteria and other particles back up the trachea away from the lungs. Rings of cartilage support the shape of the trachea even during exhaling. At the base of the trachea the windpipe splits into two bronchi that pass immediately into the two lungs. Here the right bronchus and left bronchus branch repeatedly into bronchioles, and

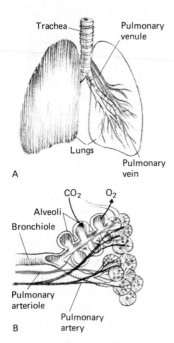

Trachea

Pulmonary venule

Lungs

Pulmonary vein

A

CO_2 O_2

Alveoli

Bronchiole

Pulmonary arteriole

Pulmonary artery

B

FIGURE 28–9

Gas exchange takes place especially in the tiny alveoli, or air sacs, which are richly supplied with capillaries.

the cleansed air is passed into millions of tiny, expandable, thin-walled clustered sacs called alveoli, in which the bronchioles terminate.

The **alveoli** constitute the bulk of lung tissue, and it is their substance that makes the lungs soft and spongy and indeed so light they can float (Figure 28–9). The lungs of an average-sized man contain an estimated 3 million of these air sacs. Their combined weight is about 2 1/2 pounds. When expanded, the total alveoli surface area amounts to about 1,000 square feet. When the chest expands or contracts—on stimulus from a respiratory center in the brain and cued by the level of carbon dioxide in the blood—it is the alveoli that are expanding or contracting. Within an alveolus the membrane is moist and oxygen dissolves into this film of moisture, allowing oxygen to pass through the wall into a rich network of capillaries that surrounds each cluster of alveoli. Layers of capillaries and alveoli cells thus lie in direct contact side by side, a double membrane almost unimaginably thin, with air moving on one side and blood flowing past on the other. Oxygen molecules pass into the blood via this virtually transparent wall and are picked up by the hemoglobin molecules on the red blood cells. The iron present in the hemoglobin locks the oxygen into place and, carried in the blood stream, the oxygen finally arrives at the body's far-flung cells, there to participate in the energy-producing reactions of cellular respiration. As much as five quarts of oxygen may be absorbed per minute by a hard-running man, but a resting man may absorb only half a pint of oxygen a minute.

Simultaneously with the passage of oxygen into the blood stream, the alveoli draw out of the blood the waste carbon dioxide that results from the combustion of carbon compounds in the cells during cellular respiration. Carbon dioxide is brought alongside the alveolus membrane and seeps out from the blood stream by diffusion, just as the oxygen seeps in. From the lungs, the carbon dioxide makes its way out of the body along the same route that the oxygen followed on its way in. While an excess of carbon dioxide may be poisonous to the body, a small amount is retained in the blood. This amount is vital to life, for it not only maintains the proper degree of acidity in body fluids but it also controls the internal breathing mechanism. Blood passing through the *respiratory center* in the brain has its carbon dioxide concentration monitored; if we hold our breath or otherwise voluntarily cause our carbon dioxide concentration to accumulate in the blood, the respiratory center is stimulated to dictate to the lungs an involuntary response to renew breathing.

The actual mechanics of lung respiration is what we commonly call *breathing* (Figure 28–10). This mechanical process generally involves muscular contractions of the rib cage and the diaphragm, a

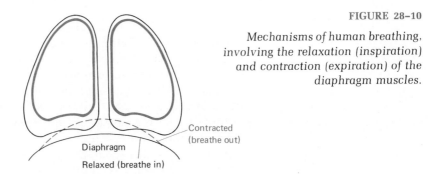

FIGURE 28–10

Mechanisms of human breathing, involving the relaxation (inspiration) and contraction (expiration) of the diaphragm muscles.

muscular partition separating the chest and abdominal cavities. When the rib muscles are contracted and thus pull the rib cage upward and outward, the increase in chest volume reduces the air pressure within the chest below the pressure of the outside atmosphere and thus draws air into the lungs. The normally upward-arched diaphragm is pulled downward by contraction of the diaphragm muscles at the same time.

The exhaling of air is normally a passive process. The chest and diaphragm muscles simply relax, allowing the diaphragm to arch back upward from its temporary downward position and the rib cage to fall back to its resting position. At the same time, the chest volume is reduced. The elasticity of the alveoli increases the pressure inside the lungs above that of the outside atmosphere and expels the air.

In general, then, the chief functions of the lung, gill, and tracheae are (1) the transport of oxygen to all the cells in the body from the outside atmosphere, and (2) the excretion of waste carbon dioxide from the body tissues to the outside atmosphere, through the connection of the circulatory system with this system of respiratory structures. The mechanical act of breathing is only one link in the chain of events involved in respiration. As we have seen, the ultimate importance of breathing is that it brings oxygen in contact with the circulatory system, which can then carry oxygen to satisfy the metabolic demands of our numerous body tissues.

Summary

Respiratory systems for gas exchange are necessary in multicellular organisms for the supply of oxygen to tissues and the elimination of carbon dioxide produced by cellular respiration. Plants have leaf stomata and stem lenticels that allow diffusion of gases into the internal tissues where moist intercellular spaces allow extensive gas exchange. Aquatic lower animals employ primarily gills for respiratory exchange; these thin flaps of tissue are richly supplied with blood capillaries. Terrestrial insects utilize an elaborate system of branching tracheal tubes to provide oxygen to all the cells of the body. Most vertebrates use lungs, which in the advanced forms are chambers with a division of the inner surface into many tiny, richly vascularized sacs, the alveoli. In human beings the air passes through a filtering process in the nose, nasal cavities, and trachea before it enters the two bronchi and lungs. En route, it crosses the pharynx (where the food passage also crosses) and goes through the larynx where the vocal chords are located. Gas exchange takes place between the inner surfaces of the alveoli and adjacent capillaries. Breathing is a mechanical process that involves muscular contractions of the rib cage and the diaphragm. Their rate is normally regulated without conscious effort by a respiratory center in the brain, which is cued by the CO_2 concentration in the blood. We can voluntarily hold our breath for a while, but eventually the respiratory center will override this voluntary control and cause breathing to begin again involuntarily.

29 The regulation of body temperature

Though animals inhabit every corner of the globe, active life is restricted to a narrow range of temperature—from a lower limit of about −1°C, the temperature of the Arctic Ocean waters, to an upper limit of about 50°C, the temperature of certain hot springs where some animals and blue-green algae manage to survive. This relatively narrow range of temperature (compared with the great range throughout the universe, from absolute zero to millions of degrees centigrade) still encompasses livable temperatures over most of the surface of the earth and in the oceans, at least during parts of the year. Rather than an absolute continuum of types of adjustment to this temperature range, though, animals have developed two principal ways of handling the regulation of body temperature in relation to environmental temperatures.

Over 2,000 years ago, Aristotle proposed that higher and lower animals be distinguished according to whether they were *warm-blooded* or *cold-blooded*. These two categories are still recognized today, although their definitions have changed greatly. The majority of animal species are cold-blooded, but the mammals and birds are warm-blooded. The basic distinction lies in the ability of the respective animal to control its temperature. Cold-blooded, or **poikilothermic** (which means "changeable temperature" and is actually more

descriptive), animals have about the same temperature as their surroundings — whether water, air, or soil — and if the temperature of the
surroundings changes, the body temperature of the animals also
changes. Warm-blooded, or **homeothermic,** birds and mammals are
able to maintain a quite constant body temperature in spite of wide
variations in the surrounding environmental temperature.

Temperature regulation in poikilothermic animals

Poikilotherms, also called ectotherms, include most of the zoologi-
cal world, as we have seen (Chapter 16), with over a million species.
Their body temperatures generally fluctuate with the surroundings,
but they do not lack control over the temperature. Many species have
a large number of *behavioral* regulators that aid in changing their
body warmth to some extent. Within the temperature range permitting an active and normal life for a poikilothermic animal, a change
in temperature will greatly affect metabolic processes, slowing them
with increasingly lower temperatures. Nevertheless, poikilotherms
are able to tolerate temperatures above or below their limit for activity by going into an inactive state; however, a tolerance for increased
temperature is usually quite limited.

The rate of metabolism in a poikilotherm rises with an increase in
temperature. Usually, the rate about doubles for a temperature rise of
10°C. Thus when a temperature increase amounts to 20° or 30°, the
animal rapidly will use up its stored carbohydrates and other molecules in increased metabolic activity. In particular, many enzymatic
actions are accelerated at greatly increased rates by increased temperatures.

Most terrestrial animals, whether poikilotherms or homeotherms
have a lethal limit of about 45°C. In fact, some aquatic arctic species
may die if the temperature is raised to only 10°C above zero. These
higher temperatures chiefly affect proteins, which become inac-
tivated and denatured through the increase in the temperature. Also,
because of the increased rate of metabolism that goes on for at least a
short time, there is a greatly increased need for oxygen; when it cannot be met, the animal dies.

When the temperature of a poikilotherm (or other organism)
drops, the speed of physiological processes also decreases, and oxygen consumption goes down as the animal becomes lethargic or
even torpid. A number of animals can survive a temperature drop to
below freezing, and some, such as mosquito larvae in the Arctic, are
known to tolerate repeated freezings and thawings. The growth of ice

FIGURE 29–1

Behavioral thermoregulation in reptiles. When this Colorado collared lizard gets quite warm from the sun, it raises its body above the rock to allow cooling air to circulate across its belly.

crystals in a cell disrupts the organization of most animals, however, and the adaptations used by these mosquito larvae and similar organisms to withstand the disintegrating effects of ice formation are not understood as yet.

As we noted, behavioral activities help the poikilotherm animal to achieve some control over its body temperature. Organisms as diverse as butterflies and reptiles will seek out sunlit areas to bask in during the early morning hours (Figure 29–1). The radiation from the sun, especially its warming effect on the thin membranes of the butterfly wing and the surface skin of the lizard, increases the temperature of the blood above that of the surroundings and the circulatory system carries this warm blood throughout the body. On a cool day with the outside temperature only 20°C, a butterfly is able to increase its internal muscle temperature to 35°C in six minutes of sunning.

Later in the day, when air temperatures are high, these insects and reptiles may keep out of the sun for a while, hiding in the shade of rocks or bushes and thus cooling themselves, preventing overheating of their bodies. If a desert lizard or snake does not hide during the heat of the day, the body temperature may reach a lethal level at about 45°C in only 10 or 15 minutes of exposure to the open sun.

The many arctic butterflies and moths, as well as most temperate and tropical species, have an insulating layer of scales over the thorax and other body portions that apparently helps to maintain body heat. Nocturnal moths, which do not have the advantage of solar radiation for warming up to flight activity, have a particularly thick layer of scales and may warm up their bodies initially by the special warm-up behavior of vibrating their wings and associated thoracic muscles. Some dragonflies have an internal series of air sacs lying just beneath the surface of the thorax; these restrict the outward flow of heat after the body is warmed in the sun.

Perhaps the most noted case of generation of body heat by a so-called poikilotherm is that of the brooding python. Female pythons can warm their clutch of eggs by violently contracting their trunk muscles. This method of periodic muscle activity also results in *regulating* the temperature of the egg clutch at 31° to 32°C, when the outside air temperature ranges from 26° to 30°C. Thus cold-blooded vertebrates and invertebrates have the ability, given a small capacity for heat production and appropriate behavioral mechanisms, to regulate their body temperature almost as closely as that of any bird or mammal, as long as the outside environmental temperature does not fluctuate much through the day or from day to day.

Temperature regulation in homeothermic animals

Homeotherms, or endotherms, which include 8,600 bird species and more than 10,000 mammal species, rely on internal heat production to regulate their body temperature. They have the capacity for maintaining an almost constant body temperature. In most human beings, for instance, the normal body temperature remains at about 37°C (98.6°F), with a slight decrease at night and a slight increase in the daytime, amounting to a 24-hour change of about 1°C. The physiological receptors available in birds and mammals to regulate body temperature (produced by metabolic activity) are quite diverse. *Insulation* in the form of feathers and fur is particularly important to arctic and temperate zone animals (Figure 29–2). They are able to maintain their temperature in a neutral range without particularly

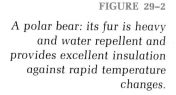

FIGURE 29–2

A polar bear: its fur is heavy and water repellent and provides excellent insulation against rapid temperature changes.

increasing their metabolic activities in cold weather. An Alaskan husky covered with a heavy winter coat can curl up in the snow and go to sleep, yet he is so well insulated that metabolism increases only when the air temperature drops below −30°C. Aquatic arctic animals such as seals and walrus, and deep-diving oceanic animals such as whales, depend on a thick heavy layer of fat and blubber for insulation. In those arctic animals that leave the water to rest on the ice, fur is still needed to protect the outer part of the skin from subfreezing temperatures. In completely aquatic mammals, though, such as whales and porpoises, hair is superfluous and the skin is essentially hairless.

A typical man will be comfortable without clothes in a narrow range of temperatures lying between 27° and 31°C. The heat produced by his normal metabolic activities is dissipated in the surrounding air, and his body temperature remains constant (Figure 29–

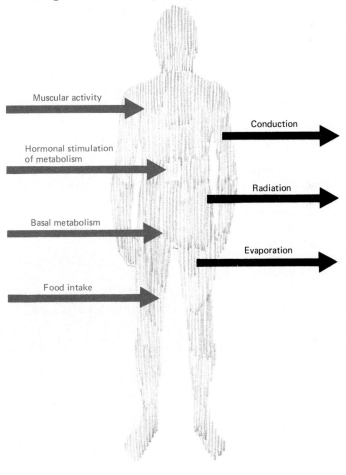

Muscular activity

Hormonal stimulation
of metabolism

Basal metabolism

Food intake

Conduction

Radiation

Evaporation

FIGURE 29–3

Factors influencing
temperature regulation in
human beings.

3). However, if the air temperature drops below 27°C, his body loses heat faster than it is being produced and the only way he can keep his body temperature at the same level is to produce more heat. This additional heat is produced by involuntary muscle contractions, which we call *shivering*. The colder it is, the more violently he shivers in order to produce the additional heat needed to keep up with the rate of cooling.

Above an environmental temperature of 31°C, the air is not cool enough to remove the metabolically produced heat of the body and thus the body begins *sweating* in order to evaporate water, cooling the blood and the surface layers of the skin (Figure 29–4). We can insulate our bodies with clothes to keep from shivering in the cold and can remove them in warm surroundings to sweat more readily. The circulatory system in the body itself also assists in maintaining a constant temperature by *reducing* the amount of blood circulating to the skin in cold temperatures and *expanding* the blood vessels in warm temperatures so as to put as much blood as possible in the skin areas to cool it faster.

Shivering and sweating are the main human mechanisms to control the body temperature. However, animals that are not capable of sweating substantially, such as dogs, evaporate water from the tongue and upper respiratory tract by an extremely rapid rate of res-

FIGURE 29–4

A highly magnified view of human skin, showing the glands, hairs, and circulatory elements involved in temperature regulation. The sweat glands secrete a salty liquid that evaporates and cools the skin surface. Hairs above the skin trap air, conserving heat near the skin surface. In cold conditions, the arterioles constrict and reduce the flow of heat-carrying blood to the exposed body surface. In hot weather, the arterioles expand and allow an increased flow of blood through the skin capillaries. Well below the skin surface is a layer of subcutaneous fat that serves as insulation, retaining body heat in the underlying muscle tissues.

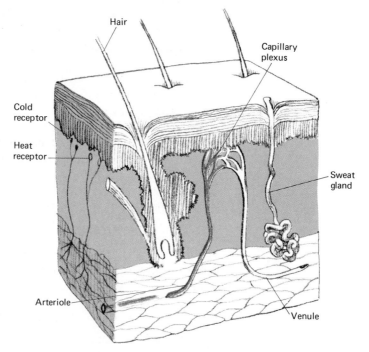

Hair

Capillary plexus

Cold receptor

Heat receptor

Sweat gland

Arteriole

Venule

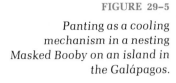

FIGURE 29–5

Panting as a cooling mechanism in a nesting Masked Booby on an island in the Galápagos.

piration called *panting*. Normally, a dog will breathe some 15 to 30 times per minute, but in panting, this rate of respiration rises to over 300 times per minute. Evaporation is facilitated by this increased passage of air volume over the wet surfaces of the throat and tongue. This rapid respiration does not include the bulk of the lung because otherwise the loss of CO_2 from the blood into the lungs would be excessive and the acidity level in the blood would change (Chapters 27 and 28). Cattle and sheep are among the many animals that are capable of both sweating and panting.

Another method of cooling the body is used by kangaroos, rabbits, and cats. They *lick* the fur of the belly and limbs and cool down by the *evaporation* of the moisture. This is a relatively inefficient method compared with sweating or panting, however.

Small mammals in general do not possess true sweat glands and they avoid panting. They also avoid exposure to high temperatures; cooling by evaporation would dehydrate them because they have a relatively high surface area compared with body volume. Accordingly, they avoid the heat by living in underground burrows, especially during diurnal hours.

Birds also lack sweat glands. They pant to some extent but the extensive *air sac system* connected with their lungs is probably the principal way by which the heat is dissipated from the body, at least in flight. Panting may be particularly well observed in nesting birds on oceanic islands, where a parent must spend long hours in the full sun exposed to high heat stress (Figure 29–5).

Hibernation and estivation

Hibernation and estivation are physiological adaptations to marginal temperature conditions and environmental stress caused by a relative lack of humidity in the atmosphere. In **hibernation,** an animal is able to escape the cold temperatures of the temperate-zone winter by remaining dormant during this period, when food is less readily available. If it was active at these times, the low external temperature would require a high metabolic rate to maintain body temperature, whether the animal was a poikilotherm or a homeotherm.

The physiologist defines hibernation of warm-blooded animals as a state in which the body temperature is greatly decreased and the metabolism, heart rate, and respiration are considerably reduced (Figure 29–6). Three orders of mammals contain species that hibernate through cold winters. These are the insectivores, such as shrews and moles; the bats; and the rodents, which include such animals as ground squirrels, hamsters, hedgehogs, and woodchucks. Bears do not actually hibernate, although they may sleep through a major part of the winter. Their body temperature drops only a few degrees and thus they can be awakened without a long period of rewarming. In contrast, hibernating rodents or bats are completely torpid.

During the cold period, when the body temperature drops to a low level, the animal can sleep for months while it slowly uses its reserves of fats built up during the preceding summer and fall. Thus the hibernating homeotherm reaches a state comparable to that of a cold-blooded animal, for its body temperature remains close to that of the surroundings and rises and falls with it. When the outside temperature increases in the following spring, the animal warms up again and returns to the homeotherm state. During hibernation, the heart rate may go down to only a few beats per minute, respiratory movements become very slow, and the oxygen consumption may drop to one-thirtieth to one-hundredth of normal.

Even birds may hibernate during cold winters; for example, a California desert poorwill (in the nighthawk family, Caprimulgidae) hibernates in rock crevices for the winter months. Hummingbirds experience a *daily* temperature drop. While feeding during the daylight hours, largely on flower nectar, their body temperature is

FIGURE 29–6

A hibernating ground squirrel in its underground chamber. Curled in deep torpor, a hibernating squirrel is not disturbed even if dug up and handled.

quite high, and they have a very high metabolic rate. However, at night the body temperature drops, along with the metabolic rate. Because of the small body size of the hummingbird, this decrease in metabolism is essential or it would burn up all of its body reserves by the time that morning came.

In **estivation** an animal that inhabits desert areas that are seasonally hot or the tropical regions affected by a severe dry season goes into a state of suspended animation, similar to that of hibernation. Generally, the reptile or mammal goes underground into a cool burrow where humidity may be maintained at a reasonably high level, and the body temperature falls to the level of the surrounding soil. At this lowered body temperature, the reserves of fat and carbohydrates are used up at a slower rate and the animal passes the unfavorable season without the need for regular activity. Many insects in tropical dry forest areas pass the dry season in a state of estivation. Adult tropical butterflies may hang motionless on trees and shrubs most of every day to reduce their metabolic needs for flower nectar at a season when such food sources are scarce (Figure 29–7).

FIGURE 29–7

Estivating adult pierid butterflies (Eurema daira) during the dry season in northwestern Costa Rica.

Summary

Animals have developed two principal ways of regulating their body temperature (and hence metabolic rates) in relation to environmental temperatures. The cold-blooded, or poikilothermic, groups have changeable body temperatures that largely depend on the temperature of their surroundings, although many species of this type do have behavioral features that aid in adjusting their body warmth. These behavioral traits include basking, seeking shade, and rapid wing or body muscle vibrations. Warm-blooded, or homeothermic, animals (the birds and mammals), rely on internal heat production to regulate their body temperature. They may maintain an almost constant body temperature with the aid of metabolic heat, insulation, shivering, sweating, panting, licking and subsequent cooling by evaporation, air sacs (in birds), and behavioral traits such as hiding in underground burrows during the warm daylight hours. In hibernation, a temperate-zone animal is able to escape the cold winter temperatures by remaining dormant, greatly decreasing body temperature, metabolism, heart rate, and respiration. Hummingbirds exhibit nocturnal dormancy every day. In estivation, the desert or dry-tropical-region animal goes into a state of suspended animation similar to that of hibernation, but to escape unfavorably hot or dry periods of the year.

30 The integration and coordination of bodily processes: hormones

The processes of digestion and excretion, internal transport, respiration and gas exchange, and the regulation of tissue temperature show that plants and animals must integrate and coordinate a complex range of metabolic activities to produce a viable, functioning organism capable of growth and reproduction. The normal development of a plant or animal depends on the interplay of a number of internal and external factors. In earlier sections of this book we have looked at the impact of environmental and genetic factors at the population and individual levels. Now we shall begin our consideration of some of the internal factors that affect coordination, growth, and the differentiation of specialized tissues in the higher plants and animals. Chief among these internal controlling factors in plants and animals are the hormones.

Plant hormones

Hormones, whether found in plants or animals, are substances produced in one part of the body and transported to another part where they have a particular effect on the tissues involved. In fact, the word *hormone* comes from the Greek word meaning "to excite."

However, many hormones also have inhibitory influences. Thus hormones serve as chemical messengers in the body, complementing the electrical messages sent via the nervous system in animals. In plants, most of the hormones function as growth regulators, controlling the development of the shoot stem, leaves, and flowers, as well as fruits. The best-studied hormones include the auxins, gibberellins, cytokinins, and florigen. They do not act alone but frequently together to produce a particular effect on the target organ.

Auxins

The auxins were among the first hormones to be investigated, and some of the first thorough work was done by Darwin and his son, Francis, published in 1881 in *The Power of Movement in Plants*. Working with young seedlings of grasses and oats, they observed that if the tip of the hollow cylindrical sheath that encloses the young shoot is covered by a hollow tube painted black, it does not bend toward light coming in from the side (as from a window or a laterally placed lamp). Shoots with their tips freely exposed or enclosed in transparent glass tubes bend strongly toward light coming in from the side (Figure 30–1). The Darwins concluded that a material produced in the tip of the shoot sheath must be transmitted from the upper to the lower part of the plant, causing the shoot to bend in the strongly evidenced positive phototropic response, for if the tip was removed from the plant by surgery, the sheath failed to bend at all.

Some 45 years later, a Dutch plant physiologist, Fritz Went, demonstrated that the growth stimulus moving downward from the tip

FIGURE 30–1

Auxin and light in the shoot tip of an oat seedling. The distribution of the auxin is indicated by dots. The difference in auxin concentration on the two sides actually starts at the tip, which is the light-sensitive site, and moves down the shoot as auxin is transported toward the base. As a result of this distribution, the shaded side grows more rapidly and the shoot bends toward the light source.

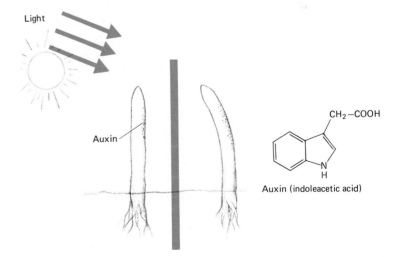

Light

Auxin

CH_2-COOH

Auxin (indoleacetic acid)

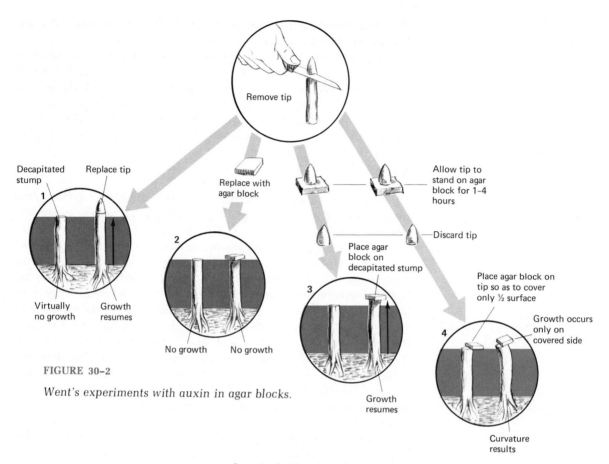

FIGURE 30–2

Went's experiments with auxin in agar blocks.

was a chemical. He placed cut shoot tips on blocks of agar (a gelatinous material made from seaweed and similar to gelatin in consistency), and within an hour a colorless substance had moved into the agar; when he put the blocks of agar (now lacking the tips) on the cut ends of shoot stumps, the stumps behaved as if their tips had been replaced. They would then start growth again and bend to the side in response to light; if the agar block was placed on only one side of the total breadth of the shoot, the shoot would bend toward that side, even in total darkness (Figure 30–2).

These and subsequent experiments made it possible to isolate the chemical stimulus, which became known as **auxin,** a term derived from the Greek word *auxein,* meaning "to increase." Since that time, many different chemicals have been found in plants that fall in the auxin class. The most common of these natural auxins is *indoleacetic acid* (abbreviated IAA), which has been found in the plants of many families. IAA and other auxins are produced in the sheath tips of grasses, in the apical meristems of shoots, and in roots. Auxins also appear in young leaves, fruits, and plant embryos.

Auxins cause the shoot to elongate. The hormone is produced in the rapidly dividing cells of the meristem and moves downward by diffusion (and probably by transport in the phloem), stimulating cell elongation in the lower part of the shoot. The hormone stimulates only cells directly under the point of release. However, the work of Winslow Briggs, among other researchers, more than a decade ago demonstrated that the lateral transport of auxins away from light reduces the auxin supply on the lighted side of a stem and thus is responsible for the phototropic bending response. The dark side grows more rapidly than the illuminated side of the plant. This asymmetrical growth promotes bending toward the light source.

In the roots, auxins promote the growth of small roots coming out from the main roots. Synthetic auxins are now commonly employed to induce the formation of adventitious roots in nursery cuttings. Auxins also affect leaf drop. When the production of auxin in the leaf is diminished, an **abscission** layer—a layer of weak and thin-walled cells across the base of the **petiole** (leaf stem)—becomes drier and disintegrates until the leaf is held on to the tree by only a few strands of vascular tissue. Abscission layers are also formed in the stems of fruits such as oranges and apples; and after the fruit has ripened, this layer breaks, allowing the fruit to drop. Commercial fruit growers will spray orchards with synthetic auxins in order to keep the ripe fruit on trees until it can be harvested.

Auxins apparently work basically by increasing the plasticity of the cell wall. Thus the cell is able to enlarge, owing to the pressure of water within its vacuole, and it will continue to take up water until the wall exerts sufficient resistance to stop cell enlargement.

Gibberellins

Gibberellins were discovered in 1926 by a Japanese worker, E. Kurosawa, who was studying a disease of rice plants called "foolish seedling disease." Rice plants with this disease grew rapidly but were spindly and sickly, tending to fall over. Kurosawa discovered that a fungus, *Gibberella fujikoroi*, which is parasitic on rice seedlings, produces a chemical causing these symptoms. Kurosawa named this chemical substance **gibberellin,** and it was soon isolated and chemically identified in the 1930s. Since that time gibberellins have been found in many higher plants and more than 20 different types are known (Figure 30–3). They have a dramatic effect on stem elongation in plants. They stimulate cell division as well as cell elongation and affect leaves as well as stems. In simple genetic mu-

FIGURE 30–3

The effects of gibberellin. The cabbage plant on the right was treated with a dose of this plant hormone, and though it is the same age as the cabbage on the left it shows great stem elongation and earlier maturation to the flowering stage.

tants that cannot produce gibberellins, the entire plant is dwarfed. If gibberellins are added to the plant, these dwarf mutants will grow to the size of normal tall plants and be indistinguishable from them. Gibberellins also stimulate flowering in many plants that require exposure to long days (as in the summer) to flower or to cold weather, as the temperate-zone plants that flower in their second year of growth. In other species, gibberellins promote the maintenance of juvenile characteristics and inhibit flowering as long as they remain at a certain concentration. Gibberellins have proved useful with commercial growers of celery because they induce rapid growth that promotes tender stocks. They may also be used to artificially stimulate flowering and break dormancy in some horticulturally valuable species.

Cytokinins

Another group of growth-regulating hormones in plants was discovered in 1941 when a Dutch plant physiologist, Johannes van Overbeek, found that coconut milk (which provides nutrients for the developing coconut palm embryo) contained a potent growth factor. Other investigators found that this growth factor was able to induce previously differentiated cells, such as adult carrot cells or tobacco stem cells, to begin division again. These **cytokinins** have now been found in about 40 different species of higher plants, especially in the actively dividing tissues. Their chemistry and activity are still under intensive investigation. Apparently, they interact with auxins to produce the total growth pattern of the plant.

Florigen

The hormonal control of flowering in plants is an active area of current research in plant physiology. Considerable evidence has shown the existence of a moving chemical substance, probably a hormone, that translocates from the leaves to the stem buds and induces flowering. This hypothetical hormone has been given the name **florigen.** It was first discovered by observations on the effect of photoperiodism on flowering (Figure 30–4). The length of the day critically determines the timing of flowering of most angiosperm species. Those that flower in the spring were found to have flowering induced by a short light period, whereas those that flower in the early summer needed a long day period. Short-day plants include asters, goldenrod, ragweed, chrysanthemums, and poinsettias—plants that flower when the day length is below some critical value (e.g., 12 hours), as in the spring or fall. Long-day plants include the clovers

FIGURE 30-4

Photoperiodism and the hormonal control of flowering in plants: responses of typical summer-flowering (long day) plants and spring-flowering (short-day) plants to a short day length and a long day length.

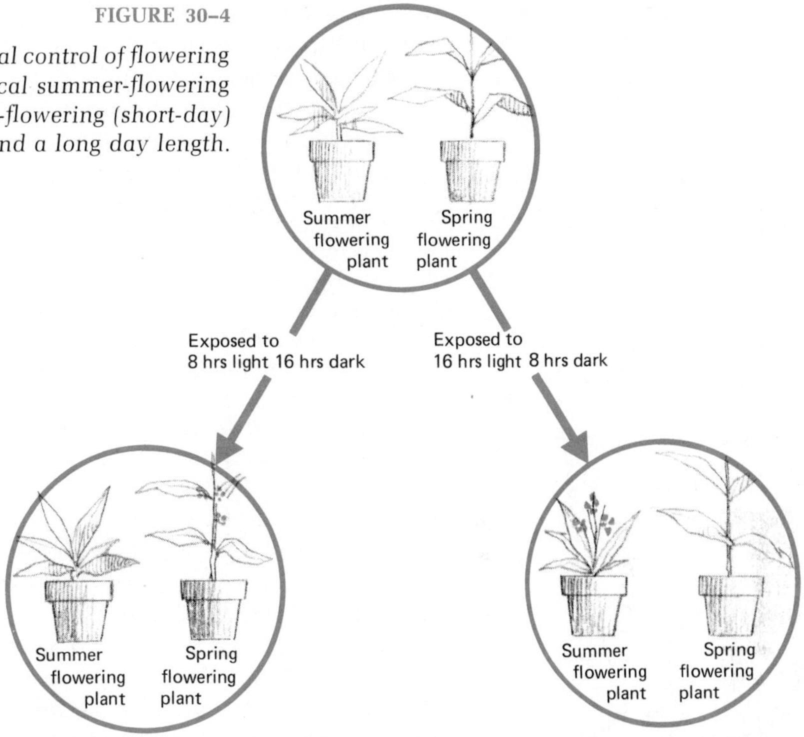

and black-eyed susans, the common summer roadside flowers. Apparently, the organism times its production of this hypothetical hormone by timing the relative length of light and dark conditions it detects in the daily cycle. Even a single leaf exposed to a single long-day photoperiod is enough to stimulate a long-day flowering plant to begin flowering. However, a flower-inducing hormone has not yet been isolated and identified, even though many workers are actively working on the problem at present.

Animal hormones

Like plants, both invertebrate and vertebrate animals have a wide variety of hormonal mechanisms. In the lower animals, only hormones in the arthropods have been reasonably well studied, particularly those of insects. V. B. Wigglesworth did much of the pioneering work on insect hormones in the 1930s in England while studying the metamorphosis of insects. He discovered that a *brain hormone* stimulates a pair of glands in the insect's thorax area (where the legs are

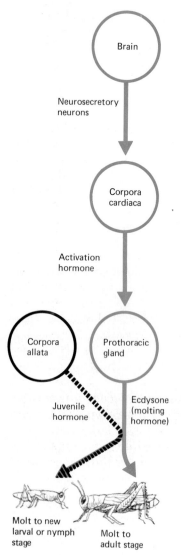

Molt to new
larval or nymph
stage

Molt to
adult stage

FIGURE 30–5

The hormonal control of molting and metamorphosis in insects. Activation hormone from neurosecretory neurons in the brain is passed along the axons to the corpora cardiaca (located behind the brain, next to the aorta), from which the hormone is liberated. The flow of activation hormone is controlled by increased sensory nervous impulses from stretch receptors located in the insect body wall. As the larva or nymph grows to a certain size, the increased amount of activation hormone in the blood stimulates the prothoracic gland to release molting hormone (ecdysone). As long as the copora allata glands release juvenile hormone simultaneously, the insect will molt to a new larval or nymph stage. But when the immature insect has molted to the stage just prior to adulthood, juvenile hormone stops and the next molt is to the adult stage.

attached) to secrete another hormone, often called *ecdysone*, or *molting hormone*, that causes the insect to shed its exoskeleton and develop a larger new one in its place (Figure 30–5).

Insects go through a number of younger stages before molting to an adult. At this last point, Wigglesworth found that a third hormone is involved, *juvenile hormone*, which is produced by a pair of glands (the corpora allata) located close to the brain. As long as juvenile hormone is present in high concentration at the time of molting, the animal simply molts to another immature stage. However, when the juvenile concentration drops in the final larval stage, the larva will molt into the pupal stage. In the pupa there is no juvenile hormone and when it molts an adult emerges.

The close physical association between hormonal system and nervous system in these animals is even more pronounced in the vertebrates. We shall look at the mammalian hormonal system in human beings in detail to achieve some appreciation of the diverse ways in which chemical control regulates the many functions of a complex multicellular animal.

Hormones and endocrine glands

Hormones are produced by tissues and organs called **endocrine glands.** *Endocrine* means "secreting internally." The hormones are secreted by these endocrine tissues directly into the blood and no special ducts or tubes carry the hormone product of the endocrine gland to their target tissues. Instead, the circulatory system accomplishes the transport task through picking the hormones up in the capillary network present in the endocrine gland and carrying them to the capillaries in the target tissues. Thus endocrine glands are often also called the ductless glands.

The pancreas and its hormones

In 1889 the two German physiologists, Joseph von Mering and Oscar Minkowski, began experimenting on the role of **pancreas glands** in dogs. They discovered that the animals that had their pancreas removed soon began to show many physiological deficiencies, and after a short time they died. Although the dogs had continued to have hearty appetites after the operation, they lost weight, lacked energy, and in general showed many symptoms of starvation, including the start of metabolic breakdown of stored fat and cellular protein for energy needed to keep alive.

These two investigators also noticed the curious fact that ants gathered in the kennels where the sick dogs were kept and were particularly attracted to the places where the dogs had urinated. Yet there were no such gatherings of ants in the kennel enclosures of the healthy dogs. The two Germans discovered that the blood sugar (glucose) level in the sick animals was abnormally high. In fact, it was so high that the animals were excreting large amounts of glucose in their urine. Despite the abundance of blood sugar, then, these animals were unable to use it and were "starving," just as do people who have the disease *diabetes mellitus*. Later experiments of these workers showed clearly that the absence of certain parts of the pancreas was associated with this disease.

In addition to the digestive secretions produced by the pancreas, there are scattered clumps of cells called **islet cells,** or *islets of Langerhans,* in the pancreas that produce a hormone needed for glucose utilization by the other cells of the body (Figure 30–6). The actual hormone, a small protein called **insulin,** was isolated in 1922 by F. G. Banting and C. H. Best, working with J. J. R. MacLeod at the University of Toronto. The following year, Banting and MacLeod received the Nobel prize for this important work. They discovered that if insulin hormone extracts were injected into dogs showing diabetic conditions, the dogs showed marked improvement. Today, millions of people who once would have become invalids and died prematurely are able to lead relatively normal lives with periodic injections of insulin extracted from natural sources.

The actual function of insulin seems to make cells more permeable to glucose. The diabetic cannot use glucose because the sugar cannot get into these cells, and thus the cell must turn to other molecular sources for energy—breakdown of the stored fats and proteins.

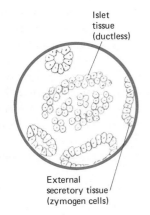

Islet
tissue
(ductless)

External
secretory tissue
(zymogen cells)

FIGURE 30–6

Human pancreas cells. Islet cells (which have no ducts) secrete the hormone insulin into the circulating blood. Zymogen cells secrete digestive enzymes into the pancreatic duct, which carries the enzymes to the duodenum.

Because the concentration of blood sugar consequently rises to such high levels that it cannot be reabsorbed by the kidney tubules, large amounts of glucose escape through the Bowman's capsules (discussed in Chapter 26) and flow out the ureter in the urine. An excessive amount of water also must be discharged with this sugar, and this loss of body fluid itself can become so great in the untreated diabetic that it can lead to circulatory collapse and death.

The thyroid and its hormones

Another major endocrine organ is the **thyroid.** Human beings have a single united thyroid gland, but most vertebrates have two thyroid glands located in the neck (Figure 30-7). Several thyroid hormones have important functions. Chief among these is the hormone **thyroxin,** which controls and stimulates oxidative metabolism in the body. If the thyroid is removed by surgery or is damaged in some way and secretes too little hormone, a condition called *hypothyroidism* develops. In newborn children, this condition creates dwarflike creatures, called *cretins*, with extremely low intelligence, who seldom achieve a mental age of more than 4 or 5 years. This condition can be cured by injecting thyroxin hormone into the body. These symptoms of hypothyroidism all result from a slowing down of the metabolic reactions of the body cells. Thus thyroid hormone seems to act to speed up the metabolism.

If the thyroid is overactive, a condition known as *hyperthyroidism* results. The person has higher than normal body temperature, perspires profusely, loses weight, becomes irritable, has generally weak muscular control, and has high blood pressure. Hyperthyroidism also causes the eyeballs to protrude. To treat this condition, part of the thyroid gland may be removed by surgery or it may be destroyed

FIGURE 30-7

The thyroid gland produces thyroxin, which helps regulate the overall metabolic rate of the body. Follicles in the gland store the secreted hormone.

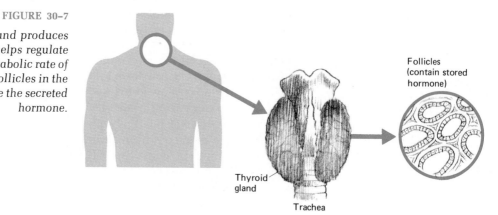

Follicles
(contain stored
hormone)

Thyroid
gland

Trachea

by radiation, or drugs that inhibit the activity of the thyroid may be administered.

The role of the thyroid in controlling metabolic activity via the hormone thyroxin was first discovered when it was observed that iodine was needed in the diet to maintain metabolic activity. In people, or animals such as certain fish, who do not get sufficient iodine in their diet, enlargements of the thyroid occur, a condition known as *goiter*. There is also a general slowdown in metabolism, including a slower-than-normal heartbeat, physical lethargy, mental dullness, loss of hair, and obesity. The discovery of the chemical structure of the hormone thyroxin showed why iodine was important in the diet, for it is basically an amino acid containing four atoms of iodine. Our understanding of how this hormone works in speeding up oxidative metabolism is still very rudimentary.

The adrenal glands and their hormones

Another important hormone-producing endocrine tissue is found in the two **adrenal glands,** which lie very near the kidneys. Each adrenal gland is actually composed of two parts, which in lower animals are separated into different glands (Figure 30–8).

The inner adrenal **medulla** secretes two hormones, **adrenalin** (also called epinephrine) and **noradrenalin** (norepinephrine). Adrenalin causes the acceleration of the heartbeat, a rise in blood pressure, increased secretion of insulin from the pancreas, increased conversion of the animal starch called glycogen into the glucose sugar in the liver, and the release of this glucose into the blood from the liver. The animal may consume more oxygen and release more red blood cells into the circulatory system from the spleen; it may have greater resistance to fatigue, have its hair become erect, have goose pimples

FIGURE 30–8

The adrenal gland, located on top of the kidney, is a dual endocrine organ. The medulla secretes noradrenalin and adrenalin, and the cortex secretes several other hormones, including cortisone.

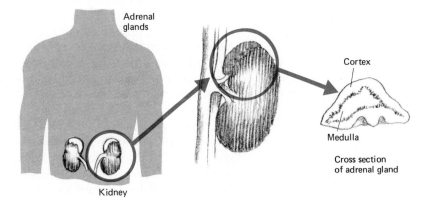

FIGURE 30–9

A section of the human brain, showing the location of the pituitary gland and its several divisions.

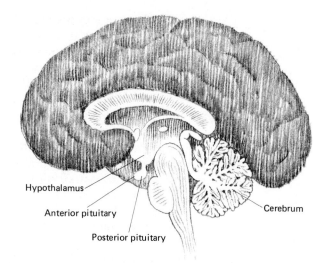

and dilated pupils, and exhibit many other effects. These are responses to stress, which may be caused by anger, fear, pain, or other highly emotional states. Noradrenalin has very similar affects. Both these hormones help mobilize the body's defenses in emergencies.

The **adrenal cortex,** the outer barklike area of an adrenal gland, produces a great number of hormones, over 50 of which have been isolated and identified so far. As a result of the production of these hormones, the adrenal glands are essential to life, and death soon results if they are removed. One of these hormones is the well-known **cortisone;** together with several related steroid hormones, they act as antagonists to insulin, promoting the formation of more glucose and increasing the blood sugar level. A person who has Addison's disease, such as the late President Kennedy, has adrenal cortices that are insufficiently active; to alleviate these symtoms and restore blood sugar levels to normal, the hormones in this group can be administered by regular injection.

Other hormones produced by the adrenal glands stimulate the cells around the tubules of the kidney to regulate the amount of potassium and sodium as well as chloride and water in the blood. This affects the volume as well as the pressure of the blood, obviously. Thus many of the adrenal hormones act together with insulin and other important elements in the intricate regulation of the body's internal fluid environment.

A third group of adrenal cortex hormones are similar in chemical structure and functioning to the sex hormones produced by the gonads. They are particularly important in controlling the development of secondary sexual characteristics in the male, such as deepening of the voice, maturation of the genital organs, and growth of facial hair. If these hormones are secreted to excess levels in the female, she will begin to develop masculine characteristics. Cortical tumors in young girls may result in such effects on their general features, suppressing feminine characteristics.

The pituitary gland and its hormones

Lying just below the brain, the knoblike **pituitary gland** consists of two main sections that are distinct in both developmental origin and function (Figure 30–9). The *anterior pituitary lobe* in a sense is the master control gland of the body. It produces at least seven hormones of great importance (Figure 30–10). They include a *growth hormone* that controls the normal processes of growth. An oversupply of this hormone during childhood will result in a giant; an un-

FIGURE 30–10

The target organs and functions of the six hormones from the anterior pituitary gland: prolactin, FSH, LH, ACTH, TSH, and growth hormone.

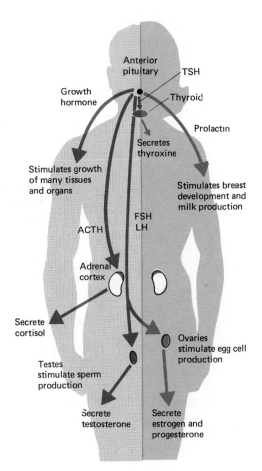

Anterior pituitary

TSH

Growth hormone

Thyroid

Prolactin

Secretes thyroxine

Stimulates growth of many tissues and organs

Stimulates breast development and milk production

FSH LH

ACTH

Adrenal cortex

Secrete cortisol

Ovaries stimulate egg cell production

Testes stimulate sperm production

Secrete testosterone

Secrete estrogen and progesterone

dersupply will result in a midget. Another hormone, **prolactin,** stimulates milk production by the female mammary glands, following the birth of a baby. A third hormone, *melanocyte-stimulating hormone,* causes darkening of the skin by stimulating the spreading of dark pigment molecules in specialized pigment-containing cells in the epidermal layer of the skin. Four other hormones exert controlling effects on other endocrine organs, including the thyroid, the adrenal cortex, and the gonads. Without the anterior pituitary, these other endocrine glands stop functioning. Thus the master control-switch activity of the pituitary is vital to the well-being of the body.

The *posterior pituitary lobe* and the adjacent part of the brain called the *hypothalamus* seem to function together to exert nervous system control over the master gland, the anterior pituitary.

The gonads and their hormones

In the human male and female genital systems, the **gonads,** or sex organs, produce a number of important hormones in addition to sperm and egg cells (Figure 30–11). The key male sex hormone, **testosterone,** is produced by the interstitial cells lying between the spermatocyte cells undergoing meiosis and producing sperm. These interstitial cells begin producing the male sex hormone when several

FIGURE 30–11

Gonadal hormones, with their target areas and effects on secondary sexual characteristics in a man and a woman.

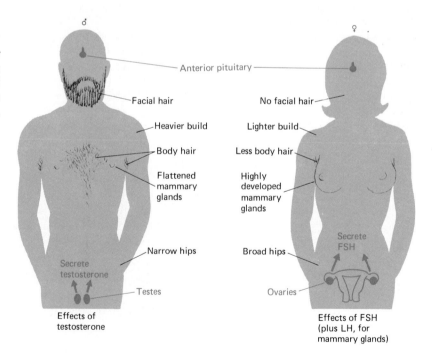

Hormonal control mechanisms

The endocrine system and the nervous system constitute the two great communication systems of the body. Rather than having their own separate network of conduits throughout the body as nerve impulses do, hormones serve as blood-borne messengers that regulate cell function in specific ways. They help control many bodily functions: in particular, reproduction, organic metabolism, energy balance, and mineral metabolism. The ability to reproduce is dependent on a normally functioning endocrine system; however, no other bodily function absolutely requires hormonal control, and, for this reason, the endocrine system is not strictly essential for life. A life without it, though, would be extremely precarious and abnormal; the individual would be unable to adapt to environmental alteration or stress, and his or her physical and mental abilities would be drastically impaired. Survival would require the constant attention bestowed on a greenhouse orchid in Alaska.

The concentration of a hormone in the blood depends on several general factors: the rate of secretion, transport phenomena in the blood, and the rates of hormone inactivation and excretion from the liver or kidneys. No hormone is secreted at a constant rate. Intracellular synthesis of the hormone depends on a rather irregular supply of chemical precursors either from the meals taken in by the body or from other body cells. The release of the hormone by the secretory cells into the blood may depend on emotional state or other variable parameters. The rate of removal from the blood affects the concentration of a hormone in the plasma as much as the rate of secretion. These continuous destructive processes are located primarily in the liver and the kidneys. Because for many hormones urinary excretion is directly proportional to glandular secretion, physiologists often use the rate of excretion of a hormone as an indicator of active secretory rate. Also, transport in the blood occurs not only as free molecular forms; many of the hormone molecules that circulate in the blood are bound to various plasma proteins. This phenomenon can be very influential when the carrier protein, usually synthesized by the liver, is in excess or deficit.

Finally, hormones exert their effects by altering the rates at which particular cellular processes proceed. Hormones never initiate a process; they merely *alter its rate.* For example, the absence of insulin secreted from the pancreas results in markedly reduced uptake of the sugar glucose by body cells, but not cessation of uptake. In this case, the hormone insulin is required to favorably increase the activity of a crucial enzyme involved in the process that synthesizes the starch glycogen by stringing glucose molecules together. There are additional ways besides alteration of enzyme activity that hormones accelerate or decelerate cellular processes, but the other principal category is alteration of the rate of transport of a substance through the cellular membrane, either by stimulation or inhibition. Thus, for instance, a second and separate effect of insulin is to somehow increase the rate of glucose transport through cell membranes.

hormones from the anterior pituitary are secreted at the onset of puberty. The flow of testosterone into the circulatory system, starting at this time, stimulates the development of secondary sexual characteristics such as the growth of body hair, deepening of the voice, the development of larger and stronger trunk muscles, and maturation of the various glands associated with sperm and semen production. If the testes are removed through castration before puberty, the secondary sexual characteristics will never appear.

In the human female, the ovaries produce gametes and secrete sex hormones. The two female sex hormones produced by mature ovaries are **estrogen** and **progesterone.** These, like testosterone in the male, stimulate development of the female secondary sexual characteristics, such as an increase in the width of the pelvis, the onset of the menstrual cycle, the growth of pubic hair, and the development of the breasts. The complex rhythmic variations in the secretions of these hormones and their connection with the secretions from the pituitary gland will be explored in Chapter 34. For now let us note that the quantity of these two hormones, estrogen and progesterone, in the blood has profound effects through a monthly cycle in releasing the egg from a follicle in the ovary, in preparing the uterus to receive the embryo, and in preparing the body for a new cycle if no fertilization occurs during a particular 28-day period. Hormonal changes in the female also take place during pregnancy, the birth process (parturition), and lactation, as well as at the end of a woman's reproductive life, the **menopause,** a period lasting a year or two and usually coming between the ages of 45 and 55.

FIGURE 30–12

A general view of endocrine glands in human beings.

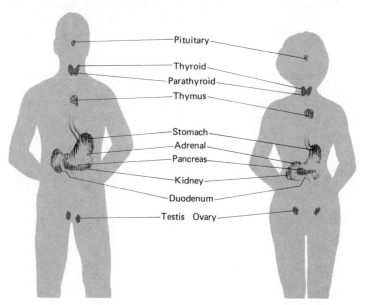

Other endocrine glands

There are many other essential human endocrine glands, such as the **thymus** in the neck, which produces a hormone important in stimulating the body's defenses against foreign elements in the blood; the **parathyroids,** located on the surface of the thyroid, which help regulate the calcium-phosphate balance between the blood and the other tissues; the **pineal gland,** associated with the forebrain, the exact functions of which are still unknown; and other endocrine tissues located in the stomach and upper small intestine (Figure 30–12). Undoubtedly a great many more hormones remain to be discovered, even in human beings, the best studied animal for hormonal integration and coordination of the bodily processes.

Summary

Hormones represent a major class of internal chemical factors that affect coordination, growth, and differentiation of specialized tissues in the higher plants and animals. They serve as chemical messengers to remote areas of the body. In plants, most hormones function as growth regulators. Auxins, such as indoleacetic acid (IAA), cause the shoot to elongate and new roots to form, and their lack causes leaves and ripe fruits to drop. Gibberellins stimulate cell division as well as cell elongation in plants. Cytokinins act as division-stimulating agents and interact with auxins to produce the total growth pattern of the plant. A hypothetical hormone called florigen is believed to induce flowering in plants. Invertebrate and vertebrate animals also have a wide variety of hormonal mechanisms. Metamorphosis in insects is controlled by ecdysone, or molting hormone, a brain hormone, and a juvenile hormone. Endocrine glands of great importance in human beings include the pancreas (which secretes insulin), the thyroid (which secretes thyroxin), the adrenal glands (which secrete adrenalin and noradrenalin from the medullar portion and cortisone and a host of other vital hormones from the other cortex), the pituitary gland (which secretes master hormones for growth, milk production, skin darkening, and control of other endocrine glands), the gonads (which secrete testosterone, estrogen, and progesterone), the thymus, the parathyroids, the pineal gland, and glands in the stomach and small intestine.

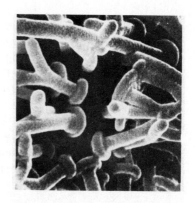

31 The integration and coordination of bodily processes: the nervous system

Just as plants are unique in having photosynthesis and cell walls, animals have developed irritability as a property of their cells and body systems to a far greater degree than any plant species. Plants do have behavior, which is usually integrated and coordinated by relatively slow hormonal reactions; bending toward light, in the form of phototropism, and the relatively rapid closing of the leaf trap of the insectivorous Venus flytrap are examples of plant behavior that are coordinated through chemical means and through the passive as well as active processes connected with diffusion.

Of course, instantaneous response is not needed in most plants, which lead a sedentary life; most animals, however, are mobile and require a more immediate response to any stimulus. The evolution of rapid nervous control, or a system analogous to it, was essential for coordination of muscular movements in the higher multicellular animals. Thus in the invertebrates above the sponge level and the vertebrates, we observe a progressive development or evolution of nervous systems. However, even in the ciliate protozoans a network of fibriles connects the bases of the cilia and coordinates their beating (Figure 31–1). There are also specific areas such as eye spots in protozoans that are specialized for stimulus reception.

Thus, even the unicellular animals exhibit a high development of the capacity to respond to stimuli, namely, *irritability*, which we have already defined (Chapter 20) as a universal characteristic of protoplasm. This irritability involves three components: *stimulus reception*, *conduction* of the signal, and *response* by the animal.

At the minimum level in all animal groups above the level of the sponge, simple nervous pathways exist that provide conduction of the signal and coordination of a response. As we have noted, however, even the single-celled animals are capable of at least stimulus reception and conduction through the slow conductivity of the cytoplasm or the cell membrane itself; on reaching the target area of the cell, a response ensues.

The nervous pathways of the higher animals are composed of at least three separate cells. The initial reception of a stimulus is achieved in a **receptor** cell, which becomes excited in some way when light, sound, pressure, chemicals, or other stimuli impinge on it. The stimulus generated in this receptor cell is passed on to a **conductor** cell, which is specialized for transmitting the impulse over a considerable distance, ending up at an **effector** cell, which is specialized for a response. In most animals this effector cell is a muscle cell or gland cell. Progressively more complex nervous systems introduce more conductor cells between the receptor and effector, and interconnections between different pathways allow flexibility in response to the incoming stimulus from the receptor. Let us now look at the basic structural units involved in the nervous system plan in invertebrates and vertebrates.

FIGURE 31–1

Nervous systems in lower invertebrates: (A) the network of nervous fibrils in a ciliate protozoan; (B) the nerve network in a Hydra.

A
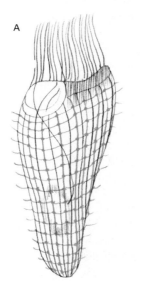

B
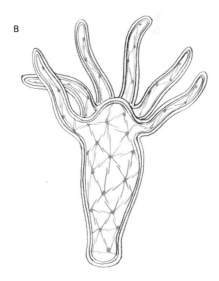

The neuron

The basic unit of the nervous system is the individual nerve cell, or **neuron.** However, it should be noted that even in human beings, only 10 percent or so of the cells in the nervous system are neurons. The remainder are specialized surrounding cells that probably sustain the neurons metabolically and support them physically. The typical neuron consists of a number of long cylindrical processes of membrane that extend from a large central mass of cytoplasm, the **cell body,** where the nucleus is located. Information is transmitted between the various parts of the body over the long neuron processes.

Different nerve-cell processes carrying information between the same two areas, or even carrying the same types of information, are usually grouped together in bundles. These bundles are called *nerves* if they are part of the peripheral nervous system and *pathways,* or tracks, if they occur within the central nervous system. The brain and spinal cord together form the *central nervous system,* which in vertebrates is protected by being housed inside the bony skull and vertebral column or backbone (Figure 31–2). The nerve

FIGURE 31–2

The central nervous system (brain and spinal cord) and peripheral nervous system of the human trunk, viewed from behind.

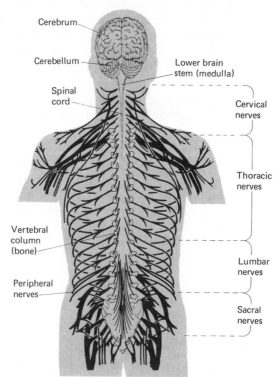

Cerebrum

Cerebellum

Lower brain stem (medulla)

Spinal cord

Cervical nerves

Thoracic nerves

Vertebral column (bone)

Peripheral nerves

Lumbar nerves

Sacral nerves

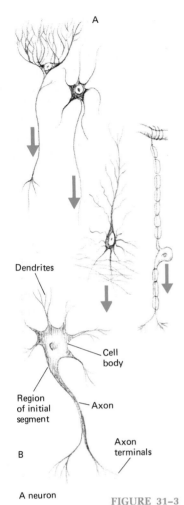

Dendrites

Cell
body

Region
of initial
segment

Axon

Axon
terminals

A

B

A neuron

FIGURE 31-3

Neuron cells: (A) diagrammatic representations of three of the many types of neurons found in the brain and one sensory neuron (right side); (B) the various structural parts of a typical neuron cell (diagrammatic).

cells or parts of the nerve cells that lie outside the central nervous system are known collectively as the *peripheral nervous system.*

The neuron can be divided structurally into three or four parts, each associated with a particular function. The **dendrites** form a series of branched outgrowths from the neuron's cell body; these are essentially a receptor extension of the cell membrane. Here the cell receives signals from the junctions of the dendrites with other neurons or specialized receptor cells. Each junction, or tiny gap, between two nerve cells is called a **synapse.** A neuron may have thousands of synaptic junctions on the surface of its dendrites and cell body that enable the cell to receive information from hundreds of other nerve cells (Figure 31–3).

The *cell body,* which contains the nucleus and many of the organelles such as mitochondria involved in metabolic processes, handles the maintenance of the neuron and its growth in early development. The output response of a neuron travels out the long **axon** from the cell body, and it depends on the number and type of input signals fed into the cell body and axon through the synaptic junctions of the dendrites. Frequently, the single long axon fiber coming out of the cell is covered with a thick fatty sheath, called **myelin,** which gives the characteristic white color to nerve fibers. Finally, the impulse arrives at the *axon terminals,* which are enlarged areas at the end of the long, usually single axon process extending from the cell body. These areas are responsible for transmitting a signal from the neuron to the cell touched by the axon terminal. This next cell may be a gland cell, a muscle cell, or another neuron.

A neuron that picks up an impulse from a receptor cell is called a **sensory neuron;** those sending their impulses to effector cells are called **motor neurons.** The neurons lying between the sensory and motor nerve cells are called *association neurons.* These neurons do not actually touch one another; a tiny gap, the *synapse,* is always present. About 97 percent of all nerve cells in human beings are association neurons. Stimuli involving memory or language may ultimately involve thousands of association neurons in the brain, whereas simple reflexes coordinated at the spinal-cord level may have no association neurons.

The transmission of nerve signals

The nature of the message passed along a nerve is rather mysterious to most of us. Basically, the electrical impulse that is transmitted along a nerve is produced by a change in electrical charge between the two sides of the tubular cell membrane surrounding the nerve

FIGURE 31-4

A

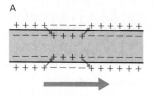

Two theories of the propagation of an action potential. (A) In unmyelinated neurons, there is a continuous shifting movement of a small area of reversed charges along the membrane. This ionic exchange of positive and negative ions exists very briefly at any one spot. (B) In myelinated neurons, node-to-node saltatory conduction occurs; since the amount of sodium and potassium ion exchange is thus greatly reduced, the net amount of work required of the nerve cell for a response to stimulus is much lower. The bottom arrows show the direction of travel of the impulse in each case.

B

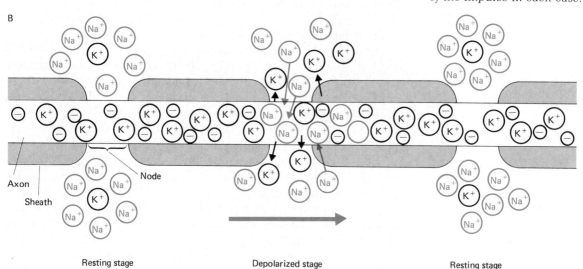

Resting stage   Depolarized stage   Resting stage

cell cytoplasm. When at rest, the membrane of a nerve cell is electrically polarized; the inside of the axon is negative in relation to the outside. This polarization of the nerve membrane is caused by the distribution of ions of certain elements. The inside of a nerve cell, like all cells, is rich in potassium and poor in sodium ions. The active transport process located in the membrane *pumps* sodium ions *out* of the cell and pumps potassium ions *into* the cell. The fluids in the environment outside the cell, however, are rich in sodium and poor in potassium ions. While resting, a nerve membrane is relatively much more permeable to potassium and as a result small amounts of potassium tend to diffuse outward, leaving the cell and building up a positive charge on the outside of the membrane. Thus the resting membrane becomes polarized; the outside is positive and the inside is negative.

During the passage of an impulse along the axon, however, the membrane polarity reverses. Because of the stimulation or excite-

PLATE 19

Animal courtship takes on
an almost infinite variety of
intricate forms and modes.
African hippos face each
other with mouths opened
in giant "yawns" during the
premating sequence of ac-
tivities. Male giraffes exhibit
literal "necking," with one
male gently rubbing his
head or neck against the
body of a second male. This
gentle necking or sparring
may increase to serious
fighting thrusts, but ap-
parently under wild
conditions, the normal
course of events is that prolonged bouts of "necking"
establish a dominance hierarchy among males. This
conjectured function is believed to decide which
male has the right to mount
a female in heat.

PLATE 20

Animal courtship displays reach their greatest level of ritualized development among birds. Both male and female members of a pair will proceed through a series of highly stylized movements, postures, and calls — one bird's display act calling forth the appropriate response from the other bird. Four parts of the courtship display of the Black-browed Albatross appear on this page. At upper left on the facing page is a pair of Great Frigate birds on Tower Island in the Galápagos; the males inflate a red throat pouch during courtship. The display of the Blue-footed Booby of the western Pacific includes "sky-pointing," where one or both birds point their bill upward, lower their wings toward the ground, and emit a piercing whistling call. The polygamous peacocks gather in an arena at mating time, where they display to any female that approaches. The tail is spread in an arresting sight, the plumes are rattled, and the wings are quivered.

PLATE 21

Auditory stimuli are particularly important in the courtship of amphibians and birds. A male toad calling on a spring evening may be heard for a considerable distance, and females attracted to his calling site will never worry about his warts or looks in general. The final stages of successful courtship involve mating in all animals with internal fertilization of the female by the male, such as land snails and desert tortoises (Gopherus agassiz). In fishes and amphibians, among others, fertilization is external. Above, a large male Brown Trout (Salmo trutta) approaches a spawning female from above, preparing to deposit his sperm-containing milt over the eggs.

ment of that area of the membrane, the membrane suddenly becomes more permeable to sodium than to potassium, and sodium enters the cell faster than potassium leaves. This causes the inside to become momentarily more positive than the outside (Figure 31–4). Thus the nerve impulse basically consists of a temporary reversal of charge on either side of the cell membrane, and the nerve impulse passes like a wave along the membrane to the end of the axon. This wave of reverse polarity traveling down the nerve fiber is called the **action potential.** The actual magnitude of the action potential is determined by the concentration of sodium and potassium ions on the two sides of the cell membrane. Normally, these concentrations do not vary to any great extent; as a result, every time the nerve is excited, the action potential is essentially the same size. When the minimum amount of stimulation has been given to the nerve, the action potential is generated no matter how large the stimulus might eventually become. In other words, the nerve fiber behaves in an *all-or-none way.*

We do not understand yet how a nerve cell membrane changes its permeability in response to a stimulus, but it is this property that distinguishes it from ordinary cell membranes. Fundamentally, then, the messages that are carried from one part of the body to another in these nerve impulses are transmitted as patterns of impulses. impinging on effectors and other receiving stations like the dots and dashes of Morse code. The code in the nervous system, however, depends both on the number of nerve impulses arriving at a given place during a particular unit of time and on the specific axons that carry the impulse. The function of the central nervous system is to interpret these patterns and convey its controlling commands to muscles, glands, and other effectors by a similar code.

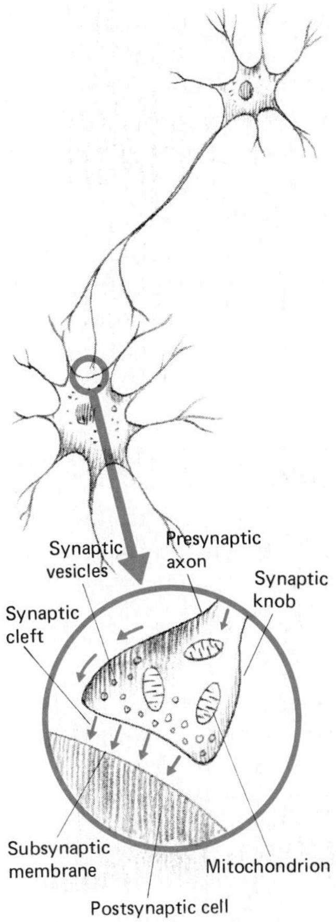

The synapse

The specialized junction between two neurons, the synapse, is crossed through the electrical activity in one nerve influencing the excitability of the second nerve (Figure 31–5). If the excitability of the receiving second nerve is temporarily increased, the synapse is said to be *excitatory.* If the excitability is temporarily decreased, the synapse is said to be *inhibitory.* The two types of synapses aid in controlling nervous transmissions, depending on their role and placement in the overall circuit.

In a synaptic junction, the axon terminal of the first (or presynaptic) neuron, which conducts the action potential toward the synapse,

FIGURE 31–5

The transmission of the nerve impulse across a synapse.

Synaptic vesicles

Presynaptic axon

Synaptic knob

Synaptic cleft

Subsynaptic membrane

Mitochondrion

Postsynaptic cell

ends in a slight swelling, the synaptic knob. The second (or post-synaptic) neuron conducts its action potential away from the synapse. The synaptic cleft, a narrow extracellular space, then separates the two neurons and prevents direct propagation of the action potential from the presynaptic neuron to the postsynaptic neuron.

The explanation of how this gap is crossed also reveals why information transfer across the synapse operates in only one direction. A chemical agent is stored in small membrane-enclosed vesicles in the synaptic knob. When an action potential in the presynaptic neuron reaches the axon terminal and depolarizes this synaptic knob, small quantities of the chemical transmitter substance are released from the synaptic knob into the synaptic cleft. Once released from these vesicles, the transmitter diffuses across the synaptic gap and combines with reactive sites on the membrane of the postsynaptic cell next to the synaptic knob. This combination of a transmitter with the reactive sites causes changes in the permeability properties of the membrane and generates a new action potential in the postsynaptic cell. The delay between excitation of the synaptic knob and the membrane potential changes in the postsynaptic cell is less than one-thousandth of a second and is called a *synaptic delay*. Even with a number of synapses in a chain of neurons, however, nerve impulses are able to travel at several hundred feet per second through the body. The excitation of the postsynaptic cell is terminated when the transmitter is chemically transformed into an ineffective substance, diffuses away from the reactive sites, or is reabsorbed by the synaptic knob.

Reflex arcs

The simplest form of a nervous coordination of bodily processes is the *reflex arc*, which consists of several neurons linking a receptor and an effector. The reflex reactions they control are usually relatively automatic and rapid. The most basic reflex arc involves only a sensory neuron and a motor neuron. The knee-jerk reflex arc is a classic example (Figure 31–6). Stretch receptors at the base of the knee cap include the terminal branches of a dendrite of a sensory neuron. When tapped, these receptors send an impulse back to the cell body of the sensory neuron, which is located in a ganglion just outside the spinal cord near its dorsal surface. The axon of the sensory neuron enters the dorsal side of the spinal cord, where it synapses with the dendrites or directly on the cell body of a motor neuron. The axon of the motor neuron comes out at the ventral side

FIGURE 31–6

*The human knee-jerk reflex:
an example of a reflex arc.*

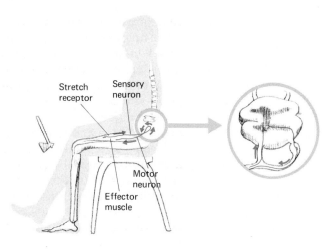

of the spinal cord and runs several feet back through the upper leg to the effector cells composing muscle fibers, which contract and cause the leg to jerk. The brain is not involved at any point, and the only function of the central nervous system is to provide a meeting place for the sensory nerve axon and the dendrites (or directly on the cell body) of the motor neuron lying in the gray matter of the spinal cord. The result is an exceptionally fast response to an environmental stimulus. Higher nervous processes such as decision making, learning, and memory must involve hundreds or even thousands of additional association neurons interspersed between the sensory neurons and the motor neurons that will produce the response to a stimulus received by the receptor.

The human brain: its structure and functioning

The basic divisions of the central nervous system are the spinal cord and the brain. The spinal cord is made up of a single cylinder of nervous tissue less than half an inch in diameter. The central area of the spinal cord is *gray matter*, largely filled with the cell bodies of association neurons and motor neurons, dendrites, and the entering axons of sensory and other afferent (inward-message-carrying) neurons. The peripheral *white matter* region of the spinal cord is made up of nerve axons that transmit action potentials between different levels of the spinal cord or between areas of the spinal cord and the brain. The fatty myelin coating of these fibers gives the region its whitish appearance and hence its common name. These fibers are organized into bundles called pathways, or tracts, and are made up of nerve cell processes that transmit the same types of functional information. The spinal white matter contains many such tracts, some of them descending to convey information from the brain to the spinal cord and others ascending to transmit in the opposite direction.

Nervous pathways that require integration between association neurons in the brain first enter the brain stem area. The **brain stem** is only one of the three primary divisions of the brain, which also includes the cerebellum and the cerebrum (Figure 31–7). The brain stem, composed of the *medulla, pons, and midbrain,* is literally the stalk of the brain, through which pass all the nerve fibers relaying input and output signals between the spinal cord and the higher brain centers. Here lie the cell bodies of motor neurons that control the skeletal muscles of the head and the fibers that enervate the smooth muscle and glands of the head, chest, and abdominal organs. Running through the entire brain stem is a core of tissue called the *reticular formation.* This interconnected network of nerve cells includes important integrating centers for respiratory, circulatory, sensory, and motor systems, as well as states of consciousness such as sleeping and waking. The chief role of the **cerebellum,** located in an area the size of a small fist at the back side (dorsal surface) of the brain stem, is the unconscious coordination of muscle movements.

FIGURE 31–7

The basic divisions of the human brain.

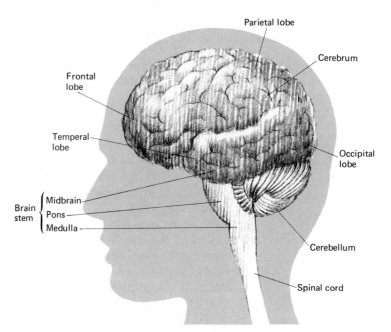

The organization of the body's activity centers in the sensory and motor areas of the cerebral cortex. Compare the sensory and motor functions of various parts of the body. Note, for example, that more cortex is devoted to the motor function of the hands than to their sensory function, whereas more sensory area than motor area is devoted to the face areas (especially the lips).

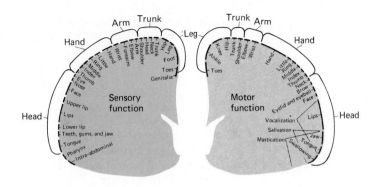

The **cerebrum** is the remaining large part of the brain. Specific areas of the cerebrum have been found to be associated with sight, hearing, speaking ability, body movements from the thumb to the whole forearm or leg, swallowing, chewing, eyelid movements, and the thought processes underlying speech. Its outer, highly convoluted portion, the **cortex,** which forms a shell about one-quarter of an inch thick and contains about 14 billion neurons, is one of the principal areas where integration of neural mechanisms is carried out (Figure 31–8). The cortex reaches its highest development in human beings and hence it is popularly called the "site of the mind and the intellect." However, we do not have any real understanding of the detailed functioning of the cortex, though its several sections, or lobes, have been studied in part. The cortex is an area of gray matter because of the predominance of cell bodies located there.

Another portion of the cerebrum, the **thalamus,** is an important integrating center for all sensory input (except smell) on its way to the cortex. It also contains a significant portion of the reticular system coming up from the brain stem. The **hypothalamus,** which lies below the thalamus, is a tiny area of only 5 to 6 cubic centimeters, yet it is the most important control center for many reflexes in the body and the visceral functions. It coordinates the nervous system tracts that control sensations of hunger, thirst, pain, pleasure, and hostility, as well as regulating body temperature, water balance, blood pressure, and reproductive behavior. One can generate these feelings and responses in an experimental animal by stimulating various areas of the hypothalamus with electrodes. Often these centers are located only fractions of an inch apart.

Overall, our understanding of how the human brain functions is still very elementary, and the phenomena of learning and memory and the role of sleep in the human brain are areas of active research at present.

Muscles

Muscles fascinate almost everyone. Their differing distribution over the human body makes girls attractive to guys and vice versa. In composition, muscle cells include high concentrations of special protein molecules that can *contract* by converting chemical energy, such as that stored in ATP, into the mechanical energy of movement.

Three different types of human muscle cells can be identified on the basis of structural and contractile properties: (1) skeletal muscle, (2) smooth muscle, and (3) cardiac muscle. Most skeletal muscle is attached to the bones of the body, and its contraction is responsible for their movements. Skeletal muscle also plays a role in controlling the voluntary release of urine and feces. It is the longest bulk of tissue in the body, accounting for 40 to 45 percent of the total body weight. Smooth muscle surrounds such hollow chambers in the body as the stomach and intestinal tract, the bladder, blood vessels, and the uterus. Cardiac muscle forms the contractile tissue of the heart.

Over 600 muscles can be identified in the human body. Each is composed of many cylindrical muscle cells, of about 10 to 100 microns in diameter and up to 30 cm or more in length. A single muscle cell is known as a *muscle fiber*. The term *muscle* refers to a number of muscle fibers bound together by connec-

tive tissue; blood vessels and nerve fibers pass through the network of connective tissue to the muscle. In some muscles, the individual fibers are as long as the muscle itself; however, most fibers are shorter than the muscle, and their ends are attached to the connective tissue network. Generally, each end of the intact muscle is attached to a bone by fibrous protein (collagen) fibers known as *tendons*. The tendons do not contract, but they do have great strength and act as a structural framework to which the muscle fibers and bones are attached. The forces generated by the contracting muscle fibers are transmitted by the connective tissue and tendons in the bones; opposing sets of muscles move the bones and hence the body in a remarkable variety of ways.

Learning and memory

As we have seen, the nervous system serves as an integrative system for all incoming sensory information, as well as initiating motor responses. Increased learning abilities are found in organisms with more highly developed neural systems (Chapter 18). The restriction of the highest type of learning — reasoning — to mammals illustrates this correlation. Thus it seems likely that learning, with its attendant information storage (memory), is physically located within the nervous system.

The changes within the nervous system that are caused by experience are called *memory traces*, or *engrams*. Much interest has focused specifically on the cerebral cortex as an important structure in learning. Yet 98 percent of this structure can be removed in an animal without causing significant impairment of the animal's ability to learn. Summarizing the search for the engram, the great neuropsychologist Karl Lashley concluded in 1950: "I sometimes feel in reviewing the evidence on the localization of the memory trace that the necessary conclusion is that learning is not possible at all. Nevertheless, in spite of such evidence against it, learning does sometimes occur." The picture has changed somewhat in the last quarter century through intensive research and there are numerous examples of detectable changes within the nervous system due to experience. Such findings have led to two kinds of theories about the mechanisms of learning: physical structural change and chemical change theories.

Structural theories essentially propose that memory is stored by means of physical changes in the functional morphology of the nervous system. Probably the simplest model of learning is that neural pathways that are used *often* will develop a facilitated transmission (with primarily excitory synapses), whereas those that are not used may develop an inhibited transmission (perhaps with primarily inhibitory kinds of synapses). Another structural theory emphasizes changes in the physical presence of hairlike projections along the dendrites of cells in the cortex. If the nervous input to the cortex is reduced, the number of these dendritic spines also greatly decreases. Stimulation of a particular pathway via learning would presumably result in an increase of the number of these dendritic spines and thus increase the surface area on these nerve cells available for synaptic connections with other neurons.

Biochemical theories propose that memory is stored in molecules within the nerve cells of the brain. Learning may thus result in the manufacture of a new "memory molecule," an increase in the rate of synthesis of existing molecules, or perhaps even a change in

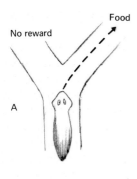

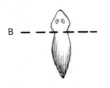

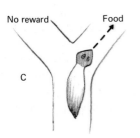

FIGURE 31–9

Learning in planarian flatworms: (A) a planarian learns to turn right in a Y-shaped tube to reach a food source; (B) after the planarian learned this task, its head was cut off; (C) when a new head was generated, the planarian still retained the memory of turning right to reach the food.

location of existing molecules. One of the first experiments to involve a biochemical mechanism of learning postulated that RNA was the memory molecule. If planarians were trained to do a specific task such as turning to the right in a T-shaped maze, and then were fed to cannibalistic untrained planarians, the cannibals showed a significant increase in the rate of learning the task when *they* were placed in the T-maze (Figure 31–9). This result indicated that some chemical within the trained animals had probably been changed in the learning process and that this information was transmitted to the untrained group by feeding. Because of its central role in protein synthesis, RNA was suspected of being the chemical involved. Extracts of RNA were then taken from another trained group following learning. When this RNA extract was administered to the untrained group, learning proceeded at a faster rate than in a group that was fed RNA extracted from the untrained planarians.

Other experiments have indicated that proteins may be the memory molecule. Drugs that inhibit protein synthesis have been shown to disrupt memory storage, whereas RNA-synthesis inhibitors do not block storage. However, since proteins are composed of amino acids that are coded by the triplet code present in transfer RNA and messenger RNA, the storage of information in proteins might result from a change (induced by learning) in the ratio of purine and pyrimidine bases (Chapter 24) present in the RNA of the organism. In one experiment, rats that had learned to climb an inclined wire possessed an altered RNA base ratio on the neurons that transmitted sensory information on body balance from the semicircular canals in the ears to the brain. Part of the problem in these and other studies has been that the experimental tests may interfere with at least three processes: (1) the storage of memory, (2) the retrieval of this information, and (3) the expression of the information in a later behavior.

The storage process seems to have a considerable time component. Having learned a particular item, the organism is thought to store this information in a *short-term* memory system. Within a few hours, this information may be then either discarded or converted to *long-term* storage through the process of consolidation. Thus high-voltage electrical shocks will interfere with the storage of events that have occurred within an hour before the treatment, but these shocks do not affect memories of events that have taken place before this time. Memory storage may actually involve a *multistaged* temporal process and not simply be short-term or long-term. Research on the human brain and its functioning in memory represents one of the most exciting frontiers of biology at present, and we may expect rapid advances in our knowledge of the mechanisms of learning and memory over the next few decades.

The function of sleep

The mechanisms involved in sleeping, a periodic activity that involves virtually all animals on earth, are as yet little understood, despite intensive research. We also have little information yet as to *why* we sleep. Most theories proposed to answer this question are basically biochemical. Sleep is thought to be due either to the build-up of a noxious substance that the body rids itself of during this inactive resting time or to the need to replenish overworked organs with the necessary nutrients for proper functioning during the next principal activity period. In this view, sleep is necessary because it performs a recovery or restorative function, which, of course, would ultimately take place on the cellular level, especially in the cells of the nervous system.

An alternative theory that does not exclude the possibility of a restorative role places sleep in an ecological perspective and proposes that sleep has an adaptive function in behavioral control, the behavior in this case being a lack of response to stimuli. Implicit in this concept is the assumption that an organism has a limited capacity to respond efficiently to stimuli and must cease at a certain point to avoid exhaustion or inefficient energy expenditure. Sleep serves as an adaptation to cause the nonresponding event. Furthermore, the placement of sleep within the 24-hour day has been determined evolutionarily by environmental factors such as the danger of predators and food availability, and is now an integral part of the organism's daily rhythm.

Summary

Animals are unique among organisms in their highly developed capacity for irritability, the ability to respond to stimuli. The nervous system plays a key role in the integration and coordination of bodily processes in response to external and internal stimuli. Irritability involves stimulus reception by specialized receptor cells, conduction of the signal by neurons, and response of the animal via effector cells. The central nervous system includes the brain and spinal cord; the peripheral nervous system includes all other nerve cells. The nerve cell, or neuron, is the basic unit in the nervous system. It is composed of dendrites, a cell body, and one or more axons. A sensory neuron picks up impulses from receptor cells; motor neurons send the impulse on to effector cells. Association neurons usually connect sensory and motor neurons. The nervous impulse, or action potential, is transmitted as a change in electrical charge distribution of ions between the two sides of a nerve membrane. The synaptic gaps between adjacent axon terminals and the dendrites of the next cell are crossed by chemical transmitter substances released at the axon end. Reflex arcs of several neurons allow the most rapid nervous responses to stimuli. More complicated nervous processes involve great numbers of association neurons interspersed between the sensory neurons and the motor neurons characteristic of the simplest reflex arcs. The spinal cord is composed of the central gray matter, with many cell bodies present, and the peripheral white matter, made up of myelin-coated nerve axons. The brain is divided into the brain stem (containing the medulla, pons, and midbrain), the cerebellum (controlling the unconscious coordination of muscle movements), and the cerebrum (integrating most neural mechanisms). Learning and memory (information storage) are seated in the nervous system. Various structural and biochemical change theories have been proposed to account for the mechanisms of learning. The function and mechanisms of sleep are as yet unknown, and, like learning, sleep is currently of great research interest to biologists and psychologists.

32 Growth and development in plants

To biologists and nonbiologists alike, the process of development in higher plants and animals is extraordinary: Starting with a single microscopic egg cell that has been fertilized by a sperm, a series of cellular divisions produces first an undifferentiated ball of cells and eventually, through a long series of developmental stages, the magnificently complicated, yet carefully integrated body of the mature organism. In this chapter we shall look at the life history of a flowering plant, an **angiosperm,** from the developing fertilized egg inside the seed coat to the mature plant with its specialized structures and functions. An appreciation of the broad outlines of growth and development in plants aids us in understanding our own development, as well as opening experimental approaches to basic problems in developmental specialization. The differentiation of even plant cells to perform special tasks raises many questions about control mechanisms for changes induced in any cell, which can be applied to such problems as cancer, where these developmental controls are loosed.

The seed and early embryonic growth

The sperm cell reaches the egg in the typical flower by landing on top of the columnar tissue above the egg-containing chamber. There, the pollen grain germinates and sends a tube of protoplasm down the center of this female flower part, allowing the two sperm nuclei to

FIGURE 32–1

The female pistil (single, long, white flower part with a somewhat broadened stigma head) and the male stamens covered with pollen of an Easter lily (Lilium longiflorum).

travel down to the ovary in tissue of their own genotype (hence avoiding potential immunological problems with the surrounding maternal tissues of a different genotype) and unite with an egg in the center of the female tissues of the flower (Figure 32–1). The union of the sperm and the egg in fertilization gives rise to the diploid *zygote,* which undergoes a series of mitotic divisions and develops into a tiny embryo. Tissues surrounding the embryo differentiate into a food storage tissue called **endosperm,** which is enclosed with the embryo inside a tough, protective seed coat. The complete structure of embryo, endosperm, and seed coat is called a **seed.**

The seed matures on the plant, and the embryo then normally enters a period of dormancy. If the seed is inside a fruit, an animal will normally eat this fruit, catch it on its fur, or otherwise disperse the contained seed or seeds to another area. Wind-blown seeds are carried by air currents to places where the seed may lie on the ground for a while. Eventually, if the seed has reached an area where there is at least seasonally sufficient moisture (even several years into the future, as in desert regions), the seed germinates and the embryo resumes active growth. At this point, the embryo in most species continues to absorb the endosperm and also absorbs a great amount of water, which may increase the volume of the seed as much as 200 percent. The additional water activates enzymes, and the metabolic rate of the embryo increases considerably.

The structure of the young plant

The growing embryo soon bursts the seed coat, with its root and shoot determining the lower and upper ends of the future mature organism (Figure 32–2). The division of cells in the young embryo has determined the polarity, or orientation, of the embryo before the root and stem even differentiated, of course. Attached to the side of the embryonic sprout are one or two **cotyledons** ("seed leaves"), which primarily function to absorb the remaining endosperm and break it down for the use of the growing embryo. They often also function as photosynthetic organs (leaves). The portion of the stem above the cotyledons is called the **epicotyl** ("above the leaf") and the portion of the plant below the point of attachment of the cotyledons is called the **hypocotyl** ("below the leaf"), the part of the plant that will form the primary root.

The hypocotyl promptly turns downward on leaving the seed, no matter what the orientation of the seed may be, in response to gravity. The emerging epicotyl then can grow on nutrients from the

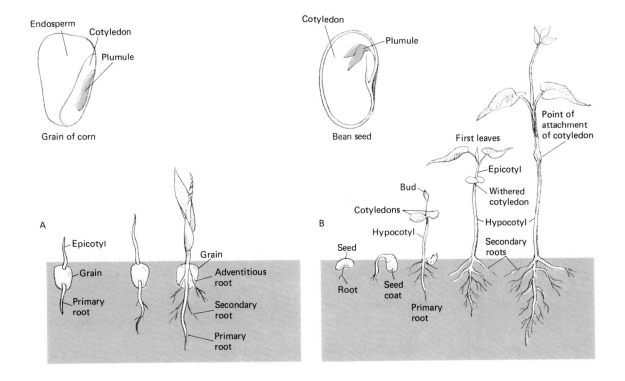

FIGURE 32–2

The structure and germination of a seed of a monocot (corn) and a dicot (bean) flowering plant. (A) Stages in the germination of a corn kernel, with one seed leaf, or cotyledon; (B) the germination of a bean seed, with two cotyledons.

young root system, which anchors the plant to the substrate and absorbs water and minerals. The epicotyl begins to elongate and put out the first true leaves. Both the vertical growth of the stem and the side growth of the leaves are due to the activities of the apical meristems; this type of growth is called **primary growth.** It elongates the plant body through cell multiplication and cell elongation.

At regular intervals along the stem, a localized region of the apical meristem will give rise to a series of swellings that will function as leaf primordia—the beginnings of the leaves. The areas where each leaf primordium arises from the stem is called a **node,** and the length of stem between two nodes is called an **internode.**

Inside the stem, the first vascular tissues form behind the apical meristem, the outer cells differentiating into primary phloem and the inner areas differentiating into the first xylem elements. As we have seen (Chapter 20), the vascular tissues of the shoot are arranged in different ways in different types of plants, but basically they serve as a collection of conduits leading to and from the leaves, branches, stems, and roots. The epidermis cells also arise at this time to serve as a protective covering over the stem, leaves, and later flowers, seeds, and fruits. On the outside of the root, the epidermis serves

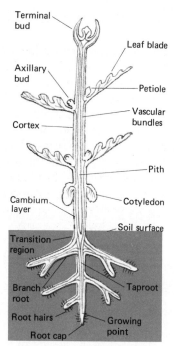

FIGURE 32–3

The structure of the young plant.

chiefly to absorb water and minerals. Within the stem of woody plants but outside the xylem lies the lateral **cambium,** a cylindrical wall of meristematic cells that promotes lateral growth. This lateral cambium layer, then, generates the *secondary growth* (width, rather than length) of a higher plant (Figure 32–3).

Leaf and bud development

With the increasing separation of the nodes by the elongating internodes, mitotic activity begins in the leaf primordia and young leaves arise on the side of the apical stem. A marginal meristem forms along the sides of the primordium and increases the lateral growth of the leaf, making it progressively flatter and broader. Another important difference between leaf growth and stem growth is that the growth of a leaf is *determinate:* it ends at a defined point. The growth of roots and stems is *indeterminate,* continuing indefinitely during the life of the plant.

As the leaf gets broader and is exposed to the sun, specialized cells filled with chloroplasts develop in the interior cells of the leaf. Light passing through the epidermal layers hits these chloroplasts, allowing photosynthesis and the production of carbohydrates. The last part of the leaf to differentiate is usually the *petiole,* the narrow stalk attaching the base of the leaf to the stem. Leaves with petioles have essentially just small extensions of the stem, which contain vascular tissue.

The term *bud* is used to describe the compound structure consisting of the apical meristem and a series of unelongated internodes enclosed within the leaf primordia at the tip of the stem. On the outside of the bud are protective, overlapping scales. Lateral, or auxiliary, buds may arise in the angle between the base of each leaf and the internode above it. Some of these will produce branch stems later. The buds at the apical end of the stem and the lateral buds will continue producing new undifferentiated tissue for the rest of the life of that particular plant, whether it lasts only a year (annuals), takes two years to reach maturity (biennials), or requires more than two seasons (perennials) to bridge the gap between seed germination and the formation of the next set of seeds. The vegetative structures of perennial plants survive year after year and may flower every season after they once reach maturity. A ground bromeliad in the Bolivian Andes (*Puya raimondii*) takes about 150 years to reach maturity and finally have its flowering period.

Differentiation and the problem of morphogenesis

More than cell division and cell elongation alone are needed to produce a new plant. In Chapter 20 it was stated that all the cells produced by the apical meristems are essentially alike, yet these cells soon begin to change in form; some becoming complicated xylem and phloem conductive tissues, and others becoming supportive and epidermal tissues. A few will eventually become the germ tissue to produce gametes that will form the next generation. The process by which a cell changes from its unspecialized immature form to a particular mature type is called **differentiation.** In plants, immediately behind the apical meristems and the zone of elongation in the stem and root is a zone of cell differentiation and maturation. Here the mature tissues of the plant begin to take shape, and structures such as leaves and side branches as well as other organs such as flowers and fruits will eventually differentiate in the process of **morphogenesis,** or the generation of form.

Investigators have made numerous attempts to understand the factors involved in differentiation by using galls and other plant tumor growths (Figure 32–4). Like cancers and benign tumors in animals, the cells of a plant tumor continue to divide until they make up large masses of undifferentiated tissue that in many cases can destroy the host organisms. Thus a crown gall tumor, which is initiated when a damaged plant is infected by a soil bacterium, *Agrobacterium tumefaciens,* causes a plant cancer that can develop into a large growth of undifferentiated cells that may reach 100 pounds or more. In plants,

FIGURE 32–4

Tumorlike gall growths in plants: (A) mite-caused galls on the leaf of a silver maple; (B) a wasp-caused gall on an oak stem.

hormones such as auxins and cytokinins stimulate this growth, and apparently these growth-promoting hormones are synthesized by the tumor cells themselves, allowing growth to continue indefinitely in the cancer tissues. Normal cells in these wounded areas cannot synthesize their own growth-promoting substances; thus if auxins and cytokinins are not added to these normal plant cells, their growth will stop, even in the laboratory where abundant nutrients can be fed to them under rigidly controlled conditions.

Differentiation, then, depends on a balance of hormones and their interaction in tissues, in order to promote growth and to allow differentiation. A high concentration of oxygen promotes root formation in tobacco pith cultures, and a high concentration of cytokinin promotes bud formation. Undifferentiated callus tissue growing on a synthetic laboratory medium containing a simple salt mixture, sucrose, and plant growth factors can be stimulated to differentiate into xylem or phloem by adjusting the concentration ratios of the sucrose, auxin, and a recently discovered hormone, kinetin. These three compounds are known to be present at the appropriate sites in the intact plant, and their relative quantities can be adjusted by selective transport processes and selective metabolic destruction through enzymes at the cell membrane or in the cytoplasm of the differentiating xylem and phloem. In live growing plants cytokinin is known to promote the growth of lateral buds; auxin has the opposite effect. Cytokinins stimulate protein synthesis in leaves and prevent senescence.

The delicate and changing balance of hormones produced by growing plants, then, has broad effects influencing differentiation. These effects depend on the target tissues and the chemical environment in which these tissues find themselves. Experimental analysis of growth and development in plants is still in an active, exploratory phase of investigation, and great advances may be expected in the future years.

Summary

Growth and development in flowering plants begins with the fertilization of the egg deep in the ovary. The embryo is enclosed with the endosperm inside a protective seed coat; the complete structure is called a seed. Germination involves the absorption of a great amount of water, which activates enzymes and mitotic growth of the young embryo. The new sprout bears one or two cotyledon seed leaves on its sides, which aid in metabolizing the stored food of the endosperm. The future stem portion is called the epicotyl, and the future primary root is called the hypocotyl. Primary growth (stem elongation and the formation of buds and leaves) is due to the activities of apical meristems; secondary growth (increase in diameter) is generated by the lateral cambium meristematic cells. The growth of a leaf is determinate; the growth of roots and stems is indeterminate. Differentiation involves the process by which a cell changes from its unspecialized immature form to a particular mature type. The principal zones of cell differentiation in plants are located immediately behind the zone of elongation, advancing along behind the apical meristems. Differentiation and morphogenesis in plants are particularly influenced by hormones such as auxins, cytokinins, and kinetin, and by sucrose and oxygen concentrations.

33 Growth and development in animals

The growth of a six-foot, 180-pound adult from a 7-pound infant over a period of 18 years is remarkable indeed, but these changes in visible growth, with an increase in height of about four times and in weight of more than 20 times, are infinitesimal compared with those that take place in the first two months the human being spends in its mother's womb. There, in eight brief weeks, it increases in length about 240 times and in weight about 1 million times, growing from a single, fertilized egg into a miniature baby. The growing embryo develops only by absorbing raw materials from its mother's body, with which it builds the structures needed for life.

Early work with development

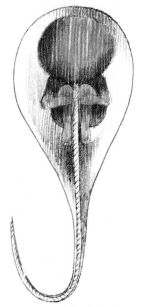

FIGURE 33–1

Preformationism: the homunculus, or little man, as interpreted in the human sperm cell by an early member of the "spermist" school.

The process of development seemed so extraordinary to early investigators studying embryos under the microscope that they could not believe the evidence of their own observations. It seemed more reasonable to adopt the point of view known as *preformationism*, namely, that the parts of the embryo do not develop but are already present at the moment of conception and simply increase in size in

FIGURE 33–2

A half embryo produced in Roux's experiment of killing one blastomere at the two-cell stage of a frog egg with a hot needle.

A

B

C

D

the growing embryo (Figure 33–1). An eighteenth-century Swiss scientist, Charles Bonnet, succinctly stated this point of view:

All the constituent parts of the body are so directly, so variously, so manifoldly intertwined as regards their functions . . . their relationship is so tight and so indivisible, that they must have originated all together at one and the same time. The artery implies the veins, their operation implies the nerves, which in their turn imply the brain and that by consequence the heart, in every single condition a whole row of other conditions.

Scientists throughout the eighteenth century felt, then, that the tiny sperm or the egg cell contained an adult human being in miniature, which simply unfolded during the process of development. Whole schools of biologists argued over whether the preformed infant was present in the egg or the sperm cell. Yet in 1759 a young German zoologist, Kaspar Friedrich Wolff, had observed the development of chick embryos and reported in a publication that the fertilized egg did not contain a minuscule chicken but that the embryo changed radically and formed as it grew. However, his idea that the structures of the embryo develop in succession—the concept of *epigenesis*—did not achieve general acceptance until early in the nineteenth century. Even at that time, though, the study of development (called *embryology*) was still a largely descriptive science, and it was not until the late 1880s that it began to be transformed into a truly experimental science.

Experimental embryology

Near the end of the nineteenth century, a pioneer German experimental embryologist, Wilhelm Roux, began working with fertilized frog eggs and observing their development (Figure 33–2). Roux found that after the first cleavage of the fertilized egg, each of the daughter cells could continue to develop as a half embryo if the other side was killed. Another experimenter, Hans Driesch, then worked with sea urchin eggs and discovered that if he broke individual cells apart, using embryos that had reached the 2-, 4-, 8-, or even the 16-cell stage, the individual cells still developed into complete embryos (Figure 33–3). Clearly, even after the fertilized egg had split

FIGURE 33–3

Driesch's experiment with sea urchin eggs. The separation of the first two blastomeres resulted in the development of two whole embryos.

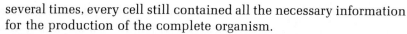

FIGURE 33–4

Spemann's experimental development of his concept of the organizer effect. Diagram of operation: (A) the insertion of a graft into the blastocoele of an amphibian early gastrula stage; (B) the position of the graft after completion of gastrulation. Result (below): the induction of a secondary embryo on the host's side by means of a transplanted piece of blastopore lip.

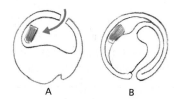

A   B

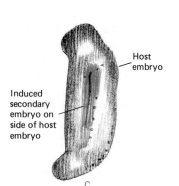

Host embryo

Induced secondary embryo on side of host embryo

C

several times, every cell still contained all the necessary information for the production of the complete organism.

The apparent contradiction between Roux's work with frogs and Driesch's experiments with sea urchins was explained by the experiments of another German, Hans Spemann. Spemann showed that the development of the fertilized egg and the cells that result from it is determined not merely by the genetic material present in its nucleus, but also by an interplay of chemical and physical factors between this material and the rest of the cell or developing embryo. Each successive step in cell differentiation and the development of morphological structures is influenced by surrounding cells and depends on the steps taken earlier.

The channeling of development

Spemann received the Nobel prize in medicine in 1935 for his "discovery of the organizer effect in embryonic development." This basic discovery has since been amplified by the work of many developmental biologists—namely, that an embryonic cell is influenced by the other cells in its environment, which gradually lead it to act on only one of the many possible sets of instructions that it contains (Figure 33–4). Once these instructions have been issued and the animal cell is differentiated into having a specific character, it cannot reverse itself and go back to being an undifferentiated cell. The developmental process has gone too far to be changed.

We know from our study of genetics and cell division that every cell in the body receives the same set of genes as was contained in the original fertilized egg. Thus the process of development involves a progressive shutting down of parts of the potentially active genome and an activation of other parts. At the first cell division of one fertilized egg into two daughter cells, all systems controlled by the genes seem to still be acting, at least in such animals as the sea urchin. As Driesch's experiments with sea urchin eggs showed, the two daughter cells retain the capacity for developing into an intact embryo. In fact, we saw that this capacity extends to the 16-cell stage, after four cell divisions; but beyond this stage, the genes available for developmental activities begin to shut down and each part of the developing embryo becomes channeled in a certain direction of development, to form a spine, skin, digestive system, and so on.

In other organisms, such as was shown by Roux in his experiments with frog eggs, the partial shutdown of genetic instructions occurs much earlier. After the fertilized egg is divided into two cells (in the frog), each daughter cell is only capable of developing into half an embryo, either the right or left side. Spemann's major discovery was that at every stage of embryonic development, structures that are already present act as *organizers*, inducing the emergence of the structures that are next on the developmental timetable. In the fertilized frog or newt egg, for example, there is a gray crescent area at the top of the egg (Figure 33–5). As the egg begins to divide, a group of cells eventually derived from this gray crescent of the egg becomes the "primary organizer," stimulating other cells to shape themselves into specific tissues. The organizer cells themselves become the **notochord** and the body muscular segments, or **somites.** The notochord and the somites, in turn, induce the outer ectoderm above them to form the neural tube, from which in turn the brain and the spinal cord will develop.

In the developing human embryo, the primary organizer is a structure in the embryo called the *primitive streak*. This streak is a group of ectodermal cells that emerges on one end of the embryo at about the twelfth day after conception. The first embryonic structures — the notochord and somites — are built from the mesoderm, or middle layers of tissues that come from the primitive streak. Thus in both newts and human beings, the notochord and its associated structures organize the rest of the embryo.

In this brief overview, we can see that the process of development and associated regulating factors involve an embryonic cell being influenced by the other cells and its environment. These factors gradually lead the cell to act on only a limited portion of the total set of genetic instructions that it contains. It is beyond the scope of this text to discuss in detail the many experiments exploring the chemical and physical nature of the inducing process that flows from the organizer areas to undifferentiated parts of the embryo, but we will look briefly at the ways in which the different tissue layers are set up, and the basic patterns of differentiation and morphogenesis that can be seen in animal embryos (Figure 33–6).

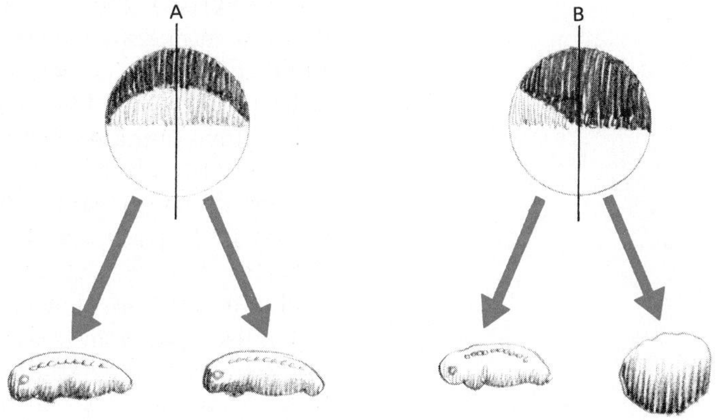

FIGURE 33–5

Following fertilization and prior to cleavage, the frog egg bears a "gray crescent" of intermediate pigmentation between the heavily pigmented upper hemisphere and the very lightly pigmented lower hemisphere. One can divide a zygote after cleavage has begun and get different results, depending on whether the half receives gray crescent material. (A) If the plane of division gives each half some gray crescent material, each half will develop into a complete though miniature larva; (B) but if a half has no gray crescent material, it will not develop into anything but a shapeless ball of cells.

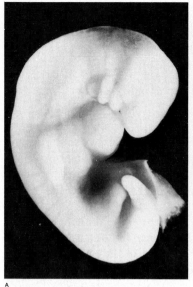

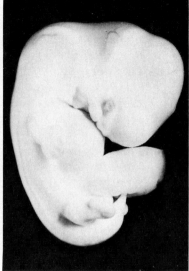

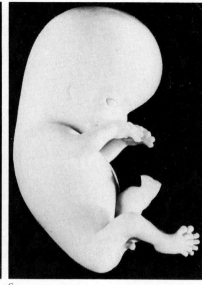

A B C

FIGURE 33–6

Early human embryos (above): (A) four weeks; (B) six weeks; (C) eight weeks. The development of a chick embryo (facing page): (A) 4 days; (B) 5 days; (C) 8 days; (D) 11 days; (E) 12 days; (F) 14 days; (G) 15 days; (H) 19 days; (I) 21 days—ready to hatch by breaking out of the confining egg shell.

Early embryonic development

The development of an embryo starts after it has been fertilized by a sperm, the male gamete. Usually just the penetration of the sperm into the membrane and cytoplasm of the egg is sufficient to stimulate development. The actual fusion with the egg of the male pronucleus, carrying its paternal set of chromosomes to be contributed to the zygote, is not necessary to initiate the first cleavage furrow. In fact, many animals normally reproduce **parthenogenically** (that is, without fertilization of the egg). In many species following entrance of the sperm, a fertilization membrane is raised off the surface of the ovum, which presumably prevents additional sperm from reaching the egg surface. The nucleus of the ovum in some species is usually arrested in an early stage of meiosis well before the sperm enters, and it completes its reduction divisions at this time of penetration in order to be able to fuse with the male pronucleus.

The development of the embryo from a zygote now goes through a series of relatively rapid cleavages (Figure 33–7). This series of mitotic divisions, without concomitant growth of the cytoplasm, results in a grapelike cluster of smaller cells, called the **morula,** which is about the same size as the single egg from which it was originally derived. Development to this stage has required a great deal of metabolic energy and as in the later stages, this energy may be supplied to the developing system in one of several ways.

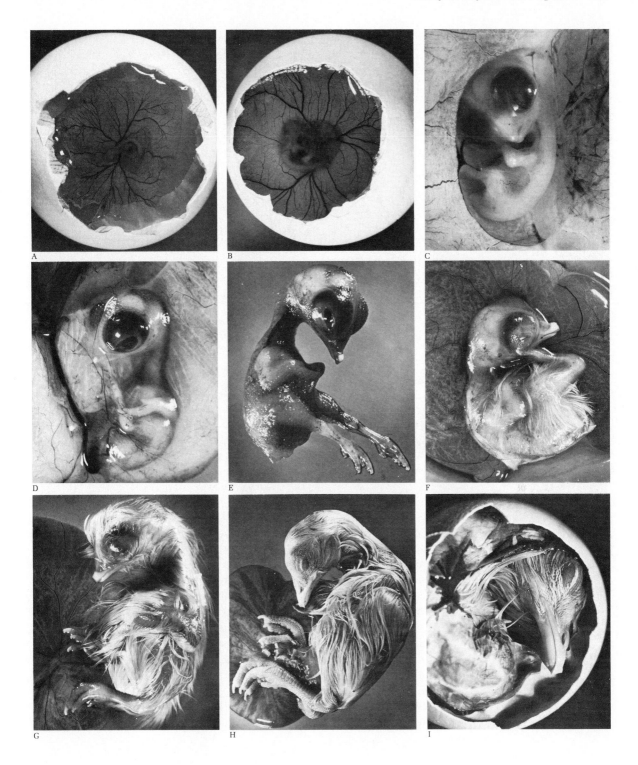

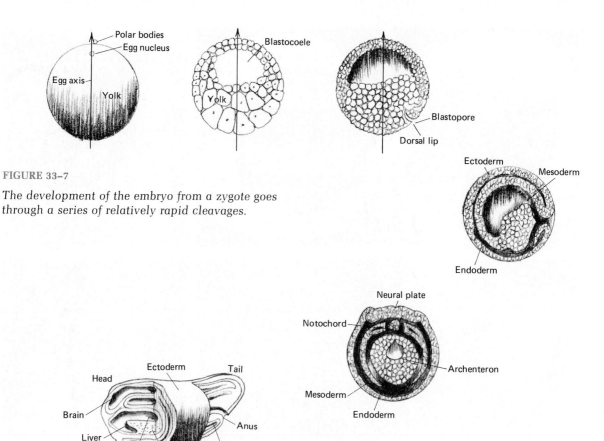

FIGURE 33–7

The development of the embryo from a zygote goes through a series of relatively rapid cleavages.

Energy for the growing embryo

Many animals produce eggs with relatively little stored food. The embryos develop quickly into immature feeding stages called **larvae,** which continue their embryonic development while feeding (Figure 33–8). Eventually they undergo metamorphosis into the adult form. Thus the eggs of species with this type of development have relatively little stored yolk and undergo development rapidly to the larval form. Starfish or sea urchin zygotes develop into free-swimming larvae in about 24 hours; frog eggs, which have considerably more yolk, require about 10 days to hatch into feeding larvae. Some tropical butterfly eggs hatch into tiny larvae in only three days. Metamorphosis into the adult in all these cases may occur far into the future.

FIGURE 33–8

Larval types: (A) a bullfrog tadpole;
(B) a starfish brachiolaria larva; (C) an east-
ern black swallowtail butterfly caterpillar;
(D) a sea urchin larva.

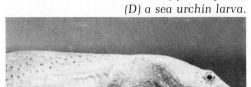

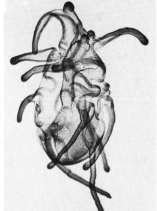

A   B

FIGURE 33–9

The structure of a bird egg,
showing the effect of a great
amount of yolk on the physical location of the developing
embryo.

C   D

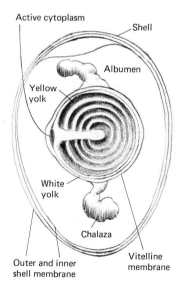

Active cytoplasm

Shell

Albumen

Yellow
yolk

White
yolk

Chalaza

Outer and inner
shell membrane

Vitelline
membrane

Reptiles and birds supply metabolic energy in another way (Figure
33–9). The eggs are relatively large and contain immense quantities
of stored food in the form of the **yolk.** Sufficient food is available inside the protective shell for the embryo to be self-sufficient until it
has grown into a miniature adult. This mode of reproduction tends
to reduce the hazards to the developing embryos and hence fewer
eggs need be laid; however, a certain amount of maternal care such
as incubating the eggs and feeding the young may be necessary, as
with birds, or eggs may have to be deposited in a well-hidden sun-
warmed pit, as with turtles and other reptiles.

The nourishment of mammalian embryos involves yet another
mechanism. Mammals produce yolkless eggs and retain the embryos
inside the uterus of the mother until a miniature adult has formed
that can survive outside the mother, though with continued maternal care for at least a short time. Within the uterus a special struc-

FIGURE 33–10

The embryo in the uterus of a placental mammal, surrounded by embryonic membranes and attached to the placenta.

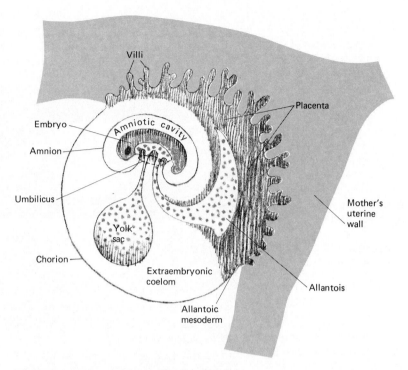

FIGURE 33–10

The embryo in the uterus of a placental mammal, surrounded by embryonic membranes and attached to the placenta.

ture, the **placenta,** serves as an exchange surface where fetal blood can come into intimate contact with the maternal blood and exchange gases, food materials, and metabolic wastes (Figure 33–10).

Yolk and the pattern of cleavage

The pattern of the cleavages that will take place through the original fertilized egg depends largely on the amount of yolk contained in the zygote. In eggs that contain relatively little yolk, such as those of the starfish, the entire egg cleaves into more or less equal-sized cells. The frog and newt zygotes contain a moderate amount of yolk concentrated particularly at one end of the cell, called the vegetal pole. While cleavage still continues to completion everywhere, the cells that form at the animal pole or top end of the egg contain relatively little yolk; hence they divide more rapidly and are smaller than the vegetal pole cells. In other words, the yolk displaces the horizontal cleavage planes toward the animal pole, which divides the amount of cytoplasm there into smaller units than at the vegetal pole.

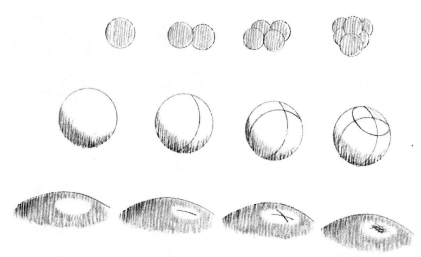

FIGURE 33–11

The effect of yolk on the pattern of cleavage in an amphioxus egg containing virtually no yolk (top), in a frog egg where there is considerable yolk in the bottom half (middle figure), and in the chicken egg (bottom), which is almost entirely yolk and consequently cleavage is restricted to a small area at the top of the inert yolk mass. The diagrams represent corresponding stages in all three animals (1, 2, 4, and 8-cell stages).

In reptiles and birds, the huge amounts of yolk present in the egg are too much to be subdivided by cleavage furrows, and only a small region of relatively yolk-free cytoplasm, which lies just under the plasma membrane, cleaves to form the early embryo. The entire yolk of an egg in these groups represents the zygote, a single cell. However, the area of active division is restricted to the little region at the top of the cell, floating above the mass of the yolk (Figure 33–11).

As cleavage continues beyond the morula stage, the zygote is broken up into smaller and smaller cells, and these become arranged in a hollow sphere called a **blastula.** The cavity in this hollow sphere is the **blastocoele.** At the blastula stage of development, the cells of the embryo rearrange themselves to form the rudiments of the major organ systems of the larva or adult to come. While cleavage continues, it is no longer the most conspicuous pattern or phenomenon in development. Instead, the establishment of shape and pattern (morphogenesis) of the developing embryo is established by a series of complex movements.

Gastrulation and the emergence of tissues

The stage of development at which the three primary germ layers of tissue are derived is called **gastrulation** (Figure 33–12). Basically, the blastula is pushed in at one end in a process called **invagination,** which continues until a two-layered embryo has formed. This pattern of gastrulation occurs in egg cells that have little yolk, such as

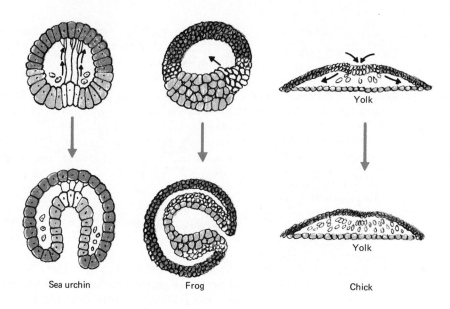

Sea urchin   Frog   Chick

FIGURE 33–12

Gastrulation in three kinds of eggs, shown in cross section. While the mechanisms of gastrulation vary from species to species, they always produce an embryonic stage that has three layers of cells. The ectoderm (dark color) gives rise to skin and nervous tissue; the endoderm (light color) is the source of the digestive tract; the mesoderm (no color) gives rise to muscle, bone, and various internal organs. To accomplish this creation of three basic tissue layers, the sea urchin egg undergoes a deep invagination of the blastula wall, similar to the shape created by a thumb pushing into a soft tennis ball. In frogs, the cells at the lip of the blastopore (point of invagination) slide inside as a continuous sheet. In chicken eggs, the blastula floats as a flattened blastoderm layer on the surface of one small area of the giant yolk. During gastrulation, cells flow inward through the "primitive streak" area at the center line of the blastoderm and spread inside the upper and lower surfaces, becoming the mesoderm.

those of a starfish. The outer layer of the new embryo is the **ectoderm,** which provides the skin and neural tissues; the inner layer is the **endoderm,** which lines the primitive gut cavity, or archenteron. This early gut cavity continues to elongate until its anterior end touches the ectoderm and breaks through to form the mouth, leaving the original rear opening to this primitive gut as the anus. Pouches push out (or evaginate) from each side of the inner layer and are cut off as vesicles. These represent the **mesoderm,** or middle tissue layer, which forms the muscles, blood, other connective tissue, and the notochord.

In animals whose eggs have more yolk, simple invagination at the vegetal pole is not possible because of the large mass of yolk. Instead, cells from the animal pole region proliferate and grow over the vegetal cells. At the same time, they turn in (involute) at one particular region, which becomes the blastopore. The resulting primitive gut cavity, or archenteron, is pushed dorsally, right under the ectoderm. The roof of this gut cavity splits off as a sheet and becomes mesoderm, and the gut closes in under this new mesoderm to become an intact tube again. The mesoderm lying on the dorsal midline differentiates into a rod of turgid cells, the notochord, the primary structural support in early chordate embryos.

In reptile and bird eggs, there is so much yolk that neither invagination of the vegetal pole nor involution around the edges of the yolky vegetal pole can occur. In the bird embryo, most of the lower layer of the blastula becomes endoderm, and most of the upper layer becomes ectoderm. In the center of the primitive streak along the future longitudinal axis of the embryo, cells from the upper layer involute along this midline to form the mesoderm, then lying between the ectoderm and endoderm.

The process of gastrulation is a turning point in morphogenesis because at this stage the three primary germ layers are determined and begin their developmental pathways toward forming the various parts of the body.

From the ectoderm comes the epidermal layers of the skin, hair, the eye lens, many glands, and the epithelium of the mouth and anus portions of the gastrointestinal tract. The ectoderm also gives rise to the nervous tissue in the form of the neural tube, which will differentiate into the spinal cord and brain as well as sending out lateral processes of nerve tissues.

The endoderm gives rise to the epithelial lining of the gut (the ectoderm being restricted to the lining of the nasal cavity, the mouth, and the anal canal). The endoderm also gives rise to other structures derived from the digestive tract, such as the lungs, liver, pancreas, bladder, and thyroid.

The mesoderm gives rise to the primitive notochord and the musculature of the body, including skeletal muscles as well as heart muscle and those of other organs. The mesoderm also gives rise to connective tissue, including blood and blood cells, and bone as well as various cartilage structures in the body, such as tendons and rib connections.

Progressive restrictions on differentiation

An example of the sequential differentiation that must occur before an adult cell type is established is shown in Figure 33–13. As development proceeds, the potential pathway of a particular cell is progressively more limited. Thus in the development of an amylase-producing pancreatic cell, the developing parental cells faced a number of choices at different times during development. The zygote could have given rise to mesoderm or ectoderm instead of endoderm. Once the endoderm was formed, it could have given rise to the hindgut or foregut instead of the midgut. Once the midgut was formed, the cell was further restricted to either the pancreas or the duodenum. In the pancreas, our prospective amylase-producing cell became an acinar cell instead of an islet cell. The acinar cells in the pancreas can differentiate into one of at least four different cell types, only one of which will produce amylase. This enzyme, like the amylase produced by the salivary glands, can digest starch, but it will do so only in the duodenum instead of in the mouth or stomach, as with salivary amylase. Though every cell in the body possesses the gene to produce pancreatic amylase, only the acinar cells at the proper location in the pancreas are able to activate the DNA switch and cause their RNA and ribosomal machinery to produce the amylase polypeptide molecule.

Once the process of gastrulation is completed, it is clear that the masses of cells have been brought into the proper position for subsequent differentiation and morphogenesis of all the structures that will go to form the adult. While the process involved is an extension of ones we have already observed at work, namely, cell cleavage

FIGURE 33–13.

Progressive restrictions ("switching stages") in the differentiation of an amylase-producing cell of the human pancreas.

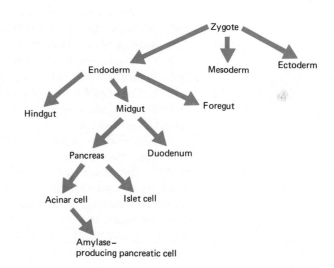

and cell differentiation under the influence of organizers, developmental biologists still have a great task ahead of them to explain these processes of organ formation and the organization of the great diversity of tissues in the adult body. One can look, for instance, at the complexity of nerves, muscles, sensory cells, and ectodermal tissue in the human eye, and appreciate more fully the large number of sensitive mechanisms controlling differentiation and morphogenesis that must smoothly unfold, without error, in order to produce a normally functioning adult organism. These considerations make developmental biology one of the most exciting frontiers of modern science; it should be a leading field well into the future.

Summary

The complexity of development is such that early biologists working with embryos adopted a preformationalist view, namely, that the parts of the embryo did not develop but rather were already present at the moment of conception and simply increased in size in the growing embryo. Wolff introduced the alternative concept of epigenesis, which states that the structure of the embryo progressively developed from an undifferentiated egg. Roux, Driesch, and Spemann provided experimental evidence that the development of the fertilized egg is determined by the genetic material present in its nucleus and by an interplay of chemical and physical factors between the genes and the rest of the cell or developing embryo. Early embryonic development involves a series of cleavages that yield the morula and then the blastula stages. Metabolic energy is usually derived from stored yolk or actively feeding embryonic stages (larvae). In mammals, the embryonic states are maintained by a uterine environment and a special exchange structure (the placenta) with the maternal circulatory system. Yolk greatly influences the pattern of cleavages to the blastula stage in different types of animals. At this point, gastrulation creates the three primary germ layers of tissue (ectoderm, mesoderm, and endoderm) through invagination at the vegetal pole, involution around the edges of the yolky vegetal pole, or involution along the midline of a primitive streak in the blastula. Throughout differentiation and morphogenesis, all the cells are fated to become specific tissues, and they experience progressive restrictions on their potency to be other types of cells. Thus unlike plants, there are no real areas set aside to continue production of undifferentiated cells throughout the animal's lifetime. Most mature animal tissues do retain the capacity to produce more cells like themselves through mitosis, but they cannot reverse the steps they took earlier along their particular developmental pathway of differentiation.

34 Reproduction: the passing of the torch

One of the basic and characteristic patterns of evolution has been the tendency of organisms to invade and inhabit every available niche in the ecosystems of the earth. In our observations on ecology, especially the phenomena involved in dispersion and population growth, we noted that even slow-breeding elephants are potentially capable of producing a large number of offspring within a relatively short time span. Indeed, if it were necessary to name a single characteristic that distinguishes life from nonliving things, it would be reproduction: the drive to produce more of one's own kind.

Asexual and sexual advantages

In the higher animals, reproduction is almost exclusively sexual; among the lower organisms and higher plants, reproduction is carried on by both sexual and asexual means. *Asexual reproduction* produces organisms that are genetically identical to their single parent, whereas *sexual reproduction*, involving the union of gametes from two different parents through the processes of meiosis and fertilization, allows for considerable shuffling and reorganization of the genetic material. Thus sexual reproduction generates variability on which evolution may operate to better adapt the organism to its environment.

This is not to say that asexual reproduction is a disadvantage to a species, for it allows relatively rapid production of new individuals, each a copy of the original parent, which may indeed be well adapted to prevailing conditions. Asexual reproduction may occur through vegetative budding, through runners sent out from the main plant body along the ground that take root elsewhere, as in strawberry plants, or below the ground by means of underground stems called **rhizomes** (Figure 34–1). Thus, potatoes, irises, and many lawn grasses (e.g., devil grass) reproduce by rhizomes. Many tropical rainforest plants are able to develop roots from sections of stems or leaves that touch the ground or moist pockets of humus material caught in tree crevices. A great many plants reproduce asexually by means of bulbs, such as lilies, onions, and tulips.

FIGURE 34–1

Common types of asexual reproduction in plants.

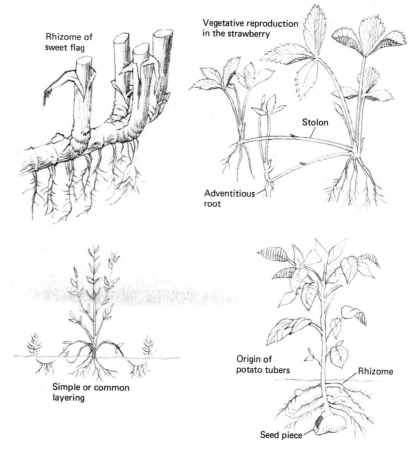

Rhizome of sweet flag

Vegetative reproduction in the strawberry

Stolon

Adventitious root

Simple or common layering

Origin of potato tubers

Rhizome

Seed piece

The regular use of asexual reproduction in fruit trees by commercial growers is well known. A particularly good genotype is propagated by grafting small twigs from that original tree onto the stumps of trees with less desirable fruit characteristics but with well-established root systems. Horticulturally valuable roses are also propagated in this way. Pineapples, bananas, seedless grapes, and other commercially important plants are sterile and can be propagated only by vegetative means.

The alternation of generations

In general, however, the advantages of sexual reproduction to diploid plants and animals are such that most species have sexual reproduction during at least part of their life. In plants the evolutionary value of sexual reproduction has been such that vascular plants have a life cycle in which there is a definitive **alternation of generations** (Figure 34–2). A *haploid*, or *gametophyte*, generation alternates with a *diploid*, or *sporophyte*, generation. Such an alternation of generations may be found even among certain algae, but at this evolutionary level the haploid and diploid plants are indistinguishable. In higher plants, the gametophyte generation is quite reduced in size compared with the sporophyte generation.

In any alternation of generations the sporophyte plant is composed of cells that are diploid. At the time of reproduction, then, certain cells in the sporophyte body undergo meiosis, resulting in the production of haploid spores. The haploid spore may be either dropped off the plant or held on the plant. The haploid spore will germinate, the location depending on the plant group involved, and give rise to a haploid gametophyte plant.

In most plants, the gametophyte generation is a smaller and more delicate structure than the sporophyte, and, in fact, it is often directly dependent on the sporophyte if the spores were held on the sporophyte body. The gametophyte is responsible for producing the gametes and since all the cells of the gametophyte are already haploid, no meiotic reductional division is necessary to produce the correct chromosome number in these gametes. Thus when sex organs are produced on the haploid gametophyte through mitosis and tissue differentiation, the result is the production of haploid sperm and eggs. Through various means, these germ cells can then combine in fertilization and give rise to a new diploid sporophyte plant.

This simple outline of the details of alternation of generations in

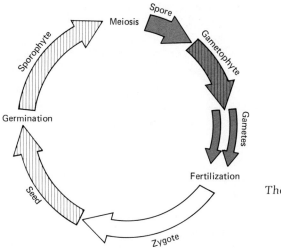

FIGURE 34–2

The alternation of generations in plants.

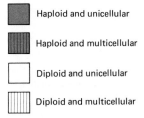

Haploid and unicellular

Haploid and multicellular

Diploid and unicellular

Diploid and multicellular

the haploid-phase and diploid-phase reproductive cycle in plants should not conceal the fact that there is great variability in the types of life cycles among the different divisions and classes. As a more or less "typical" example of sexual reproduction in plants that have well-developed sporophyte and gametophyte phases, let us look at the reproductive cycle of the fern.

Reproduction in the fern

The ferns are a large class of vascular plants found throughout the world in every conceivable ecological niche (Figure 34–3). They are usually small plants occupying shaded and moist locations, though tree ferns can grow to dozens of feet in height and even rocky desert canyons have numerous ferns of modest stature but persistent existence. The sporophyte generation in the fern is represented by the plant body that most of us are familiar with — namely, large leaves, or fronds as they are usually called, that rise from an underground stem, the rhizome, and have roots growing down from the rhizome into the ground.

If we look at almost any of the common ferns of North America, we will see that the **spore**-producing bodies (*sporangia*) are produced on the lower side of the leaf. Often the sporangia are clustered into groups called *sori*, which look like brown patches scattered regularly on the underside of the frond. Spore mother cells are found in the upper part of each sporangium and they subsequently divide by

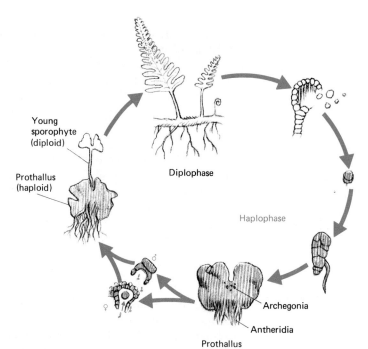

FIGURE 34–3

The reproductive cycle of a fern, showing the dominant diploid (2n) stages and the short-lived haploid (1n) stages. Sperm are produced in the antheridia organs and eggs in the archegonia organs on the gametophyte plant, a haploid "prothallus." The sperm swim to the egg via water on the plant body. The fertilized egg grows into the young sporophyte that eventually produces spores that float out on air currents to moist places, germinate, and grow into prothalli to start the sexual cycle again.

meiosis, producing a large number of haploid spores. As the sporangium matures, it breaks open and throws the spores out into the air currents. The spores are the principal means of dispersal for all ferns (a few have long rhizome runners that form new individuals elsewhere).

When a spore lands on a favorable substrate, such as a moist soil surface, it begins cell division immediately and produces a heart-shaped gametophyte plant. This gametophyte, or prothallus, is a flattened green body that is photosynthetic and has rhizoids on its lower surface to anchor it to the ground and supply the delicate plant with water. Both male and female sex organs are produced on the same gametophyte plant. At maturity the sperm swims out of the male sex organs and through surface water covering the gametophyte to enter the female sex organ and fertilize the egg. The zygote then develops a rhizome itself, sprouts leaves, and grows up into the new sporophyte, destroying the old gametophyte in the process.

Reproduction in flowering plants

In the flowering plants, or angiosperms, the gametophyte generation is reduced to a few parts of the flower, which is the compound organ for sexual reproduction in these plants. The flower is a transitory structure, lasting only long enough for one season of reproduction, in contrast to the life-long maintenance of the sex organs in male and female animals.

In an evolutionary sense, the flower is a modified stem tip. During vegetative growth, the stem produces the usual green leaves. At the seasonal time for flowering, however, a series of physiological and morphological changes take place in the growing tip, producing the flower parts. There are four basic sets of flower parts (Figure 34–4). The outer whorl of flower parts is called the *calyx* and is composed of subunits called **sepals,** which look like slightly modified green leaves and protect the developing flower bud. After production of the sepals, the petals, stamens, and pistils are initiated and mature in the bud. The *petals,* collectively called the **corolla,** are usually brightly colored and advertise the presence of the flowers to potential pollinators. Within the corolla are located the **stamens,** the male reproductive structures, which are composed of long stalks called *filaments* and tipped by enlarged *anther* organs where the pollen grains are actually produced. Here in the form of pollen grains is the immature male gametophyte generation. In the center of the flower is one (or more) *pistil,* which contains the female gametophyte.

The pistil typically consists of three parts: a hollow base, the *ovary,* in which the ovules are contained; a slender stalk, the *style,* which is hollow and will form a channel for the pollen tube to grow to the ovary; and the *stigma,* a sticky landing platform for the pollen, which is located at the end of the style. Each ovule in the ovary encloses the female gametophyte, which contains a single egg cell. When the sperm fertilizes the egg cell, the ovule tissues will develop

FIGURE 34–4

The four basic sets of flower parts: sepals, petals, stamens, and pistil.

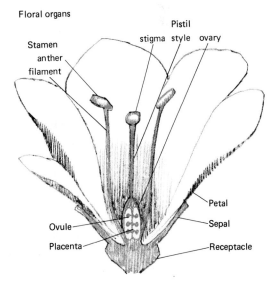

FIGURE 34-5

Reproduction in angiosperms: (A) the reproductive cycle, including the very brief haplophase of the gametes; (B) germination of a pollen grain on the stigma of the pistil and growth of a pollen tube down the style to an ovule in the ovary.

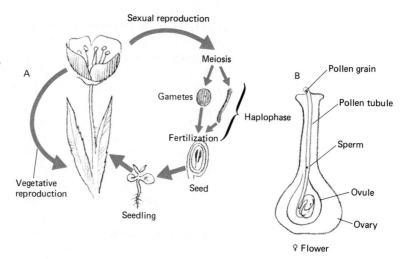

into a *seed* containing the young embryo. Some flowering plants have male flowers separate from female flowers; in other species, male and female flowers may be present on the same plant. The flower described above and illustrated in Figure 34–4 is one in which male and female flower parts are located in the same flower.

The process of fertilization involves pollen grains being carried from the anther to the stigma, usually of another flower, where they germinate. They may be carried by the wind, by insect vectors, or even by birds and mammals such as bats, which visit flowers for nectar and for the protein content of the pollen itself. On the top of the stigma, the pollen grain germinates and develops a pollen tube, which grows down through the hollow center of the style toward the ovule. Two sperm nuclei reach the female gametophyte within the ovule. One fuses with the nucleus of the egg cell present there and the other fuses with two extra nuclei in the gametophyte to produce a unique triploid ($3n$) tissue (Figure 34–5).

From this extraordinary $3n$ nucleus, a $3n$ tissue (the endosperm) is produced, which, like yolk in the animal embryo, nourishes the young developing plant embryo. However, in the entire diversity of life, this phenomenon of triple fusion has been observed only among the angiosperms. In some angiosperms such as corn and wheat, the endosperm is not completely used up in the course of development and is found in the mature seed. Its rich nutritional value supplies food for most of the world's human population in the form of meal and other substances made from the corn kernels.

After fertilization in the angiosperm, the enlarged base of the carpels, which is the ovary, develops into a **fruit.** Inside, the ovules develop into the seeds, each containing its own embryonic sporophyte. The diverse kinds of fruits that we see are largely evolutionary derivatives of the ovary, developed to ensure a means of seed dispersal. Some fruits have membranous wings that carry the seeds considerable distances on the wind. Brightly colored and highly edible fruits are very attractive to birds and other animals, which eat them and carry the seeds for some distance, as they pass unharmed through the digestive tract. Seeds with outside spurs and hooks travel readily on fur and feathers, to be carried by birds or mammals to other areas.

The unique combination of these fruits, which increase the efficiency of distribution, the flower, which increases the efficiency of fertilization, and the angiosperm's more efficient *conducting systems* make the flowering plants the most successful of all the land plants, not only in terms of diversity and numbers of species and individuals but also their coevolutionary effects on the existence and adaptations of other organisms such as major herbivore groups.

Reproductive cycles in animals

The complexities of types of reproduction among the animals are as great as the diversity of body shapes and modes of life of the various phyla and classes. Asexual reproduction is widely distributed among the invertebrates (Figure 34–6). It may occur by **fission,** especially in protozoans. Among the flagellates, fission is usually bi-

FIGURE 34–6

Several kinds of asexual reproduction in animals.

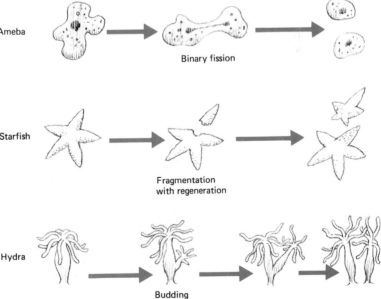

Ameba

Binary fission

Starfish

Fragmentation with regeneration

Hydra

Budding

nary, the individual dividing lengthwise. Some dinoflagellates reproduce by multiple fission, as do various other protozoans. **Budding** as a means of reproduction is usual in sponges, cnidarians, some protozoans, and in certain other invertebrates. Often the buds remain attached to the original parental individual, and colonies form around that parent in which all members may be alike, or are differentiated as colonial individuals fitted for the performance of different functions.

Fragmentation, which involves splitting the body and regenerating missing parts, is another interesting method of asexual reproduction. This is characteristic of many of the fresh-water flatworms, nemerteans, annelids, and echinoderms. Fragmentation and regeneration are usually combined with some sort of fission process. Still another means of asexual reproduction is by special reproductive cells or groups of cells that are capable of developing into a complete individual. These are called **gemmules** and are diploid cells, not haploid gametes. Natural parthenogenesis (growth of an unfertilized egg) is common among rotifers, some gastropod mollusks, certain crustaceans, and insects.

Asexual reproduction in animals

Obvious advantages and strong disadvantages occur with the types of invertebrate asexual reproduction. The chief *advantage* lies in the minimization of chance, which tends to play a large part in sexual reproduction in controlling the meeting of gametes and even of males and females. Chance is avoided if one simply releases a carbon copy of oneself into the environment. Thus with asexual reproduction, perpetuation of the line is assured. All the reproductive cells are parts derived by fission, fragmentation, or budding, and they have equal genetic opportunity for developing into new members of the species. Additionally, it is an advantage in some environments, such as the deep sea trenches, to be able to reproduce without another individual of the same species.

The chief *disadvantage* in asexual reproduction is the same for animals as it is for plants: the loss of the possibility for variation in one's descendants. Offspring produced asexually must inevitably contain the same chromosomal composition and genetic endowment as the parental individual they came from, with the only possibility of variation coming from mutation, a relatively rare event in nature. Thus in an invertebrate population reproducing asexually for many generations, there will be few variants on which natural selection can act.

The types of sexual reproduction

The process of sexual reproduction is found in every invertebrate phylum and in some groups it seems to be the only means of reproduction. Special cells, or gametes, are produced and these unite into a compound cell, the fertilized egg, or zygote. Frequently among invertebrates the gametes may be quite similar in structure, so the distinction of egg and sperm does not always hold. Sometimes the two kinds of gametes are even produced by the same individual. However, the most general situation involves the occurrence of cross fertilization, that is, union of the gametes from two different individuals. This union may take place outside the bodies of the gamete-producing individuals as *external fertilization,* or it may occur inside the female as *internal fertilization,* a more precise means of ensuring union of the gametes. In general, these gametes are produced by specialized organs present in the normal adult phase of that particular species.

An alternation of major body forms, as in alternation of generation in plants, is found in cnidarians (Figure 34–7). Certain of these coelenterates have a polyp form alternating with a motile medusa form during their life cycle. The medusas in the usual coelenterate species are male or female (in separate individuals), and they produce sperm and eggs that unite in a fertilized zygote. The zygote develops into a larval stage that eventually attaches itself to a substrate and grows into a branched colony of polyps by budding. Part of the polyp colony then produces more buds that detach themselves and swim off as medusas, which look like small jellyfish.

FIGURE 34–7

Alternation of body forms cnidarians. Both stages are diplophase; as in other animals the haplophase is represented by the gametes only. When the fertilized egg begins cleaving, it develops into a ciliated swimming larva that soon settles onto the substrate and metamorphoses into an adult polyp. The adult polyp colony eventually buds off free-swimming medusas asexually, which mature to produce eggs and sperm.

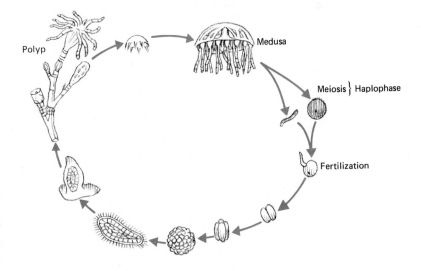

Polyp Medusa Meiosis } Haplophase Fertilization

Aside from the alternation of body forms, this complex cycle, which is sometimes called an alternation of generations, is not comparable to the alternation of sporophyte and gametophyte generations in plants. In the reproductive cycle of these coelenterates, both the polyp and medusa are diploid, and, as in other animals, the haploid phase is represented only by the gametes. The fertilization union of the haploid egg and sperm create a young, embryonic diploid stage, which becomes a ciliated swimming larva; then the larva metamorphoses into a polyp adult. The adult colony of diploid polyps has both feeding polyps and reproductive polyps. The latter will asexually bud off diploid medusas, which also go through a maturation process before they are capable of producing the haploid sperm and eggs.

Reproductive cycles in mammals and human beings

Most mammals undergo one or more cycles of potential reproductive activity during a year. In some species such as the primates and domestic dogs and cattle, the males remain sexually active at all times; the females may enter a period of heat only several times during the year, at which time they will accept the male. But in most mammals, both the male and the female produce gametes (Figures 34–8 and 34–9) and prepare to incubate an embryo only during a short annual breeding season.

In either case, the uterus of the female undergoes a gradual

FIGURE 34–8

The parts of the male reproductive tract: lateral view.

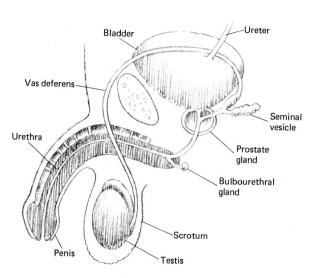

FIGURE 34–9

The organization of the female reproductive tract. The fingerlike fimbriae projecting from the ends of the oviducts trap the egg after a monthly release from one of the ovaries. The hymen is a fold of tissue partially across the lower end of the vagina; it does not close off the vaginal opening completely. The clitoris is a small, very sensitive stalk of tissue similar in structure to the penis.

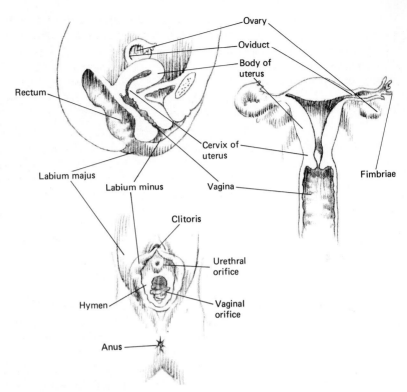

increase in vascularization and when the uterine lining is supplied with a great number of blood vessels, the ovary releases one or more eggs. If these are fertilized, the uterine lining is ready to nourish a developing embryo. If no embryo develops, the vascularization of the uterine lining will be reabsorbed in part and sloughed off in part, and after a while the cycle is repeated.

The human menstrual cycle

In the monthly menstrual cycle of a woman who has passed puberty, about one ovum a month matures in the ovary (Figure 34–10). This egg cell is surrounded by follicle cells that synthesize nutrients and pass them to the cytoplasm of the developing ovum. (All of the eggs that a female will release during her entire lifetime are present at birth in the ovaries, with their attendant follicle cells.) The follicle enlarges and the ovum becomes suspended in a fluid-filled cavity. Meanwhile, the nucleus of the ovum has begun meiosis. The mature

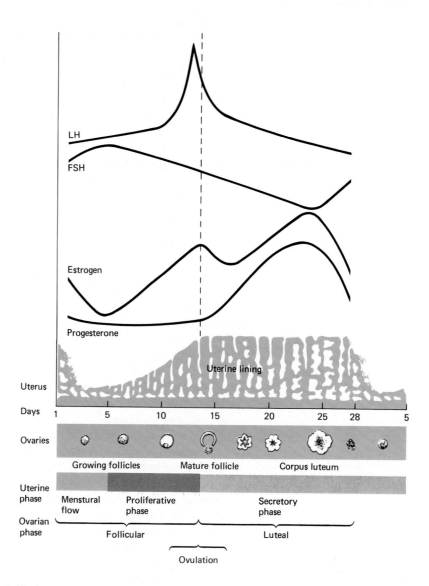

FIGURE 34–10

The human menstrual cycle, including hormone concentrations, uterine lining change, and ovarian events.

follicle soon presses against the surface of the ovary and ruptures, liberating the ovum in the act of ovulation. The empty follicle now fills with a plug of yellow connective tissue cells during the next seven days, as it forms a **corpus luteum.**

The newly released egg is drawn into the upper end of the oviduct by a current created by the sweeping cilia lining the oviduct. The upper ends of these paired oviducts in mammals are called **Fallopian tubes;** the lower ends of the oviducts have fused to form a **uterus.** The uterus develops a richly vascularized lining and serves the function of holding, protecting, and nourishing a developing embryo. The lower end of this single-chambered uterus in human beings connects to the **vagina,** which passes to the exterior. The vagina receives the penis of the male during copulation and it is also capable of marked dilation so as to become large enough to permit the infant to pass through it at the time of birth.

After ovulation occurs, the ruptured follicle in the ovary is transformed into the new structure called the corpus luteum. The corpus luteum will persist for several months if the egg is fertilized. In most cases, however, the egg is not fertilized and the corpus luteum then degenerates about 14 days after ovulation.

In the uterus, cyclic changes are also taking place about every 28 days. The reproductive cycle in the female receives its name — menstrual cycle — because of the physical changes evident here rather than the internal changes evident in the ovaries. Day one of the menstrual cycle begins with the onset of menstruation, which will last for several days. This process, involving the discharge of the detached portions of the lining of the uterus, called the **endometrium,** and blood, occurs if the egg is not fertilized. Some 28 days earlier, the lining of the uterus had begun to thicken and soften, as the numbers of blood vessels and small glands increased, in order to support and nourish the embryo if the egg was fertilized on its travel to the uterus through the Fallopian tubes. However, if pregnancy does not result, endometrium development stops when the corpus luteum begins to degenerate. This development causes a local spasm of blood vessels that starves the cells of the thickened portions of the endometrium for nutrients and oxygen. When these cells die, the thickened portions of the endometrium become detached and are released from the uterus, along with a small amount of blood (about 50 to 250 ml) through the vagina. It takes about three to five days to complete this discharge. The "cramps" sometimes felt at this time are caused by contractions of the muscular walls of the uterus.

At the time of a menstrual discharge (the start of the cycle), the pituitary gland begins to secrete increasing amounts of *follicle-stimulating hormone* (FSH). When this hormone reaches the ovary, it stimulates the maturation of an egg-containing follicle. The follicle secretes another hormone, estrogen, which, as we have noted earlier, stimulates the development and maintenance of the secondary sexual characteristics that first appear at puberty. In the uterus, estrogen stimulates the onset of the thickening and vascularization of the endometrium lining. Increasing concentrations of estrogen inhibit the production of FSH (thus preventing additional eggs from maturing) and stimulate the pituitary to secrete a second hormone, *luteinizing hormone* (LH).

The high concentration of estrogen and the rising amount of LH apparently stimulates **ovulation** on about the fourteenth day of the cycle. The ovum is swept into the oviduct; if it is fertilized there, a developing embryo will reach the uterus in about four more days. If it is not fertilized, the ovum disintegrates within a day or two.

Meanwhile in the ovary, LH in the blood has stimulated the ruptured follicle to differentiate into a corpus luteum, which begins to secrete *progesterone,* the so-called pregnancy hormone. Progesterone causes the uterine lining to undergo the final preparation for receiving an embryo and maintains the lining in a richly vascularized condition. It also stimulates the production of glandular cells in the mammary glands, which can secrete milk. At this time, increased concentrations of progesterone in the blood also inhibit the production of both FSH and LH by the pituitary.

With progesterone inhibiting the pituitary gland, the corpus luteum, maintained by LH, reaches its maximum development after about seven days. When the corpus luteum degenerates, after about 24 days following the onset of the preceding menstruation, the source of progesterone and much of the estrogen production is halted. As a consequence, the uterine lining, which had been maintained by the stimulating effects of estrogen and progesterone, is sloughed off along with bleeding. Now that high concentrations of progesterone and estrogen are no longer present, the pituitary is no longer inhibited from producing FSH and so it begins a new cycle, starting the maturation of another egg-containing follicle.

If, on the other hand, the egg is fertilized and develops, pregnancy results and the embryo is implanted in the uterine wall on about the seventh day of development. In ways not completely understood, this results in the continuing production of LH. Probably, the uterus signals the pituitary via the nervous system to continue to produce LH despite high concentrations of progesterone. In this event the corpus luteum is maintained and it continues to produce increasing amounts of progesterone so that the vascular uterine lining persists and a *placenta* develops. The placenta brings the fetal blood into intimate contact with the maternal circulation, permitting the exchange of oxygen, nutrients, carbon dioxide, and metabolic wastes. By the end of the third month of pregnancy, the placenta takes over the production of progesterone. This is the most frequent time for miscarriages, for if the corpus luteum ceases to produce progesterone before the placenta begins its secretion, the uterine lining is no longer maintained and the fetus can be lost.

The menstrual cycle continues uninterrupted, as we have seen, unless pregnancy occurs. The menstrual cycle is finally terminated by *menopause* when a woman reaches the age of 45 to 55; the uterine cycles become irregular at first and then after a few months to perhaps several years, the cycles cease. The ovaries do not respond any longer to FSH (though it is still produced during this time). Estrogen is no longer produced, and ovulation ceases. The physiological and psychological adjustments that are characteristic of the menopause period must be made because of the absence of estrogen in the blood.

The perpetuation of life

We have thus gone full circle in our story of life. At the individual level, we have seen how sex cells are produced in the body, how they unite, how the young embryo develops into the mature organism, and how the mature organism produces another set of sex cells in still another turning of the cycle of life. We have also seen the adaptations of life to a broad range of physical conditions across the face of the earth and through time, as revealed by the historical record of evolution. The intimate and diverse adaptations made by life at the community, population, individual, and cellular levels are remarkable indeed, yet all are manifestations resulting from different aspects of the same biological phenomenon: *reproduction*, the perpetuation of one's own kind.

Reproduction is essential for the survival of life on earth, the perpetuation of existing species, and the potential creation of new species. It forms the means by which an individual parent produces carriers of its own characteristics, transmitting to them a genetic constitution in which specific directions are encoded for the development of these characteristics that will distinguish the offspring as individuals and yet as the same species. Reproduction provides mechanisms for the long-term continuity of a species and yet also for variation within this taxonomic and biological unit. From the variants that arise through various natural forces such as background radiation, the individuals that have the greatest degree of success in meeting the conditions of their immediate environments and in adapting themselves to new conditions are those best fitted to survive, and hence those that will become the parents of future generations and possibly the ancestors of new species.

Success in reproduction, then, is the ultimate criterion for the continued perpetuation of a species and of an individual's genotype. The total diversity of life on earth arises from these basic internal properties and external environmental agents acting on the biological material of our planet.

Summary

Reproduction is the drive to produce more of one's own kind, the most significant characteristic that distinguishes life from nonliving things. Like the ancient Greek runners in an Olympic Games relay race, the parents pass the torch in the form of a package of genes to a fresh generation, always pressing forward toward evolutionary survival, a finish line that is frequently reached but that can never be crossed because of the constantly changing environment. Asexual reproduction produces organisms that are genetically identical to their single parent, and it may occur by fission, budding, runners, rhizomes, spores, fragmentation, gemmules, and parthenogenesis. Its chief advantages include relatively rapid production of new individuals that may be well adapted to prevailing conditions and the minimization or elimination of the element of chance, required for the meeting of two sexes and two gametes. Its chief disadvantage is the lack of variability generated among the offspring. Sexual reproduction involves the union of gametes from two different parents through the chance processes of meiosis and fertilization, introducing considerable shuffling and reorganization of the genetic material. In vascular plants an alternation of haploid gametophyte and diploid sporophyte generations is commonplace. This type of life cycle allows sexual reproduction to occur at regular intervals even in ferns, where wind-carried diploid spores are the principal means of dispersal for the plant; but haploid gametes allow the advantages of meiotic reshuffling of the genotype. The flowering plants have highly evolved pollination and seed-dispersal systems that ensure wide gene exchange and dispersal, making the angiosperms the most successful of all land plants. In animals sexual reproduction can occur with external or internal fertilization (union of the gametes). No true alternation of generations occurs, but there are cycles of reproductive activity during the course of a year. The complicated human menstrual cycle is an excellent example of the interplay of internal and external factors in achieving the basic goal of organic life: reproduction, the perpetuation of one's own kind.

EPILOGUE: THE END OF ALL

When the moon shall have faded out from the sky, and the sun shall shine at noonday a dull cherry-red and the seas shall be frozen over, and the ice-cap shall have crept downward to the equator from either pole, and no keels shall cut the waters, nor wheels turn in mills, when all cities shall have long been dead and crumbled into dust, and all life shall be on the very last verge of extinction on this globe; then, on a bit of lichen, shall be seated a tiny insect, preening its antennae in the glow of the worn-out sun, representing the sole survival of animal life on this our earth—a melancholy "bug."

ALFRED, LORD TENNYSON

Suggestions for further reading

The books listed below have been chosen because they are interestingly written, particularly for the general reader who wishes to further pursue these topics. They are readily available in most libraries or are paperback books that are generally stocked in college and university bookstores.

Part one: The earth as a biosphere

Ehrenfeld, David W. *Biological Conservation*. New York: Holt, Rinehart & Winston, 1970, 226 pp.

Ehrlich, Paul R., and Anne H. Ehrlich. *Population, Resources, Environment: Issues in Human Ecology*. San Francisco: Freeman, 1972, 509 pp.

Farb, Peter. *Ecology*. New York: Time-Life Books, 1970, 192 pp.

Leopold, Aldo. *A Sand County Almanac, With Essays on Conservation from Round River*. New York: Oxford University Press, 1966, 295 pp.

Milne, Lorus, and Margery Milne. *The Nature of Life: Earth, Plants, Animals, Man and Their Effect on Each Other*. New York: Crown, 1972, 320 pp.

Part two: Life at the community level

Cheng, Thomas C. *Symbiosis: Organisms Living Together*. New York: Pegasus, 1970, 250 pp.

Emmel, Thomas C. *Behavior and Ecology: A Book of Readings*. Dubuque, Iowa: Kendall/Hunt, 1970, 221 pp.

Whittaker, Robert H. *Communities and Ecosystems*. New York: Macmillan, 1975, 385 pp.

Wickler, Wolfgang. *Mimicry in Plants and Animals*. New York: McGraw-Hill, 1968, 255 pp.

Part three: Biology at the population level

Emmel, Thomas C. *An Introduction to Ecology and Population Biology*. New York: Norton, 1973, 208 pp.

Emmel, Thomas C., ed. *Genetics and Evolution: A Book of Readings*. Dubuque, Iowa: Kendall/Hunt, 1970, 156 pp.

Howell, F. Clark. *Early Man*. New York: Time-Life Books, 1965, 200 pp.

Moore, Ruth. *Evolution*. New York: Time-Life Books, 1964, 192 pp.

Wilson, Edward O., ed. *Readings from Scientific American: Ecology, Evolution, and Population Biology*. San Francisco: Freeman, 1974, 319 pp.

Part four: The biology of individual organisms

Ayers, Charlotte J. *Biology of Sex*. New York: Wiley, 1974, 280 pp.

Golanty, Eric. *Human Reproduction*. New York: Holt, Rinehart & Winston, 1975, 225 pp.

Hanawalt, Philip C., and Robert H Haynes, eds. *Readings from Scientific American: The Chemical Basis of Life: An Introduction to Molecular and Cell Biology*. San Francisco: Freeman, 1973, 405 pp.

Kennedy, Donald. *Readings from Scientific American: Cellular and Organismal Biology*. San Francisco: Freeman, 1974, 355 pp.

Nourse, Alan E. *The Body*. New York: Time-Life Books, 1971, 200 pp.

Tanner, James M., and Gordon R. Taylor. *Growth*. New York: Time-Life Books, 1971, 198 pp.

Glossary of biological terms

abdomen In vertebrates, the part of the body trunk containing the visceral organs except for the heart and lungs; in lower animals, the posterior portion of the body containing the reproductive organs and rear portion of the digestive tract

abortion The premature expulsion of the developing fetus from the uterine cavity

abscission In plants, the dropping of leaves, roots, or other plant parts at the end of the growing season as the result of formation of a layer of specialized cells

absorption The transport of small food molecules and water from the gastrointestinal tract through the cell membranes of the intestinal cells to the blood and lymph vessels

acellular organism An organism that lacks division into cells, as the body of a slime mold

action potential A rapid change in the membrane potential of a nerve cell, resulting in transmission of a nerve impulse

active transport The movement of a substance across a cell membrane by a process requiring the expenditure of energy

adaptation The acquisition of characters by an organism that help it to survive; also, any genetically controlled characteristic such as structure, physiology, or behavior that aids in fitting the organism to its particular environment

adaptive polymorphism The different adaptively advantageous forms that occur within a single population of a species

adaptive radiation The evolution of a number of specialized forms from a relatively primitive and unspecialized type of organism; for example, Darwin's finches in the Galápagos Islands

adaptive strategy The way an animal or plant acquires energy and maintains itself

adenosine diphosphate (ADP) A lower-energy form of ATP, with only two phosphate groups instead of three as in ATP

adenosine triphosphate (ATP) An organic compound with three phosphate groups that functions as energy-releasing currency for organisms

adrenal cortex The outer, barklike area of the adrenal gland

adrenal gland An endocrine gland located near the kidneys in vertebrates; the source of adrenalin and many steroid hormones

adrenal medulla The inner portion of the adrenal gland

adrenalin A hormone produced by the adrenal medulla that stimulates a number of body reactions

aerobe A microorganism that requires oxygen for respiration

aggregation A group of individuals, not necessarily all of the same species

alga A simple photosynthetic plant that lacks multicellular sex organs

allele One of several alternative gene forms at a particular position (locus) on homologous chromosomes

allopatric species Species that live in different geographic areas; see sympatric species

alternation of generations A reproductive cycle in which a haploid phase, the gametophyte, gives rise to gametes that fuse and form a zygote that grows into a diploid phase, the sporophyte, which in turn produces spores that give rise to new gametophytes

alveolus One of the air sacs in the lungs

amino acid One of the organic acids that are the building blocks of protein molecules

amphibian One of the class of vertebrates between fish and reptiles that must return to the water to lay their eggs

amylase A starch-digesting enzyme

anaerobe A microorganism that is able to live without free oxygen

anaphase The stage in mitosis or meiosis in which the chromatids of each chromosome separate and move to opposite spindle poles

anemia A lack of sufficient hemoglobin, resulting in poor oxygen-transporting capacity

aneuploidy The condition in which the nuclei do not contain an exact multiple of the haploid number of chromosomes

angiosperm One of a group of plants whose seeds are born within a covered ovary or fruit

annual A plant that completes its life cycle within a single growing season

antenna One of the long, paired appendages on the head of many arthropods that contain sensory cells

antibody A protein that destroys or inactivates an antigen

antigen A foreign substance, usually a protein, that stimulates the body to produce antibodies against it

anus The posterior (rear) opening of the digestive tract to the outside of the body

aorta The main artery in the blood-circulating system

appendix A small projection from the terminal point of the cecum

arteriole A small artery

artery A blood vessel that carries blood away from the heart

asexual reproduction Cellular reproduction that involves only the production of new daughter cells from the equal division of a mother cell

atom The smallest unit of an element that still retains characteristic properties of that substance

ATP See adenosine triphosphate

atrium Either of the two upper parts of the mammalian heart

autosome A chromosome that is not a sex chromosome

autotroph A "self-feeder"; a plant that can photosynthesize the needed carbohydrate food from simple inorganic materials

auxin Any of the plant hormones that promotes cell elongation

axon The part of a neuron that carries impulses away from the cell body

bacteriophage A virus that parasitizes a bacterial cell; also called a phage

biennial A plant that requires two years to complete its reproductive cycle

bilateral symmetry A kind of symmetry in which an organism has two similar sides that are virtually mirror images of each other

bile A secretion from the gall bladder, which empties into the duodenum and helps in the digestion of fats

binary fission Cell reproduction by division into two essentially equal parts.

biodegradable material Material that is capable of being broken down by microorganisms or other living things

biogenesis The origin of living organisms from other living organisms

biogeochemical cycle The cycling of materials in

characteristic paths from the earth through living systems and back to the earth

biogeography The study of the geographic distribution of living plants and animals

biological magnification The concentration of persistent chemicals such as DDT as they are passed along a food chain

biomass The total weight of all organic matter in a particular habitat or area

biome One of about a dozen broad groupings of plants and animals spread across the earth at characteristic latitudes and altitudes

biosphere The entire area of land, water, and air on the surface of the earth that is occupied by living things.

biota The plants and animals of a particular area

biotic potential The maximum possible population growth rate under ideal conditions

blastula An animal embryo after rapid cleavage and before gastrulation, which usually consists of a hollow sphere formed of a single layer of cells

botany The study of plants

brain stem One of the three primary divisions of the brain; composed of the medulla, pons, and midbrain

Brownian movement The random movement of microscopic particles suspended in a gas or liquid

budding The asexual formation of a miniature version of a parental plant or animal body

calorie A unit of heat; the amount of heat energy required to raise the temperature of one gram of water 1° C (cal.); also the amount of heat required to raise the temperature of 1,000 grams of water 1° C (Cal.)

calyx The outer whorl of flower parts (sepals) found in angiosperms

cambium The principal lateral meristem of vascular plants, which gives rise to most secondary tissues

capillary One of the smallest blood vessels, connecting arteries with veins; the exchange of oxygen and other materials takes place between the blood and the tissues through capillary walls

carbohydrate An organic compound composed of carbon, hydrogen, and oxygen; the hydrogen and oxygen atoms are attached in a 2 to 1 ratio; examples are sugars, starch, cellulose, and glycogen

carnivore An organism that eats animals to obtain its food energy

carotenoid A yellow, orange, or red pigment found in plants and some algae

carpel A leaflike floral part that encloses the ovule or ovules of the angiosperm

carrying capacity The maximum population size that a given environment can support indefinitely

cartilage A specialized type of skeletal connective tissue in vertebrates, with a rubbery intercellular matrix

catalyst A substance that accelerates the rate of a chemical reaction but is not used up in the process; enzymes are catalysts

cecum A side pocket of the digestive tract

cell The fundamental unit of animal and plant life, consisting of a nucleus, cytoplasm, and numerous specialized organelles

cell body The central mass of cytoplasm in a neuron

cell wall A nonliving protective coat around a plant cell

cellulose A complex insoluble carbohydrate that is a major constituent of most plant cell walls

central body The centrosome, a region of specialized cytoplasm outside the nucleus of most animal cells, which contains two dark bodies called centrioles

central nervous system The portion of the nervous system that contains the brain and spinal cord in vertebrates and the main controlling nerve cords in invertebrates

centriole A cytoplasmic organelle, located just outside the nucleus in animal cells, that organizes the spindle fibers during mitosis and meiosis

centromere A section of the chromosome to which a spindle fiber attaches during mitosis or meiosis; also called kinetochore

cerebellum The enlarged part of the dorsal hindbrain of vertebrates that controls muscular activity and coordination

cerebrum The principal portion of the forebrain in vertebrates, forming the chief coordination center of the nervous center

character displacement The rapid divergence of characters in sympatric species that will minimize competition between them

chemosynthesis The autotrophic synthesis of organic materials by using energy derived from inorganic molecules rather than light, as in photosynthesis

chiasmata The points of contact between a homologous pair of chromosomes in meiosis, at which there is usually an exchange of genetic material

chlorophyll The green pigments of plant cells that are necessary for photosynthesis

chloroplast A small plastid body containing chlorophyll, which is the site of photosynthesis in green plant cells

chromatid A single chromosomal strand; one of the two daughter strands of a duplicated chromosome, which are joined by a single centromere

chromatin The long threads of nucleoprotein that form the unraveled chromosome

chromosome One of the DNA bodies in the cell nucleus that contains a linear sequence of genes

chyme A thick mixture of food particles and secreted fluids in the stomach

cilium A short, hairlike locomotory organelle on the surface of a cell

clear-cutting The lumbering off all the trees within a particular area

cleavage A successive cell division of a zygote

climax community A relatively stable community reached in a successional series

codon A unit of genetic coding, comprising three adjacent nucleotides on a molecule of m-RNA that form the code for a single amino acid

coevolution The simultaneous evolutionary adjustments made by two or more closely associated groups of plants and/or animals

colon The large intestine of vertebrates, leading to the rectum

commensalism A symbiosis of two species in which one benefits and the other is neither benefited nor harmed

community The ecological unit composed of all the plant and animal populations living in a given area

competition The interaction between two or more individuals or by two or more populations resulting from the utilization of the same limited and mutually required resource

conductor A specialized cell that transmits a nervous impulse from the receptor cell to other conductor cells or to an effector cell

conifer An evergreen tree or shrub, such as pine, fir, and spruce

connective tissue A type of animal tissue whose cells are irregularly distributed through a relatively large amount of intercellular material (matrix) that connects, supports, or surrounds other tissues and organs

continental drift The drifting movement of our present continents apart from a single land source area during the past 600 million years or so

contraceptive A means of preventing conception (successful union of a sperm and egg cell); examples are condoms, pills, and diaphragms

corolla The petals of a flower

corpus luteum The ruptured mature follicle near the surface of the human ovary, after release of an egg

cortex The outer, highly convoluted portion of the cerebrum (forebrain), where most higher thinking processes are centered; the bark or outer layers of a plant stem and root

cortisone A hormone produced by the adrenal cortex

cotyledon A seed leaf on the embryonic sprout of a seed plant

cross-over The exchange of parts of two homologous chromosomes at meiosis

cuticle A waxy layer on the outer surface of leaves or arthropods

cytochrome One of several iron-containing pigments that serve as electron carriers in electron transport chains, involved in cellular respiration and photosynthesis

cytokinesis The division of the cytoplasm of a cell

cytokinin One of the growth hormones in plants that promote cell division

cytoplasm The living matter of a cell, excluding the nucleus

dark reactions The part of photosynthesis that involves the reduction of CO_2 by hydrogen to form carbohydrate molecules, a process not requiring light

deciduous plant A plant that sheds its leaves each year or season decomposer An organism that feeds on and breaks down dead organic matter

deduction Reasoning from accepted general principles to specific examples

defoliant A chemical herbicide that causes plants to lose their leaves

demographic transition A decline in birth rate following a drop in death rate, as in the population of western Europe from 1650 to 1850

demographic transposition A predicted reversal of the present relative positions of the death and birth rates in major underdeveloped countries when they reach the point of being unable to supply the subsistence-level needs of a significant proportion of their population

dendrite A nerve fiber, usually highly branched, that conducts impulses toward the nerve cell body

density The number of individuals within a particular unit of space

density-dependent factor An influence on population growth whose effect is directly correlated with the size of the population in a particular area

density-independent factor An influence on population growth whose effect is not changed by the size of the population in a particular area

deoxyribonucleic acid (DNA) A nucleic acid in the cell that is thought to be the genetic material in all organisms except certain viruses, where RNA serves as the genetic material

detritus Organic debris

differential reproduction Reproduction in which a better-adapted organism leaves more living offspring than another organism of the same or different species

differentiation The developmental process by which a relatively unspecialized immature cell becomes a more specialized cell or tissue

diffusion The movement of suspended or dissolved particles from one place to another as a result of the random movement of individual particles from a more concentrated to a less concentrated region

digestion The breaking up of a large molecule into its building-block units

diploid cell A cell that has two of each type of chromosome (2n) except for the sex chromosomes

disaccharide A double sugar; one composed of two simple sugar units

dispersion The spatial arrangement of the individuals that compose a population

DNA See deoxyribonucleic acid

dominance hierarchy A "peck order" of social rankings within a population

dominant gene A gene allele that exerts its full phenotypic effect despite the presence of another allele of the same gene, whose effect is thus masked by the dominant allele

duodenum The first portion of the small intestine of vertebrates

ecology The study of the interrelationships between organisms and their environment

ecosystem The physical environment and the biological community that occurs in a given area

ecotone A transition zone or boundary area between two major types of plant communities

ectoderm The outermost layer of body tissue, especially in an embryo

effector The part of an organism that produces a response to stimuli

embryo The early developmental stage of a plant or animal produced from a fertilized egg

emigration A one-way movement out of a particular area

endocrine gland A ductless gland whose hormone secretions are released into the circulatory system

endoderm The innermost tissue layer of an animal embryo

endometrium The glandular lining of the uterus in mammals that is sloughed off in menstruation

endoplasmic reticulum A network of membrane-bounded channels in the cytoplasm of a cell

endosperm A nutritive material in the seeds of flowering plants

energy flow The transfer of energy from one organism to another in an ecosystem

enzyme A protein molecule that catalyzes a chemical reaction

epicotyl The upper part of the axis of a plant embryo that will form the shoot

epidermis The outermost portion of the skin of an animal

epiglottis The flap of tissue that covers the glottis while food is being swallowed

epiphyte A plant that grows on the branches or trunks of other plants

epistasis The masking of one gene by another, nonallelic gene

epithelial tissue The tissue that covers external body surfaces and some internal surfaces

erythrocyte A red blood cell containing hemoglobin

esophagus In mammals, the forward part of the digestive tract leading from the pharynx to the stomach

estivation A state of dormancy similar to hibernation, but carried out to survive a hot, dry period rather than a cold period

estrogen A female sex hormone in vertebrates

ethology The study of behavior in its natural context, as opposed to strictly laboratory observations and empirical testing

eutrophication The promotion of excessive plant and animal growth (and resultant oxygen depletion) by the addition of substantial amounts of nutrients to aquatic ecosystems

evolution The origin of new forms over time and the continuous genetic adjustment of existing forms from generation to generation

exoskeleton A skeleton covering the outside of the body such as in arthropods

Fallopian tube In human females, an oviduct; one of the two reproductive tubes that carry the egg from the ovary to the uterus

fauna The animal life of a given area or geologic time period

feces The indigestible waste discharged from the digestive tract

fertilization The fusion of two gametes, normally the egg and sperm

fetus An embryo in its later stages of development, still contained inside the egg shell or uterus; the developing human being from about the third month after conception until birth

fiber One of the long, thin extensions of the cell body of a nerve cell

fibroblast A relatively undifferentiated cell found in loose connective tissue that is thought to give rise to fibers

fission Asexual reproduction in cells or organisms by division of the body into two more or less equal parts

fitness The relative ability to leave offspring, as in evolution

flagellum A fine, long, threadlike structure that can be used in propulsion and feeding by a cell body

flora The plant life of a given area or geologic time period

florigen The flowering hormone in plants

flower The reproductive structure of an angiosperm

follicle An enveloping jacket of cells around a potential egg cell in an ovary

food chain A sequence of organisms including producers, consumers, and decomposers, through which energy and materials move within an ecosystem

food web The total collection of feeding relationships in a biotic community

fossil The preserved remains or impressions of a previously existing organism

founder effect The situation in which a single organism (or a few) founds a new population, and the alleles for the genes it is carrying determine the genetic constitution of the population

fragmentation A method of sexual reproduction that involves splitting the body into several portions, each of which regenerates the missing parts

free energy The energy available for doing work in a chemical system

fruit A mature ovary or cluster of ovaries, formed from the ovule in an angiosperm, which contains the seeds

gamete A sexual reproductive cell such as the egg or sperm

gametophyte A haploid plant generation

ganglion A cluster of cell bodies of neurons

gastrula A two-layered and eventually three-layered embryo

gastrulation The process by which the blastula with its single layer of cells turns into a three-layered embryo made up of ectoderm, mesoderm, and endoderm

gemmule One of the special diploid reproductive cells or groups of cells that are capable of developing into a complete individual

gene A unit of inheritance on the chromosome, usually controlling the structure of one protein

gene frequency The relative abundance of a particular allele of a gene in a population

gene pool The collection of all of the alleles of all of the genes in a population

genetic drift In small populations, the influence of random events on the gene frequencies of subsequent generations, such that one allele may "drift" to fixation and be the only representative of that gene locus

genome The total collection of genes in an individual or species

genotype The genetic constitution of an organism; inherited characteristics that may or may not be displayed

genus A group of related species

germination The sprouting of a spore or seed

gibberellin One of a group of plant growth hormones whose most characteristic effect is stem elongation in plants

gill A thin-walled tissue projection from the body of an animal, specialized for gas exchange

gland A cell or organ producing one or more secretions that are discharged to the outside of the gland

globular protein A protein in which the polypeptide chain is folded three-dimensionally to form a globular shape

glomerulus The cluster of capillaries enclosed by the Bowman's capsule in the kidney

glottis The ventral opening in the pharynx for the air channel

glucose A six-carbon sugar that is important in cellular metabolism

glycogen A complex carbohydrate analogous to plant starch, which is stored in animals

Golgi apparatus An organelle in cells that serves as a collecting and packaging center for secretory products

gonad A testis or ovary

grana The stacks of membranes bearing the chlorophyll molecules in chloroplasts

growth curve The manner and speed of population increase

gymnosperm One of the higher plants with vascular tissues and with naked seeds borne on the open leaf or in cones; for example; conifers

habitat The kind of place where a particular organism normally lives

habituation The progressive waning of response to an insignificant stimulus that is repeatedly presented

haploid cell A cell that has only one of each type of chromosome (1n)

hemoglobin A reddish, iron-containing pigment in the blood that is able to transport oxygen and carbon dioxide

herbaceous plant A plant whose stem remains soft and succulent, rather than woody

herbivore An animal that eats plants to obtain its food energy

heterotroph A plant that requires organic nutrients from the environment; it is unable to manufacture its own organic compounds; see autotroph

heterozygous organism An organism that has two different alleles of a given gene

hibernation A state of winter dormancy, in which an animal greatly reduces its metabolism and active behavior

histone Any of several proteins found in the nucleus that are complexed, at one time or another, with DNA

homeotherm A warm-blooded animal, capable of self-regulation of body temperature

homologous chromosome One of the chromosomes that bear genes for the same characters

homology Correspondence in structure and function

homozygous organism An organism that has two copies of the same allele of a given gene

hormone A chemical substance secreted in one part of the body that affects other parts of the body

hybrid An offspring of two different varieties or of two different species

hyphae The bodies of a fungus

hypothalamus A portion of the cerebrum that lies below the thalamus; the most important control center for many reflexes in the body and the visceral functions

hypocotyl The lower portion of the axis of a plant embryo that will form the root

immigration A one-way movement into a particular area

imprinting A rapid learning association with a particular stimulus (e.g., the shape of the mother) soon after birth

inbreeding Mating and reproduction among close relatives

induction Reasoning from specific facts to the general principle

inorganic compound A chemical compound that does not contain carbon

instinct Genetically programed behavior; that is, unlearned patterned behavior

insulin A hormone produced by the islets of Langerhans in the pancreas that helps regulate carbohydrate metabolism, especially the conversion of glucose into glycogen

invagination A folding or protruding inward.

invertebrate An organism that lacks a backbone

ileum In human beings, the longest and final section of the small intestine, about five feet long

ion An electrically charged atom

irritability A general property of all organisms, involving the ability to respond to stimuli or changes in the environment

islet cell One of a clump of cells in the pancreas that produce the hormone insulin

isolating mechanism One of the mechanisms —geographic, ecological, seasonal, behavioral, or genetic—that prevents gene exchange between different species living in the same area

jejunum In human beings, the middle section of the small intestine, about three feet long

karyotype The collection of the chromosomes found in the diploid cells of an individual

kidney In vertebrates, the organ that regulates the balance of water in dissolved substances in the blood and controls the excretion of nitrogen wastes in the form of urine

larva An immature animal that is morphologically very different from the adult and must go through radical transformations to attain the adult form

larynx The voice box, between the pharynx and the trachea

laterite A leached tropical soil with a high iron and aluminum oxide concentration

laterization The process of formation of a hard, impermeable reddish soil crust in cleared tropical forest areas

leaching The removal of minerals and other elements from the soil by the downward movement of water

learning Any of the several processes that produce adaptive change in individual behavior as the result of experience

lek A displaying group formed by the attraction of males (and subsequently females) to a communal mating ground

lenticel One of the openings that allow gas exchange through the bark of stems and roots

leukocyte A white blood cell

lichen A plant composed of a symbiotic alga and fungus

ligament A tough, fibrous band of connective tissue that connects bones or supports internal organs

light reactions The part of photosynthesis that involves the trapping of light energy by chlorophyll molecules, which is used to split water

linkage The inheritance of certain genes as a unit because they are physically located on the same chromosome

lipid Any of a variety of organic fat or fatlike compounds including oils, waxes, steroids, and carotenes

locus The position of a gene on a chromosome

lumen An opening or cavity, such as the central cavity of the gut

lymph A colorless fluid derived from tissue fluid and transported in special ducts to the blood

lysosome An organelle in the cell in which digestive enzymes are stored

macroevolution The appearance of new species in time

macromolecule A molecule of very high molecular weight, including the proteins, polysaccharides, and nucleic acids

matrix The substance of a structure, such as the constituents of a bone

medulla The lower part of the brain, largely responsible for motor neuron coordination

medusa The free-swimming stage in the life cycle of a coelenterate

meiosis The process of nuclear division that gives rise to the sex cells, in which the number of chromosomes is reduced by half

menopause The end of a woman's reproductive life

meristematic **tissue** The undifferentiated plant tissue that functions primarily in the production of new cells by mitosis

mesoderm The middle tissue layer of an animal embryo

messenger RNA (m-RNA) The form of RNA that carries genetic information from the gene to the ribosome, where it determines the order of the amino acids as a polypeptide is formed

metabolism The sum of the various chemical activities going on within a cell or a whole organism

metamorphosis The transformation of an entire animal into an adult

microevolution The development of new races from already existing species through the gradual accumulation of genetic differences between geographically isolated populations

migration A periodic movement between two points

mimicry The resemblance in form, color, or behavior of an organism (the mimic) to other protected organisms (models), resulting in protection or concealment for the mimic

mitochondrion An organelle in the cell in which aerobic respiration takes place

mitosis The usual process of nuclear division in which chromosomes are exactly duplicated and two identical daughter cells are formed

molecular weight The sum of the atomic weights of the constituent atoms in a molecule

molecule The smallest possible unit of a compound substance, consisting of two or more atoms

molting The shedding of part or all of the outer covering, as in arthropods

morphogenesis The establishment of size, form, and structure in an organism

morphology The study of form and structure, at any level of organization

mortality The death rate

morula An early embryonic stage composed of a cluster of mitotically produced cells, which is about the same size as the single egg from which it was originally derived

motor neuron A neuron leading from the central nervous system to skeletal muscle or another effector

mucosa A thick, largely glandular layer of cells that line the cavity of the gut

mutagen Any agent that increases the mutation rate of a gene

mutant A mutated gene or an organism carrying a mutated gene

mutation Any relatively stable heritable change in the genetic material, usually a change of a gene from one allelic form to another

mutualism A symbiosis in which both species benefit

mycelium The mass of hyphae forming the body of a fungus

myelin A thick, fatty sheath on nerve fibers

natality The birth rate

natural selection The nonrandom reproduction of genotypes, resulting in differential survival and hence evolutionary change

nectar A fluid containing many sugars that attracts insects to plants for the purpose of pollination

nephron The functional unit of a vertebrate kidney, consisting of a Bowman's capsule, convoluted tubule, and cluster of capillaries

nerve A bundle or group of nerve fibers

nerve fiber A long process of a neuron, either the dendrite or axon end

neuron A nerve cell

nervous system All the nerve cells of a body

niche The functional role and ecological position of an organism in its ecosystem; its "way of life," unique to each species

nitrogen fixation The incorporation of nitrogen from the atmosphere into inorganic nitrogen compounds, which are more generally usable by green plants

node In plants, the point on a stem where a leaf or a bud is attached

noradrenalin In human beings, a hormone secreted by the medulla portion of the adrenal gland

notochord In the lower chordates and in the embryos of the higher vertebrates, a flexible solid supportive rod running longitudinally along the back of the animal

nucleic acid An organic acid consisting of nucleotides; the principal types are deoxyribonucleic acid (DNA) and ribonucleic acid (RNA)

nucleolus A dense body within the nucleus, containing mostly RNA

nucleoplasm The cellular protoplasm found inside the nuclear membrane

nucleotide A single unit of a nucleic acid, consisting of a five-carbon sugar with a phosphate group and a purine or a pyrimidine nitrogen base attached

nucleus A large, membrane-bounded structure in the cell that contains the chromosomes; also, the central part of an atom; or a group of nerve cell bodies in the central nervous system

nutrient A substance usable in metabolic activity

olfaction The sense of smell

ontogeny The development or life history of an organism

oocyte A cell that is a precursor to an egg cell

oogenesis The formation of egg cells in the ovaries by meiosis

organ A body part composed normally of several tissues grouped together into a structural and functional unit

organic compound A compound made up of carbon atoms and other elements

organelle A structurally distinct part of a cell, with a particular function

organism Any individual living creature

organizer The part of an embryo that is capable of inducing undifferentiated cells to follow a particular course in development toward specialization

osmosis The movement of water by diffusion through a semipermeable membrane

outbreeding Mating and reproduction among unrelated individuals

ovary The female reproductive organ, in which egg cells are produced

oviduct The tube transporting the eggs from the ovary to the uterus or to the outside; in human beings, their upper portions are known as the Fallopian tubes

ovulation The release of an egg from the ovary in animals

ovule In seed plants, a structure wrapped around the egg cell that becomes the seed after fertilization of the egg

ovum The egg cell, or female gamete

oxidation The loss of an electron by an atom

paleontology The study of the life of past geologic periods, particularly fossils

pancreas A large glandular organ in vertebrates located near the stomach, which secretes digestive enzymes and also produces hormones

panmixis Random mating in a population

parasitism A symbiotic relationship in which one species benefits at the expense of the other; the parasite benefits at the expense of the host

parathyroid One of the endocrine glands that help regulate the calcium-phosphate balance between the blood and the other tissues

parthenogenesis The development of an egg without fertilization

pathogen A disease-producing organism

pepsin An enzyme released into the stomach that digests proteins

peptide bond A bond between two amino acids

periodicity Having a particular time of activity during the 24-hour day

peristalsis Alternating waves of contraction and relaxation passing along the digestive tract

permeable membrane A membrane that permits certain substances to pass through

peroxisome A membrane-bound cellular vesicle that contains oxidative enzymes

pesticide One of the chemicals that kill pests

petiole The stock of a leaf

pharynx The part of the digestive tract between the mouth cavity and the esophagus

phenotype The physical manifestation of a genetic trait; such displayed characteristics as eye color, height; see genotype

pheromone A chemical substance that acts as an external hormone, influencing the behavior or physiology of other organisms of the same species

phloem A vascular tissue in plants that transports organic materials up and down the plant body

phosphorylation The addition of a phosphate group to a molecule such as AMP, ADP, or NAD

photoperiodism An activity response by an organism to the duration and timing of light and dark periods during the daily cycle

photosynthesis The autotrophic synthesis of organic materials, in which the source of energy is light from the sun

phototropism A turning response to light, mediated by auxins

phylogeny The evolutionary history of an organism or group of organisms

physiology The study of the functioning of organ systems

pigment A substance that absorbs light

pineal gland An endocrine gland associated with the forebrain

pistil The female reproductive organ of the flower, comprising the ovary, the style, and the stigma

pith A soft tissue located in the center of a stem

pituitary An endocrine gland located at the base of the vertebrate brain, which controls many of the other endocrine glands of the body through its secretion of master hormones

placenta A structure formed in part from the inner lining of the uterus and in part from the fetal components, which aids in the exchange of materials between the fetus and the mother in mammals

plasma The blood fluid, minus the cells and platelets

plasma membrane The outer boundary membrane of a cell

plastid An organelle or plant cell that functions in photosynthesis and/or nutrient storage

pleiotropy In a gene, the quality of having more than one phenotypic effect

poikilotherm A cold-blooded animal, incapable of internally regulating body temperature except by behavior

pollination The transfer of pollen from the male reproductive organs (anthers) to the end (stigma) of the female reproductive organ (pistil) in flowering plants, or an analogous transfer of pollen from male cone to female cone in gymnosperms

polymer A large molecule composed of many similar smaller molecules

polymorphism The simultaneous occurrence of several phenotypes in a population

polyp The sedentary stage in the life cycle of a coelenterate

polypeptide A chain of amino acids linked together by peptide bonds

plate tectonics The concept that the earth's surface is composed of a number of huge moving plates, each including not only a continental mass but portions of ocean basins as well

polyploidy The condition of having more than two complete sets of chromosomes per cell.

polysaccharide A carbohydrate composed of many simple sugars; for example, starch and cellulose

population A group of individuals that belong to the same species

predation The feeding of free-living organisms on other organisms

primary growth In plants, the growth originating from the apical meristem of the shoots and roots; see secondary growth

primary production The energy accumulated by plants and stored in the form of carbohydrates

producer A green plant that produces its own food

progesterone One of the principal female sex hormones in vertebrates

prolactin A pituitary hormone that stimulates milk production by mammary glands

purine Any of several double-ringed nitrogenous bases important in the composition of nucleic acids

pylorus The valve junction between the stomach and the small intestine

pyrimidine Any of several single-ringed nitrogenous bases important in the composition of nucleic acids

protein A complex organic compound composed of one or more polypeptide chains, each made up of many amino acids joined by peptide bonds

protoplasm The living substance in cells

race A subspecies of a species

radial symmetry A type of symmetry in which the body parts are arranged regularly around the central point rather than on two sides of a longitudinal plane

radiation, adaptive The evolutionary divergence of species into different niches or adaptive zones

receptor A specialized cell that initially receives a stimulus

recessive gene A gene allele that does not express its phenotype in the presence of another allele of the same gene

rectum The terminal portion of the large intestine

reduction An energy-storing process involving the addition of electrons to a substance by the addition of hydrogen

reflex A functional pathway of the nervous system from receptor cell to effector

reflex arc A complete nervous pathway consisting of a sensory neuron and a motor neuron, but not requiring the participation of the brain

reinforcement Reward for a particular behavior

releaser A stimulus that triggers behavioral responses

respiration In cellular form, the release of energy by the oxidation of food molecules; on an organismic scale, the taking in of oxygen and the release of CO_2 in breathing

rhizome An underground stem

ribonucleic acid (RNA) A nucleic acid in which the sugar component is ribose and one of the nitrogenous bases is uracil

ribosomal RNA (r-RNA) The RNA found in ribosomes

ribosome A small organelle in the cytoplasm of a cell that functions in protein synthesis

RNA See ribonucleic acid

saliva The fluid secreted by three pairs of salivary glands in the mouth

saltational evolution The origin of new types of organisms by "jumping," that is, large-scale mutations

sap The water and dissolved materials moving in the xylem and less commonly applied to solutions moving in the phloem

saprophyte A fungus or bacterium that lives on dead organic matter

secondary growth The lateral growth of a plant stem (increase in width)

seed A reproductive unit of plants that consists of an embryo and stored food enclosed in a protective hard coat

segregation The separation of two alleles into different gametes during mitosis (Mendel's first law)

sensory neuron A neuron leading from a receptor cell to the central nervous system

semen The fluid product of the male reproductive system, including sperm

sepal One of the subunits of the calyx, the outermost whorl of flower parts in angiosperms

sessile organism An organism that is attached and not free to move around; for example, a plant

sex-linked gene One of the genes located on the sex chromosomes

sexual reproduction Reproduction involving meiosis in the union of gametes

somatic cell One of the body cells, except the germ cells

somite One of the muscular body segments of the embryo

speciation The process of formation of new species

species A group of organisms that are capable of interbreeding and are reproductively isolated from all other such groups of populations; a species has some heritable difference that distinguishes it from all other related species

sperm A male gamete

spermatogenesis The meiotic divisions and maturation of meiotic products in the testes leading to the formation of mature sperm cells

sphincter A ring-shaped muscle that can close a tube by contracting

spindle A structure of fibrils with which the chromosomes are associated in mitosis and meiosis

spore An asexual reproductive cell

sporophyte A diploid plant that produces spores

stamen A male sexual part of a flower that bears the pollen

starch A polymer of glucose

sterilization The act or process of killing all living cells; also, the prevention of gametes being discharged in sexual reproduction

stimulus Any environmental change that is detected by a receptor

stomata Openings that allow gas exchange through the epidermis of plant leaves

stratification The distribution in vertical layers of the living organisms in a community

subspecies A distinctive geographic subunit of a species

substrate The environmental base on which an organism lives, such as soil, or a substance acted on in chemical reactions

succession A series of progressive changes in the plant and animal life of an area

sucrose A double sugar, composed of one glucose and one fructose unit; common table sugar

survivorship The percentage of individuals in a population still living at various times after birth

symbiosis The living together of two organisms in intimate relationship

sympatric species Species that have the same geographic range; see allopatric species

synapse The space lying in the junction between two neurons

synapsis The pairing of homologous chromosomes during meiosis

taxis The continuous orientation by an organism to a specific stimulus in its environment

taxonomy The study of the classification of organisms

tendon A type of connective tissue that attaches muscle to bone

territory A particular area defended by an individual against intrusion by other individuals, particularly of the same species

testis The male gamete-producing organ

testosterone A male hormone secreted by the testis in higher vertebrates and stimulating the development and maintenance of male secondary sexual characteristics, as well as the production of sperm

thalamus A portion of the cerebrum serving as an important integrating center for all sensory input (except smell) on its way to the cortex

thorax In insects, the three leg-bearing segments of the body lying between the head and the abdomen; in vertebrates, the portion of the body trunk containing the heart and lungs

thrombocyte One of the blood platelets

thymus A glandular organ that is essential for developing immunological capabilities in vertebrates

thyroid A vertebrate endocrine gland located in the neck, which secretes an iodine-containing hormone (thyroxine) that increases the metabolic rate

thyroxin A thyroid hormone that controls and stimulates oxidative metabolism in the body

tissue A group of similar cells organized into a structural and functional unit

trachea In vertebrates, the part of the respiratory system running from the pharynx into the beginning of the bronchus going to each lung; that is, the wind pipe; tracheae are the multiple air-conducting tubes that form the breathing systems of insects

transfer RNA (t-RNA) A small type of RNA molecule that carries a specific kind of amino acid molecule to a ribosome-messenger RNA complex for incorporation into a polypeptide molecule

transformation A genetic change produced by the incorporation into a cell of DNA from another cell

translocation The movement of organic materials from one place to another within the plant body, largely through the phloem; also, the physical dislocation of a section of one chromosome to become attached to another, nonhomologous chromosome.

transpiration The loss of water vapor from the stomata in the leaves in plants

trophic level The position of a species in the food chain

tropism A turning response to a stimulus

urea The principal chemical form of disposal of nitrogenous wastes in mammals and some other vertebrates, formed in the liver by a combination of ammonia and carbon dioxide

ureter A duct carrying urine from the kidney to the bladder in higher vertebrates

urethra The tube carrying urine from the bladder to the exterior in mammals

urine The liquid waste filtered from the blood by the kidney and excreted by the bladder

uterus The chamber of the female reproductive tract in which the mammalian embryo undergoes much of its development and nourishment

vacuole A membrane-bounded chamber in a cell

vagina The part of the female reproductive duct below the uterus in mammals, which receives the male penis during copulation

vas deferens In male mammals, a tube bearing sperm from the testis to the urethra

vascular bundle In plants, a group of conducting tissues (xylem and phloem)

vein A blood vessel that transports blood toward the heart

ventricle A muscular chamber of the heart that receives blood from an atrium and pumps blood out of the heart

vertebral column The backbone found in nearly all vertebrates, which supports the body and protects the spinal cord

villus A highly vascularized, fingerlike process arising from the intestinal lining, which serves to increase the absorptive surface area of the small intestine

virus A submicroscopic noncellular particle, composed of a nucleic acid core and a protein shell, which is parasitic and reproduces only within a host cell

xylem A vascular tissue in plants that transports water and dissolved minerals upward from the roots

yolk The stored nutritive material in an egg that will nourish the embryo

zoology The study of animals

zygote The diploid (2n) fertilized egg cell that results from the fusion of male and female gametes

Acknowledgments for illustrations

Cover: Photos by Manfred Kage and A. Lloyd from Peter Arnold; NASA; George Holton from Photo Researchers; Grant Heilman (3).

All drawings by Vantage Art, Inc.

2-14 Adapted from Eibl-Eibesfeldt, *Ethnology*, Holt, Rinehart and Winston, 1970.
3-4 Adapted from Boughey, *Ecology of Populations*, 2nd ed., Macmillan, 1973.
7-16 Adapted from R. W. Ficken, et al., *Science* 173 (1971), A.A.A.S.
8-2 Adapted from Paul A. Colenvaux, *Introduction to Ecology*, Wiley, 1973 (from MacArthur, 1958 and Hutchinson, 1965).
8-4 Adapted from Paul A. Colenvaux, *Introduction to Ecology*, Wiley, 1973 (from G. E. Hutchinson, *The Ecological Theatre and the Evolutionary Play*, Yale Univ. Press, 1965, p. 139).
8-5 Adapted from A. G. Fischer, "Latitudinal variations in organic diversity," *Evolution*, 14, 1960.
9-5 Adapted from Archie Carr, *American Scientist*, Vol. 50, No. 3, 1962, Fig. 1 (p. 360).
10-4 Adapted from Allee, et al., *Principles of Animal Ecology*, Saunders, 1949, p. 311.
10-5 Adapted from Andrewartha and Birch, *Distribution and Abundance of Animals*, Univ. of Chicago Press, 1954.
10-6 Adapted from Eugene P. Odum, *Fundamentals of Ecology*, Saunders, 1959.
11-1 Adapted from *Population Bulletin*, Vol. 18, No. 1. Courtesy of the Population Reference Bureau, Inc., Washington, D.C.
13-3 Adapted from Robert C. Stebbins in *University of California Publication of Zoology*, Vol. 48, (1949).
13-5 Adapted from Erlich, Holm, and Parnell, *Process of Evolution*, McGraw-Hill, 1974, Figs. 7-1 and 7-2.
13-8 Adapted from Robert I. Bowman in *University of California Publication of Zoology*, Vol. 58 (1961).
14-6 Adapted from Robert Jastrow, *Stars, Planets and Life*, William Heinemann Ltd., 1967, p. 134.
15-16 Adapted from I. Carlquist, *Island Life*, Natural History Press, 1965, pp. 370–371.
17-10 Adapted from Simpson and Beck, *Life: An Introduction to Biology*, 2nd ed., Harcourt Brace Jovanovich, 1965.
18-6 Adapted from N. Tinbergen, *The Study of Instinct*, Oxford Univ. Press, 1951.
19-5 Adapted from N. Tinbergen, "The Evolution of Behavior in Gulls," *Scientific American*, Dec., 1960.
19-6 Adapted from studies of James E. Lloyd, University of Florida.
19-7 Adapted from Jan van Lawich-Goodall, Behavior of Free-living Chimpanzees in the Gombe Stream Reserve," *Animal Behavior Monographs*, Vol. 1, Part 3, 1968.
19-12 Adapted from K. von Frisch, *Bees: Their Vision, Chemical Senses & Language*, Cornell Univ. Press, 1950.
33-2 Adapted from Balinsky, *An Introduction to Embryology*, Saunders, 1960 (after Roux, from Morgan, 1927).
33-3 Adapted from Balinsky, *An Introduction to Embryology*, Saunders, 1960 (after Driesch, from Gabriel and Fogel, 1955).
33-4 Adapted from Balinsky, *An Introduction to Embryology*, Saunders, 1960 (after Spemann and H. Mangold, from Spemann, 1938).

Picture Credits

Legend: t–top, b–bottom, c–center, l–left, r–right
Chapter 1: pp. 3, 5, Grant Heilman; p. 4, NASA; pp. 5, 9, all Grant Heilman.
Chapter 2: pp. 11, Grant Heilman; p. 20 tl, br Charles L. Hogue; tr Dr. Boyce A. Drummond; p. 21, b Grant Heilman; t Robert C. Hermes/Photo Researchers; p. 22, tl Grant Heilman; bl Thomas C. Emmel; r Isabelle Conant/Photo Researchers; p. 23, t Grant Heilman; b Leonard Lee Rue/Photo Researchers; p. 24, tl Joseph Van Wormer/Photo Researchers; bl A. Lowry/Photo Researchers; c Hal Harrison for Grant Heilman r L. N. Stone/Photo Researchers; p. 25, l Grant Heilman; r Hall Harrison for Grant Heilman; p. 24, tl Donald Burrows/Taurus Photos; tr Hal Harrison for Grant Heilman; bl Grant Heilman; p. 27, tl Thomas C. Emmel; bl Patricia Witherspoon/Photo Researchers; r Leonard Lee Rue/Monkmeyer.
Chapter 3: p. 33, Dr. Boyce A. Drummond; p. 37, tl Dr. Boyce A. Drummond; tr Grant Heilman; bl Leonard Lee Rue/Monkmeyer; br Charles L. Hogue; p. 43, Grant Heilman.
Chapter 4: pp. 45, 57, Grant Heilman.
Chapter 5: p. 61, Fred Lyon/Photo Researchers; p. 62, Klaus D. Francke/Peter Arnold; p. 64, Monkmeyer; p. 65, Karl Weidman/Photo Researchers; p. 67, Grant Heilman; p. 68, 1 r UPI; p. 69, Environment Magazine; p. 70, Grant Heilman; p. 73, Pro Pix/Monkmeyer; p 75, Fred Lyon/Photo Researchers; p. 79, l Grant Heilman, r Gerhard E. Gscheidle/Peter Arnold; p. 81, Thomas C. Emmel; p. 84, l Grant Heilman; r Bob Harrington/Michigan Dept. of Natural Resources.
Chapter 6: pp. 91, 94, Jen and Des Bartlett/Photo Researchers; p. 99, all Grant Heilman.
Chapter 7: p. 107, Mark N. Brultan/Photo Researchers; p. 108, Wayne Fitch; p. 111, Thomas C. Emmel; p. 112, l Alfred M. Bailey/Photo Researchers; r Thomas C. Emmel; p. 113, Mark N. Brultan/Photo Researchers; p. 114, l Jeanne White/Photo Researchers; r Duryea Morton/Photo Researchers; p. 115, l Jeanne White/Photo Researchers; c Spencer/Monkmeyer; r Hal Harrison for Grant Heilman; p. 116, Thomas C. Emmel; p. 118, l Julian P. Donahue; c Grant Heilman; r John H. Gerard/Photo Researchers; p. 119, l c Runk/Schoenberger for Grant Heilman; r Thomas C. Emmel; p. 120, l Runk/Schoenberger for Grant Heilman; tr Walter C. Auffenberg; br Grant Heilman; p. 122, James E. Lloyd.
Chapter 8: p. 124, Leonard Lee Rue/Monkmeyer; p. 125, Carl Frank/Photo Researchers.
Chapter 9: p. 137, Monkmeyer; p. 140, Leonard Lee Rue/Monkmeyer; p. 141, Woods Hole Oceanographic Institution; pp. 142, 147, Thomas C. Emmel; p. 150, Monkmeyer.
Chapter 10: p. 154, Grant Heilman.

Chapter 11: p. 171, Studio Laborie, Bergerac, France.
Chapter 12: p. 188, Grant Heilman.
Chapter 13: p. 206, Grant Heilman; p. 207, lr American Museum of Natural History; p. 211, l Bettmann Archive; r American Museum of Natural History.
Chapter 14: p. 223, Leonard Lee Rue/Monkmeyer; p. 224, Russ Kinne/Photo Researchers; p. 227, t AMNH; ctl Grant Heilman; cbl Russ Kinne/Photo Researchers; cr John H. Gerard/Photo Researchers; bl Russ Kinne/Photo Researchers; bc Runk/Schoenberger for Grant Heilman; br American Museum of Natural History; p. 229, both Wide World; p. 231, t American Museum of Natural History; b Chicago Natural History Museum; p. 233, Runk-Schoenberger for Grant Heilman; p. 234, tl Hal H. Harrison/Photo Researchers; tr Laurence Pringle/Photo Researchers; bl Runk-Schoenberger for Grant Heilman; br Grant Heilman; p. 235, l Grant Heilman; r Russ Kinne/Photo Researchers; pp. 237, 238, 240, 243, 244, American Museum of Natural History; p. 246, tl tr, bl Leonard Lee Rue/Monkmeyer; br Van de Poll/Monkmeyer.
Chapter 15: p. 248, Gordon Smith/Photo Researchers; p. 249, Russ Kinne/Photo Researchers; p. 250, l Gordon Smith/Photo Researchers; c, r Grant Heilman; p. 251, t American Museum of Natural History; b Trustees of the British Museum; pp. 252, 257, 260, American Museum of Natural History.
Chapter 16: p. 262, N. E. Beck Jr./Photo Researchers; p. 265, l Bettmann Archive; r Culver Pictures; p. 270, Carl Struewe/Monkmeyer; pp. 271, 272, 273, 274, Runk/Schoenberger for Grant Heilman; p. 275, t, br Runk/Schoenberger for Grant Heilman; bl R. F. Head/Photo Researchers; p. 276, Runk/Schoenberger for Grant Heilman; p. 277, l Runk/Schoenberger for Grant Heilman; r Grant Heilman; p. 279, l, tr Runk/Schoenberger for Grant Heilman; br Russ Kinne/Photo Researchers; p. 281, tl, tr, br Hal Harrison for Grant Heilman; cl, cr, bl Grant Heilman; p. 282, l General Biological Supply House; c Carl Struwe/Monkmeyer; r Runk/Schoenberger for Grant Heilman; 283; p. 284 l Runk/Schoenberger for Grant Heilman; br N. E. Beck Jr./Photo Researchers; p. 285, both Grant Heilman; p. 286, tl Grant Haist/Photo Researchers; tr, bl Grant Heilman; br Jack Dermid/Photo Researchers.
Chapter 17: p. 284, Pro Pix/Monkmeyer; p. 289, both P. R. Chapman; p. 290, t Grant Heilman; b De Wys Inc.; p. 293, all U.S. Forest Service, Dept. of Agriculture; p. 303, American Museum of Natural History.
Chapter 18: p. 308, Miguel Castro/Photo Researchers; p. 309, Caisse Nationale des Monuments Historiques; p. 313, l Derek Bayes/Time-Life Books; c Peter Arnold; r Monkmeyer; p. 318, Ray Ellis/Photo Researchers; p. 323, l Miguel Castro/Photo Researchers; r Thomas McAvoy/Time-Life Picture Agency.
Chapter 19: p. 325, Leonard Lee Rue/Monkmeyer; p. 326, l Dr. Boyce A. Drummond; r George Holton/Photo Researchers; p. 327, l George Holton/Photo Researchers; tr Dr. Boyce A. Drummond; br Grant Heilman; p. 329, Leonard Lee Rue/Monkmeyer; p. 333, t Sol Mednick; b C. L. Hogue; p. 334, A. W. Ambler/Photo Researchers (National Audubon Society); p. 336, Francisco Erize, p. 337, Dr. Boyce A. Drummond; p. 339 (19-16), Timothy Ransom; p. 340, Eric Hasking/Photo Researchers; p. 341, R. Mittermeier; p. 343, Toni Angermayer/Photo Researchers.
Chapter 20: p. 347, Leonard Lee Rue/Monkmeyer
Chapter 21: p. 365, t Laboratory for Cell Biology, Harvard University; (b) Culver Pictures; p. 366, l Rare Book Division, New York Public Library; r Grant Heilman; p. 372, both, Lester V. Bergman and Associates; p. 373, Laboratory for Cell Biology, Harvard University; p. 375, l M.C.D. Biology, University of Colorado; r Grant Heilman; p. 376, M.C.D. Biology, University of Colorado; p. 377, Grant Heilman; p. 378, Laboratory for Cell Biology, Harvard University; p. 380, t T. E. Weier; b U.S.D.A.; p. 381, l A. L. Houwink, r Lester V. Bergman and Associates.
Chapter 22: pp. 385, 390, Turtox; p. 391, Dr. William T. Jackson, Dartmouth College.

Chapter 23: p. 402, M. W. F. Tweedie/Photo Researchers; p. 403, Bettmann Archive.
Chapter 24: p. 426, Manfred Kage/Peter Arnold.
Chapter 25: p. 443, Grant Heilman; p. 449, Dr. Melvin, Calvin Lawrence Radiation Laboratory, University of California, Berkeley; p. 451, Grant Heilman.
Chapter 26: p. 458, Max Tharp/Monkmeyer.
Chapter 27: p. 475, Grant Heilman; p. 481, Bettmann Archive; p. 488, all Grant Heilman.
Chapter 28: pp. 492, 495, Ripon Microslides Laboratory.
Chapter 29: p. 503, Steve McCutcheon; p. 505, H. E. Stork/Photo Researchers; p. 506, Steve McCutcheon; pp. 509, 511, Thomas C. Emmel.
Chapter 30: p. 512, Grant Heilman; p. 515, S. H. Wittmer/Michigan State University Agricultural Experimental Station.
Chapter 31: p. 528, UPI.
Chapter 32: pp. 543, 544, 547, all Grant Heilman.
Chapter 33: p. 550, Grant Heilman; p. 554, Carnegie Institution, Dept of Embryology, Davis Division; p. 555, all Grant Heilman; p. 557, tl George Porter/Photo Researchers; tr, bl Grant Heilman; br Carolina Biological Supply Co.
Chapter 34: p. 546, Paul Conklin/Monkmeyer.

Color Plates:

1: t. Thomas C. Emmel; cl, Shostal; cr, Bucky Reeves from National Audubon Society/Photo Researchers; bl, W. E. Harvey from National Audubon Society/Photo Researchers; br, Leonard Lee Rue III from National Audubon Society/Photo Researchers.
2: left page, Dr. Boyce A. Drummon; right page, 1 both Thomas C. Emmel; tr, George Halton/Photo Researchers.
3: left page, all Shostal; right page, t and bl, Shostal; br, Frederick Ayer/Photo Researchers.
4: both pages, Grant Heilman.
5: left page, t, Shostal; bl, J. S. Flannery/Bruce Coleman; br, Shostal; right page, t, Joe Rychetnik/Photo Researchers; b, both Shostal.
6: left page, tl, Frank E. Gunnel/Monkmeyer; tr, Alford W. Cooper/Photo Researchers; bl, Shostal; br, Herbert Lanks/Shostal; right page, t, Alan Pitcairn/Grant Heilman; b, Shostal.
7: both Grant Heilman.
8: t, both Grant Heilman; c and b, Dr. Boyce A. Drummond.
9: left page, br, Grant Heilman; all others Dr. Boyce A. Drummond; right page, tl, N. Smythe from National Society Photo Researchers; tr and bl, Dr. Boyce A. Drummond; cr, Grant Heilman.
10: tl, Grant Heilman; tr, Tom McHugh/Photo Researchers (Steinhart Aquarium); cl, Shostal; bl, Dr. Boyce A. Drummond.
11: Thomas C. Emmel.
12: both pages, all Thomas C. Emmel.
13: all, Thomas C. Emmel.
14: t, Manfred Kage/Peter Arnold; b, Runk-Schoenberger/Grant Heilman.
15: all Grant Heilman.
16: left page, all Grant Heilman; right page, tl, Grant Heilman; tr and br, Shostal; bl, John R. MacGregor/Peter Arnold.
17: left page, all Grant Heilman; right page, all Grant Heilman except bl, W. H. Hodge/Peter Arnold.
18: t, Runk/Schoenberger/Grant Heilman; bl, Shostal; br, Grant Heilman.
19: t, Lee Lyon/Bruce Coleman; b, Mohamed Amin/Bruce Coleman.
20: left page, all Francisco Erize/Bruce Coleman; right page, tl, Jen Des Bartlett/Bruce Coleman; tr, Sven Gillsater; b, Eric Hosking/Bruce Coleman.
21: tl, David Hughes/Bruce Coleman; tr, S. C. Bisserot/Bruce Coleman; bl, J. R. Simon/Bruce Coleman; br, Oxford Scientific Films/Bruce Coleman.

Index

Page numbers in italics refer to the illustrations on those pages.